AF324123

PROBABILISTIC INEQUALITIES

SERIES ON CONCRETE AND APPLICABLE MATHEMATICS

Series Editor: Professor George A. Anastassiou
Department of Mathematical Sciences
The University of Memphis
Memphis, TN 38152, USA

Published

Vol. 1 Long Time Behaviour of Classical and Quantum Systems
edited by S. Graffi & A. Martinez

Vol. 2 Problems in Probability
by T. M. Mills

Vol. 3 Introduction to Matrix Theory: With Applications to Business
and Economics
by F. Szidarovszky & S. Molnár

Vol. 4 Stochastic Models with Applications to Genetics, Cancers,
Aids and Other Biomedical Systems
by Tan Wai-Yuan

Vol. 5 Defects of Properties in Mathematics: Quantitative Characterizations
by Adrian I. Ban & Sorin G. Gal

Vol. 6 Topics on Stability and Periodicity in Abstract Differential Equations
by James H. Liu, Gaston M. N'Guérékata & Nguyen Van Minh

Vol. 7 Probabilistic Inequalities
by George A. Anastassiou

Series on Concrete and Applicable Mathematics – Vol. 7

PROBABILISTIC INEQUALITIES

George A Anastassiou

University of Memphis, USA

World Scientific

NEW JERSEY · LONDON · SINGAPORE · BEIJING · SHANGHAI · HONG KONG · TAIPEI · CHENNAI

Published by

World Scientific Publishing Co. Pte. Ltd.

5 Toh Tuck Link, Singapore 596224

USA office: 27 Warren Street, Suite 401-402, Hackensack, NJ 07601

UK office: 57 Shelton Street, Covent Garden, London WC2H 9HE

British Library Cataloguing-in-Publication Data
A catalogue record for this book is available from the British Library.

ISBN-13 978-981-4280-78-5
ISBN-10 981-4280-78-X

Printed by FuIsland Offset Printing (S) Pte Ltd, Singapore

Dedicated to my wife Koula

Preface

In this monograph we present univariate and multivariate probabilistic inequalities regarding basic probabilistic entities like expectation, variance, moment generating function and covariance. These are built on recent classical form real analysis inequalities also given here in full details. This treatise relies on author's last twenty one years related research work, more precisely see [18]-[90], and it is a natural outgrowth of it. Chapters are self-contained and several advanced courses can be taught out of this book. Extensive background and motivations are given per chapter. A very extensive list of references is given at the end. The topics covered are very diverse. Initially we present probabilistic Ostrowski type inequalities, other various related ones, and Grothendieck type probabilistic inequalities. A great bulk of the book is about Information theory inequalities, regarding the Csiszar's f-Divergence between probability measures. Another great bulk of the book is regarding applications in various directions of Geometry Moment Theory, in particular to estimate the rate of weak convergence of probability measures to the unit measure, to maximize profits in the stock market, to estimating the difference of integral means, and applications to political science in electorial systems. Also we develop Grüss type and Chebyshev-Grüss type inequalities for Stieltjes integrals and show their applications to probability. Our results are optimal or close to optimal. We end with important real analysis methods with potential applications to stochastic. The exposed theory is destined to find applications to all applied sciences and related subjects, furthermore has its own theoretical merit and interest from the point of view of Inequalities in Pure Mathematics. As such is suitable for researchers, graduate students, and seminars of the above subjects, also to be in all science libraries.

The final preparation of book took place during 2008 in Memphis, Tennessee, USA.

I would like to thank my family for their dedication and love to me, which was the strongest support during the writing of this monograph.

I am also indebted and thankful to Rodica Gal and Razvan Mezei for their heroic and great typing preparation of the manuscript in a very short time.

George A. Anastassiou
Department of Mathematical Sciences
The University of Memphis, TN
U.S.A.
January 1, 2009

Contents

Chapter 1

Introduction

This monograph is about probabilistic inequalities from the theoretical point of view, however with lots of implications to applied problems. Many examples are presented. Many inequalities are proved to be sharp and attained. In general the presented estimates are very tight.

This monograph is a rare of its kind and all the results presented here appear for the first time in a book format. It is a natural outgrowth of the author's related research work, more precisely see [18]-[90], over the last twenty one years.

Several types of probabilistic inequalities are discussed and the monograph is very diverse regarding the research on the topic.

In Chapters 2-4 are discussed univariate and multivariate Ostrowski type and other general probabilistic inequalities, involving the expectation, variance, moment generating function, and covariance of the engaged random variables. These are applications of presented general mathematical analysis classical inequalities.

In Chapter 5 we give results related to the probabilistic version of Grothendieck inequality. These inequalities regard the positive bilinear forms.

Next, in Chapters 6-15 we study and present results regarding the Csiszar's f-Divergence in all possible directions. The Csiszar's discrimination is the most essential and general measure for the comparison between two probability measures. Many other divergences are special cases of Csiszar's f-divergence which is an essential component of information theory.

The probabilistic results derive as applications of deep mathematical analysis methods. Many Csiszar's f-divergence results are applications of developed here Ostrowski, Grüss and integral means inequalities. Plenty of examples are given.

In Chapter 16 we present the basic moment theory, especially we give the geometric moment theory method. This basically follows [229]. It has to do with finding optimal lower and upper bounds to probabilistic integrals subject to prescribed moment conditions.

So in Chapters 17-24 we apply the above described moment methods in various studies. In Chapters 17-19 we deal with optimal lower and upper bounds to the expected convex combinations of the Adams and Jefferson rounding rules of a nonnegative random variable subject to given moment conditions. This theory has applications to political science.

In Chapters 20-22, we estimate and find the Prokhorov radius of a family of distributions close to Dirac measure at a point, subject to some basic moment constraints. The last expresses the rate of weak convergence of a sequence of probability measures to the unit measure at a point.

In Chapter 23 we apply the geometric moment theory methods to optimal portfolio management regarding maximizing profits from stocks and bonds and we give examples. The developed theory provides more general optimization formulae with applications to stock market.

In Chapter 24 we find, among others, estimates for the differences of general integral means subject to a simple moment condition.

The moment theory part of the monograph relies on [28], [23], [89], [85], [88], [90], [86], [44] and [48].

In Chapter 25 we develop Grüss type inequalities related to Stieltjes integral with applications to expectations.

In Chapter 26 we present Chebyshev-Grüss type and difference of integral means inequalities for the Stieltjes integral with applications to expectation and variance.

Chapters 27-28 work as an Appendix of this book. In Chapter 27 we discuss new expansion formula, while in last Chapter 28 we study the integration by parts on the multivariate domain.

Both final Chapters have potential important applications to probability.

The writing of the monograph was made to help the reader the most. The chapters are self-contained and several courses can be taught out of this book. All background needed to understand each chapter is usually found there. Also are given, per chapter, strong motivations and inspirations to write it.

We finish with a rich list of 349 related references. The exposed results are expected to find applications in most applied fields like: probability, statistics, physics, economics, engineering and information theory, etc.

Chapter 2

Basic Stochastic Ostrowski Inequalities

Very general univariate and multivariate stochastic Ostrowski type inequalities are given, involving $\| \cdot \|_\infty$ and $\| \cdot \|_p$, $p \geq 1$ norms of probability density functions. Some of these inequalities provide pointwise estimates to the error of probability distribution function from the expectation of some simple function of the involved random variable. Other inequalities give upper bounds for the expectation and variance of a random variable. All are presented over finite domains. At the end of chapter are presented applications, in particular for the Beta random variable. This treatment relies on [31].

2.1 Introduction

In 1938, A. Ostrowski [277] proved a very important integral inequality that made the foundation and motivation of a lot of further research in that direction. Ostrowski inequalities have a lot of applications, among others, to probability and numerical analysis. This chapter presents a group of new general probabilistic Ostrowski type inequalities in the univariate and multivariate cases over finite domains. For related work see S.S. Dragomir et al. ([174]). We mention the main result from that article.

Theorem 2.1. ([174]) *Let X be a random variable with the probability density function $f\colon [a,b] \subset \mathbb{R} \to \mathbb{R}_+$ and with cumulative distribution function $F(x) = Pr(X \leq x)$. If $f \in L_p[a,b]$, $p > 1$, then we have the inequality:*

$$\left| Pr(X \leq x) - \left(\frac{b - E(X)}{b - a} \right) \right|$$

$$\leq \left(\frac{q}{q+1} \right) \|f\|_p (b-a)^{1/q} \left[\left(\frac{x-a}{b-a} \right)^{\frac{1+q}{q}} + \left(\frac{b-x}{b-a} \right)^{\frac{1+q}{q}} \right]$$

$$\leq \left(\frac{q}{q+1} \right) \|f\|_p (b-a)^{1/q}, \quad \text{for all } x \in [a,b], \text{ where } \frac{1}{p} + \frac{1}{q} = 1, \tag{2.1}$$

E stands for expectation.

3

Later Brnetić and Pečarić improved Theorem 2.1, see [117]. This is another related work here. We would like to mention

Theorem 2.2 ([117]). *Let the assumptions of Theorem 2.1 be fulfilled. Then*

$$\left| Pr(X \leq x) - \left(\frac{b - E(X)}{b - a} \right) \right| \leq \left(\frac{(x-a)^{q+1} + (b-x)^{q+1}}{(q+1)(b-a)^q} \right)^{1/q} \cdot \|f\|_p, \qquad (2.2)$$

for all $x \in [a, b]$.

2.2   One Dimensional Results

We mention

Lemma 2.3. *Let X be a random variable taking values in $[a, b] \subset \mathbb{R}$, and F its probability distribution function. Let f be the probability density function of X. Let also $g \colon [a, b] \to \mathbb{R}$ be a function of bounded variation, such that $g(a) \neq g(b)$. Let $x \in [a, b]$. We define*

$$P(g(x), g(t)) = \begin{cases} \dfrac{g(t) - g(a)}{g(b) - g(a)}, & a \leq t \leq x, \\[4mm] \dfrac{g(t) - g(b)}{g(b) - g(a)}, & x < t \leq b. \end{cases} \qquad (2.3)$$

Then

$$F(x) = \frac{1}{(g(b) - g(a))} \int_a^b F(t) dg(t) + \int_a^b P(g(x), g(t)) f(t) dt. \qquad (2.4)$$

Proof. Here notice that $F' = f$ and apply Proposition 1 of [32]. $\qquad \square$

Remark 2.4 (on Lemma 2.3). i) We see that the expectation

$$E(g(X)) = \int_a^b g(t) dF(t) = g(b) - \int_a^b F(t) dg(t), \qquad (2.5)$$

proved by using integration by parts.

ii) By (2.4) we obtain

$$\frac{g(b)}{(g(b) - g(a))} - F(x) = \frac{1}{(g(b) - g(a))} \left(g(b) - \int_a^b F(t) dg(t) \right)$$
$$- \int_a^b P(g(x), g(t)) f(t) dt.$$

Hence by (2.5) we get

$$\frac{g(b)}{(g(b) - g(a))} - F(x) = \frac{1}{(g(b) - g(a))} E(g(X))$$
$$- \int_a^b P(g(x), g(t)) f(t) dt.$$

And we have proved that

$$F(x) - \left(\frac{g(b) - E(g(X)))}{g(b) - g(a)} \right) = \int_a^b P(g(x), g(t)) f(t) dt. \tag{2.6}$$

Now we are ready to present

Theorem 2.5. *Let X be a random variable taking values in $[a, b] \subset \mathbb{R}$, and F its probability distribution function. Let f be the probability density function of X. Let also $g \colon [a, b] \to \mathbb{R}$ be a function of bounded variation, such that $g(a) \neq g(b)$. Let $x \in [a, b]$, and the kernel P defined as in (2.3). Then we get*

$$\left| F(x) - \left(\frac{g(b) - E(g(X)))}{g(b) - g(a)} \right) \right|$$

$$\begin{cases} \left(\int_a^b |P(g(x), g(t))| dt \right) \cdot \|f\|_\infty, & \textit{if } f \in L_\infty([a, b]), \\[2mm] \sup_{t \in [a, b]} |P(g(x), g(t))|, & \\[2mm] \left(\int_a^b |P(g(x), g(t))|^q dt \right)^{1/q} \|f\|_p, & \textit{if } f \in L_p([a, b]), p, q > 1 \colon \frac{1}{p} + \frac{1}{q} = 1. \end{cases} \tag{2.7}$$

Proof. From (2.6) we obtain

$$\left| F(x) - \left(\frac{g(b) - E(g(X)))}{g(b) - g(a)} \right) \right| \leq \int_a^b |P(g(x), g(t))| f(t) dt. \tag{2.8}$$

The rest are obvious. $\qquad\square$

In the following we consider the moment generating function

$$M_X(x) := E(e^{xX}), \qquad x \in [a, b];$$

and we take $g(t) := e^{xt}$, all $t \in [a, b]$.

Theorem 2.6. *Let X be a random variable taking values in $[a, b] \colon 0 < a < b$. Let F be its probability distribution function and f its probability density function of X.*

Let $x \in [a,b]$. Then

$$\left| F(x) - \left(\frac{e^{xb} - M_X(x)}{e^{xb} - e^{xa}} \right) \right|$$

$$\leq \begin{cases} \dfrac{\|f\|_\infty}{(e^{xb} - e^{xa})} \cdot \left\{ \dfrac{1}{x}(2e^{x^2} - e^{xa} - e^{xb}) + e^{xa}(a - x) + e^{xb}(b - x) \right\}, \\ \qquad \text{if } f \in L_\infty([a,b]), \\ \dfrac{\max\{(e^{x^2} - e^{xa}), (e^{xb} - e^{x^2})\}}{(e^{xb} - e^{xa})}, \\ \left(\dfrac{\|f\|_p}{e^{xb} - e^{xa}} \right) \cdot \left[\displaystyle\int_a^x (e^{xt} - e^{xa})^q dt + \int_x^b (e^{xb} - e^{xt})^q dt \right]^{1/q}, \\ \qquad \text{if } f \in L_p \in ([a,b]), \ p,q > 1 : \frac{1}{p} + \frac{1}{q} = 1, \\ \qquad \text{and in particular,} \\ \dfrac{\|f\|_2}{(e^{xb} - e^{xa})} \cdot \Big[e^{2xb}(b - x) + e^{2xa}(x - a) + \dfrac{3}{2x}e^{2xa} \\ \qquad - \dfrac{3}{2x}e^{2xb} - \dfrac{2e^{xa+x^2}}{x} + \dfrac{2e^{xb+x^2}}{x} \Big]^{1/2}, \quad \text{if } f \in L_2([a,b]). \end{cases}$$
(2.9)

Proof. Application of Theorem 2.5. □

Another important result comes:

Theorem 2.7. *Let X be a random variable taking values in $[a,b] \subset \mathbb{R}$ and F its probability distribution function. Let f be the probability density function of X. Let $x \in [a,b]$ and $m \in \mathbb{N}$. Then*

$$\left| F(x) - \left(\frac{b^m - E(X^m)}{b^m - a^m} \right) \right|$$

$$\leq \begin{cases} \left(\dfrac{\|f\|_\infty}{b^m - a^m} \right) \cdot \left\{ \dfrac{(2x^{m+1} - a^{m+1} - b^{m+1})}{m + 1} + a^{m+1} + b^{m+1} - a^m x - b^m x \right\}, \\ \qquad \text{if } f \in L_\infty([a,b]), \\ \dfrac{\max(x^m - a^m, b^m - x^m)}{b^m - a^m}, \\ \left(\dfrac{\|f\|_p}{b^m - a^m} \right) \cdot \left[\displaystyle\int_a^x (t^m - a^m)^q dt + \int_x^b (b^m - t^m)^q dt \right]^{1/q}, \\ \qquad \text{if } f \in L_p([a,b]); \ p,q > 1 : \frac{1}{p} + \frac{1}{q} = 1, \\ \qquad \text{and in particular,} \\ \left(\dfrac{\|f\|_2}{b^m - a^m} \right) \cdot \Big\{ \dfrac{2x^{m+1}}{m + 1}(b^m - a^m) + x(a^{2m} - b^{2m}) \\ \qquad + \dfrac{2m^2}{(m + 1)(2m + 1)}(b^{2m+1} - a^{2m+1}) \Big\}^{1/2}, \qquad \text{if } f \in L_2([a,b]). \end{cases}$$
(2.10)

Proof. Application of Theorem 2.5, when $g(t) = t^m$. □

An interesting case follows.

Theorem 2.8. *Let f be the probability density function of a random variable X. Here X takes values in $[-a, a]$, $a > 0$. We assume that f is 3-times differentiable on $[-a, a]$, and that f'' is bounded on $[-a, a]$. Then it holds*

$$\left| E(X^2) + \left(\frac{2(f(a) + f(-a)) + a(f'(a) - f'(-a))}{6} \right) \cdot a^3 \right|$$

$$\leq \frac{41a^4}{120} \cdot \| 6f'(t) + 6tf''(t) + t^2 f'''(t) \|_\infty. \tag{2.11}$$

Proof. Application of Theorem 4 from [32]. $\qquad\square$

The next result is regarding the variance $\mathrm{Var}(X)$.

Theorem 2.9. *Let X be a random variable that takes values in $[a, b]$ with probability density function f. We assume that f is 3-times differentiable on $[a, b]$. Let $EX = \mu$. Define*

$$p(r, s) := \begin{cases} s - a, & a \leq s \leq r \\ s - b, & r < s \leq b, \end{cases} \quad r, s \in [a, b].$$

Then

$$\left| \mathrm{Var}(X) + \left((b - \mu)^2 f(b) - (\mu - a)^2 f(a) \right) \cdot \left(\mu - \frac{(a + b)}{2} \right) \right.$$

$$+ \left\{ (b - \mu)f(b) + (\mu - a)f(a) + \frac{1}{2}[(b - \mu)^2 f'(b) - (\mu - a)^2 f'(a)] \right\}$$

$$\left. \cdot \left[\mu^2 + (a + b)^2 - (a + b)\mu - \left(\frac{5a^2 + 5b^2 + 8ab}{6} \right) \right] \right|$$

$$\leq \begin{cases} \dfrac{\| f^{(3)} \|_p}{(b - a)^2} \cdot \displaystyle\int_a^b \int_a^b |p(\mu, s_1)| \, |p(s_1, s_2)| \cdot \| p(s_2, \cdot) \|_q ds_1 ds_2, \\ \qquad \text{if } f^{(3)} \in L_p([a, b]), \ p, q > 1: \frac{1}{p} + \frac{1}{q} = 1, \\ \dfrac{\| f^{(3)} \|_1}{(b - a)^2} \cdot \displaystyle\int_a^b \int_a^b |p(\mu, s_1)| \, |p(s_1, s_2)| \cdot \| p(s_2, \cdot) \|_\infty ds_1 ds_2, \\ \qquad \text{if } f^{(3)} \in L_1([a, b]). \end{cases} \tag{2.12}$$

Proof. Application of Lemma 1([32]), for $(t - \mu)^2 f(t)$ in place of f, and $x = \mu \in [a, b]$. $\qquad\square$

2.3 Multidimensional Results

Background 2.10. Here we consider random variables $X_i \geq 0$, $i = 1, \ldots, n$, $n \in \mathbb{N}$, taking values in $[0, b_i]$, $b_i > 0$. Let F be the joint probability distribution function of $(X_1, \ldots, X_n)$ assumed to be a continuous function. Let f be the probability density function. We suppose that $\frac{\partial^n F}{\partial x_1 \cdots \partial x_n}$ $(= f)$ exists on $\times_{i=1}^n [0, b_i]$ and is integrable. Clearly, the expectation

$$E\left(\prod_{i=1}^n X_i\right) = \int_0^{b_1} \int_0^{b_2} \cdots \int_0^{b_n} x_1 \cdots x_n f(x_1, x_2, \ldots, x_n) dx_1 \cdots dx_n. \qquad (2.13)$$

Consider on $\times_{i=1}^n [0, b_i]$ the function

$$G_n(x_1, \ldots, x_n) := 1 - \left(\sum_{j=1}^n F(b_1, b_2, \ldots, b_{j-1}, x_j, b_{j+1}, \ldots, b_n)\right)$$

$$+ \left(\sum_{\substack{\ell=1 \\ i<j}}^{\binom{n}{2}} F(b_1, b_2, \ldots, b_{i-1}, x_i, \ldots, x_j, b_{j+1}, \ldots, b_n)_\ell\right)$$

$$+ (-1)^3 \left(\sum_{\substack{\ell^*=1 \\ i<j<k}}^{\binom{n}{3}} F(b_1, b_2, \ldots, x_i, \ldots, x_j, \ldots, x_k, b_{k+1}, \ldots, b_n)_{\ell^*}\right)$$

$$+ \cdots + (-1)^n F(x_1, \ldots, x_n). \qquad (2.14)$$

Above ℓ counts the (i, j)'s, while ℓ^* counts the (i, j, k)'s, etc. One can easily establish that

$$\frac{\partial^n G_n(x_1, \ldots, x_n)}{\partial x_1 \cdots \partial x_n} = (-1)^n f(x_1, \ldots, x_n). \qquad (2.15)$$

We give

Theorem 2.11. *According to the above assumptions and notations we have*

$$E\left(\prod_{i=1}^n X_i\right) = \int_0^{b_1} \cdots \int_0^{b_n} G_n(x_1, \ldots, x_n) dx_1 \cdots dx_n. \qquad (2.16)$$

Proof. Integrate by parts G_n with respect to x_n, taking into account that

$$G_n(x_1, x_2, \ldots, x_{n-1}, b_n) = 0.$$

Then again integrate by parts with respect to variable x_{n-1} the outcome of the first integration. Then repeat, successively, the same procedure by integrating by parts with respect to variables $x_{n-2}, x_{n-3}, \ldots, x_1$. Thus we have recovered formula (2.13). That is deriving (2.16). $\qquad \square$

Let $(x_1, \ldots, x_n) \in \times_{i=1}^n [0, b_i]$ be fixed. We define the kernels $p_i \colon [0, b_i]^2 \to \mathbb{R}$:

$$p_i(x_i, s_i) := \begin{cases} s_i, & s_i \in [0, x_i], \\ s_i - b_i, & s_i \in (x_i, b_i], \end{cases} \qquad (2.17)$$

for all $i = 1, \ldots, n$.

Theorem 2.12. *Based on the above assumptions and notations we derive that*

$$\tilde{\theta}_{1,n} := (-1)^n \cdot \int_{\times_{i=1}^n [0,b_i]} \prod_{i=1}^n p_i(x_i, s_i) f(s_1, s_2, \ldots, s_n) ds_1 ds_2 \cdots ds_n$$

$$= \left\{ \left(\prod_{i=1}^n b_i \right) \cdot G_n(x_1, \ldots, x_n) \right\}$$

$$- \left[\sum_{i=1}^{\binom{n}{1}} \left(\prod_{\substack{j=1 \\ j \neq i}}^n b_j \right) \int_0^{b_i} G_n(x_1, \ldots, s_i, \ldots, x_n) ds_i \right]$$

$$+ \left[\sum_{\ell=1}^{\binom{n}{2}} \left(\prod_{\substack{k=1 \\ k \neq i,j}}^n b_k \right) \left(\int_0^{b_i} \int_0^{b_j} G_n(x_1, \ldots, s_i, \ldots, s_j, \ldots, x_n) ds_i ds_j \right)_{(\ell)} \right]$$

$$- + \cdots - + \cdots + (-1)^{n-1}$$

$$\left[\sum_{j=1}^{\binom{n}{n-1}} b_j \int_{\substack{\times_{i=1}^n [0,b_i] \\ i \neq j}} G_n(s_1, \ldots, x_j, \ldots, s_n) ds_1 \cdots \widehat{ds_j} \cdots ds_n \right]$$

$$+ (-1)^n \cdot E \left(\prod_{i=1}^n X_i \right) =: \tilde{\theta}_{2,n}. \tag{2.18}$$

The above ℓ counts all the (i,j)'s: $i < j$ and $i, j = 1, \ldots, n$. Also $\widehat{ds_j}$ means that ds_j is missing.

Proof. Application of Theorem 2 ([30]) for $f = G_n$, also taking into account (2.15) and (2.16). $\qquad \square$

As a related result we give

Theorem 2.13. *Under the notations and assumptions of this Section 2.3, additionally we assume that $\|f(s_1, \ldots, s_n)\|_\infty < +\infty$. Then*

$$|\tilde{\theta}_{2,n}| \leq \frac{\|f(s_1, \ldots, s_n)\|_\infty}{2^n} \cdot \left\{ \prod_{i=1}^n [x_i^2 + (b_i - x_i)^2] \right\}, \tag{2.19}$$

for any $(x_1, \ldots, x_n) \in \times_{i=1}^n [0, b_i]$, $n \in \mathbb{N}$.

Proof. Based on Theorem 2.12, and Theorem 4 ([30]). $\qquad \square$

The L_p analogue follows

Theorem 2.14. *Under the notations and assumptions of this Section 2.3, additionally we suppose that $f \in L_p(\times_{i=1}^n [0, b_i])$, i.e., $\|f\|_p < +\infty$, where $p, q > 1$: $\frac{1}{p} + \frac{1}{q} = 1$. Then*

$$|\tilde{\theta}_{2,n}| \leq \frac{\|f\|_p}{(q+1)^{n/q}} \cdot \left\{ \prod_{i=1}^n [x_i^{q+1} + (b_i - x_i)^{q+1}] \right\}^{1/q}, \tag{2.20}$$

for any $(x_1, \ldots, x_n) \in \times_{i=1}^n [0, b_i]$.

Proof. Based on Theorem 2.12, and Theorem 6 ([30]). $\qquad\qquad\qquad\qquad\qquad$ $\square$

2.4 Applications

i) **On Theorem 2.6**: Let X be a random variable uniformly distributed over $[a, b]$, $0 < a < b$. Here the probability density function is

$$f(t) = \frac{1}{b - a}. \tag{2.21}$$

And the probability distribution function is

$$F(x) = P(X \leq x) = \frac{x - a}{b - a}, \qquad a \leq x \leq b. \tag{2.22}$$

Furthermore the moment generating function is

$$M_X(x) = \frac{e^{bx} - e^{ax}}{(b - a)x}, \qquad a \leq x \leq b. \tag{2.23}$$

Here $\|f\|_\infty = \frac{1}{b-a}$, and $\|f\|_p = (b - a)^{\frac{1}{p} - 1}$, $p > 1$. Finally the left-hand side of (1.9) is

$$\left| \left(\frac{x - a}{b - a} \right) - \left(\frac{e^{bx} - \left(\frac{e^{bx} - e^{ax}}{x(b-a)} \right)}{e^{bx} - e^{ax}} \right) \right|. \tag{2.24}$$

Then one finds interesting comparisons via inequality (2.9).

ii) **On Theorem 2.7** (see also [105], [107], [117], [174]): Let X be a Beta random variable taking values in $[0, 1]$ with parameters $s > 0$, $t > 0$. It has probability density function

$$f(x; s, t) := \frac{x^{s-1}(1 - x)^{t-1}}{B(s, t)}; \qquad 0 \leq x \leq 1, \tag{2.25}$$

where

$$B(s, t) := \int_0^1 \tau^{s-1}(1 - \tau)^{t-1} d\tau,$$

the Beta function. We know that $E(X) = \frac{s}{s+t}$, see [105]. As noted in [174] for $p > 1$, $s > 1 - \frac{1}{p}$, and $t > 1 - \frac{1}{p}$ we find that

$$\|f(\cdot; s, t)\|_p = \frac{1}{B(s, t)} \cdot [B(p(s - 1) + 1, p(t - 1) + 1)]^{1/p}. \tag{2.26}$$

Also from [107] for $s, t \geq 1$ we have

$$\|f(\cdot, s, t)\|_\infty = \frac{(s - 1)^{s-1}(t - 1)^{t-1}}{B(s, t)(s + t - 2)^{s+t-2}}. \tag{2.27}$$

One can find easily for $m \in \mathbb{N}$ that

$$E(X^m) = \frac{\prod_{j=1}^{m}(s+m-j)}{\prod_{j=1}^{m}(s+t+m-j)}. \tag{2.28}$$

Here $F(x) = P(X \le x)$, P stands for probability.
So we present

Theorem 2.15. *Let X be a Beta random variable taking values in $[0,1]$ with parameters s,t. Let $x \in [0,1]$, $m \in \mathbb{N}$. Then we obtain*

$$\left| P(X \ge x) - \left(\frac{\prod_{j=1}^{m}(s+m-j)}{\prod_{j=1}^{m}(s+t+m-j)} \right) \right|$$

$$\le \begin{cases} \left(\dfrac{(s-1)^{s-1}(t-1)^{t-1}}{B(s,t)(s+t-2)^{s+t-2}} \right) \cdot \left\{ \dfrac{(2x^{m+1}-1)}{m+1} + 1 - x \right\}, & \text{when } s,t \ge 1, \\[2ex] \max(x^m, 1 - x^m), \text{ when } s,t > 0, \\[2ex] \dfrac{1}{B(s,t)} \cdot [B(p(s-1)+1, p(t-1)+1)]^{1/p} \cdot \left[\dfrac{x^{mq+1}}{mq+1} + \displaystyle\int_x^1 (1-t^m)^q dt \right]^{1/q}, \\[1ex] \qquad \text{when } p,q > 1 : \dfrac{1}{p} + \dfrac{1}{q} = 1, \text{ and } s,t > 1 - \dfrac{1}{p}, \\[2ex] \dfrac{1}{B(s,t)} \cdot [B(2s-1, 2t-1)]^{1/2} \\[1ex] \cdot \left(\dfrac{2x^{m+1}}{m+1} - x + \dfrac{2m^2}{(m+1)(2m+1)} \right)^{1/2}, & \text{when } s,t \ge \dfrac{1}{2}. \end{cases} \tag{2.29}$$

The case of $x = \frac{1}{2}$ is especially interesting. We find

Theorem 2.16. *Let X be a Beta random variable taking values in $[0,1]$ with*

parameters s,t. Let $m \in \mathbb{N}$. Then

$$\left| P\left(X \geq \frac{1}{2} \right) - \left(\frac{\prod\limits_{j=1}^{m} (s+m-j)}{\prod\limits_{j=1}^{m} (s+t+m-j)} \right) \right|$$

$$\leq \begin{cases} \left(\dfrac{(s-1)^{s-1}(t-1)^{t-1}}{B(s,t)(s+t-2)^{s+t-2}} \right) \cdot \left(\dfrac{\left(\frac{1}{2^m} - 1 \right)}{m+1} + \dfrac{1}{2} \right), \\ \qquad \text{when } s,t \geq 1, \\[2mm] 1 - \dfrac{1}{2^m}, \quad \text{when } s,t > 0, \\[2mm] \dfrac{1}{B(s,t)} \cdot [B(p(s-1)+1, p(t-1)+1)]^{\frac{1}{p}} \cdot \left[\dfrac{1}{(mq+1)2^{mq+1}} + \displaystyle\int_{\frac{1}{2}}^{1} (1-t^m)^q dt \right]^{\frac{1}{q}}, \\ \qquad \text{when } p,q > 1 : \dfrac{1}{p} + \dfrac{1}{q} = 1, \ \text{and } s,t > 1 - \dfrac{1}{p}, \\[2mm] \dfrac{1}{B(s,t)} \cdot [B(2s-1, 2t-1)]^{\frac{1}{2}} \cdot \left(\dfrac{1}{(m+1) \cdot 2^m} - \dfrac{1}{2} + \dfrac{2m^2}{(m+1)(2m+1)} \right)^{\frac{1}{2}}, \\ \qquad \text{when } s,t > \dfrac{1}{2}. \end{cases} \qquad (2.30)$$

2.5   Addendum

We present here a detailed proof of Theorem 2.11, when $n = 3$.

Theorem 2.17. *Let the random vector (X,Y,Z) taking values on $\prod_{i=1}^{3}[0,b_i]$, where $b_i > 0$, $i = 1,2,3$. Let F be the cumulative distribution function of (X,Y,Z), assumed to be continuous on $\prod_{i=1}^{3}[0,b_i]$. Assume that there exists $f = \frac{\partial^3 F}{\partial x \partial y \partial z} \geq 0$ a.e, and is integrable on $\prod_{i=1}^{3}[0,b_i]$. Then f is a joint probability density function of (X,Y,Z). Furthermore we have the representation*

$$E(XYZ) = \int_0^{b_1} \int_0^{b_2} \int_0^{b_3} G_3(x,y,z)\,dxdydz, \qquad (2.31)$$

where

$$G_3(x,y,z) := 1 - F(x,b_2,b_3) - F(b_1,y,b_3) + F(x,y,b_3)$$

$$-F(b_1,b_2,z) + F(x,b_2,z) + F(b_1,y,z) - F(x,y,z). \qquad (2.32)$$

Proof. For the definition of the multidimensional cumulative distribution see [235], $p.\,97$.

We observe that

$$\int_0^{b_3} \left(\int_0^{b_2} \left(\int_0^{b_1} f(x,y,z)dx \right) dy \right) dz$$

$$= \int_0^{b_3} \left(\int_0^{b_2} \left(\int_0^{b_1} \frac{\partial^3 F(x,y,z)}{\partial x \partial y\, dz} dx \right) dy \right) dz$$

$$= \int_0^{b_3} \left(\int_0^{b_2} \frac{\partial^2}{\partial y \partial z} \left(F(b_1,y,z) - F(o,y,z) \right) dy \right) dz \tag{2.33}$$

$$= \int_0^{b_3} \frac{\partial}{\partial z} \left[F(b_1,b_2,z) - F(0,b_2,z) - F(b_1,0,z) + F(0,0,z) \right] dz$$

$$= F(b_1,b_2,b_3) - F(0,b_2,b_3) - F(b_1,0,b_3) + F(0,0,b_3) - F(b_1,b_2,0)$$

$$+ F(0,b_2,0) + F(b_1,0,0) - F(0,0,0) = F(b_1,b_2,b_3) = 1, \tag{2.34}$$

proving that f is a probability density function.

We know that

$$E\left(XYZ\right) = \int_0^{b_1} \int_0^{b_2} \int_0^{b_3} xyz f(x,y,z) dx dy dz.$$

We notice that $G_3(x,y,b_3) = 0$.

So me have

$$\int_0^{b_3} G_3(x,y,z) dz = z G_3(x,y,z) \Big|_0^{b_3} - \int_0^{b_3} z\, d_z G_3(x,y,z)$$

$$= -\int_0^{b_3} z\, d_z G_3(x,y,z) = \int_0^{b_3} x\alpha_1(x,y,z) dz, \tag{2.35}$$

where

$$\alpha_1(x,y,z) := \frac{\partial F(b_1,b_2,z)}{\partial z} - \frac{\partial F(x,b_2,z)}{\partial z} - \frac{\partial F(b_1,y,z)}{\partial z} + \frac{\partial F(x,y,z)}{\partial z}. \tag{2.36}$$

Next take

$$\int_0^{b_2} \alpha_1(x,y,z) dy$$

$$= \int_0^{b_2} \left[\frac{\partial F(b_1,b_2,z)}{\partial z} - \frac{\partial F(x,b_2,z)}{\partial z} - \frac{\partial F(b_1,y,z)}{\partial z} + \frac{\partial F(x,y,z)}{\partial z} \right] dy$$

$$= y \left[\frac{\partial F(b_1,b_2,z)}{\partial z} - \frac{\partial F(x,b_2,z)}{\partial z} - \frac{\partial F(b_1,y,z)}{\partial z} + \frac{\partial F(x,y,z)}{\partial z} \right] \Big|_0^{b_2}$$

$$- \int_0^{b_2} y \left[\frac{-\partial^2 F(b_1,y,z)}{\partial y \partial z} + \frac{\partial^2 F(x,y,z)}{\partial y \partial z} \right] dy = \int_0^{b_2} y\alpha_2(x,y,z) dy, \tag{2.37}$$

where

$$\alpha_2(x,y,z) := \frac{\partial^2 F(b_1,y,z)}{\partial y \partial z} - \frac{\partial^2 F(x,y,z)}{\partial y \partial z}. \tag{2.38}$$

Finally we see that

$$\int_0^{b_1} \alpha_2(x,y,z)dx$$

$$= \int_0^{b_1} \left(\frac{\partial^2 F(b_1,y,z)}{\partial y \partial z} - \frac{\partial^2 F(x,y,z)}{\partial y \partial z} \right) dx$$

$$= x \left(\frac{\partial^2 F(b_1,y,z)}{\partial y \partial z} - \frac{\partial^2 F(x,y,z)}{\partial y \partial z} \right) \cdot \Big|_0^{b_1} - \int_0^{b_1} x \left(- \frac{\partial^3 F(x,y,z)}{\partial x \partial y \partial z} \right) dx$$

$$= \int_0^{b_1} x \frac{\partial^3 F(x,y,z)}{\partial x \partial y \partial z} dx = \int_0^{b_1} x f(x,y,z)dx. \tag{2.39}$$

Putting things together we get

$$\int_0^{b_1} \left(\int_0^{b_2} \left(\int_0^{b_3} G_3(x,y,z)dz \right) dy \right) dx$$

$$= \int_0^{b_1} \left(\int_0^{b_2} \left(\int_0^{b_3} z\alpha_1(x,y,z)dz \right) dy \right) dx$$

$$= \int_0^{b_3} z \left(\int_0^{b_1} \left(\int_0^{b_2} \alpha_1(x,y,z)dy \right) dx \right) dz$$

$$= \int_0^{b_3} z \left(\int_0^{b_1} \left(\int_0^{b_2} y\,\alpha_2(x,y,z)dy \right) dx \right) dz$$

$$= \int_0^{b_3} z \left(\int_0^{b_2} y \left(\int_0^{b_1} \alpha_2(x,y,z)dx \right) dy \right) dz$$

$$= \int_0^{b_3} z \left(\int_0^{b_2} y \left(\int_0^{b_1} x f(x,y,z)dx \right) dy \right) dz$$

$$= \int_0^{b_1} \int_0^{b_2} \int_0^{b_3} xyz\, f(x,y,z)dx dy dz = E(XYZ), \tag{2.40}$$

proving the claim. $\qquad\square$

Remark 2.18. Notice that

$$\frac{\partial G_3(x,y,z)}{\partial z} = - \frac{\partial F(b_1,b_2,z)}{\partial z} + \frac{\partial F(x,b_2,z)}{\partial z}$$

$$+ \frac{\partial F(b_1,y,z)}{\partial z} - \frac{\partial F(x,y,z)}{\partial z},$$

$$\frac{\partial^2 G_3(x,y,z)}{\partial y \partial z} = \frac{\partial^2 F(b_1,y,z)}{\partial y \partial z} - \frac{\partial^2 F(x,y,z)}{\partial y \partial z},$$

and

$$\frac{\partial^3 G_3(x,y,z)}{\partial x \partial y \partial z} = - \frac{\partial^3 F(x,y,z)}{\partial x \partial y \partial z} = -f(x,y,z).$$

That is giving us

$$\frac{\partial^3 G_3(x,y,z)}{\partial x \partial y \partial z} = -f(x,y,z). \tag{2.41}$$

Chapter 3

Multidimensional Montgomery Identities and Ostrowski type Inequalities

Multidimensional Montgomery type identities are presented involving, among others, integrals of mixed partial derivatives. These lead to interesting multidimensional Ostrowski type inequalities. The inequalities in their right-hand sides involve L_p-norms ($1 \le p \le \infty$) of the engaged mixed partials. An application is given also to Probability. This treatment is based on [33].

3.1 Introduction

In 1938, A. Ostrowski [277] proved the following inequality:

Theorem 3.1. *Let* $f: [a,b] \to \mathbb{R}$ *be continuous on* $[a,b]$ *and differentiable on* (a,b) *whose derivative* $f': (a,b) \to \mathbb{R}$ *is bounded on* (a,b), *i.e.,* $\|f'\|_\infty = \sup\limits_{t \in (a,b)} |f'(t)| < +\infty$. *Then*

$$\left| \frac{1}{b-a} \int_a^b f(t)dt - f(x) \right| \le \left[\frac{1}{4} + \frac{(x - \frac{a+b}{2})^2}{(b-a)^2} \right] \cdot (b-a)\|f'\|_\infty, \qquad (3.1)$$

for any $x \in [a,b]$. *The constant* $\frac{1}{4}$ *is the best possible.*

Since then there has been a lot of activity about these inequalities with interesting applications to Probability and Numerical Analysis.

This chapter has been greatly motivated by the following result of S. Dragomir et al. (2000), see [176] and [99], p. 263, Theorem 5.18. It gives a multidimensional Montgomery type identity.

Theorem 3.2. *Let* $f: [a,b] \times [c,d] \to \mathbb{R}$ *be such that the partial derivatives* $\frac{\partial f(t,s)}{\partial t}$, $\frac{\partial f(t,s)}{\partial s}$, $\frac{\partial^2 f(t,s)}{\partial t \partial s}$ *exist and are continuous on* $[a,b] \times [c,d]$. *Then for all* $(x,y) \in [a,b] \times [c,d]$, *we have the representation*

$$f(x,y) = \frac{1}{(b-a)(d-c)} \left\{ \int_a^b \int_c^d f(t,s)dsdt + \int_a^b \int_c^d p(x,t)\frac{\partial f(t,s)}{\partial t}dsdt \right.$$

$$\left. + \int_a^b \int_c^d q(y,s)\frac{\partial f(t,s)}{\partial s}dsdt + \int_a^b \int_c^d p(x,t)q(y,s)\frac{\partial^2 f(t,s)}{\partial t \partial s}dsdt \right\}, \qquad (3.2)$$

15

where $p\colon [a,b]^2 \to \mathbb{R}$, $q\colon [c,d]^2 \to \mathbb{R}$ and are given by

$$p(x,t) := \begin{cases} t - a, & \text{if } t \in [a,x] \\ t - b, & \text{if } t \in (x,b] \end{cases}$$

and

$$q(y,s) := \begin{cases} s - c, & \text{if } s \in [c,y] \\ s - d, & \text{if } s \in (y,d]. \end{cases}$$

In this chapter we give more general multidimensional Montgomery type identities and derive related new multidimensional Ostrowski type inequalities involving $\|\cdot\|_p$ ($1 \le p \le \infty$) norms of the engaged mixed partial derivatives. At the end we give an application to Probability. That is finding bounds for the difference of the expectation of a product of three random variables from an appropriate linear combination of values of the related joint probability distribution function.

3.2   Results

We present a multidimensional Montgomery type identity which is a generalization of Theorem 5.18, p. 263 in [99], see also [176]. We give the representation.

Theorem 3.3. *Let $f\colon \times_{i=1}^3 [a_i,b_i] \to \mathbb{R}$ such that the partials $\frac{\partial f(s_1,s_2,s_3)}{\partial s_j}$, $j = 1,2,3$; $\frac{\partial^2 f(s_1,s_2,s_3)}{\partial s_k \partial s_j}$, $j < k$, $j,k \in \{1,2,3\}$; $\frac{\partial^3 f(s_1,s_2,s_3)}{\partial s_3 \partial s_2 \partial s_1}$ exist and are continuous on $\times_{i=1}^3 [a_i,b_i]$. Let $(x_1,x_2,x_3) \in \times_{i=1}^3 [a_i,b_i]$. Define the kernels $p_i\colon [a_i,b_i]^2 \to \mathbb{R}$:*

$$p_i(x_i,s_i) = \begin{cases} s_i - a_i, & s_i \in [a_i,x_i] \\ s_i - b_i, & s_i \in (x_i,b_i], \end{cases}$$

for $i = 1,2,3$. Then

$$
f(x_1,x_2,x_3)
$$
$$
= \frac{1}{\prod\limits_{i=1}^3 (b_i - a_i)} \left\{ \int_{a_1}^{b_1} \int_{a_2}^{b_2} \int_{a_3}^{b_3} f(s_1,s_2,s_3) ds_3 ds_2 ds_1 \right.
$$
$$
+ \sum_{j=1}^3 \left(\int_{a_1}^{b_1} \int_{a_2}^{b_2} \int_{a_3}^{b_3} p_j(x_j,s_j) \frac{\partial f(s_1,s_2,s_3)}{\partial s_j} ds_3 ds_2 ds_1 \right)
$$
$$
+ \sum_{\substack{\ell=1 \\ j<k}}^3 \left(\int_{a_1}^{b_1} \int_{a_2}^{b_2} \int_{a_3}^{b_3} p_j(x_j,s_j) p_k(x_k,s_k) \cdot \frac{\partial^2 f(s_1,s_2,s_3)}{\partial s_k \partial s_j} ds_3 ds_2 ds_1 \right) (\ell)
$$
$$
+ \int_{a_1}^{b_1} \int_{a_2}^{b_2} \int_{a_3}^{b_3} \left(\prod_{i=1}^3 p_i(x_i,s_i) \right) \frac{\partial^3 f(s_1,s_2,s_3)}{\partial s_3 \partial s_2 \partial s_1} ds_3 ds_2 ds_1 \left. \right\}. \tag{3.3}
$$

Above ℓ counts (j,k): $j < k$; $j,k \in \{1,2,3\}$.

Proof. Here we use repeatedly the following identity due to Montgomery (see [265], p. 565), which can be easily proved by using integration by parts

$$g(u) = \frac{1}{\beta - \alpha} \int_\alpha^\beta g(z)dz + \frac{1}{\beta - \alpha} \int_\alpha^\beta k(u,z)g'(z)dz,$$

where $k \colon [\alpha, \beta]^2 \to \mathbb{R}$ is defined by

$$k(u,z) := \begin{cases} z - \alpha, & \text{if } z \in [\alpha, u] \\ z - \beta, & \text{if } z \in (u, \beta], \end{cases}$$

and g is absolutely continuous on $[\alpha, \beta]$.

First we see that

$$f(x_1, x_2, x_3) = A_0 + B_0,$$

where

$$A_0 := \frac{1}{b_1 - a_1} \int_{a_1}^{b_1} f(s_1, x_2, x_3)ds_1,$$

and

$$B_0 := \frac{1}{b_1 - a_1} \int_{a_1}^{b_1} p_1(x_1, s_1) \frac{\partial f(s_1, x_2, x_3)}{\partial s_1} ds_1.$$

Furthermore we have

$$f(s_1, x_2, x_3) = A_1 + B_1,$$

where

$$A_1 := \frac{1}{b_2 - a_2} \int_{a_2}^{b_2} f(s_1, s_2, x_3)ds_2,$$

and

$$B_1 := \frac{1}{b_2 - a_2} \int_{a_2}^{b_2} p_2(x_2, s_2) \frac{\partial f(s_1, s_2, x_3)}{\partial s_2} ds_2.$$

Also we find that

$$f(s_1, s_2, x_3) = \frac{1}{b_3 - a_3} \int_{a_3}^{b_3} f(s_1, s_2, s_3)ds_3$$
$$+ \frac{1}{b_3 - a_3} \int_{a_3}^{b_3} p_3(x_3, s_3) \frac{\partial f(s_1, s_2, s_3)}{\partial s_3} ds_3.$$

Next we put things together, and we derive

$$A_1 = \frac{1}{(b_2 - a_2)(b_3 - a_3)} \int_{a_2}^{b_2} \int_{a_3}^{b_3} f(s_1, s_2, s_3)ds_3 ds_2$$
$$+ \frac{1}{(b_2 - a_2)(b_3 - a_3)} \int_{a_2}^{b_2} \int_{a_3}^{b_3} p_3(x_3, s_3) \frac{\partial f(s_1, s_2, s_3)}{\partial s_3} ds_3 ds_2.$$

And

$$A_0 = \frac{1}{\prod_{i=1}^{3}(b_i - a_i)} \int_{a_1}^{b_1} \int_{a_2}^{b_2} \int_{a_3}^{b_3} f(s_1, s_2, s_3) ds_3 ds_2 ds_1$$

$$+ \frac{1}{\prod_{i=1}^{3}(b_i - a_i)} \int_{a_1}^{b_1} \int_{a_2}^{b_2} \int_{a_3}^{b_3} p_3(x_3, s_3) \frac{\partial f(s_1, s_2, s_3)}{\partial s_3} ds_3 ds_2 ds_1$$

$$+ \frac{1}{(b_1 - a_1)(b_2 - a_2)} \int_{a_1}^{b_1} \int_{a_2}^{b_2} p_2(x_2, s_2) \frac{\partial f(s_1, s_2, x_3)}{\partial s_2} ds_2 ds_1.$$

Also we obtain

$$\frac{\partial f(s_1, s_2, x_3)}{\partial s_2} = \frac{1}{b_3 - a_3} \int_{a_3}^{b_3} \frac{\partial f(s_1, s_2, s_3)}{\partial s_2} ds_3$$

$$+ \frac{1}{b_3 - a_3} \int_{a_3}^{b_3} p_3(x_3, s_3) \frac{\partial^2 f(s_1, s_2, s_3)}{\partial s_3 \partial s_2} ds_3.$$

Therefore we get

$$A_0 = \frac{1}{\prod_{i=1}^{3}(b_i - a_i)} \int_{a_1}^{b_1} \int_{a_2}^{b_2} \int_{a_3}^{b_3} f(s_1, s_2, s_3) ds_3 ds_2 ds_1$$

$$+ \frac{1}{\prod_{i=1}^{3}(b_i - a_i)} \int_{a_1}^{b_1} \int_{a_2}^{b_2} \int_{a_3}^{b_3} p_3(x_3, s_3) \frac{\partial f(s_1, s_2, s_3)}{\partial s_3} ds_3 ds_2 ds_1$$

$$+ \frac{1}{\prod_{i=1}^{3}(b_i - a_i)} \int_{a_1}^{b_1} \int_{a_2}^{b_2} \int_{a_3}^{b_3} p_2(x_2, s_2) \frac{\partial f(s_1, s_2, s_3)}{\partial s_2} ds_3 ds_2 ds_1$$

$$+ \frac{1}{\prod_{i=1}^{3}(b_i - a_i)} \int_{a_1}^{b_1} \int_{a_2}^{b_2} \int_{a_3}^{b_3} p_2(x_2, s_2) p_3(x_3, s_3)$$

$$\cdot \frac{\partial^2 f(s_1, s_2, s_3)}{\partial s_3 \partial s_2} ds_3 ds_2 ds_1.$$

Similarly we obtain that

$$
B_0 = \frac{1}{\prod\limits_{i=1}^{3}(b_i - a_i)} \int_{a_1}^{b_1} \int_{a_2}^{b_2} \int_{a_3}^{b_3} p_1(x_1, s_1)\frac{\partial f(s_1, s_2, s_3)}{\partial s_1} ds_3 ds_2 ds_1
$$

$$
+ \frac{1}{\prod\limits_{i=1}^{3}(b_i - a_i)} \int_{a_1}^{b_1} \int_{a_2}^{b_2} \int_{a_3}^{b_3} p_1(x_1, s_1)p_3(x_3, s_3)\frac{\partial^2 f(s_1, s_2, s_3)}{\partial s_3 \partial s_1} ds_3 ds_2 ds_1
$$

$$
+ \frac{1}{\prod\limits_{i=1}^{3}(b_i - a_i)} \int_{a_1}^{b_1} \int_{a_2}^{b_2} \int_{a_3}^{b_3} p_1(x_1, s_1)p_2(x_2, s_2)\frac{\partial^2 f(s_1, s_2, s_3)}{\partial s_2 \partial s_1} ds_3 ds_2 ds_1
$$

$$
+ \frac{1}{\prod\limits_{i=1}^{3}(b_i - a_i)} \int_{a_1}^{b_1} \int_{a_2}^{b_2} \int_{a_3}^{b_3} p_1(x_1, s_1)p_2(x_2, s_2)p_3(x_3, s_3)
$$

$$
\cdot \frac{\partial^3 f(s_1, s_2, s_3)}{\partial s_3 \partial s_2 \partial s_1} ds_3 ds_2 ds_1.
$$

We have proved (3.3). $\square$

A generalization of Theorem 3.3 to any $n \in \mathbb{N}$ follows:

Theorem 3.4. *Let $f \colon \times_{i=1}^{n}[a_i, b_i] \to \mathbb{R}$ such that the partials*

$$
\frac{\partial f}{\partial s_j}(s_1, s_2, \ldots, s_n),\ j = 1, \ldots, n;\quad \frac{\partial^2 f}{\partial s_k \partial s_j},\ j < k,
$$

$$
j, k \in \{1, \ldots, n\};\quad \frac{\partial^3 f}{\partial s_r \partial s_k \partial s_j},\ j < k < r,
$$

$$
j, k, r \in \{1, \ldots, n\}; \ldots;\quad \frac{\partial^{n-1} f}{\partial s_n \cdots \partial \hat{s}_\ell \cdots \partial s_1},\ \ell \in \{1, \ldots, n\};\quad \frac{\partial^n f}{\partial s_n \partial s_{n-1} \cdots \partial s_1}
$$

*all exist and are continuous on $\times_{i=1}^{n}[a_i, b_i]$, $n \in \mathbb{N}$. Above $\partial \hat{s}_\ell$ means ∂s_ℓ is missing.
Let $(x_1, \ldots, x_n) \in \times_{i=1}^{n}[a_i, b_i]$. Define the kernels $p_i \colon [a_i, b_i]^2 \to \mathbb{R}$:*

$$
p_i(x_i, s_i) := \begin{cases} s_i - a_i, & s_i \in [a_i, x_i], \\ s_i - b_i, & s_i \in (x_i, b_i], \end{cases} \quad \text{for } i = 1, 2, \ldots, n. \tag{3.4}
$$

Then

$$f(x_1, x_2, \ldots, x_n) = \frac{1}{\left(\prod_{i=1}^{n} (b_i - a_i)\right)}$$

$$\cdot \left\{ \int_{\times_{i=1}^{n}[a_i,b_i]} f(s_1, s_2, \ldots, s_n) ds_n ds_{n-1} \cdots ds_1 \right.$$

$$+ \sum_{j=1}^{n} \left(\int_{\times_{i=1}^{n}[a_i,b_i]} p_j(x_j, s_j) \frac{\partial f(s_1, s_2, \ldots, s_n)}{\partial s_j} ds_n \cdots ds_1 \right)$$

$$+ \sum_{\substack{\ell_1 = 1 \\ j < k}}^{\binom{n}{2}} \left(\int_{\times_{i=1}^{n}[a_i,b_i]} p_j(x_j, s_j) p_k(x_k, s_k) \frac{\partial^2 f(s_1, s_2, \ldots, s_n)}{\partial s_k \partial s_j} ds_n \cdots ds_1 \right)_{(\ell_1)}$$

$$+ \sum_{\substack{\ell_2 = 1 \\ j < k < r}}^{\binom{n}{3}} \left(\int_{\times_{i=1}^{n}[a_i,b_i]} p_j(x_j, s_j) p_k(x_k, s_k) p_r(x_r, s_r) \right.$$

$$\left. \cdot \frac{\partial^3 f(s_1, \ldots, s_n)}{\partial s_r \partial s_k \partial s_j} ds_n \cdots ds_1 \right)_{(\ell_2)}$$

$$+ \cdots + \sum_{\ell=1}^{\binom{n}{n-1}} \left(\left(\int_{\times_{i=1}^{n}[a_i,b_i]} p_1(x_1, s_1) \cdots \widehat{p_\ell(x_\ell, s_\ell)} \cdots p_n(x_n, s_n) \right. \right.$$

$$\left. \left. \cdot \frac{\partial^{n-1} f(s_1, \ldots, s_n)}{\partial s_n \cdots \partial \hat{s}_\ell \cdots \partial s_1} ds_n \cdots \widehat{ds_\ell} \cdots ds_1 \right) \right.$$

$$\left. + \int_{\times_{i=1}^{n}[a_i,b_i]} \left(\prod_{i=1}^{n} p_i(x_i, s_i) \right) \cdot \frac{\partial^n f(s_1, \ldots, s_n)}{\partial s_n \cdots \partial s_1} ds_n \cdots ds_1 \right\}. \tag{3.5}$$

Above ℓ_1 counts (j,k): $j < k$; $j,k \in \{1,2,\ldots,n\}$, also ℓ_2 counts (j,k,r): $j < k < r$; $j,k,r \in \{1,2,\ldots,n\}$, etc. Also $\widehat{p_\ell(x_\ell, s_\ell)}$ and $\widehat{ds_\ell}$ means that $p_\ell(x_\ell, s_\ell)$ and ds_ℓ are missing, respectively.

Proof. Similar to Theorem 3.3. $\qquad\qquad\square$

Next we present some Ostrowski type [277] inequalities. In particular the result that follows generalizes Theorem 5.21, p. 266 from [99], also the corresponding result from [176].

Theorem 3.5. *Let $f \colon \times_{i=1}^{3} [a_i, b_i] \to \mathbb{R}$ as in Theorem 3.3. Then*

$$\left| f(x_1, x_2, x_3) - \frac{1}{\prod_{i=1}^{3} (b_i - a_i)} \int_{a_1}^{b_1} \int_{a_2}^{b_2} \int_{a_3}^{b_3} f(s_1, s_2, s_3) ds_3 ds_2 ds_1 \right|$$

$$\leq (M_{1,3} + M_{2,3} + M_{3,3}) / \left(\prod_{i=1}^{3} (b_i - a_i) \right). \tag{3.6}$$

Here we have

$$
M_{1,3} := \begin{cases}
\sum_{j=1}^{3} \left(\left\| \frac{\partial f}{\partial s_j} \right\|_{\infty} \right. \\
\quad \left. \cdot \left(\prod_{\substack{i=1 \\ i \neq j}}^{3} (b_i - a_i) \right) \cdot \left(\frac{(x_j - a_j)^2 + (b_j - x_j)^2}{2} \right) \right), \\
\quad \text{if } \frac{\partial f}{\partial s_j} \in L_{\infty}\left(\times_{i=1}^{3} [a_i, b_i] \right), \ j = 1,2,3; \\[2mm]
\sum_{j=1}^{3} \left(\left\| \frac{\partial f}{\partial s_j} \right\|_{p_j} \cdot \left(\prod_{\substack{i=1 \\ i \neq j}}^{3} (b_i - a_i)^{1/q_j} \right) \right. \\
\quad \left. \cdot \left[\frac{(x_j - a_j)^{q_j+1} + (b_j - x_j)^{q_j+1}}{q_j+1} \right]^{1/q_j} \right), \\
\quad \text{if } \frac{\partial f}{\partial s_j} \in L_{p_j}\left(\times_{i=1}^{3} [a_i, b_i] \right), \ p_j > 1, \\
\qquad \frac{1}{p_j} + \frac{1}{q_j} = 1, \ \text{for } j = 1,2,3; \\[2mm]
\sum_{j=1}^{3} \left(\left\| \frac{\partial f}{\partial s_j} \right\|_1 \cdot \left(\frac{b_j - a_j}{2} + \left| x_j - \frac{a_j + b_j}{2} \right| \right) \right), \\
\quad \text{if } \frac{\partial f}{\partial s_j} \in L_1\left(\times_{i=1}^{3} [a_i, b_i] \right), \ \text{for } j = 1,2,3.
\end{cases}
\tag{3.7}
$$

Also we have

$$
M_{2,3} := \begin{cases}
\sum_{\substack{\ell=1 \\ j<k}}^{3} \left(\left\| \frac{\partial^2 f}{\partial s_k \partial s_j} \right\|_{\infty} \cdot \left(\prod_{\substack{i=1 \\ i \neq j,k}}^{3} (b_i - a_i) \right) \right. \\
\quad \left. \cdot \frac{((x_j - a_j)^2 + (b_j - x_j)^2)((x_k - a_k)^2 + (b_k - x_k)^2)}{4} \right)_{(\ell)}, \\
\quad \text{if } \frac{\partial^2 f}{\partial s_k \partial s_j} \in L_{\infty}\left(\times_{i=1}^{3} [a_i, b_i] \right), k, j \in \{1,2,3\}; \\[2mm]
\sum_{\substack{\ell=1 \\ j<k}}^{3} \left(\left\| \frac{\partial^2 f}{\partial s_k \partial s_j} \right\|_{p_{kj}} \cdot \left(\prod_{\substack{i=1 \\ i \neq j,k}}^{3} (b_i - a_i)^{1/q_{kj}} \right) \right. \\
\quad \cdot \left[\frac{(x_j - a_j)^{q_{kj}+1} + (b_j - x_j)^{q_{kj}+1}}{q_{kj}+1} \right]^{1/q_{kj}} \\
\quad \left. \left[\frac{(x_k - a_k)^{q_{kj}+1} + (b_k - x_k)^{q_{kj}+1}}{q_{kj}+1} \right]^{1/q_{kj}} \right)_{(\ell)}, \\
\quad \text{if } \frac{\partial^2 f}{\partial s_k \partial s_j} \in L_{p_{kj}}\left(\times_{i=1}^{3} [a_i, b_i] \right), p_{kj} > 1, \frac{1}{p_{kj}} + \frac{1}{q_{kj}} = 1, \\
\qquad \text{if } j, k \in \{1,2,3\}; \\[2mm]
\sum_{\substack{\ell=1 \\ j<k}}^{3} \left(\left\| \frac{\partial^2 f}{\partial s_k \partial s_j} \right\|_1 \cdot \left(\frac{b_j - a_j}{2} + \left| x_j - \frac{a_j + b_j}{2} \right| \right) \right. \\
\quad \left. \cdot \left(\frac{b_k - a_k}{2} + \left| x_k - \frac{a_k + b_k}{2} \right| \right) \right)_{(\ell)}, \\
\quad \text{if } \frac{\partial^2 f}{\partial s_k \partial s_j} \in L_1\left(\times_{i=1}^{3} [a_i, b_i] \right), \ \text{for } j, k \in \{1,2,3\}.
\end{cases}
\tag{3.8}
$$

And finally,

$$M_{3,3} := \begin{cases} \left\| \dfrac{\partial^3 f(s_1, s_2, s_3)}{\partial s_3 \partial s_2 \partial s_1} \right\|_\infty \cdot \dfrac{\prod\limits_{j=1}^{3} \left((x_j - a_j)^2 + (b_j - x_j)^2 \right)}{8}, \\[2mm] \quad \text{if } \dfrac{\partial^3 f(s_1, s_2, s_3)}{\partial s_3 \partial s_2 \partial s_1} \in L_\infty \left(\times_{i=1}^{3} [a_i, b_i] \right); \\[3mm] \left\| \dfrac{\partial^3 f(s_1, s_2, s_3)}{\partial s_3 \partial s_2 \partial s_1} \right\|_{p_{123}} \prod\limits_{j=1}^{3} \\[2mm] \quad \cdot \left(\dfrac{(x_j - a_j)^{q_{123}+1} + (b_j - x_j)^{q_{123}+1}}{q_{123} + 1} \right)^{1/q_{123}}, \\[2mm] \quad \text{if } \dfrac{\partial^3 f(s_1, s_2, s_3)}{\partial s_3 \partial s_2 \partial s_1} \in L_{p_{123}} \left(\times_{i=1}^{3} [a_i, b_i] \right), p_{123} > 1, \\[2mm] \quad \dfrac{1}{p_{123}} + \dfrac{1}{q_{123}} = 1; \\[3mm] \left\| \dfrac{\partial^3 f(s_1, s_2, s_3)}{\partial s_3 \partial s_2 \partial s_1} \right\|_1 \cdot \prod\limits_{j=1}^{3} \left(\dfrac{b_j - a_j}{2} + \left| x_j - \dfrac{a_j + b_j}{2} \right| \right), \\[2mm] \quad \text{if } \dfrac{\partial^3 f(s_1, s_2, s_3)}{\partial s_3 \partial s_2 \partial s_1} \in L_1 \left(\times_{i=1}^{3} [a_i, b_i] \right). \end{cases} \tag{3.9}$$

Inequality (3.6) is valid for any $(x_1, x_2, x_3) \in \times_{i=1}^{3} [a_i, b_i]$, *where* $\| \cdot \|_p$ $(1 \leq p \leq \infty)$ *are the usual* L_p*-norms on* $\times_{i=1}^{3} [a_i, b_i]$.

Proof. We have by (3.3) that

$$\left| f(x_1, x_2, x_3) - \frac{1}{\prod\limits_{i=1}^{3} (b_i - a_i)} \int_{a_1}^{b_1} \int_{a_2}^{b_2} \int_{a_3}^{b_3} f(s_1, s_2, s_3) ds_3 ds_2 ds_1 \right|$$

$$\leq \frac{1}{\prod\limits_{i=1}^{3} (b_i - a_i)} \left(\sum_{j=1}^{3} \left(\int_{a_1}^{b_1} \int_{a_2}^{b_2} \int_{a_3}^{b_3} |p_j(x_j, s_j)| \right. \right.$$

$$\left. \cdot \left| \frac{\partial f(s_1, s_2, s_3)}{\partial s_j} \right| ds_3 ds_2 ds_1 \right)$$

$$+ \sum_{\substack{\ell=1 \\ j<k}}^{3} \left(\int_{a_1}^{b_1} \int_{a_2}^{b_2} \int_{a_3}^{b_3} |p_j(x_j, s_j)| \cdot |p_k(x_k, s_k)| \right.$$

$$\left. \cdot \left| \frac{\partial^2 f(s_1, s_2, s_3)}{\partial s_k \partial s_j} \right| ds_3 ds_2 ds_1 \right)_{(\ell)}$$

$$+ \int_{a_1}^{b_1} \int_{a_2}^{b_2} \int_{a_3}^{b_3} \left(\prod_{i=1}^{3} |p_i(x_i, s_i)| \right) \cdot \left| \frac{\partial^3 f(s_1, s_2, s_3)}{\partial s_3 \partial s_2 \partial s_1} \right| ds_3 ds_2 ds_1 \left. \right).$$

We notice the following $(j = 1, 2, 3)$

$$\int_{a_1}^{b_1} \int_{a_2}^{b_2} \int_{a_3}^{b_3} |p_j(x_j, s_j)| \cdot \left| \frac{\partial f(s_1, s_2, s_3)}{\partial s_j} \right| ds_3 ds_2 ds_1$$

$$\leq \begin{cases} \left\| \dfrac{\partial f}{\partial s_j} \right\|_\infty \cdot \int_{a_1}^{b_1} \int_{a_2}^{b_2} \int_{a_3}^{b_3} |p_j(x_j, s_j)| ds_3 ds_2 ds_1, \\ \quad \text{if } \dfrac{\partial f}{\partial s_j} \in L_\infty \left(\times_{i=1}^3 [a_i, b_i] \right); \\[2mm] \left\| \dfrac{\partial f}{\partial s_j} \right\|_{p_j} \left(\int_{a_1}^{b_1} \int_{a_2}^{b_2} \int_{a_3}^{b_3} |p_j(x_j, s_j)|^{q_j} ds_3 ds_2 ds_1 \right)^{1/q_j}, \\ \quad \text{if } \dfrac{\partial f}{\partial s_j} \in L_{p_j} \left(\times_{i=1}^3 [a_i, b_i] \right), p_j > 1, \dfrac{1}{p_j} + \dfrac{1}{q_j} = 1; \\[2mm] \left\| \dfrac{\partial f}{\partial s_j} \right\|_1 \cdot \sup_{s_j \in [a_j, b_j]} |p_j(x_j, s_j)|, \text{ if } \dfrac{\partial f}{\partial s_j} \in L_1 \left(\times_{i=1}^3 [a_i, b_i] \right). \end{cases}$$

Furthermore we observe that

$$\int_{a_1}^{b_1} \int_{a_2}^{b_2} \int_{a_3}^{b_3} |p_j(x_j, s_j)| ds_3 ds_2 ds_1$$

$$= (b_1 - a_1)(\widehat{b_j - a_j})(b_3 - a_3) \left(\frac{(x_j - a_j)^2 + (b_j - x_j)^2}{2} \right);$$

$\widehat{b_j - a_j}$ means $(b_j - a_j)$ is missing, for $j = 1, 2, 3$. Also we find

$$\left(\int_{a_1}^{b_1} \int_{a_2}^{b_2} \int_{a_3}^{b_3} |p_j(x_j, s_j)|^{q_j} ds_3 ds_2 ds_1 \right)^{1/q_j}$$

$$= (b_1 - a_1)^{1/q_j} (\widehat{b_j - a_j})^{1/q_j} (b_3 - a_3)^{1/q_j}$$

$$\cdot \left[\frac{(x_j - a_j)^{q_j + 1} + (b_j - x_j)^{q_j + 1}}{q_j + 1} \right]^{1/q_j};$$

$(\widehat{b_j - a_j})^{1/q_j}$ means $(b_j - a_j)^{1/q_j}$ is missing, for $j = 1, 2, 3$. Also

$$\sup_{s_j \in [a_j, b_j]} |p_j(x_j, s_j)| = \max\{x_j - a_j, b_j - x_j\} = \frac{b_j - a_j}{2} + \left| x_j - \frac{a_j + b_j}{2} \right|,$$

for $j = 1, 2, 3$.

Putting things together we get

$$\sum_{j=1}^3 \left(\int_{a_1}^{b_1} \int_{a_2}^{b_2} \int_{a_3}^{b_3} |p_j(x_j, s_j)| \cdot \left| \frac{\partial f(s_1, s_2, s_3)}{\partial s_j} \right| ds_3 ds_2 ds_1 \right)$$

$$\leq \begin{cases} \displaystyle\sum_{j=1}^{3}\left\|\frac{\partial f}{\partial s_j}\right\|_{\infty}\cdot\left(\prod_{\substack{i=1\\i\neq j}}^{3}(b_i-a_i)\right)\cdot\left(\frac{(x_j-a_j)^2+(b_j-x_j)^2}{2}\right), \\[4mm] \quad \text{if } \dfrac{\partial f}{\partial s_j}\in L_{\infty}\left(\times_{i=1}^{3}[a_i,b_i]\right), j=1,2,3; \\[4mm] \displaystyle\sum_{j=1}^{3}\left\|\frac{\partial f}{\partial s_j}\right\|_{p_j}\left(\prod_{\substack{i=1\\i\neq j}}^{3}(b_i-a_i)^{1/q_j}\right) \\[4mm] \quad\cdot\left[\dfrac{(x_j-a_j)^{q_j+1}+(b_j-x_j)^{q_j+1}}{q_j+1}\right]^{1/q_j}, \\[4mm] \quad \text{if } \dfrac{\partial f}{\partial s_j}\in L_{p_j}\left(\times_{i=1}^{3}[a_i,b_i]\right), p_j>1, \dfrac{1}{p_j}+\dfrac{1}{q_j}=1, \text{for } j=1,2,3; \\[4mm] \displaystyle\sum_{j=1}^{3}\left\|\frac{\partial f}{\partial s_j}\right\|_1\cdot\left(\frac{b_j-a_j}{2}+\left|x_j-\frac{a_j+b_j}{2}\right|\right), \\[4mm] \quad \text{if } \dfrac{\partial f}{\partial s_j}\in L_1\left(\times_{i=1}^{3}[a_i,b_i]\right), \text{for } j=1,2,3. \end{cases} \tag{3.10}$$

Similarly working, next we find that

$$\sum_{\substack{\ell=1\\j<k}}^{3}\left(\int_{a_1}^{b_1}\int_{a_2}^{b_2}\int_{a_3}^{b_3}|p_j(x_j,s_j)|\cdot|p_k(x_k,s_k)|\cdot\left|\frac{\partial^2 f(s_1,s_2,s_3)}{\partial s_k\partial s_j}\right|ds_3ds_2ds_1\right)_{(\ell)}$$

$$\leq \begin{cases} \displaystyle\sum_{\substack{\ell=1\\j<k}}^{3}\left(\left\|\frac{\partial^2 f}{\partial s_k\partial s_j}\right\|_{\infty}\cdot\left(\prod_{\substack{i=1\\i\neq j,k}}^{3}(b_i-a_i)\right)\right. \\[4mm] \qquad\left.\cdot\dfrac{((x_j-a_j)^2+(b_j-x_j)^2)(x_k-a_k)^2+(b_k-x_k)^2}{4}\right)_{(\ell)}, \\[4mm] \quad \text{if } \dfrac{\partial^2 f}{\partial s_k\partial s_j}\in L_{\infty}\left(\times_{i=1}^{3}[a_i,b_i]\right), k,j\in\{1,2,3\}; \\[4mm] \displaystyle\sum_{\substack{\ell=1\\j<k}}^{3}\left(\left\|\frac{\partial^2 f}{\partial s_k\partial s_j}\right\|_{p_{kj}}\cdot\left(\prod_{\substack{i=1\\i\neq j,k}}^{3}(b_i-a_i)^{1/q_{kj}}\right)\right. \\[4mm] \qquad\cdot\left[\dfrac{(x_j-a_j)^{q_{kj}+1}+(b_j-x_j)^{q_{kj}+1}}{q_{kj}+1}\right]^{1/q_{kj}} \\[4mm] \qquad\left.\cdot\left[\dfrac{(x_k-a_k)^{q_{kj}+1}+(b_k-x_k)^{q_{kj}+1}}{q_{kj}+1}\right]^{1/q_{kj}}\right)_{(\ell)}, \\[4mm] \quad \text{if } \dfrac{\partial^2 f}{\partial s_k\partial s_j}\in L_{p_{kj}}\left(\times_{i=1}^{3}[a_i,b_i]\right), p_{kj}>1, \\[4mm] \qquad \dfrac{1}{p_{kj}}+\dfrac{1}{q_{kj}}=1, \text{for } j,k\in\{1,2,3\}; \end{cases} \tag{3.11a}$$

$$\sum_{\substack{\ell=1 \\ j<k}}^{3} \left(\int_{a_1}^{b_1} \int_{a_2}^{b_2} \int_{a_3}^{b_3} |p_j(x_j, s_j)| \cdot |p_k(x_k, s_k)| \cdot \left| \frac{\partial^2 f(s_1, s_2, s_3)}{\partial s_k \partial s_j} \right| ds_3 ds_2 ds_1 \right)_{(\ell)}$$

$$\leq \begin{cases} \displaystyle\sum_{\substack{\ell=1 \\ j<k}}^{3} \left(\left\| \frac{\partial^2 f}{\partial s_k \partial s_j} \right\|_1 \cdot \left(\frac{b_j - a_j}{2} + \left| x_j - \frac{a_j + b_j}{2} \right| \right) \right. \\ \qquad \left. \cdot \left(\frac{b_k - a_k}{2} + \left| x_k - \frac{a_k + b_k}{2} \right| \right) \right)_{(\ell)}, \\ \text{if } \dfrac{\partial^2 f}{\partial s_k \partial s_j} \in L_1 \left(\times_{i=1}^{3}[a_i, b_i] \right), \text{ for } j, k \in \{1, 2, 3\}. \end{cases} \tag{3.11b}$$

Finally we obtain that

$$\int_{a_1}^{b_1} \int_{a_2}^{b_2} \int_{a_3}^{b_3} \left(\prod_{i=1}^{3} |p_i(x_i, s_i)| \right) \cdot \left| \frac{\partial^3 f(s_1, s_2, s_3)}{\partial s_3 \partial s_2 \partial s_1} \right| ds_3 ds_2 ds_1$$

$$\leq \begin{cases} \left\| \dfrac{\partial^3 f(s_1, s_2, s_3)}{\partial s_3 \partial s_2 \partial s_1} \right\|_{\infty} \cdot \dfrac{\displaystyle\prod_{j=1}^{3} \left((x_j - a_j)^2 + (b_j - x_j)^2 \right)}{8}, \\ \text{if } \dfrac{\partial^3 f(s_1, s_2, s_3)}{\partial s_3 \partial s_2 \partial s_1} \in L_{\infty} \left(\times_{i=1}^{3}[a_i, b_i] \right); \\ \left\| \dfrac{\partial^3 f(s_1, s_2, s_3)}{\partial s_3 \partial s_2 \partial s_1} \right\|_{p_{123}} \cdot \displaystyle\prod_{j=1}^{3} \left(\dfrac{(x_j - a_j)^{q_{123}+1} + (b_j - x_j)^{q_{123}+1}}{q_{123} + 1} \right)^{1/q_{123}} \\ \text{if } \dfrac{\partial^3 f(s_1, s_2, s_3)}{\partial s_3 \partial s_2 \partial s_1} \in L_{p_{123}} \left(\times_{i=1}^{3}[a_i, b_i] \right), p_{123} > 1, \dfrac{1}{p_{123}} + \dfrac{1}{q_{123}} = 1; \\ \left\| \dfrac{\partial^3 f(s_1, s_2, s_3)}{\partial s_3 \partial s_2 \partial s_1} \right\|_1 \cdot \displaystyle\prod_{j=1}^{3} \left(\dfrac{b_j - a_j}{2} + \left| x_j - \dfrac{a_j + b_j}{2} \right| \right), \\ \text{if } \dfrac{\partial^3 f(s_1, s_2, s_3)}{\partial s_3 \partial s_2 \partial s_1} \in L_1 \left(\times_{i=1}^{3}[a_i, b_i] \right). \end{cases} \tag{3.12}$$

Taking into account (3.10), (3.11a), (3.11b), (3.12) we have completed the proof of (3.6). $\qquad\square$

A generalization of Theorem 3.5 follows:

Theorem 3.6. *Let* $f: \times_{i=1}^{n}[a_i, b_i] \to \mathbb{R}$ *as in Theorem 3.4,* $n \in \mathbb{N}$, *and* $(x_1, \ldots, x_n) \in \times_{i=1}^{n}[a_i, b_i]$. *Here* $\|\cdot\|_p$ ($1 \leq p \leq \infty$) *is the usual* L_p-*norm on* $\times_{i=1}^{n}[a_i, b_i]$. *Then*

$$\left| f(x_1, x_2, \ldots, x_n) - \frac{1}{\displaystyle\prod_{i=1}^{n}(b_i - a_i)} \int_{\times_{i=1}^{n}[a_i,b_i]} f(s_1, s_2, \ldots, s_n) ds_n \cdots ds_1 \right|$$

$$\leq \left(\sum_{i=1}^{n} M_{i,n} \right) \bigg/ \left(\prod_{i=1}^{n}(b_i - a_i) \right). \tag{3.13}$$

Here we have

$$
M_{1,n} := \begin{cases}
\sum_{j=1}^{n} \left(\left\| \dfrac{\partial f}{\partial s_j} \right\|_{\infty} \cdot \left(\prod_{\substack{i=1 \\ i \neq j}}^{n} (b_i - a_i) \right) \right. \\
\qquad \left. \cdot \left(\dfrac{(x_j - a_j)^2 + (b_j - x_j)^2}{2} \right) \right), \\
\quad if \ \dfrac{\partial f}{\partial s_j} \in L_{\infty} \left(\times_{i=1}^{n} [a_i, b_i] \right), j = 1, 2, \ldots, n; \\[4pt]
\sum_{j=1}^{n} \left(\left\| \dfrac{\partial f}{\partial s_j} \right\|_{p} \cdot \left(\prod_{\substack{i=1 \\ i \neq j}}^{n} (b_i - a_i)^{1/q} \right) \right. \\
\qquad \left. \cdot \left[\dfrac{(x_j - a_j)^{q+1} + (b_j - x_j)^{q+1}}{q+1} \right]^{1/q} \right), \\
\quad if \ \dfrac{\partial f}{\partial s_j} \in L_p \left(\times_{i=1}^{n} [a_i, b_i] \right), p > 1, \\
\qquad \dfrac{1}{p} + \dfrac{1}{q} = 1, for \ j = 1, 2, \ldots, n; \\[4pt]
\sum_{j=1}^{n} \left(\left\| \dfrac{\partial f}{\partial s_j} \right\|_{1} \cdot \left(\dfrac{b_j - a_j}{2} + \left| x_j - \dfrac{a_j + b_j}{2} \right| \right) \right), \\
\quad if \ \dfrac{\partial f}{\partial s_j} \in L_1 \left(\times_{i=1}^{n} [a_i, b_i] \right), for \ j = 1, 2, \ldots, n.
\end{cases} \tag{3.14}
$$

Also we have

$$
M_{2,n} := \begin{cases}
\sum_{\substack{\ell_1 = 1 \\ j < k}}^{\binom{n}{2}} \left(\left\| \dfrac{\partial^2 f}{\partial s_k \partial s_j} \right\|_{\infty} \cdot \left(\prod_{\substack{i=1 \\ i \neq j,k}}^{n} (b_i - a_i) \right) \right. \\
\quad \left. \cdot \dfrac{((x_j - a_j)^2 + (b_j - x_j)^2) \cdot ((x_k - a_k)^2 + (b_k - x_k)^2)}{4} \right)_{(\ell_1)}, \\
\quad if \ \dfrac{\partial^2 f}{\partial s_k \partial s_j} \in L_{\infty} \left(\times_{i=1}^{n} [a_i, b_i] \right), j, k \in \{1, 2, \ldots, n\}; \\[4pt]
\sum_{\substack{\ell_1 = 1 \\ j < k}}^{\binom{n}{2}} \left(\left\| \dfrac{\partial^2 f}{\partial s_k \partial s_j} \right\|_{p} \cdot \left(\prod_{\substack{i=1 \\ i \neq j,k}}^{n} (b_i - a_i)^{1/q} \right) \right. \\
\quad \cdot \left[\dfrac{(x_j - a_j)^{q+1} + (b_j - x_j)^{q+1}}{q+1} \right]^{1/q} \\
\quad \left. \cdot \left[\dfrac{(x_k - a_k)^{q+1} + (b_k - x_k)^{q+1}}{q+1} \right]^{1/q} \right)_{(\ell_1)}, \\
\quad if \ \dfrac{\partial^2 f}{\partial s_k \partial s_j} \in L_p \left(\times_{i=1}^{n} [a_i, b_i] \right), p > 1, \\
\qquad \dfrac{1}{p} + \dfrac{1}{q} = 1, for \ j, k \in \{1, 2, \ldots, n\};
\end{cases} \tag{3.15a}
$$

$$M_{2,n} := \begin{cases} \displaystyle\sum_{\substack{\ell_1=1 \\ j<k}}^{\binom{n}{2}} \left(\left\| \frac{\partial^2 f}{\partial s_k \partial s_j} \right\|_1 \cdot \left(\frac{b_j - a_j}{2} + \left| x_j - \frac{a_j + b_j}{2} \right| \right) \right. \\ \qquad \left. \left(\frac{b_k - a_k}{2} + \left| x_k - \frac{a_k + b_k}{2} \right| \right) \right)_{(\ell_1)}, \\ \quad if \ \dfrac{\partial^2 f}{\partial s_j \partial s_j} \in L_1 \left(\times_{i=1}^n [a_i, b_i] \right), for \ j, k \in \{1, 2, \dots, n\}. \end{cases} \tag{3.15b}$$

And,

$$M_{3,n} := \begin{cases} \displaystyle\sum_{\substack{\ell_2=1 \\ j<k<r}}^{\binom{n}{3}} \left(\left\| \frac{\partial^3 f}{\partial s_r \partial s_k \partial s_j} \right\|_\infty \cdot \left(\prod_{\substack{i=1 \\ i \neq j,k,r}}^n (b_i - a_i) \right) \right. \\ \qquad \left. \cdot \prod_{m \in \{j,k,r\}} \frac{((x_m - a_m)^2 + (b_m - x_m)^2)}{2^3} \right)_{(\ell_2)}, \\ \quad if \ \dfrac{\partial^3 f}{\partial s_r \partial s_k \partial s_j} \in L_\infty \left(\times_{i=1}^n [a_i, b_i] \right), j, k, r \in \{1, \dots, n\}; \\[2mm] \displaystyle\sum_{\substack{\ell_2=1 \\ j<k<r}}^{\binom{n}{3}} \left(\left\| \frac{\partial^3 f}{\partial s_r \partial s_k \partial s_j} \right\|_p \cdot \left(\prod_{\substack{i=1 \\ i \neq j,k,r}}^n (b_i - a_i)^{1/q} \right) \right. \\ \qquad \left. \cdot \prod_{m \in \{j,k,r\}} \left[\frac{(x_m - a_m)^{q+1} + (b_m - x_m)^{q+1}}{q+1} \right]^{1/q} \right)_{(\ell_2)}, \\ \quad if \ \dfrac{\partial^3 f}{\partial s_r \partial s_k \partial s_j} \in L_p \left(\times_{i=1}^n [a_i, b_i] \right), p > 1, \\ \quad \dfrac{1}{p} + \dfrac{1}{q} = 1, for \ j, k, r \in \{1, 2, \dots, n\}; \\[2mm] \displaystyle\sum_{\substack{\ell_2=1 \\ j<k<r}}^{\binom{n}{3}} \left(\left\| \frac{\partial^3 f}{\partial s_r \partial s_k \partial s_j} \right\|_1 \cdot \prod_{m \in \{j,k,r\}} \right. \\ \qquad \left. \cdot \left(\frac{b_m - a_m}{2} + \left| x_m - \frac{a_m + b_m}{2} \right| \right) \right)_{(\ell_2)}, \\ \quad if \ \dfrac{\partial^3 f}{\partial s_r \partial s_k \partial s_j} \in L_1 \left(\times_{i=1}^n [a_i, b_i] \right), for \ j, k, r \in \{1, 2, \dots, n\}. \end{cases} \tag{3.16}$$

And,

$$M_{n-1,n} := \begin{cases} \sum_{\ell=1}^{\binom{n}{n-1}} \left(\left\| \dfrac{\partial^{n-1} f}{\partial s_n \cdots \widehat{\partial s_\ell} \cdots \partial s_1} \right\|_\infty (b_\ell - a_\ell) \right. \\ \qquad \left. \cdot \dfrac{\prod\limits_{m=1,m\neq\ell}^{n} ((x_m - a_m)^2 + (b_m - x_m)^2)}{2^{n-1}} \right), \\ \quad if\ \dfrac{\partial^{n-1} f}{\partial s_n \cdots \widehat{\partial s_\ell} \cdots \partial s_1} \in L_\infty \left(\times_{i=1}^n [a_i, b_i] \right), \ell = 1, \ldots, n; \end{cases} \tag{3.17a}$$

$$M_{n-1,n} := \begin{cases} \sum_{\ell=1}^{\binom{n}{n-1}} \left(\left\| \dfrac{\partial^{n-1} f}{\partial s_n \cdots \partial \hat{s}_\ell \cdots \partial s_1} \right\|_p \cdot (b_\ell - a_\ell)^{1/q} \right. \\ \qquad \left. \cdot \left(\prod_{\substack{m=1 \\ m\neq\ell}}^{n} \left[\dfrac{(x_m - a_m)^{q+1} + (b_m - x_m)^{q+1}}{q+1} \right]^{1/q} \right) \right), \\ \quad if\ \dfrac{\partial^{n-1} f}{\partial s_n \cdots \partial \hat{s}_\ell \cdots \partial s_1} \in L_p \left(\times_{i=1}^n [a_i, b_i] \right), p > 1, \\ \qquad \dfrac{1}{p} + \dfrac{1}{q} = 1, \ell = 1, \ldots, n; \\[4pt] \sum_{\ell=1}^{\binom{n}{n-1}} \left(\left\| \dfrac{\partial^{n-1} f}{\partial s_n \cdots \partial \hat{s}_\ell \cdots \partial s_1} \right\|_1 \right. \\ \qquad \left. \cdot \prod_{\substack{m=1 \\ m\neq\ell}}^{n} \left(\dfrac{b_m - a_m}{2} + \left| x_m - \dfrac{a_m + b_m}{2} \right| \right) \right), \\ \quad if\ \dfrac{\partial^{n-1} f}{\partial s_n \cdots \partial \hat{s}_\ell \cdots \partial s_1} \in L_1 \left(\times_{i=1}^n [a_i, b_i] \right), \ell = 1, \ldots, n. \end{cases} \tag{3.17b}$$

Finally we have

$$M_{n,n} := \begin{cases} \left\| \dfrac{\partial^n f}{\partial s_n \cdots \partial s_1} \right\|_\infty \cdot \dfrac{\prod\limits_{j=1}^{n} ((x_j - a_j)^2 + (b_j - x_j)^2)}{2^n}, \\ \quad if\ \dfrac{\partial^n f}{\partial s_n \cdots \partial s_1} \in L_\infty \left(\times_{i=1}^n [a_i, b_i] \right); \\[4pt] \left\| \dfrac{\partial^n f}{\partial s_n \cdots \partial s_1} \right\|_p \cdot \sum_{j=1}^{n} \left(\dfrac{(x_j - a_j)^{q+1} + (b_j - x_j)^{q+1}}{q+1} \right)^{1/q}, \\ \quad if\ \dfrac{\partial^n f}{\partial s_n \cdots \partial s_1} \in L_p \left(\times_{i=1}^n [a_i, b_i] \right), p > 1, \dfrac{1}{p} + \dfrac{1}{q} = 1; \\[4pt] \left\| \dfrac{\partial^n f}{\partial s_n \cdots \partial s_1} \right\|_1 \cdot \prod_{j=1}^{n} \left(\dfrac{b_j - a_j}{2} + \left| x_j - \dfrac{a_j + b_j}{2} \right| \right), \\ \quad if\ \dfrac{\partial^n f}{\partial s_n \cdots \partial s_1} \in L_1 \left(\times_{i=1}^n [a_i, b_i] \right). \end{cases} \tag{3.18}$$

Proof. Similar to Theorem 3.5. $\qquad\qquad\qquad\qquad\qquad\qquad\qquad\qquad\square$

3.3   Application to Probability Theory

Here we consider the random variables $X_i \geq 0$, $i = 1, 2, 3$, taking values in $[0, b_i]$, $b_i > 0$. Let F be the joint probability distribution function of (X_1, X_2, X_3) assumed to be in $C^3(\times_{i=1}^{3}[0, b_i])$. Let f be the corresponding probability density function, that is $f = \frac{\partial^3 F}{\partial x_1 \partial x_2 \partial x_3}$. Clearly the expectation

$$E(X_1 X_2 X_3) = \int_0^{b_1} \int_0^{b_2} \int_0^{b_3} x_1 x_2 x_3 f(x_1, x_2, x_3) dx_1 dx_2 dx_3. \qquad (3.19)$$

Consider on $\times_{i=1}^{3}[0, b_i]$ the function

$$G_3(x_1, x_2, x_3) := 1 - F(x_1, b_2, b_3) - F(b_1, x_2, b_3) - F(b_1, b_2, x_3) + F(x_1, x_2, b_3)$$
$$+ F(x_1, b_2, x_3) + F(b_1, x_2, x_3) - F(x_1, x_2, x_3). \qquad (3.20)$$

We see that

$$\frac{\partial^3 G_3(x_1, x_2, x_3)}{\partial x_1 \partial x_2 \partial x_3} = -f(x_1, x_2, x_3). \qquad (3.21)$$

In [31], Theorem 6, we established that

$$E(X_1 X_2 X_3) = \int_0^{b_1} \int_0^{b_2} \int_0^{b_3} G_3(x_1, x_2, x_3) dx_1 dx_2 dx_3. \qquad (3.22)$$

Obviously here we have $F(x_1, b_2, b_3) = F_{X_1}(x_1)$, $F(b_1, x_2, b_3) = F_{X_2}(x_2)$, $F(b_1, b_2, x_3) = F_{X_3}(x_3)$, where F_{X_1}, F_{X_2}, F_{X_3} are the probability distribution functions of X_1, X_2, X_3. Furthermore, $F(x_1, x_2, b_3) = F_{X_1, X_2}(x_1, x_2)$, $F(x_1, b_2, x_3) = F_{X_1, X_3}(x_1, x_3)$, $F(b_1, x_2, x_3) = F_{X_2, X_3}(x_2, x_3)$, where F_{X_1, X_2}, F_{X_1, X_3}, F_{X_2, X_3} are the joint probability distribution functions of (X_1, X_2), (X_1, X_3), (X_2, X_3), respectively. That is

$$G_3(x_1, x_2, x_3) = 1 - F_{X_1}(x_1) - F_{X_2}(x_2) - F_{X_3}(x_3) + F_{X_1, X_2}(x_1, x_2)$$
$$+ F_{X_1, X_3}(x_1, x_3) + F_{X_2, X_3}(x_2, x_3) - F(x_1, x_2, x_3). \qquad (3.23)$$

We set by $f_1 = F'_{X_1}$, $f_2 = F'_{X_2}$, $f_3 = F'_{X_3}$, $f_{12} = \frac{\partial^2 F_{X_1, X_2}}{\partial x_1 \partial x_2}$, $f_{13} = \frac{\partial^2 F_{X_1, X_3}}{\partial x_1 \partial x_3}$, and $f_{23} = \frac{\partial^2 F_{X_2, X_3}}{\partial x_2 \partial x_3}$, the corresponding existing probability density functions.

Then we have

$$\frac{\partial G_3(x_1, x_2, x_3)}{\partial x_1} = -f_1(x_1) + \frac{\partial F_{X_1, X_2}(x_1, x_2)}{\partial x_1}$$
$$+ \frac{\partial F_{X_1, X_3}(x_1, x_3)}{\partial x_1} - \frac{\partial F(x_1, x_2, x_3)}{\partial x_1},$$

$$\frac{\partial G_3(x_1, x_2, x_3)}{\partial x_2} = -f_2(x_2) + \frac{\partial F_{X_1, X_2}(x_1, x_2)}{\partial x_2}$$
$$+ \frac{\partial F_{X_2, X_3}(x_2, x_3)}{\partial x_2} - \frac{\partial F(x_1, x_2, x_3)}{\partial x_2},$$

$$\frac{\partial G_3(x_1, x_2, x_3)}{\partial x_3} = -f_3(x_3) + \frac{\partial F_{X_1, X_3}(x_1, x_3)}{\partial x_3}$$
$$+ \frac{\partial F_{X_2, X_3}(x_2, x_3)}{\partial x_3} - \frac{\partial F(x_1, x_2, x_3)}{\partial x_3}. \qquad (3.24)$$

Furthermore we find that

$$\frac{\partial G_3(x_1, x_2, x_3)}{\partial x_1 \partial x_2} = f_{12}(x_1, x_2) - \frac{\partial^2 F(x_1, x_2, x_3)}{\partial x_1 \partial x_2},$$

$$\frac{\partial G_3(x_1, x_2, x_3)}{\partial x_2 \partial x_3} = f_{23}(x_2, x_3) - \frac{\partial^2 F(x_1, x_2, x_3)}{\partial x_2 \partial x_3}, \qquad (3.25)$$

and

$$\frac{\partial G_3(x_1, x_2, x_3)}{\partial x_1 \partial x_3} = f_{13}(x_1, x_3) - \frac{\partial^2 F(x_1, x_2, x_3)}{\partial x_1 \partial x_3}.$$

Then one can apply (3.6) for $a_1 = a_2 = a_3 = 0$ and $f = G_3$.

The left-hand side of (3.6) will be

$$\left| G_3(x_1, x_2, x_3) - \frac{1}{b_1 b_2 b_3} \int_0^{b_1} \int_0^{b_2} \int_0^{b_3} G_3(s_1, s_2, s_3) ds_1 ds_2 ds_3 \right|$$

$$= \left| \left(1 - F(x_1, b_2, b_3) - F(b_1, x_2, b_3) \right.\right.$$

$$- F(b_1, b_2, x_3) + F(x_1, x_2, b_3) + F(x_1, b_2, x_3)$$

$$\left.\left. + F(b_1, x_2, x_3) - F(x_1, x_2, x_3) \right) - \frac{1}{b_1 b_2 b_3} E(X_1 X_2 X_3) \right|, \qquad (3.26)$$

by using (3.20) and (3.21). Here $(x_1, x_2, x_3) \in \times_{i=1}^3 [0, b_i]$. Next one can apply (3.21), (3.24), (3.25) in the right side of (3.6).

Chapter 4

General Probabilistic Inequalities

In this chapter we use an important integral inequality from [79] to obtain bounds for the covariance of two random variables (1) in a general setup and (2) for a class of special joint distributions. The same inequality is also used to estimate the difference of the expectations of two random variables. Also we present about the attainability of a related inequality. This chapter is based on [80].

4.1 Introduction

In [79] was proved the following result:

Theorem 4.1. *Let $f \in C^n(B)$, $n \in \mathbb{N}$, where $B = [a_1, b_1] \times \cdots \times [a_n, b_n]$, a_j, $b_j \in \mathbb{R}$, with $a_j < b_j$, $j = 1, ..., n$. Denote by ∂B the boundary of the box B. Assume that $f(x) = 0$, for all $x = (x_1, ..., x_n) \in \partial B$ (in other words we assume that $f(\cdots, a_j, \cdots) = f(\cdots, b_j, \cdots) = 0$, for all $j = 1, ..., n$). Then*

$$\int_B |f(x_1, ..., x_n)| \, dx_1 \cdots dx_n \leq \frac{m(B)}{2^n} \int_B \left| \frac{\partial^n f(x_1, ..., x_n)}{\partial x_1 \cdots \partial x_n} \right| dx_1 \cdots dx_n, \qquad (4.1)$$

where $m(B) = \prod_{j=1}^{n} (b_j - a_j)$ is the n-th dimensional volume (i.e. the Lebesgue measure) of B.

In the chapter we give probabilistic applications of the above inequality and related remarks.

4.2 Make Applications

Let $X \geq 0$ be a random variable. Given a sequence of positive numbers $b_n \to \infty$, we introduce the events

$$B_n = \{X > b_n\}.$$

Then (as in the proof of Chebychev's inequality)

$$E[X \, 1_{B_n}] = \int_{B_n} X \, dP \geq b_n P(B_n) = b_n P(X > b_n), \qquad (4.2)$$

31

where, as usual, P is the corresponding probability measure and E its associated expectation. If $E[X] < \infty$, then monotone convergence implies (since $X1_{B_n^c} \nearrow X$)

$$\lim_n E\left[X1_{B_n^c}\right] = E[X].$$

Thus

$$E[X1_{B_n}] = E[X] - E\left[X1_{B_n^c}\right] \to 0.$$

Hence by (4.2)

$$b_n P(X > b_n) \to 0, \tag{4.3}$$

for any sequence $b_n \to \infty$.

The following proposition is well-known, but we include it here for the sake of completeness:

Proposition 4.2. *Let $X \geq 0$ be a random variable with (cumulative) distribution function $F(x)$. Then*

$$E[X] = \int_0^\infty [1 - F(x)]dx \tag{4.4}$$

(notice that, if the integral in the right hand side diverges, it should diverge to $+\infty$, and the equality still makes sense).

Proof. Integration by parts gives

$$\int_0^\infty [1 - F(x)]\, dx =$$

$$\lim_{x \to \infty} x[1 - F(x)] + \int_0^\infty x dF(x) = \lim_{x \to \infty} xP(X > x) + E[X].$$

If $E[X] < \infty$, then (4.4) follows by (4.3). If $E[X] = \infty$, then the above equation gives that

$$\int_0^\infty [1 - F(x)]\, dx = \infty,$$

since $xP(X > x) \geq 0$, whenever $x \geq 0$. $\qquad\qquad\square$

Remark 4.3. If $X \geq a$ is a random variable with distribution function $F(x)$, we can apply Proposition 4.2 to $Y = X - a$ and obtain

$$E[X] = a + \int_a^\infty [1 - F(x)]\, dx. \tag{4.5}$$

We now present some applications of Theorem 4.1.

Proposition 4.4. *Consider two random variables X and Y taking values in the interval $[a, b]$. Let $F_X(x)$ and $F_Y(x)$ respectively be their distribution functions,* *which are assumed absolutely continuous. Their corresponding densities are $f_X(x)$ and $f_Y(x)$ respectively. Then*

$$|E[X] - E[Y]| \leq \frac{b-a}{2} \int_a^b |f_X(x) - f_Y(x)|\, dx. \tag{4.6}$$

Proof. We have that $F_X(b) = F_Y(b) = 1$, thus formula (4.5) gives us

$$E[X] = a + \int_a^b [1 - F_X(x)]dx \quad \text{and} \quad E[Y] = a + \int_a^b [1 - F_Y(x)]dx.$$

It follows that

$$|E[X] - E[Y]| \leq \int_a^b |F_X(x) - F_Y(x)|\, dx.$$

Since $F_X(x)$ and $F_Y(x)$ are absolutely continuous with densities $f_X(x)$ and $f_Y(x)$ respectively, and since $F_X(a) - F_Y(a) = F_X(b) - F_Y(b) = 0$, Theorem 4.1, together with the above inequality, imply (4.6). $\qquad\square$

Theorem 4.5. *Let X and Y be two random variables with joint distribution function $F(x,y)$. We assume that $a \leq X \leq b$ and $c \leq Y \leq d$. We also assume that $F(x,y)$ possesses a (joint) probability density function $f(x,y)$. Then*

$$|cov(X,Y)| \leq \frac{(b-a)(d-c)}{4} \int_c^d \int_a^b |f(x,y) - f_X(x)f_Y(y)|\, dxdy, \tag{4.7}$$

where

$$cov(X,Y) = E[XY] - E[X]E[Y]$$

is the covariance of X and Y, while $f_X(x)$ and $f_Y(y)$ are the marginal densities of X and Y respectively.

Proof. Integration by parts implies

$$E[XY] = aE[Y] + cE[X] - ac + \int_c^d \int_a^b [1 - F(x,d) - F(b,y) + F(x,y)]\, dxdy \tag{4.8}$$

(in fact, this formula generalizes to n random variables). Notice that $F(x,d)$ is the (marginal) distribution function of X alone, and similarly $F(b,y)$ is the (marginal) distribution function of Y alone. Thus, by (4.5)

$$E[X] = a + \int_a^b [1 - F(x,d)]\, dx \quad \text{and} \quad E[Y] = c + \int_c^d [1 - F(b,y)]\, dy,$$

which derives

$$(E[X] - a)(E[Y] - c) = \int_c^d \int_a^b [1 - F(x,d)][1 - F(b,y)]\, dxdy,$$

or

$$E[X]E[Y] = aE[Y] + cE[X] - ac +$$

$$\int_c^d \int_a^b \left[1 - F(x,b) - F(a,y) + F(x,d)F(b,y)\right] dx\,dy. \tag{4.9}$$

Subtracting (4.9) from (4.8) we get

$$\operatorname{cov}(X,Y) = E\left[XY\right] - E\left[X\right]E\left[Y\right] = \int_c^d \int_a^b \left[F(x,y) - F(x,d)F(a,y)\right] dx\,dy.$$

Hence

$$\left|\operatorname{cov}(X,Y)\right| \le \int_c^d \int_a^b \left|F(x,y) - F(x,d)F(b,y)\right| dx\,dy.$$

Now $F(a,y) = F(x,c) = 0$ and $F(b,d) = 1$. It follows that $F(x,y) - F(x,d)F(b,y)$ vanishes on the boundary of $B = (a,b) \times (c,d)$. Thus Theorem 4.1, together with the above inequality, imply (4.7). $\qquad\square$

In Feller's classical book [194], par. 5, page 165, we find the following fact: Let $F(x)$ and $G(y)$ be distribution functions on $\mathbb{R}$, and put

$$U(x,y) = F(x)G(y)\left\{1 - \alpha\left[1 - F(x)\right]\left[1 - G(y)\right]\right\}, \tag{4.10}$$

where $-1 \le \alpha \le 1$. Then $U(x,y)$ is a distribution function on $\mathbb{R}^2$, with marginal distributions $F(x)$ and $G(y)$. Furthermore $U(x,y)$ possesses a density if and only if $F(x)$ and $G(y)$ do.

Theorem 4.6. *Let X and Y be two random variables with joint distribution function $U(x,y)$, as given by (4.10). For the marginal distributions $F(x)$ and $G(y)$, of X and Y respectively, we assume that they posses densities $f(x)$ and $g(y)$ (respectively). Furthermore, we assume that $a \le X \le b$ and $c \le Y \le d$. Then*

$$\left|\operatorname{cov}(X,Y)\right| \le |\alpha|\,\frac{(b-a)(d-c)}{16}. \tag{4.11}$$

Proof. The (joint) density $u(x,y)$ of $U(x,y)$ is

$$u(x,y) = \frac{\partial^2 U(x,y)}{\partial x \partial y} = f(x)g(y)\left\{1 - \alpha\left[1 - 2F(x)\right]\left[1 - 2G(y)\right]\right\}.$$

Thus

$$u(x,y) - f(x)g(y) = -\alpha\left[1 - 2F(x)\right]\left[1 - 2G(y)\right]f(x)g(y)$$

and Theorem 4.5 implies

$$\left|\operatorname{cov}(X,Y)\right| \le$$

$$|\alpha|\,\frac{(b-a)(d-c)}{4}\left[\int_a^b |1 - 2F(x)|\,f(x)dx\right]\left[\int_c^d |1 - 2G(y)|\,g(y)dy\right],$$

or

$$\left|\operatorname{cov}(X,Y)\right| \le |\alpha|\,\frac{(b-a)(d-c)}{4}\,E\left[|1 - 2F(X)|\right]E\left[|1 - 2G(Y)|\right]. \tag{4.12}$$

Now, $F(x)$ and $G(y)$ have density functions, hence $F(X)$ and $G(Y)$ are uniformly distributed on $(0,1)$. Thus

$$E\left[|1 - 2F(X)|\right] = E\left[|1 - 2G(Y)|\right] = \frac{1}{2}$$

and the proof is done by using the above equalities in (4.12). $\qquad\square$

4.3 Remarks on an Inequality

The basic ingredient in the proof of Theorem 4.1 is the following inequality (also shown in [79]):
Let $B = (a_1, b_1) \times \cdots \times (a_n, b_n) \subset \mathbb{R}^n$, with $a_j < b_j$, $j = 1, ..., n$. Denote by $\overline{B}$ and ∂B respectively the (topological) closure and the boundary of the open box B. Consider the functions $f \in C_0(\overline{B}) \cap C^n(B)$, namely the functions that are continuous on $\overline{B}$, have n continuous derivatives in B, and vanish on ∂B. Then for such functions we have

$$|f(x_1, ..., x_n)| \le \frac{1}{2^n} \int_B \left| \frac{\partial^n f(x_1, ..., x_n)}{\partial x_1 \cdots \partial x_n} \right| dx_1 \cdots dx_n,$$

true for all $(x_1, ..., x_n) \in B$. In other words

$$\|f\|_\infty \le \frac{1}{2^n} \int_B \left| \frac{\partial^n f(x_1, ..., x_n)}{\partial x_1 \cdots \partial x_n} \right| dx_1 \cdots dx_n, \tag{4.13}$$

where $\| \cdot \|_\infty$ is the supnorm of $C_0(\overline{B})$.

Remark 4.7. Suppose we have that $a_j = -\infty$ for some (or all) j and/or $b_j = \infty$ for some (or all) j. Let $\widehat{B} = \overline{B} \cup \{\infty\}$ be the one-point compactification of $\overline{B}$ and assume that $f \in C_0(\widehat{B}) \cap C^n(B)$. This means that, if $a_j = -\infty$, then $\lim_{s_j \to -\infty} f(x_1, ..., s_j, ..., x_n) = 0$, for all $x_k \in (a_k, b_k)$, $k \ne j$, and also that, if $b_j = \infty$, then $\lim_{s_j \to \infty} f(x_1, ..., s_j, ..., x_n) = 0$, for all $x_k \in (a_k, b_k)$, $k \ne j$.
Then the proof of (4.13), as given in [79], remains valid.

Remark 4.8. In the case of an unbounded domain, as described in the previous remark, although $\|f\|_\infty < \infty$, it is possible that

$$\int_B \left| \frac{\partial^n f(x_1, ..., x_n)}{\partial x_1 \cdots \partial x_n} \right| dx_1 \cdots dx_n = \infty.$$

For example, take $B = (0, \infty)$ and

$$f(x) = \int_0^x \left(e^{i\xi^2} - \frac{\sqrt{\pi}}{2} e^{i\pi/4} e^{-\xi} \right) d\xi.$$

Before discussing sharpness and attainability for (4.13) we need some notation. For $c = (c_1, ..., c_n) \in B$, we introduce the intervals

$$I_{j,0}(c) = (a_j, c_j) \qquad \text{and} \qquad I_{j,1}(c) = (c_j, b_j), \qquad j = 1, ..., n.$$

Theorem 4.9. *For $f \in C_0(\overline{B}) \cap C^n(B)$, inequality (4.13) is sharp. Equality is attained if and only if there is a $c = (c_1, ..., c_n) \in B$ such that the n-th mixed derivative $\partial^n f(x_1, ..., x_n)/\partial x_1 \cdots \partial x_n$ does not change sign in each of the 2^n "subboxes" $I_{1,\varepsilon_1}(c) \times \cdots \times I_{n,\varepsilon_n}(c)$.*

Proof. ($\Leftarrow$) For an arbitrary $c = (c_1, ..., c_n) \in B$ we have

$$f(c_1, ..., c_n) = (-1)^{\varepsilon_1 + \cdots + \varepsilon_n} \int_{I_{1,\varepsilon_1}(c)} \cdots \int_{I_{n,\varepsilon_n}(c)} \frac{\partial^n f(x_1, ..., x_n)}{\partial x_1 \cdots \partial x_n} dx_1 \cdots dx_n, \tag{4.14}$$

where each ε_j can be either 0 or 1. If c is as in the statement of the theorem, then

$$|f(c_1,...,c_n)| = \int_{I_{1,\varepsilon_1}(c)} \cdots \int_{I_{n,\varepsilon_n}(c)} \left| \frac{\partial^n f(x_1,...,x_n)}{\partial x_1 \cdots \partial x_n} \right| dx_1 \cdots dx_n. \qquad (4.15)$$

Adding up (4.15) for all 2^n choices of $(\varepsilon_1,...,\varepsilon_n)$ we find

$$2^n |f(c_1,...,c_n)| = \sum_{\varepsilon_1,...,\varepsilon_n} \int_{I_{1,\varepsilon_1}(c)} \cdots \int_{I_{n,\varepsilon_n}(c)} \left| \frac{\partial^n f(x_1,...,x_n)}{\partial x_1 \cdots \partial x_n} \right| dx_1 \cdots dx_n, \qquad (4.16)$$

or

$$|f(c_1,...,c_n)| = \frac{1}{2^n} \int_B \left| \frac{\partial^n f(x_1,...,x_n)}{\partial x_1 \cdots \partial x_n} \right| dx_1 \cdots dx_n.$$

This forces (4.13) to become an equality. In fact we also must have

$$|f(c_1,...,c_n)| = \|f\|_\infty,$$

i.e. $|f(x)|$ attains its maximum at $x = c$.

($\Rightarrow$) Conversely, let $c \in B$ be a point at which $|f(c)| = \|f\|_\infty$ (such a c exists since f is continuous and vanishes on ∂B and at infinity). If for this c, the sign of the mixed derivative $\partial^n f(x_1,...,x_n)/\partial x_1 \cdots \partial x_n$ changes inside some "sub-box" $I_{1,\varepsilon_1}(c) \times \cdots \times I_{n,\varepsilon_n}(c)$, then (4.14) implies

$$\|f\|_\infty = |f(c_1,...,c_n)| < \int_{I_{1,\varepsilon_1}(c)} \cdots \int_{I_{n,\varepsilon_n}(c)} \left| \frac{\partial^n f(x_1,...,x_n)}{\partial x_1 \cdots \partial x_n} \right| dx_1 \cdots dx_n.$$

Thus forces $f, (4.13)$ become a strict inequality. $\qquad\square$

4.4 $L_q, q > 1$, Related Theory

We give

Theorem 4.10. *Let $f \in AC([a,b])$ (absolutely continuous): $f(a) = f(b) = 0$, and $p, q > 1 : \frac{1}{p} + \frac{1}{q} = 1$. Then*

$$\int_a^b |f(x)| dx \le \frac{(b-a)^{1+\frac{1}{p}}}{2(p+1)^{1/p}} \left(\int_a^b |f'(y)|^q dy \right)^{1/q}. \qquad (4.17)$$

Proof. Call $c = \frac{a+b}{2}$. Then

$$f(x) = \int_a^x f'(y) dy, \quad a \le x \le c,$$

and

$$f(x) = -\int_x^b f'(y) dy, \quad c \le x \le b.$$

For $a \leq x \leq c$ we observe,

$$|f(x)| \leq \int_a^x |f'(y)|dy,$$

and

$$\int_a^c |f(x)|dx \leq \int_a^c \int_a^x |f'(y)|dydx$$

(by Fubini's theorem)

$$= \int_a^c \left(\int_y^c |f'(y)|dx \right) dy = \int_a^c (c-y)|f'(y)|dy$$

$$\leq \left(\int_a^c (c-y)^p dy \right)^{1/p} \|f'\|_{L^q(a,c)}.$$

That is,

$$\int_a^c |f(x)|dx \leq \frac{(b-a)^{1+\frac{1}{p}}}{2^{1+\frac{1}{p}}(p+1)^{1/p}} \|f'\|_{L^q(a,c)}. \tag{4.18}$$

Similarly we obtain

$$|f(x)| \leq \int_x^b |f'(y)|dy$$

and

$$\int_c^b |f(x)|dx \leq \int_c^b \int_x^b |f'(y)|dydx$$

(by Fubini's theorem)

$$= \int_c^b \int_c^y |f'(y)|dxdy = \int_c^b (y-c)|f'(y)|dy$$

$$\leq \left(\int_c^b (y-c)^p dy \right)^{1/p} \|f'\|_{L^q(c,b)}.$$

That is, we get

$$\int_c^b |f(x)|dx \leq \frac{(b-a)^{1+\frac{1}{p}}}{2^{1+\frac{1}{p}}(p+1)^{1/p}} \|f'\|_{L^q(c,b)}. \tag{4.19}$$

Next we see that

$$\left(\int_a^b |f(x)|dx \right)^q = \left[\int_a^c |f(x)|dx + \int_c^b |f(x)|dx \right]^q$$

$$\leq \left(\frac{(b-a)^{1+\frac{1}{p}}}{2^{1+\frac{1}{p}}(p+1)^{1/p}}\right)^{q}\left(\|f'\|_{L^{q}(a,c)}+\|f'\|_{L^{q}(c,b)}\right)^{q}$$

$$\leq \left(\frac{(b-a)^{1+\frac{1}{p}}}{2^{1+\frac{1}{p}}(p+1)^{1/p}}\right)^{q} 2^{q-1}\left(\|f'\|_{L^{q}(a,c)}^{q}+\|f'\|_{L^{q}(c,b)}^{q}\right)$$

$$= \left(\frac{(b-a)^{1+\frac{1}{p}}}{2^{1+\frac{1}{p}}(p+1)^{1/p}}\right)^{q} 2^{q-1}\|f'\|_{L^{q}(a,b)}^{q}.$$

That is we proved

$$\int_{a}^{b}|f(x)|dx \leq \frac{(b-a)^{1+\frac{1}{p}}}{2^{1+\frac{1}{p}}(p+1)^{1/p}}2^{1-\frac{1}{q}}\|f'\|_{L^{q}(a,b)}, \tag{4.20}$$

which gives

$$\int_{a}^{b}|f(x)|dx \leq \frac{(b-a)^{1+\frac{1}{p}}}{2(p+1)^{1/p}}\|f'\|_{L^{q}(a,b)}, \tag{4.21}$$

proving (4.17). $\qquad\square$

Remark 4.11. We notice that $\lim\limits_{p\to\infty} ln(p+1)^{1/p} = 0$, that is $\lim\limits_{p\to\infty}(p+1)^{1/p} = 1$. Let $f \in C'([a,b])$, $f(a) = f(b) = 0$. Then by Theorem 4.1 we obtain

$$\int_{a}^{b}|f(x)|dx \leq \frac{b-a}{2}\int_{a}^{b}|f'(y)|dy, \tag{4.22}$$

which is (4.17) when $p = \infty$ and $q = 1$. When $p = q = 2$, by Theorem 4.10, we get

$$\int_{a}^{b}|f(x)|dx \leq \frac{(b-a)^{3/2}}{\sqrt{12}}\left(\int_{a}^{b}(f'(y))^{2}dy\right)^{1/2}. \tag{4.23}$$

The L_{q}, $q > 1$, analog of Proposition 4.4 follows

Proposition 4.12. *Consider two random variables X and Y taking values in the interval $[a,b]$. Let $F_X(x)$ and $F_Y(x)$ respectively be their distribution functions which are assumed absolutely continuous. Call their corresponding densities $f_X(x)$, $f_Y(x)$, respectively. Let $p,q > 1 : \frac{1}{p} + \frac{1}{q} = 1$. Then*

$$|E(X) - E(Y)| \leq \frac{(b-a)^{1+\frac{1}{p}}}{2(p+1)^{1/p}}\left(\int_{a}^{b}|f_X(y) - f_Y(y)|^{q}dy\right)^{1/q}. \tag{4.24}$$

Proof. Notice that

$$|E(X) - E(Y)| \leq \int_{a}^{b}|F_X(x) - F_Y(x)|dx. \tag{4.25}$$

Then use Theorem 4.10 for $f = F_X - F_Y$, which fulfills the assumptions there. $\qquad\square$

Observe that (4.6) is the case of (4.24) when $p = \infty$ and $q = 1$.

Chapter 5

About Grothendieck Inequalities

We give here the analog of Grothendieck inequality for positive linear forms. We find upper and lower bounds of L_p $0 < p \leq \infty$, type, which all lead to sharp inequalities.This treatments is based on [68].

5.1 Introduction

We mention the famous Grothendieck inequality [204].

Theorem 5.1. (Grothendieck 1956) *Let K_1 and K_2 be compact spaces. Let $u : C(K_1) \times C(K_2) \to \mathbb{R}$ be a bounded bilinear form. Then there exist probability measures μ_1 and μ_2 on K_1 and K_2, respectively, such that*

$$|u(f,g)| \leq K_G^{\mathbb{R}} \|u\| \left(\int_{K_1} f^2 d\mu_1 \right)^{1/2} \left(\int_{K_2} g^2 d\mu_2 \right)^{1/2}, \tag{5.1}$$

for all $f \in C(K_1)$ and $g \in C(K_2)$, where $K_G^{\mathbb{R}}$ is a universal constant.

We have that

$$1.57\ldots = \frac{\pi}{2} \leq K_G^{\mathbb{R}} \leq \frac{\pi}{2 \ln(1 + \sqrt{2})} = 1.782\ldots. \tag{5.2}$$

See [204], [240], for the left hand side and right hand side bounds, respectively.

Still $K_G^{\mathbb{R}}$ precise value is unknown.

The complex case is studied separately, see [209]. Here we present analogs of Theorem 5.1 for the case of a positive linear form, which is of course a bounded linear one.

We use

Theorem 5.2. (W. Adamski (1991), [3]) *Let X be a pseudocompact space and Y be an arbitrary topological space, then, for any positive bilinear form $\Phi : C(X) \times C(Y) \to \mathbb{R}$, there exists a uniquely determined measure μ on the product of the Baire $\sigma-$ algebras of X and Y such that*

$$\Phi(f,g) = \int_{X \times Y} f \otimes g d\mu \tag{5.3}$$

holds for all $f \in C(X)$ and all $g \in C(Y)$.

Note. Above the tensor product $(f \otimes g)(x, y) = f(x) \cdot g(y)$, X pseudocompact means all continuous functions $f : X \to \mathbb{R}$ are bounded. Positivity of Φ means that for $f \geq 0$, $g \geq 0$ we have $\Phi(f, g) \geq 0$.

From (5.3) we see that $0 \leq \mu(X \times Y) = \Phi(1, 1) < \infty$.

If $\Phi(1, 1) = 0$, then $\Phi \equiv 0$, the trivial case.

Also $|\Phi(f, g)| \leq \Phi(1, 1) \|f\|_{\infty} \|g\|_{\infty}$, with $\|\Phi\| = \Phi(1, 1)$, in the case that X, Y are both pseudocompact, so that Φ is a bounded bilinear form.

5.2   Main Results

We give

Theorem 5.3. *Let X, Y be pseudocompact spaces, $\Phi : C(X) \times C(Y) \to \mathbb{R}$ be a positive bilinear form and $p, q > 1 : \frac{1}{p} + \frac{1}{q} = 1$. Then there exist uniquely determined probability measures μ_1, μ_2 on the Baire σ-algebras of X and Y, respectively, such that*

$$|\Phi(f, g)| \leq \|\Phi\| \left(\int_X |f(x)|^p \, d\mu_1 \right)^{1/p} \left(\int_Y |g(y)|^q \, d\mu_2 \right)^{1/q}, \tag{5.4}$$

for all $f \in C(X)$ and all $g \in C(Y)$.

Inequality (5.4) is sharp, namely it is attained, and the constant 1 is the best possible.

Proof. By Theorem 5.2 we have

$$\Phi(f, g) = \int_{X \times Y} f(x) g(y) \, d\mu(x, y), \tag{5.5}$$

and μ is a unique measure on the product of the Baire σ- algebras of X and Y. Without loss of generality we may suppose that $\Phi \not\equiv 0$. Denote by $m := \mu(X \times Y)$, so it is $0 < m < \infty$, and consider the measures $\mu^*(A) := \mu(A \times Y)$, for any A in the Baire σ-algebra of X, and $\mu^{**}(B) := \mu(X \times B)$, for any B in the Baire σ-algebra of Y. Here μ^*, μ^{**} are uniquely defined. Notice that

$$\mu^*(X) = \mu(X \times Y) = \mu^{**}(Y) = m. \tag{5.6}$$

Denote the probability measures $\mu_1 := \frac{\mu^*}{m}$, $\mu_2 := \frac{\mu^{**}}{m}$.

Also it holds

$$\int_{X \times Y} |f(x)|^p \, d\mu = \int_X |f(x)|^p \, d\mu^*, \tag{5.7}$$

and

$$\int_{X\times Y} |g(y)|^q \, d\mu = \int_Y |g(y)|^q \, d\mu^{**}. \tag{5.8}$$

So from (5.5) we get

$$|\Phi(f,g)| \le \int_{X\times Y} |f(x)| \, |g(y)| \, d\mu(x,y) \tag{5.9}$$

(by Hölder's inequality)

$$\le \left(\int_{X\times Y} |f(x)|^p \, d\mu(x,y)\right)^{1/p} \left(\int_{X\times Y} |g(y)|^q \, d\mu(x,y)\right)^{1/q} \tag{5.10}$$

$$= \left(\int_X |f(x)|^p \, d\mu^*(x)\right)^{1/p} \left(\int_Y |g(y)|^q \, d\mu^{**}(y)\right)^{1/q} \tag{5.11}$$

$$= m \left(\int_X |f(x)|^p \, d\mu_1(x)\right)^{1/p} \left(\int_Y |g(y)|^q \, d\mu_2(y)\right)^{1/q},$$

proving (5.4).

Next we derive the sharpness of (5.4).

Let us assume $f(x) = c_1 > 0$, $g(y) = c_2 > 0$. Then the left hand side of (5.4) equals $c_1 c_2 m$, equal to the right hand side of (5.4). That is proving attainability of (5.4) and that 1 is the best constant. $\qquad\square$

We give

Corollary 5.4. *All as in Theorem 5.3 with $p = q = 2$. Then*

$$|\Phi(f,g)| \le \|\Phi\| \left(\int_X (f(x))^2 \, d\mu_1\right)^{1/2} \left(\int_Y (g(y))^2 \, d\mu_2\right)^{1/2}, \tag{5.12}$$

for all $f \in C(X)$ and all $g \in C(Y)$.

Inequality (5.12) is sharp and the best constant is 1.

So when Φ is positive we improve Theorem 5.1.

Corollary 5.5. *All as in Theorem 5.3. Then*

$$|\Phi(f,g)| \le \|\Phi\| \inf_{\{p,q>1:\frac{1}{p}+\frac{1}{q}=1\}} \left(\int_X |f(x)|^p \, d\mu_1\right)^{1/p} \left(\int_Y |g(y)|^q \, d\mu_2\right)^{1/q}, \tag{5.13}$$

for all $f \in C(X)$ and all $g \in C(Y)$. Inequality (5.13) is sharp and the best constant is 1.

Corollary 5.6. *All as in Theorem 5.3, but $p = 1$, $q = \infty$. Then*

$$|\Phi(f,g)| \le \|\Phi\|\,\|g\|_\infty \left(\int_X |f(x)|\,d\mu_1 \right), \tag{5.14}$$

for all $f \in C(X)$ and all $g \in C(Y)$.

Inequality (5.14) is attained when $f(x) \ge 0$ and $g(y) = c > 0$, so it is sharp.

Proof. We observe that

$$|\Phi(f,g)| \le \|g\|_\infty \left(\int_{X \times Y} |f(x)|\,d\mu(x,y) \right)$$

$$= \|g\|_\infty \left(\int_X |f(x)|\,d\mu^*(x) \right) \tag{5.15}$$

$$= m\,\|g\|_\infty \left(\int_X |f(x)|\,d\mu_1(x) \right) \tag{5.16}$$

$$= \|\Phi\|\,\|g\|_\infty \left(\int_X |f(x)|\,d\mu_1 \right).$$

Sharpness of (5.14) is obvious. $\qquad\square$

Corollary 5.7. *Let X, Y be pseudocompact spaces, $\Phi : C(X) \times C(Y) \to \mathbb{R}$ be a positive bilinear form. Then there exist uniquely determined probability measures μ_1, μ_2 on the Baire $\sigma-$ algebras of X and Y, respectively, such that*

$$|\Phi(f,g)| \le \|\Phi\| \min \left(\|g\|_\infty \left(\int_X |f(x)|\,d\mu_1 \right), \|f\|_\infty \left(\int_Y |g(y)|\,d\mu_2 \right) \right), \tag{5.17}$$

for all $f \in C(X)$ and all $g \in C(Y)$.

Proof. Clear from Corollary 5.6. $\qquad\square$

Next we give some converse results.
We need

Definition 5.8. Let X, Y be pseudocompact spaces. We define

$$C^+(X) := \{ f : X \to \mathbb{R}_+ \cup \{0\} \text{ continuous and bounded} \}, \tag{5.18}$$

and

$$C^{++}(Y) := \{ g : Y \to \mathbb{R}_+ \text{ continuous and bounded} \},$$

where $\mathbb{R}_+ := \{ x \in \mathbb{R} : x > 0 \}$.

Clearly $C^+(X) \subset C(X)$, and $C^{++}(Y) \subset C(Y)$.

So $C^{++}(Y)$ is the positive cone of $C(Y)$.

Theorem 5.9. *Let X, Y be pseudocompact spaces, $\Phi : C^+(X) \times C^{++}(Y) \to \mathbb{R}$ be a positive bilinear form and $0 < p < 1$, $q < 0 : \frac{1}{p} + \frac{1}{q} = 1$. Then there exist* *uniquely determined probability measures μ_1, μ_2 on the Baire $\sigma-$ algebras of X and Y, respectively, such that*

$$\Phi(f,g) \geq \|\Phi\| \left(\int_X (f(x))^p \, d\mu_1\right)^{1/p} \left(\int_Y (g(y))^q \, d\mu_2\right)^{1/q}, \qquad (5.19)$$

for all $f \in C^+(X)$ and all $g \in C^{++}(Y)$.

Inequality (5.19) is sharp, namely it is attained, and the constant 1 is the best possible.

Proof. We use the same notations as in Theorem 5.3. By Theorem 5.2 and reverse Hölder's inequality we obtain

$$\Phi(f,g) = \int_{X \times Y} f(x) \, g(y) \, d\mu(x,y) \geq$$

$$\left(\int_{X \times Y} (f(x))^p \, d\mu(x,y)\right)^{1/p} \left(\int_{X \times Y} (g(y))^q \, d\mu(x,y)\right)^{1/q} \qquad (5.20)$$

$$= \left(\int_X (f(x))^p \, d\mu^*(x)\right)^{1/p} \left(\int_Y (g(y))^q \, d\mu^{**}(y)\right)^{1/q}$$

$$= m \left(\int_X (f(x))^p \, d\mu_1(x)\right)^{1/p} \left(\int_Y (g(y))^q \, d\mu_2(y)\right)^{1/q}, \qquad (5.21)$$

proving (5.19).

For the sharpness of (5.19), take $f(x) = c_1 > 0$, $g(y) = c_2 > 0$. Then $L.H.S$ $(5.19) = R.H.S\ (5.19) = mc_1 c_2$. $\qquad\square$

We need

Definition 5.10. Let X, Y be pseudocompact spaces. We define

$$C^-(X) := \{f : X \to \mathbb{R}_- \cup \{0\} \text{ continuous and bounded}\}, \qquad (5.22)$$

and

$$C^{--}(Y) := \{g : Y \to \mathbb{R}_- \text{ continuous and bounded}\},$$

where $\mathbb{R}_- := \{x \in \mathbb{R} : x < 0\}$.

Clearly $C^-(X) \subset C(X)$, and $C^{--}(Y) \subset C(Y)$.

So $C^{--}(Y)$ is the negative cone of $C(Y)$.

Theorem 5.11. *Let X, Y be pseudocompact spaces, $\Phi : C^-(X) \times C^{--}(Y) \to \mathbb{R}$ be a positive bilinear form and $0 < p < 1$, $q < 0 : \frac{1}{p} + \frac{1}{q} = 1$. Then there exist uniquely determined probability measures μ_1, μ_2 on the Baire $\sigma-$ algebras of X and Y, respectively, such that*

$$\Phi(f,g) \geq \|\Phi\| \left(\int_X |f(x)|^p \, d\mu_1\right)^{1/p} \left(\int_Y |g(y)|^q \, d\mu_2\right)^{1/q}, \qquad (5.23)$$

for all $f \in C^- (X)$ and all $g \in C^{--} (Y)$.

Inequality (5.23) is sharp, namely it is attained, and the constant 1 is the best possible.

Proof. Acting as in Theorem 5.9, we have that $\Phi (f, g) \geq 0$ and

$$\Phi (f, g) = \int_{X \times Y} f (x) g (y) \, d\mu (x, y) = \int_{X \times Y} |f (x)| \, |g (y)| \, d\mu (x, y) \tag{5.24}$$

$$\geq \left(\int_{X \times Y} |f (x)|^p \, d\mu (x, y) \right)^{1/p} \left(\int_{X \times Y} |g (y)|^q \, d\mu (x, y) \right)^{1/q} \tag{5.25}$$

$$= \left(\int_X |f (x)|^p \, d\mu^* (x) \right)^{1/p} \left(\int_Y |g (y)|^q \, d\mu^{**} (y) \right)^{1/q}$$

$$= m \left(\int_X |f (x)|^p \, d\mu_1 (x) \right)^{1/p} \left(\int_Y |g (y)|^q \, d\mu_2 (y) \right)^{1/q}, \tag{5.26}$$

establishing (5.23).

For the sharpness of (5.23), take $f (x) = c_1 < 0$, $g (y) = c_2 < 0$. Then *L.H.S* (5.23) $= m c_1 c_2 = $ *R.H.S* (5.23).

Hence optimality in (5.23) is established. $\qquad\qquad \square$

We finish this chapter with

Corollary 5.12.   (to Theorem 5.9) *We have*

$$\Phi (f, g) \geq \|\Phi\| \sup_{\left\{ p, q: \ 0 < p < 1, q < 0, \frac{1}{p} + \frac{1}{q} = 1 \right\}}$$

$$\left[\left(\int_X (f (x))^p \, d\mu_1 \right)^{1/p} \left(\int_Y (g (y))^q \, d\mu_2 \right)^{1/q} \right], \tag{5.27}$$

for all $f \in C^+ (X)$ and all $g \in C^{++} (Y)$.

Inequality (5.27) is sharp, namely it is attained, and the constant 1 is the best possible.

And analogously we find

Corollary 5.13.   (to Theorem 5.11) *We have*

$$\Phi (f, g) \geq \|\Phi\| \sup_{\left\{ p, q: \ 0 < p < 1, q < 0, \frac{1}{p} + \frac{1}{q} = 1 \right\}}$$

$$\left[\left(\int_X |f (x)|^p \, d\mu_1 \right)^{1/p} \left(\int_Y |g (y)|^q \, d\mu_2 \right)^{1/q} \right], \tag{5.28}$$

for all $f \in C^- (X)$ and all $g \in C^{--} (Y)$.

Inequality (5.28) is sharp, namely it is attained, and the constant 1 is the best possible.

Chapter 6

Basic Optimal Estimation of Csiszar's f-Divergence

In this chapter are established basic sharp probabilistic inequalities that give best approximation for the Csiszar's f-divergence, which is the most essential and general measure for the discrimination between two probability measures.This treatment relies on [37].

6.1 Background

Throughout this chapter we use the following. Let f be a convex function from $(0, +\infty)$ into $\mathbb{R}$ which is strictly convex at 1 with $f(1) = 0$.

Let (X, A, λ) be a measure space, where λ is a finite or a σ-finite measure on (X, A). And let μ_1, μ_2 be two probability measures on (X, A) such that $\mu_1 \ll \lambda$, $\mu_2 \ll \lambda$ (absolutely continuous), e.g. $\lambda = \mu_1 + \mu_2$.

Denote by $p = \frac{d\mu_1}{d\lambda}$, $q = \frac{d\mu_2}{d\lambda}$ the (densities) Radon–Nikodym derivatives of μ_1, μ_2 with respect to λ. Here we suppose that

$$0 < a \le \frac{p}{q} \le b, \quad \text{a.e. on } X \text{ and } a \le 1 \le b.$$

The quantity

$$\Gamma_f(\mu_1, \mu_2) = \int_X q(x) f\left(\frac{p(x)}{q(x)}\right) d\lambda(x), \tag{6.1}$$

was introduced by I. Csiszar in 1967, see [140], and is called f-*divergence* of the probability measures μ_1 and μ_2.

By Lemma 1.1 of [140], the integral (6.1) is well-defined and $\Gamma_f(\mu_1, \mu_2) \ge 0$ with equality only when $\mu_1 = \mu_2$, see also our Theorem 6.8, etc, at the end of the chapter. Furthermore $\Gamma_f(\mu_1, \mu_2)$ does not depend on the choice of λ. The concept of f-divergence was introduced first in [139] as a generalization of Kullback's *"information for discrimination"* or I-*divergence (generalized entropy)* [244], [243] and of Rényi's *"information gain"* (I-*divergence*) *of order* α [306]. In fact the *I-divergence of order* 1 equals

$$\Gamma_{u \log_2 u}(\mu_1, \mu_2).$$

45

The choice $f(u) = (u-1)^2$ produces again a known measure of difference of distributions that is called χ^2-*divergence*. Of course the *total variation* distance

$$|\mu_1 - \mu_2| = \int_X |p(x) - q(x)| \, d\lambda(x)$$

is equal to $\Gamma_{|u-1|}(\mu_1, \mu_2)$. Here by supposing $f(1) = 0$ we can consider $\Gamma_f(\mu_1, \mu_2)$ the f-divergence as a measure of the difference between the probability measures μ_1, μ_2.

The f-divergence is in general asymmetric in μ_1 and μ_2. But since f is convex and strictly convex at 1 so is

$$f^*(u) = uf\left(\frac{1}{u}\right) \tag{6.2}$$

and as in [140] we obtain

$$\Gamma_f(\mu_2, \mu_1) = \Gamma_{f^*}(\mu_1, \mu_2). \tag{6.3}$$

In Information Theory and Statistics many other divergences are used which are special cases of the above general Csiszar f-divergence, for example Hellinger distance D_H, α-divergence D_a, Bhattacharyya distance D_B, Harmonic distance D_{Ha}, Jeffrey's distance D_J, triangular discrimination D_Δ, for all these see e.g. [102], [152].

The problem of finding and estimating the proper *distance* (or *difference* or *discrimination*) of two probability distributions is one of the major ones in Probability Theory. The above f-divergence measures in their various forms have been also applied to Anthropology, Genetics, Finance, Economics, Political science, Biology, Approximation of Probability distributions, Signal Processing and Pattern Recognition.

A great motivation for this chapter has been the very important monograph on the matter by S. Dragomir [152].

6.2   Main Results

We give the following.

Theorem 6.1. *Let* $f \in C^1([a,b])$. *Then*

$$\Gamma_f(\mu_1, \mu_2) \le \|f'\|_{\infty,[a,b]} \int_X |p(x) - q(x)| \, d\lambda(x). \tag{6.4}$$

Inequality (6.4) is sharp. The optimal function is $f^*(y) = |y - 1|^\alpha$, $\alpha > 1$ *under the condition, either*

$$\text{i)} \qquad \max(b - 1, 1 - a) < 1 \ \text{ and } \ C := \operatorname{esssup}(q) < +\infty, \tag{6.5}$$

or

$$\text{ii)} \qquad |p - q| \le q \le 1, \ \text{ a.e. on } X. \tag{6.6}$$

Proof. We see that

$$\Gamma_f(\mu_1,\mu_2) = \int_X q(x) f\left(\frac{p(x)}{q(x)}\right) d\lambda$$

$$= \int_X q(x) f\left(\frac{p(x)}{q(x)}\right) d\lambda - f(1)$$

$$= \int_X q(x) f\left(\frac{p(x)}{q(x)}\right) d\lambda - \int_X q(x) f(1)\, d\lambda$$

$$= \int_X q(x) \left(f\left(\frac{p(x)}{q(x)}\right) - f(1) \right) d\lambda$$

$$\leq \int_X q(x) \left| f\left(\frac{p(x)}{q(x)}\right) - f(1) \right| d\lambda$$

$$\leq \|f'\|_{\infty,[a,b]} \int_X q(x) \left| \frac{p(x)}{q(x)} - 1 \right| d\lambda$$

$$= \|f'\|_{\infty,[a,b]} \int_X |p(x) - q(x)|\, d\lambda,$$

proving (6.4). Clearly here f^* is convex, $f^*(1) = 0$ and strictly convex at 1.

Also we have

$$f^{*\prime}(y) = \alpha|y - 1|^{\alpha-1}\, \text{sign}(y - 1)$$

and

$$\|f^{*\prime}\|_{\infty,[a,b]} = \alpha\big(\max(b - 1, 1 - a)\big)^{\alpha-1}.$$

We have

$$\text{L.H.S.}(6.4) = \int_X (q(x))^{1-\alpha}|p(x) - q(x)|^\alpha\, d\lambda, \tag{6.7}$$

and

$$\text{R.H.S.}(6.4) = \alpha\big(\max(b - 1, 1 - a)\big)^{\alpha-1} \int_X |p(x) - q(x)|\, d\lambda. \tag{6.8}$$

Hence

$$\lim_{\alpha \to 1} \text{R.H.S.}(6.4) = \int_X |p(x) - q(x)|\, d\lambda. \tag{6.9}$$

Based on Lemma 6.3 (6.22) following we find

$$\lim_{\alpha \to 1} \text{L.H.S.}(6.4) = \lim_{\alpha \to 1} \text{R.H.S.}(6.4), \tag{6.10}$$

establishing that inequality (6.4) is sharp. $\qquad\square$

Next we give the more general

Theorem 6.2. *Let* $f \in C^{n+1}([a,b])$, $n \in \mathbb{N}$, *such that* $f^{(k)}(1) = 0$, $k = 0, 2, \ldots, n$. *Then*

$$\Gamma_f(\mu_1,\mu_2) \leq \frac{\|f^{(n+1)}\|_{\infty,[a,b]}}{(n + 1)!} \int_X (q(x))^{-n}|p(x) - q(x)|^{n+1}\, d\lambda(x). \tag{6.11}$$

Inequality (6.11) is sharp. The optimal function is $\tilde{f}(y) := |y - 1|^{n+\alpha}$, $\alpha > 1$ under the condition (i) or (ii) of Theorem 6.1.

Proof. Let any $y \in [a, b]$, then

$$f(y) - f(1) = \sum_{k=1}^{n} \frac{f^{(k)}(1)}{k!}(y - 1)^k + \mathcal{R}_n(1, y), \qquad (6.12)$$

where

$$\mathcal{R}_n(1, y) := \int_{1}^{y} \left(f^{(n)}(t) - f^{(n)}(1)\right)\frac{(y - t)^{n-1}}{(n - 1)!}\, dt, \qquad (6.13)$$

here $y \geq 1$ or $y \leq 1$.

As in [27], p. 500 we get

$$|\mathcal{R}_n(1, y)| \leq \frac{\|f^{(n+1)}\|_{\infty,[a,b]}}{(n + 1)!}|y - 1|^{n+1}, \qquad (6.14)$$

for all $y \in [a, b]$. By theorem's assumption we have

$$f(y) = f'(1)(y - 1) + \mathcal{R}_n(1, y), \quad \text{all } y \in [a, b]. \qquad (6.15)$$

Consequently we obtain

$$\int_{X} q(x)f\left(\frac{p(x)}{q(x)}\right) d\lambda(x) = \int_{X} q(x)\mathcal{R}_n\left(1, \frac{p(x)}{q(x)}\right) d\lambda(x). \qquad (6.16)$$

Therefore

$$\Gamma_f(\mu_1, \mu_2) \leq \int_{X} q(x)\left|\mathcal{R}_n\left(1, \frac{p(x)}{q(x)}\right)\right| d\lambda(x)$$

$$\overset{\text{(by (6.14))}}{\leq} \frac{\|f^{(n+1)}\|_{\infty,[a,b]}}{(n + 1)!} \int_{X} q(x)\left|\frac{p(x)}{q(x)} - 1\right|^{n+1} d\lambda(x)$$

$$= \frac{\|f^{(n+1)}\|_{\infty,[a,b]}}{(n + 1)!} \int_{X} (q(x))^{-n}|p(x) - q(x)|^{n+1}\, d\lambda(x),$$

proving (6.11). Clearly here $\tilde{f}$ is convex, $\tilde{f}(1) = 0$ and strictly convex at 1, also $\tilde{f} \in C^{n+1}([a, b])$.

Furthermore

$$\tilde{f}^{(k)}(1) = 0, \quad k = 1, 2, \ldots, n,$$

and

$$\tilde{f}^{(n+1)}(y) = (n + \alpha)(n + \alpha - 1)\cdots(\alpha + 1)\alpha|y - 1|^{\alpha-1}\left(\text{sign}(y - 1)\right)^{n+1}.$$

Thus

$$|\tilde{f}^{(n+1)}(y)| = \left(\prod_{j=0}^{n}(n + \alpha - j)\right)|y - 1|^{\alpha-1},$$

and

$$\|\tilde{f}^{(n+1)}\|_{\infty,[a,b]} = \left(\prod_{j=0}^{n}(n+\alpha-j)\right)\left(\max(b-1,1-a)\right)^{\alpha-1}. \tag{6.17}$$

Applying $\tilde{f}$ into (6.11) we derive

$$\text{L.H.S.}(6.11) = \int_X q(x)\left|\frac{p(x)}{q(x)}-1\right|^{n+\alpha} d\lambda$$

$$= \int_X (q(x))^{1-n-\alpha}|p(x)-q(x)|^{n+\alpha}\, d\lambda(x), \tag{6.18}$$

and

$$\text{R.H.S.}(6.11) = \frac{\prod_{j=0}^{n}(n+\alpha-j)(\max(b-1,1-a))^{\alpha-1}}{(n+1)!}$$

$$\cdot \int_X (q(x))^{-n}|p(x)-q(x)|^{n+1}\, d\lambda(x). \tag{6.19}$$

Then

$$\lim_{\alpha\to 1}\text{R.H.S.}(6.11) = \int_X (q(x))^{-n}|p(x)-q(x)|^{n+1}\, d\lambda(x). \tag{6.20}$$

By applying Lemma 6.3 here, see next, we find

$$\lim_{\alpha\to 1}\text{L.H.S.}(11) = \lim_{\alpha\to 1}\int_X (q(x))^{1-n-\alpha}|p(x)-q(x)|^{n+\alpha}\, d\lambda(x)$$

$$= \int_X (q(x))^{-n}|p(x)-q(x)|^{n+1}\, d\lambda(x),$$

proving the sharpness of (6.11). $\qquad\square$

Lemma 6.3. *Here $\alpha>1$, $n\in\mathbb{Z}_+$ and condition i) or ii) of Theorem 6.1 is valid. Then*

$$\lim_{\alpha\to 1}\int_X (q(x))^{1-n-\alpha}|p(x)-q(x)|^{n+\alpha}\, d\lambda(x)$$

$$= \int_X (q(x))^{-n}|p(x)-q(x)|^{n+1}\, d\lambda(x). \tag{6.21}$$

When $n=0$ we have

$$\lim_{\alpha\to 1}\int_X (q(x))^{1-\alpha}|p(x)-q(x)|^{\alpha}\, d\lambda(x) = \int_X |p(x)-q(x)|\, d\lambda(x). \tag{6.22}$$

Proof. i) Let the condition (i) of Theorem 6.1 hold. We have

$$a-1 \leq \frac{p(x)}{q(x)}-1 \leq b-1, \quad \text{a.e. on } X.$$

Then

$$\left|\frac{p(x)}{q(x)} - 1\right| \le \max(b-1, 1-a) < 1, \quad \text{a.e. on } X.$$

Hence

$$\left|\frac{p(x)}{q(x)} - 1\right|^{n+\alpha} \le \left|\frac{p(x)}{q(x)} - 1\right|^{n+1} \le \left|\frac{p(x)}{q(x)} - 1\right| < 1, \quad \text{a.e. on } X,$$

and

$$\lim_{\alpha \to 1} \left|\frac{p(x)}{q(x)} - 1\right|^{n+\alpha} = \left|\frac{p(x)}{q(x)} - 1\right|^{n+1}, \quad \text{a.e. on } X.$$

By Dominated convergence theorem we obtain

$$\lim_{\alpha \to 1} \int_X \left|\frac{p(x)}{q(x)} - 1\right|^{n+\alpha} d\lambda(x) = \int_X \left|\frac{p(x)}{q(x)} - 1\right|^{n+1} d\lambda(x).$$

Then notice that

$$0 \le \int_X q(x) \left(\left|\frac{p(x)}{q(x)} - 1\right|^{n+1} - \left|\frac{p(x)}{q(x)} - 1\right|^{n+\alpha}\right) d\lambda$$

$$\le C \int_X \left(\left|\frac{p(x)}{q(x)} - 1\right|^{n+1} - \left|\frac{p(x)}{q(x)} - 1\right|^{n+\alpha}\right) d\lambda \to 0, \quad \text{as } \alpha \to 1.$$

Consequently,

$$\lim_{\alpha \to 1} \int_X q(x) \left|\frac{p(x)}{q(x)} - 1\right|^{n+\alpha} d\lambda = \int_X q(x) \left|\frac{p(x)}{q(x)} - 1\right|^{n+1} d\lambda,$$

implying (6.21), etc.

ii) By $q \le 1$ a.e. on X we get $q^{n+\alpha} \le q^{n+\alpha-1}$ a.e. on X.

$$|p - q|^{n+\alpha} \le q^{n+\alpha-1} \quad \text{a.e. on } X.$$

Here by assumptions obviously $q > 0$, $q^{1-n-\alpha} > 0$ a.e. on X. Set

$$\varrho_1 := \left\{x \in X : |p(x) - q(x)|^{n+\alpha} > (q(x))^{n+\alpha-1}\right\},$$

then $\lambda(\varrho_1) = 0$,

$$\varrho_2 := \left\{x \in X : q(x) = 0\right\},$$

then $\lambda(\varrho_2) = 0$,

$$\varrho_3 := \left\{x \in X : (q(x))^{1-n-\alpha}|p(x) - q(x)|^{n+\alpha} > 1\right\}.$$

Let $x_0 \in \varrho_3$, then

$$(q(x_0))^{1-n-\alpha}|p(x_0) - q(x_0)|^{n+\alpha} > 1.$$

If $q(x_0) > 0$, i.e. $(q(x_0))^{n+\alpha-1} > 0$ and $(q(x_0))^{1-n-\alpha} < +\infty$ then

$$|p(x_0) - q(x_0)|^{n+\alpha} > (q(x_0))^{n+\alpha-1}.$$

Hence $x_0 \in \varrho_1$.

If $q(x_0) = 0$, then $x_0 \in \varrho_2$. Clearly $\varrho_3 \subseteq \varrho_1 \cup \varrho_2$ and since $\lambda(\varrho_1 \cup \varrho_2) = 0$ we get that $\lambda(\varrho_3) = 0$. In the case of $q(x_0) = 0$ we get from above $\infty(p(x_0))^{n+\alpha} > 1$, which is only possible if $p(x_0) > 0$. That is, we established that

$$(q(x))^{1-n-\alpha}|p(x) - q(x)|^{n+\alpha} \le 1, \quad \text{a.e. on } X.$$

Furthermore

$$\lim_{\alpha \to 1}(q(x))^{1-n-\alpha}|p(x) - q(x)|^{n+\alpha} = (q(x))^{-n}|p(x) - q(x)|^{n+1}, \quad \text{a.e. on } X.$$

Therefore by Dominated convergence theorem we derive

$$\int_X (q(x))^{1-n-\alpha}|p(x) - q(x)|^{n+\alpha}\, d\lambda \to \int_X (q(x))^{-n}|p(x) - q(x)|^{n+1}\, d\lambda,$$

as $\alpha \to 1$, proving (6.21), etc. $\qquad\square$

Note 6.4. See that $|p - q| \le q$ a.e. on X is equivalent to $p \le 2q$ a.e. on X.

Corollary 6.5. (to Theorem 6.2) *Let $f \in C^2([a, b])$, such that $f(1) = 0$. Then*

$$\Gamma_f(\mu_1, \mu_2) \le \frac{\|f^{(2)}\|_{\infty,[a,b]}}{2} \cdot \int_X (q(x))^{-1}(p(x) - q(x))^2\, d\lambda(x). \tag{6.23}$$

Inequality (6.23) is sharp. The optimal function is $\tilde{f}(y) := |y - 1|^{1+\alpha}$, $\alpha > 1$ under the condition (i) or (ii) of Theorem 6.1.

Note 6.6. Inequalities (6.4) and (6.11) have some similarity to corresponding ones from those of N.S. Barnett et al. in Theorem 1, p. 3 of [102], though there the setting is slightly different and the important matter of sharpness is not discussed at all.

Next we connect Csiszar's f-divergence to the usual first modulus of continuity ω_1.

Theorem 6.7. *Suppose that*

$$0 < h := \int_X |p(x) - q(x)|\, d\lambda(x) \le \min(1 - a, b - 1). \tag{6.24}$$

Then

$$\Gamma_f(\mu_1, \mu_2) \le \omega_1(f, h). \tag{6.25}$$

Inequality (6.25) is sharp, namely it is attained by $f^*(x) = |x - 1|$.

Proof. By Lemma 8.1.1, p. 243 of [20] we obtain

$$f(x) \le \frac{\omega_1(f, \delta)}{\delta}|x - 1|, \tag{6.26}$$

for any $0 < \delta \le \min(1 - a, b - 1)$. Therefore

$$\begin{aligned}
\Gamma_f(\mu_1, \mu_2) &\le \frac{\omega_1(f, \delta)}{\delta} \int_X q(x)\left|\frac{p(x)}{q(x)} - 1\right| d\lambda(x) \\
&= \frac{\omega_1(f, \delta)}{\delta} \int_X |p(x) - q(x)|\, d\lambda(x).
\end{aligned}$$

By choosing $\delta = h$ we establish (6.25). Notice $\omega_1(|x-1|, \delta) = \delta$, for any $\delta > 0$. Consequently

$$\Gamma_{|x-1|}(\mu_1, \mu_2) = \int_X q(x) \left| \frac{p(x)}{q(x)} - 1 \right| d\lambda(x) = \int_X |p(x) - q(x)| \, d\lambda(x)$$

$$= \omega_1 \left(|x-1|, \int_X |p(x) - q(x)| \, d\lambda(x) \right),$$

proving attainability of (6.25). $\qquad\qquad\qquad\qquad\qquad\qquad\qquad\qquad\square$

Finally we reproduce the essence of Lemma 1.1. of [140] in a much simpler way.

Theorem 6.8. *Let $f : (0 + \infty) \to \mathbb{R}$ convex which is strictly convex at 1, $0 < a \leq 1 \leq b$, $a \leq \frac{p}{q} \leq b$, a.e on X. Then*

$$\Gamma_f(\mu_1, \mu_2) \geq f(1). \tag{6.27}$$

Proof Set $b := \frac{f'_-(1)+f'_+(1)}{2} \in \mathbb{R}$, because by f convexity both $f'_\pm$ exist. Since f is convex we get in general that

$$f(u) \geq f(1) + b(u - 1), \quad \forall u \in (0, +\infty).$$

We prove this last fact: Let $\quad 1 < u_1 < u_2, \quad$ then

$$\frac{f(u_1) - f(1)}{u_1 - 1} \leq \frac{f(u_2) - f(1)}{u_2 - 1},$$

hence

$$f'_+(1) \leq \frac{f(u) - f(1)}{u - 1} \quad \text{for} \quad u > 1.$$

That is

$$f(u) - f(1) \geq f'_+(1)(u - 1) \quad \text{for} \quad u > 1.$$

Let now $u_1 < u_2 < 1$, then

$$\frac{f(u_1) - f(1)}{u_1 - 1} \leq \frac{f(u_2) - f(1)}{u_2 - 1} \leq f'_-(1),$$

hence

$$f(u) - f(1) \geq f'_-(1)(u - 1) \quad \text{for} \quad u < 1.$$

Also by convexity $f'_-(1) \leq f'_+(1)$, and

$$f'_-(1) \leq \frac{f'_-(1) + f'_+(1)}{2} \leq f'_+(1).$$

That is, we have

$$f'_-(1) \leq b \leq f'_+(1).$$

Let $u > 1$ then $f(u) \geq f(1) + f'_+(1)(u-1) \geq f(1) + b(u-1)$, that is

$$f(u) \geq f(1) + b(u-1) \quad \text{for} \quad u > 1.$$

Let $u < 1$ then $f(u) \geq f(1) + f'_-(1)(u-1) \geq f(1) + b(u-1)$, that is

$$f(u) \geq f(1) + b(u-1) \quad \text{for} \quad u < 1.$$

Consequently it holds

$$f(u) \geq f(1) + b(n-1), \quad \forall u \in (0, +\infty),$$

proving this important feature of f. Therefore $\lim\limits_{u \to +\infty} \frac{f(u)}{u} \geq b$.

We do have

$$f\left(\frac{p}{q}\right) \geq f(1) + b\left(\frac{p}{q} - 1\right), \qquad a.e. on \ X,$$

and

$$qf\left(\frac{p}{q}\right) \geq f(1)q + b(p-q), \qquad a.e. on \ X.$$

The last gives

$$\int_X qf\left(\frac{p}{q}\right) d\lambda \geq f(1),$$

which is

$$\Gamma_f(\mu_1, \mu_2) \geq f(1). \qquad \qquad \square$$

Note 6.9 (to Theorem 6.8). Since f is strictly convex at 1 we get

$$f(u) > f(1) + b(u-1), \ \forall u \in (0, \infty) - \{1\},$$

with equality at $u = 1$.

Consequently

$$f\left(\frac{p}{q}\right) > f(1) + b\left(\frac{p}{q} - 1\right), \quad p \neq q,$$

with equality if $p = q$; $a.e. \ on \ X$.

Thus

$$qf\left(\frac{p}{q}\right) \geq f(1)q + b(p-q), \quad \text{whenever} \ p \neq q; a.e. on X,$$

and

$$\Gamma_f(\mu_1, \mu_2) > f(1), \quad \text{if} \ \mu_1 \neq \mu_2. \tag{6.28}$$

Clearly

$$\Gamma_f(\mu_1, \mu_2) = f(1), \quad \text{if} \ \mu_1 \neq \mu_2. \tag{6.29}$$

Note 6.10 (to Theorem 6.8). If $f(1) = 0,$ then $\Gamma_f(\mu_1, \mu_2) > 0$, for $\mu_1 \neq \mu_2$, and $\Gamma_f(\mu_1, \mu_2) = 0$, if $\mu_1 = \mu_2$.

Chapter 7

Approximations via Representations of Csiszar's f-Divergence

The Csiszar's discrimination is the most essential and general measure for the comparison between two probability measures. Here we provide probabilistic representation formulae for it on various very general settings. Then we present tight estimates for their remainders involving a variety of norms of the engaged functions. Also are given some direct general approximations for the Csiszar's f-divergence leading to very close probabilistic inequalities. We give also many important applications. This treatment relies a lot on [43].

7.1 Background

Throughout this chapter we use the following.

Let f be a convex function from $(0, +\infty)$ into $\mathbb{R}$ which is strictly convex at 1 with $f(1) = 0$. Let $(X, \mathcal{A}, \lambda)$ be a measure space, where λ is a finite or a σ-finite measure on $(X, \mathcal{A})$. And let μ_1, μ_2 be two probability measures on $(X, \mathcal{A})$ such that $\mu_1 \ll \lambda$, $\mu_2 \ll \lambda$ (absolutely continuous), e.g. $\lambda = \mu_1 + \mu_2$.

Denote by $p = \frac{d\mu_1}{d\lambda}$, $q = \frac{d\mu_2}{d\lambda}$ the (densities) Radon-Nikodym derivatives of μ_1, μ_2 with respect to λ. Here we suppose that

$$0 < a \le \frac{p}{q} \le b, \quad \text{a.e. on } X \quad \text{and} \quad a \le 1 \le b.$$

The quantity

$$\Gamma_f(\mu_1, \mu_2) = \int_X q(x) f\left(\frac{p(x)}{q(x)}\right) d\lambda(x), \tag{7.1}$$

was introduced by I. Csiszar in 1967, see [140], and is called f-*divergence* of the probability measures μ_1 and μ_2. By Lemma 1.1 of [140], the integral (7.1) is well-defined and $\Gamma_f(\mu_1, \mu_2) \ge 0$ with equality only when $\mu_1 = \mu_2$. Furthermore $\Gamma_f(\mu_1, \mu_2)$ does not depend on the choice of λ. The concept of f-divergence was introduced first in [139] as a generalization of Kullback's "*information for discrimination*" or I-*divergence (generalized entropy)* [244], [243] and of Rényi's "*information gain*" (I-*divergence of order* α) [306]. In fact the I-*divergence of order* 1 equals

$$\Gamma_{u \log_2 u}(\mu_1, \mu_2).$$

The choice $f(u) = (u - 1)^2$ produces again a known measure of difference of distributions that is called χ^2-*divergence*. Of course the *total variation* distance $|\mu_1 - \mu_2| = \int_X |p(x) - q(x)| \, d\lambda(x)$ is equal to $\Gamma_{|u-1|}(\mu_1, \mu_2)$.

Here by supposing $f(1) = 0$ we can consider $\Gamma_f(\mu_1, \mu_2)$, the f-divergence as a measure of the difference between the probability measures μ_1, μ_2. The f-divergence is in general asymmetric in μ_1 and μ_2. But since f is convex and strictly convex at 1 so is

$$f^*(u) = uf\left(\frac{1}{u}\right) \tag{7.2}$$

and as in [140] we obtain

$$\Gamma_f(\mu_2, \mu_1) = \Gamma_{f^*}(\mu_1, \mu_2). \tag{7.3}$$

In Information Theory and Statistics many other divergences are used which are special cases of the above general Csiszar f-divergence, e.g. Hellinger distance D_H, α-divergence D_α, Bhattacharyya distance D_B, Harmonic distance D_{Ha}, Jeffrey's distance D_J, triangular discrimination D_Δ, for all these see, e.g. [102], [152]. The problem of finding and estimating the *proper distance* (*or difference or discrimination*) of two probability distributions is one of the major ones in Probability Theory.

The above f-divergence measures in their various forms have been also applied to Anthropology, Genetics, Finance, Economics, Political Science, Biology, Approximation of Probability distributions, Signal Processing and Pattern Recognition. A great motivation for this chapter has been the very important monograph on the topic by S. Dragomir [152].

7.2 All Results

We present the following

Theorem 7.1. *Let $f, g \in C^1([a, b])$ where f as in this chapter, $g' \neq 0$ over $[a, b]$. Then*

$$\Gamma_f(\mu_1, \mu_2) \leq \left\|\frac{f'}{g'}\right\|_{\infty, [a,b]} \int_X q(x) \left|g\left(\frac{p(x)}{q(x)}\right) - g(1)\right| d\lambda. \tag{7.4}$$

Proof. From [112], p. 197, Theorem 27.8, Cauchy's Mean Value Theorem, we have that

$$f'(c)(g(b) - g(a)) = g'(c)(f(b) - f(a)), \tag{7.5}$$

where $c \in (a, b)$. Then

$$|f(b) - f(a)| = \left|\frac{f'(c)}{g'(c)}\right| |g(b) - g(a)|, \tag{7.6}$$

and

$$|f(z) - f(y)| \le \left\|\frac{f'}{g'}\right\|_{\infty,[a,b]} |g(z) - g(y)|, \tag{7.7}$$

for any $y, z \in [a, b]$. Clearly it holds

$$q(x)\left|f\left(\frac{p(x)}{q(x)}\right) - f(1)\right| \le \left\|\frac{f'}{g'}\right\|_{\infty,[a,b]} q(x)\left|g\left(\frac{p(x)}{q(x)}\right) - g(1)\right|, \tag{7.8}$$

a.e. on X. The last gives (7.4). $\qquad\square$

Examples 7.2 (to Theorem 7.1).

1) Let $g(x) = \frac{1}{x}$. Then

$$\Gamma_f(\mu_1, \mu_2) \le \|x^2 f'\|_{\infty,[a,b]} \int_X \frac{q(x)}{p(x)}|p(x) - q(x)|\, d\lambda(x). \tag{7.9}$$

2) Let $g(x) = e^x$. Then

$$\Gamma_f(\mu_1, \mu_2) \le \|e^{-x} f'\|_{\infty,[a,b]} \int_X q(x)\left|e^{\frac{p(x)}{q(x)}} - e\right| d\lambda(x). \tag{7.10}$$

3) Let $g(x) = e^{-x}$. Then

$$\Gamma_f(\mu_1, \mu_2) \le \|e^x f'\|_{\infty,[a,b]} \int_X q(x)\left|e^{-\frac{p(x)}{q(x)}} - e^{-1}\right| d\lambda(x). \tag{7.11}$$

4) Let $g(x) = \ln x$. Then

$$\Gamma_f(\mu_1, \mu_2) \le \|x f'\|_{\infty,[a,b]} \int_X q(x)\left|\ln\left(\frac{p(x)}{q(x)}\right)\right| d\lambda(x). \tag{7.12}$$

5) Let $g(x) = x\ln x$, with $a > e^{-1}$. Then

$$\Gamma_f(\mu_1, \mu_2) \le \left\|\frac{f'}{1 + \ln x}\right\|_{\infty,[a,b]} \int_X p(x)\left|\ln\frac{p(x)}{q(x)}\right| d\lambda(x). \tag{7.13}$$

6) Let $g(x) = \sqrt{x}$. Then

$$\Gamma_f(\mu_1, \mu_2) \le 2\|\sqrt{x} f'\|_{\infty,[a,b]} \int_X \sqrt{q(x)}|\sqrt{p(x)} - \sqrt{q(x)}|\, d\lambda(x). \tag{7.14}$$

7) Let $g(x) = x^\alpha$, $\alpha > 1$. Then

$$\Gamma_f(\mu_1, \mu_2) \le \frac{1}{\alpha}\left\|\frac{f'}{x^{\alpha-1}}\right\|_{\infty,[a,b]} \int_X q(x)^{1-\alpha}|(p(x))^\alpha - (q(x))^\alpha|\, d\lambda(x). \tag{7.15}$$

Next we give

Theorem 7.3. *Let $f \in C^1([a, b])$. Then*

$$\left|\Gamma_f(\mu_1, \mu_2) - \frac{1}{b-a}\int_a^b f(y)\, dy\right|$$

$$\le \frac{\|f'\|_{\infty,[a,b]}}{(b-a)}\left[\left(\frac{a^2+b^2}{2}\right) - (a+b) + \int_X q^{-1}p^2\, d\lambda(x)\right]. \tag{7.16}$$

Proof. Let $z \in [a, b]$, then by Theorem 22.1, p. 498, [27] we have

$$\left| \frac{1}{b-a} \int_a^b f(y)\, dy - f(z) \right| \leq \left(\frac{(z-a)^2 + (b-z)^2}{2(b-a)} \right) \|f'\|_{\infty,[a,b]} =: A(z), \qquad (7.17)$$

which is the basic Ostrowski inequality [277]. That is

$$-A(z) \leq f(z) - \frac{1}{b-a} \int_a^b f(y)dy \leq A(z), \qquad (7.18)$$

and

$$-qA\left(\frac{p}{q}\right) \leq qf\left(\frac{p}{q}\right) - \frac{q}{b-a} \int_a^b f(y)dy \leq qA\left(\frac{p}{q}\right), \qquad (7.19)$$

a.e. on X.

Integrating (7.19) against λ we find

$$-\int_X qA\left(\frac{p}{q}\right) d\lambda \leq \Gamma_f(\mu_1, \mu_2) - \frac{1}{b-a} \int_a^b f(y)dy \leq \int_X qA\left(\frac{p}{q}\right) d\lambda. \qquad (7.20)$$

That is,

$$\left| \Gamma_f(\mu_1, \mu_2) - \frac{1}{b-a} \int_a^b f(y)dy \right| \leq \int_X qA\left(\frac{p}{q}\right) d\lambda$$

$$= \frac{\|f'\|_{\infty,[a,b]}}{2(b-a)} \int_X q\left(\left(\frac{p}{q} - a\right)^2 + \left(b - \frac{p}{q}\right)^2 \right) d\lambda$$

$$= \frac{\|f'\|_{\infty,[a,b]}}{2(b-a)} \int_X q^{-1}\left((p - aq)^2 + (bq - p)^2\right) d\lambda, \qquad (7.21)$$

proving (7.16). $\qquad\square$

Note 7.4. In general we have, if $|A(z)| \leq B(z)$, then

$$\left| \int_X A(z)d\lambda \right| \leq \int_X B(z)d\lambda. \qquad (7.22)$$

Remark 7.5. Let f as in this chapter and $f \in C^1([a,b])$. Then we have Montogemery's identity, see p. 69 of [169],

$$f(z) = \frac{1}{b-a} \int_a^b f(t)dt + \frac{1}{b-a} \int_a^b P^*(z,t)f'(t)dt, \qquad (7.23)$$

for all $z \in [a, b]$, and

$$P^*(z,t) := \begin{cases} t - a, & t \in [a, z] \\ t - b, & t \in (z, b]. \end{cases} \qquad (7.24)$$

Hence

$$qf\left(\frac{p}{q}\right) = \frac{q}{b-a} \int_a^b f(t)dt + \frac{q}{b-a} \int_a^b P^*\left(\frac{p}{q}, t\right) f'(t)dt,$$

a.e. on X.

We get the following *representation formula*

$$\Gamma_f(\mu_1, \mu_2) = \frac{1}{b-a}\left(\int_a^b f(t)dt + \mathcal{R}_1\right), \tag{7.25}$$

where

$$\mathcal{R}_1 := \int_X q(x)\left(\int_a^b P^*\left(\frac{p(x)}{q(x)}, t\right)f'(t)dt\right)d\lambda(x). \tag{7.26}$$

Remark 7.6. Let f be as in this chapter and $f \in C^{(2)}[a, b]$. Then according to (2.2), p. 71 of [169], or (4.104), p. 192 of [126] we have

$$f(z) = \frac{1}{b-a}\int_a^b f(t)dt + \left(\frac{f(b)-f(a)}{b-a}\right)\left(z - \left(\frac{a+b}{2}\right)\right)$$
$$+ \frac{1}{(b-a)^2}\int_a^b\int_a^b P^*(z, t)P^*(t, s)f''(s)dsdt, \tag{7.27}$$

for all $z \in [a, b]$, where P^* is defined by (7.24). So that

$$qf\left(\frac{p}{q}\right) = \frac{q}{b-a}\int_a^b f(t)dt + \left(\frac{f(b)-f(a)}{b-a}\right)q\left(\frac{p}{q} - \left(\frac{a+b}{2}\right)\right)$$
$$+ \frac{q}{(b-a)^2}\int_a^b\int_a^b P^*\left(\frac{p}{q}, t\right)P^*(t, s)f''(s)dsdt, \tag{7.28}$$

a.e. on X. By integration against λ we obtain the *representation formula*

$$\Gamma_f(\mu_1, \mu_2) = \frac{1}{b-a}\int_a^b f(t)dt + \left(\frac{f(b)-f(a)}{b-a}\right)\left(1 - \left(\frac{a+b}{2}\right)\right) + \frac{\mathcal{R}_2}{(b-a)^2}, \tag{7.29}$$

where

$$\mathcal{R}_2 := \int_X q\left(\int_a^b\int_a^b P^*\left(\frac{p}{q}, t\right)P^*(t, s)f''(s)dsdt\right)d\lambda. \tag{7.30}$$

Again from Theorem 2.1, p. 70 of [169], or Theorem 4.17, p. 191 of [126] we see that

$$\left|f(z) - \frac{1}{b-a}\int_a^b f(t)dt - \left(\frac{f(b)-f(a)}{b-a}\right)\left(z - \frac{a+b}{2}\right)\right|$$
$$\leq \frac{1}{2}\left\{\left[\frac{\left(z - \frac{a+b}{2}\right)^2}{(b-a)^2} + \frac{1}{4}\right]^2 + \frac{1}{12}\right\}(b-a)^2\|f''\|_{\infty, [a,b]}, \tag{7.31}$$

for all $z \in [a, b]$.

Therefore

$$\left|qf\left(\frac{p}{q}\right) - \frac{q}{b-a}\int_a^b f(t)dt - \left(\frac{f(b)-f(a)}{b-a}\right)\left(p - \frac{(a+b)}{2}q\right)\right|$$
$$\leq \frac{1}{2}\left\{q\left[\frac{\left(\frac{p}{q} - \left(\frac{a+b}{2}\right)\right)^2}{(b-a)^2} + \frac{1}{4}\right]^2 + \frac{q}{12}\right\}(b-a)^2\|f''\|_{\infty, [a,b]}, \tag{7.32}$$

a.e. on X. Using (7.22) we have established that

$$\left| \Gamma_f(\mu_1, \mu_2) - \frac{1}{b-a} \int_a^b f(t)dt - \left(\frac{f(b) - f(a)}{b-a} \right) \left(1 - \frac{a+b}{2} \right) \right|$$

$$\leq \frac{(b-a)^2}{2} \|f''\|_{\infty,[a,b]} \left\{ \int_X q \left[\frac{\left(\frac{p}{q} - \left(\frac{a+b}{2} \right) \right)^2}{(b-a)^2} + \frac{1}{4} \right]^2 d\lambda + \frac{1}{12} \right\}. \qquad (7.33)$$

From p. 183 of [32] we have

Lemma 7.7. *Let $f \colon [a,b] \to \mathbb{R}$ be 3-times differentiable on $[a,b]$. Assume that f''' is integrable on $[a,b]$. Let $z \in [a,b]$. Define*

$$P(r,s) := \begin{cases} \dfrac{s-a}{b-a}, & a \leq s \leq r \\[2mm] \dfrac{s-b}{b-a}, & r < s \leq b; \ r,s \in [a,b]. \end{cases} \qquad (7.34)$$

Then

$$f(z) = \frac{1}{b-a} \int_a^b f(s_1)ds_1 + \left(\frac{f(b) - f(a)}{b-a} \right) \int_a^b P(z,s_1)ds_1$$

$$+ \left(\frac{f'(b) - f'(a)}{b-a} \right) \int_a^b \int_a^b P(z,s_1)P(s_1,s_2)ds_1 ds_2$$

$$+ \int_a^b \int_a^b \int_a^b P(z,s_1)P(s_1,s_2)P(s_2,s_3)f'''(s_3)ds_1 ds_2 ds_3. \qquad (7.35)$$

Remark 7.8. Let f as in this chapter and $f \in C^{(3)}([a,b])$. Then from (7.35) we obtain

$$qf\left(\frac{p}{q} \right) = \frac{q}{b-a} \int_a^b f(s_1)ds_1 + \left(\frac{f(b) - f(a)}{b-a} \right) q \int_a^b P\left(\frac{p}{q}, s_1 \right) ds_1$$

$$+ \left(\frac{f'(b) - f'(a)}{b-a} \right) q \int_a^b \int_a^b P\left(\frac{p}{q}, s_1 \right) P(s_1, s_2)ds_1 ds_2$$

$$+ q \int_a^b \int_a^b \int_a^b P\left(\frac{p}{q}, s_1 \right) P(s_1,s_2)P(s_2,s_3)f'''(s_3)ds_1 ds_2 ds_3, \qquad (7.36)$$

a.e. on X. We have derived the *representation formula*

$$\Gamma_f(\mu_1, \mu_2) = \frac{1}{b-a} \int_a^b f(s_1)ds_1 + \left(\frac{f(b) - f(a)}{b-a} \right) \int_X q \left(\int_a^b P\left(\frac{p}{q}, s_1 \right) ds_1 \right) d\lambda$$

$$+ \left(\frac{f'(b) - f'(a)}{b-a} \right) \int_X q \left(\int_a^b \int_a^b P\left(\frac{p}{q}, s_1 \right) P(s_1,s_2)ds_1 ds_2 \right) d\lambda + \mathcal{R}_3, \qquad (7.37)$$

where

$$\mathcal{R}_3 := \int_X q \left(\int_a^b \int_a^b \int_a^b P\left(\frac{p}{q}, s_1 \right) P(s_1,s_2)P(s_2,s_3)f'''(s_3)ds_1 ds_2 ds_3 \right) d\lambda.$$

$$(7.38)$$

We use also

Theorem 7.9. ([32]). *Let $f : [a, b] \to \mathbb{R}$ be 3-times differentiable on $[a, b]$. Assume that f''' is measurable and bounded on $[a, b]$. Let $z \in [a, b]$. Then*

$$\left| f(z) - \frac{1}{(b-a)} \int_a^b f(s_1) ds_1 - \left(\frac{f(b) - f(a)}{b - a} \right) \left(z - \left(\frac{a + b}{2} \right) \right) \right.$$
$$\left. - \left(\frac{f'(b) - f'(a)}{2(b - a)} \right) \left[z^2 - (a + b)z + \frac{(a^2 + b^2 + 4ab)}{6} \right] \right|$$
$$\leq \frac{\|f'''\|_{\infty, [a,b]}}{(b - a)^3} \cdot A^*(z), \tag{7.39}$$

where

$$A^*(z) := \left[abz^4 - \frac{1}{3}a^2 b^3 z + \frac{1}{3}a^3 bz^2 - ab^2 z^3 - \frac{1}{3}a^3 b^2 z + \frac{1}{3}ab^3 z^2 + a^2 b^2 z^2 \right.$$
$$- a^2 bz^3 - \frac{1}{2}az^5 - \frac{1}{2}bz^5 + \frac{1}{6}z^6 + \frac{3}{4}a^2 z^4 + \frac{3}{4}b^2 z^4 + \frac{1}{3}b^2 a^4$$
$$- \frac{2}{3}a^3 z^3 - \frac{2}{3}b^3 z^3 - \frac{1}{3}b^3 a^3 + \frac{5}{12}a^4 z^2 + \frac{5}{12}b^4 z^2 + \frac{1}{3}b^4 a^2$$
$$\left. - \frac{2}{15}ba^5 - \frac{2}{15}ab^5 - \frac{1}{6}a^5 z - \frac{1}{6}b^5 z + \frac{a^6}{20} + \frac{b^6}{20} \right]. \tag{7.40}$$

Inequality (7.39) is attained by

$$f(z) = (z - a)^3 + (b - z)^3; \tag{7.41}$$

in that case both sides of inequality equal zero.

Remark 7.10. Let f be as in this chapter and $f \in C^{(3)}([a, b])$. Then from (7.39) we find

$$\left| qf \left(\frac{p}{q} \right) - \frac{q}{b - a} \int_a^b f(s_1) ds_1 - \left(\frac{f(b) - f(a)}{b - a} \right) \left(p - q \left(\frac{a + b}{2} \right) \right) \right.$$
$$\left. - \left(\frac{f'(b) - f'(a)}{2(b - a)} \right) \left(\frac{p^2}{q} - (a + b)p + \frac{(a^2 + b^2 + 4ab)q}{6} \right) \right|$$
$$\leq \frac{\|f'''\|_{\infty, [a,b]}}{(b - a)^3} qA^* \left(\frac{p}{q} \right), \tag{7.42}$$

a.e. on X, A^* is as in (7.40).

We have proved that (via (7.22)) that

$$\left| \Gamma_f(\mu_1, \mu_2) - \frac{1}{b - a} \int_a^b f(s_1) ds_1 - \left(\frac{f(b) - f(a)}{b - a} \right) \left(1 - \left(\frac{a + b}{2} \right) \right) \right.$$
$$\left. - \left(\frac{f'(b) - f'(a)}{2(b - a)} \right) \cdot \left\{ \int_X q^{-1} p^2 d\lambda - (a + b) + \frac{(a^2 + b^2 + 4ab)}{6} \right\} \right|$$
$$\leq \frac{\|f'''\|_{\infty, [a,b]}}{(b - a)^3} \int_X qA^* \left(\frac{p}{q} \right) d\lambda, \tag{7.43}$$

where A^* is as in (7.40).

We need a generalization of Montgomery identity.

Proposition 7.11. ([32]). *Let $f \colon [a,b] \to \mathbb{R}$ be differentiable on $[a,b]$, $f' \colon [a,b] \to \mathbb{R}$ is integrable on $[a,b]$. Let $g \colon [a,b] \to \mathbb{R}$ be of bounded variation and such that $g(a) \neq g(b)$. Let $z \in [a,b]$. We define*

$$P(g(z), g(t)) := \begin{cases} \dfrac{g(t) - g(a)}{g(b) - g(a)}, & a \leq t \leq z, \\[3mm] \dfrac{g(t) - g(b)}{g(b) - g(a)}, & z < t \leq b. \end{cases} \tag{7.44}$$

Then it holds

$$f(z) = \frac{1}{(g(b) - g(a))} \int_a^b f(t) dg(t) + \int_a^b P(g(z), g(t)) f'(t) dt. \tag{7.45}$$

Remark 7.12. Let $f \in C^1([a,b])$, $g \in C([a,b])$ and of bounded variation, $g(a) \neq g(b)$. The rest are as in this chapter. Then by (7.44) and (7.45) we find

$$q f\left(\frac{p}{q}\right) = \frac{q}{g(b) - g(a)} \int_a^b f(t) dg(t) + q \int_a^b P\left(g\left(\frac{p}{q}\right), g(t)\right) f'(t) dt, \tag{7.46}$$

a.e. on X.

Finally we obtain the *representation formula*

$$\Gamma_f(\mu_1, \mu_2) = \frac{1}{g(b) - g(a)} \int_a^b f(t) dg(t) + \mathcal{R}_4, \tag{7.47}$$

where

$$\mathcal{R}_4 := \int_X q \left(\int_a^b P\left(g\left(\frac{p}{q}\right), g(t)\right) f'(t) dt \right) d\lambda. \tag{7.48}$$

Based on the last we have

Theorem 7.13. *In this chapter's setting, $f \in C^1([a,b])$ and $g \in C([a,b])$ of bounded variation, $g(a) \neq g(b)$. Then*

$$\left| \Gamma_f(\mu_1, \mu_2) - \frac{1}{g(b) - g(a)} \int_a^b f(t) dg(t) \right|$$

$$\leq \|f'\|_{\infty, [a,b]} \left\{ \int_X q \left(\int_a^b \left| P\left(g\left(\frac{p}{q}\right), g(t)\right) \right| dt \right) d\lambda \right\}. \tag{7.49}$$

Examples 7.14. (to Theorem 7.13). Here $f \in C^1([a,b])$ and as in this chapter. We take $g_1(x) = e^x$, $g_2(x) = \ln x$, $g_3(x) = \sqrt{x}$, $g_4(x) = x^\alpha$, $\alpha > 1$; $x > 0$. We elaborate on

$$P(g_i(x), g_i(t)) := \begin{cases} \dfrac{g_i(t) - g_i(a)}{g_i(b) - g_i(a)}, & a \leq t \leq x, \\[3mm] \dfrac{g_i(t) - g_i(b)}{g_i(b) - g_i(a)}, & x < t \leq b, \end{cases} \qquad i = 1, 2, 3, 4. \tag{7.50}$$

Clearly $dg_1(x) = e^x dx$, $dg_2(x) = \frac{1}{x}dx$, $dg_3(x) = \frac{1}{2\sqrt{x}}dx$, $dg_4(x) = \alpha x^{\alpha-1}dx$.

We have

1)

$$P(e^x, e^t) = \begin{cases} \dfrac{e^t - e^a}{e^b - e^a}, & a \leq t \leq x, \\[3mm] \dfrac{e^t - e^b}{e^b - e^a}, & x < t \leq b, \end{cases} \tag{7.51}$$

2)

$$P(\ln x, \ln t) = \begin{cases} \dfrac{\ln t - \ln a}{\ln b - \ln a}, & a \leq t \leq x, \\[3mm] \dfrac{\ln t - \ln b}{\ln b - \ln a}, & x < t \leq b, \end{cases} \tag{7.52}$$

3)

$$P(\sqrt{x}, \sqrt{t}) = \begin{cases} \dfrac{\sqrt{t} - \sqrt{a}}{\sqrt{b} - \sqrt{a}}, & a \leq t \leq x, \\[3mm] \dfrac{\sqrt{t} - \sqrt{b}}{\sqrt{b} - \sqrt{a}}, & x < t \leq b, \end{cases} \tag{7.53}$$

4)

$$P(x^\alpha, t^\alpha) = \begin{cases} \dfrac{t^\alpha - a^\alpha}{b^\alpha - a^\alpha}, & a \leq t \leq x, \\[3mm] \dfrac{t^\alpha - b^\alpha}{b^\alpha - a^\alpha}, & x < t \leq b. \end{cases} \tag{7.54}$$

Furthermore we derive

1)

$$|P(e^{p/q}, e^t)| = \begin{cases} \dfrac{e^t - e^a}{e^b - e^a}, & a \leq t \leq p/q, \\[3mm] \dfrac{e^b - e^t}{e^b - e^a}, & \dfrac{p}{q} < t \leq b, \end{cases} \tag{7.55}$$

2)

$$\left| P\left(\ln \frac{p}{q}, \ln t \right) \right| = \begin{cases} \dfrac{\ln t - \ln a}{\ln b - \ln a}, & a \leq t \leq \dfrac{p}{q}, \\[3mm] \dfrac{\ln b - \ln t}{\ln b - \ln a}, & \dfrac{p}{q} < t \leq b. \end{cases} \tag{7.56}$$

3)

$$\left| P\left(\sqrt{\frac{p}{q}}, \sqrt{t}\right) \right| = \begin{cases} \dfrac{\sqrt{t} - \sqrt{a}}{\sqrt{b} - \sqrt{a}}, & a \le t \le \dfrac{p}{q}, \\[3mm] \dfrac{\sqrt{b} - \sqrt{t}}{\sqrt{b} - \sqrt{a}}, & \dfrac{p}{q} < t \le b, \end{cases} \tag{7.57}$$

4)

$$\left| P\left(\frac{p^\alpha}{q^\alpha}, t^\alpha\right) \right| = \begin{cases} \dfrac{t^\alpha - a^\alpha}{b^\alpha - a^\alpha}, & a \le t \le \dfrac{p}{q}, \\[3mm] \dfrac{b^\alpha - t^\alpha}{b^\alpha - a^\alpha}, & \dfrac{p}{q} < t \le b. \end{cases} \tag{7.58}$$

Next we get

1)

$$\int_a^b |P(e^{p/q}, e^t)| dt = \frac{1}{e^b - e^a}\left\{ 2e^{p/q} - e^a\left(\frac{p}{q} - a + 1\right) + e^b\left(b - \frac{p}{q} - 1\right)\right\}, \tag{7.59}$$

2)

$$\int_a^b \left| P\left(\ln\frac{p}{q}, \ln t\right)\right| dt = \left(\ln\left(\frac{b}{a}\right)\right)^{-1}\left\{ \frac{2p}{q}\left(\ln\frac{p}{q} - 1\right) - a(\ln a - 1)\right.$$
$$\left. - (\ln a)\left(\frac{p}{q} - a\right) + (\ln b)\left(b - \frac{p}{q}\right) - b(\ln b - 1)\right\}, \tag{7.60}$$

3)

$$\int_a^b \left| P\left(\sqrt{\frac{p}{q}}, \sqrt{t}\right)\right| dt = \frac{1}{\sqrt{b} - \sqrt{a}}\left\{ \frac{4}{3}\frac{p^{3/2}}{q^{3/2}} - (\sqrt{a} + \sqrt{b})\frac{p}{q} + \frac{(a^{3/2} + b^{3/2})}{3}\right\}, \tag{7.61}$$

4)

$$\int_a^b \left| P\left(\frac{p^\alpha}{q^\alpha}, t^\alpha\right)\right| dt = \frac{1}{b^\alpha - a^\alpha}\left\{ \frac{2}{(\alpha + 1)}\left(\frac{p^{\alpha+1}}{q^{\alpha+1}}\right)\right.$$
$$\left. - (a^\alpha + b^\alpha)\frac{p}{q} + \frac{\alpha(a^{\alpha+1} + b^{\alpha+1})}{(\alpha + 1)}\right\}. \tag{7.62}$$

Clearly we find

1)

$$q\int_a^b |P(e^{p/q}, e^t)| dt = \frac{1}{e^b - e^a}\left\{ 2qe^{p/q} - e^a(p - aq + q) + e^b(bq - p - q)\right\}, \tag{7.63}$$

2)

$$q \int_a^b \left| P\left(\ln \frac{p}{q}, \ln t\right)\right| dt = \left(\ln\left(\frac{b}{a}\right)\right)^{-1} \left\{ 2p\left(\ln\frac{p}{q} - 1\right)\right.$$

$$\left. - a(\ln a - 1)q - (\ln a)(p - aq) + (\ln b)(bq - p) - b(\ln b - 1)q\right\}, \quad (7.64)$$

3)

$$q \int_a^b \left| P\left(\sqrt{\frac{p}{q}}, \sqrt{t}\right)\right| dt = \frac{1}{\sqrt{b} - \sqrt{a}} \left\{ \frac{4}{3} q^{-1/2} p^{3/2} - (\sqrt{a} + \sqrt{b})p + \frac{(a^{3/2} + b^{3/2})}{3} q\right\},$$

$$(7.65)$$

4)

$$q \int_a^b \left| P\left(\frac{p^\alpha}{q^\alpha}, t^\alpha\right)\right| dt = \frac{1}{b^\alpha - a^\alpha} \left\{ \frac{2}{(\alpha + 1)}(q^{-\alpha} p^{\alpha+1})\right.$$

$$\left. - (a^\alpha + b^\alpha)p + \left(\frac{\alpha}{\alpha + 1}\right)(a^{\alpha+1} + b^{\alpha+1})q\right\}. \quad (7.66)$$

Finally we obtain

1)

$$\int_X \left(q \int_a^b |P(e^{p/q}, e^t)| dt\right) d\lambda = \frac{1}{e^b - e^a} \left\{ 2 \int_X q e^{p/q} d\lambda - e^a(2 - a) + e^b(b - 2)\right\},$$

$$(7.67)$$

2)

$$\int_X \left(q \int_a^b \left| P\left(\ln \frac{p}{q}, \ln t\right)\right| dt\right) d\lambda$$

$$= \left(\ln\left(\frac{b}{a}\right)\right)^{-1} \left\{ 2\left(\int_X p \ln \frac{p}{q} d\lambda\right) - \ln(ab) + a + b - 2\right\}, \quad (7.68)$$

3)

$$\int_X q \int_a^b \left| P\left(\sqrt{\frac{p}{q}}, \sqrt{t}\right)\right| dt$$

$$= \frac{1}{\sqrt{b} - \sqrt{a}} \left\{ \frac{4}{3} \int_X q^{-1/2} p^{3/2} d\lambda - (\sqrt{a} + \sqrt{b}) + \frac{(a^{3/2} + b^{3/2})}{3}\right\}, \quad (7.69)$$

4)

$$\int_X q \int_a^b \left| P\left(\frac{p^\alpha}{q^\alpha}, t^\alpha\right)\right| dt = \frac{1}{b^\alpha - a^\alpha} \left\{ \frac{2}{(\alpha + 1)} \int_X q^{-\alpha} p^{\alpha+1} d\lambda - (a^\alpha + b^\alpha)\right.$$

$$\left. + \left(\frac{\alpha}{\alpha + 1}\right)(a^{\alpha+1} + b^{\alpha+1})\right\}. \quad (7.70)$$

Applying (7.49) we derive

1)

$$\left| \Gamma_f(\mu_1, \mu_2) - \frac{1}{e^b - e^a} \int_a^b f(t)e^t \, dt \right|$$

$$\leq \frac{\|f'\|_{\infty,[a,b]}}{(e^b - e^a)} \left\{ 2 \int_X q e^{p/q} \, d\lambda - e^a(2-a) + e^b(b-2) \right\}, \qquad (7.71)$$

2)

$$\left| \Gamma_f(\mu_1, \mu_2) - \left(\ln\left(\frac{b}{a}\right) \right)^{-1} \int_a^b \frac{f(t)}{t} \, dt \right|$$

$$\leq \|f'\|_{\infty,[a,b]} \left(\ln\left(\frac{b}{a}\right) \right)^{-1} \left\{ 2 \int_X p \ln \frac{p}{q} \, d\lambda - \ln(ab) + (a+b-2) \right\}, \qquad (7.72)$$

3)

$$\left| \Gamma_f(\mu_1, \mu_2) - \frac{1}{2(\sqrt{b} - \sqrt{a})} \int_a^b \frac{f(t)}{\sqrt{t}} \, dt \right|$$

$$\leq \frac{\|f'\|_{\infty,[a,b]}}{(\sqrt{b} - \sqrt{a})} \left\{ \frac{4}{3} \int_X q^{-1/2} p^{3/2} \, d\lambda + \frac{(a^{3/2} + b^{3/2})}{3} - (\sqrt{a} + \sqrt{b}) \right\}, \qquad (7.73)$$

4) $(\alpha > 1)$

$$\left| \Gamma_f(\mu_1, \mu_2) - \frac{\alpha}{(b^\alpha - a^\alpha)} \int_a^b f(t)t^{\alpha-1} \, dt \right|$$

$$\leq \frac{\|f'\|_{\infty,[a,b]}}{(b^\alpha - a^\alpha)} \left\{ \frac{2}{(\alpha+1)} \left(\int_X q^{-\alpha} p^{\alpha+1} \, d\lambda \right) \right.$$

$$\left. + \left(\frac{\alpha}{\alpha+1} \right) (a^{\alpha+1} + b^{\alpha+1}) - (a^\alpha + b^\alpha) \right\}. \qquad (7.74)$$

We need to mention

Proposition 7.15. ([32]). *Let $f \colon [a,b] \to \mathbb{R}$ be twice differentiable on $[a,b]$. The second derivative $f'' \colon [a,b] \to \mathbb{R}$ is integrable on $[a,b]$. Let also $g \colon [a,b] \to \mathbb{R}$ be of bounded variation, such that $g(a) \neq g(b)$. Let $z \in [a,b]$. The kernel P is defined as in Proposition 7.11 (7.44). Then we get*

$$f(z) = \frac{1}{(g(b) - g(a))} \int_a^b f(t) \, dg(t)$$

$$+ \frac{1}{(g(b) - g(a))} \left(\int_a^b P(g(z), g(t)) \, dt \right) \left(\int_a^b f'(t_1) \, dg(t_1) \right)$$

$$+ \int_a^b \int_a^b P(g(z), g(t)) P(g(t), g(t_1)) f''(t_1) \, dt_1 \, dt. \qquad (7.75)$$

Remark 7.16. Let $f \in C^{(2)}([a,b])$, $g \in C([a,b])$ and of bounded variation, $g(a) \neq g(b)$. The rest are as in this chapter. We have the *representation formula*

$$\Gamma_f(\mu_1, \mu_2) = \frac{1}{(g(b) - g(a))} \int_a^b f(t)dg(t) + \frac{1}{(g(b) - g(a))} \left(\int_a^b f'(t_1)dg(t_1) \right)$$

$$\cdot \left(\int_X \left(q \int_a^b P\left(g\left(\frac{p}{q}\right), g(t) \right) dt \right) d\lambda \right) + \mathcal{R}_5, \tag{7.76}$$

where

$$\mathcal{R}_5 := \int_X \left(q \int_a^b \int_a^b P\left(g\left(\frac{p}{q}\right), g(t) \right) P(g(t), g(t_1)) f''(t_1) dt_1 dt \right) d\lambda. \tag{7.77}$$

Clearly

$$|\mathcal{R}_5| \leq \|f''\|_{\infty, [a,b]}$$

$$\cdot \left(\int_X \left(q \int_a^b \int_a^b \left| P\left(g\left(\frac{p}{q}\right), g(t) \right) \right| |P(g(t), g(t_1))| dt_1 dt \right) d\lambda \right). \tag{7.78}$$

Therefore we get

$$\left| \Gamma_f(\mu_1, \mu_2) - \frac{1}{(g(b) - g(a))} \int_a^b f(t)dg(t) \right.$$

$$\left. - \frac{1}{(g(b) - g(a))} \left(\int_a^b f'(t_1)dg(t_1) \right) \left(\int_X \left(q \int_a^b P\left(g\left(\frac{p}{q}\right), g(t) \right) dt \right) d\lambda \right) \right|$$

$$\leq \|f''\|_{\infty, [a,b]}$$

$$\cdot \left(\int_X \left(q \int_a^b \int_a^b \left| P\left(g\left(\frac{p}{q}\right), g(t) \right) \right| |P(g(t), g(t_1))| dt_1 dt \right) d\lambda \right). \tag{7.79}$$

We further need to use

Theorem 7.17. *([32]). Let $f : [a,b] \to \mathbb{R}$ be n-times differentiable on $[a,b]$, $n \in \mathbb{N}$. The nth derivative $f^{(n)} : [a,b] \to \mathbb{R}$ is integrable on $[a,b]$. Let $z \in [a,b]$. Define the kernel*

$$P(r,s) := \begin{cases} \dfrac{s-a}{b-a}, & a \leq s \leq r, \\[2mm] \dfrac{s-b}{b-a}, & r < s \leq b, \end{cases} \tag{7.80}$$

where $r, s \in [a,b]$. Then

$$f(z) = \frac{1}{b-a} \int_a^b f(s_1)ds_1 + \sum_{k=0}^{n-2} \left(\frac{f^{(k)}(b) - f^{(k)}(a)}{b-a} \right)$$

$$\cdot \underbrace{\int_a^b \cdots \int_a^b}_{(k+1)\text{th}-\text{integral}} P(z, s_1) \prod_{i=1}^{k} P(s_i, s_{i+1}) ds_1 ds_2 \cdots ds_{k+1}$$

$$+ \int_a^b \cdots \int_a^b P(z, s_1) \prod_{i=1}^{n-1} P(s_i, s_{i+1}) f^{(n)}(s_n) ds_1 ds_2 \cdots ds_n. \tag{7.81}$$

Here and later we agree that $\prod_{k=0}^{-1} \bullet = 0$, $\prod_{i=1}^{0} \bullet = 1$.

Remark 7.18. Let $f \in C^{(n)}([a,b])$, $n \in \mathbb{N}$ and as in this chapter. Then by Theorem 1 ([32]) we obtain the *representation formula*

$$\Gamma_f(\mu_1,\mu_2) = \frac{1}{b-a}\int_a^b f(s_1)ds_1 + \sum_{k=0}^{n-2}\left(\frac{f^{(k)}(b)-f^{(k)}(a)}{b-a}\right)$$
$$\cdot\left(\int_X q\left(\int_a^b \cdots \int_a^b P\left(\frac{p}{q},s_1\right)\prod_{i=1}^{k}P(s_i,s_{i+1})ds_1\cdots ds_{k+1}\right)d\lambda\right)$$
$$+ \mathcal{R}_6, \tag{7.82}$$

where

$$\mathcal{R}_6 := \int_X q\left(\int_a^b \cdots \int_a^b P\left(\frac{p}{q},s_1\right)\prod_{i=1}^{n-1}P(s_i,s_{i+1})f^{(n)}(s_n)ds_1\cdots ds_n\right)d\lambda. \tag{7.83}$$

Clearly we have that

$$|\mathcal{R}_6| \le \|f^{(n)}\|_{\infty,[a,b]}\int_X q\left(\int_a^b \cdots \int_a^b \left|P\left(\frac{p}{q},s_1\right)\right|\prod_{i=1}^{n-1}|P(s_i,s_{i+1})|ds_1\cdots ds_n\right)d\lambda. \tag{7.84}$$

We also need

Theorem 7.19. ([32]). *Let* $f\colon [a,b] \to \mathbb{R}$ *be* n-*times differentiable on* $[a,b]$, $n \in \mathbb{N}$. *The* nth *derivative* $f^{(n)}\colon [a,b] \to \mathbb{R}$ *is integrable on* $[a,b]$. *Let* $g\colon [a,b] \to \mathbb{R}$ *be of bounded variation and* $g(a) \neq g(b)$. *Let* $z \in [a,b]$. *The kernel* P *is defined as in Proposition 1 ([32]). Then*

$$f(z) = \frac{1}{(g(b)-g(a))}\int_a^b f(s_1)dg(s_1) + \frac{1}{(g(b)-g(a))}\sum_{k=0}^{n-2}\left(\int_a^b f^{(k+1)}(s_1)dg(s_1)\right)$$
$$\cdot\left(\int_a^b \cdots \int_a^b P(g(z),g(s_1))\prod_{i=1}^{k}P(g(s_i),g(s_{i+1}))ds_1 ds_2\cdots ds_{k+1}\right)$$
$$+ \int_a^b \cdots \int_a^b P(g(z),g(s_1))\prod_{i=1}^{n-1}P(g(s_i),g(s_{i+1}))f^{(n)}(s_n)ds_1\cdots ds_n. \tag{7.85}$$

Remark 7.20. Let $f \in C^{(n)}([a,b])$, $n \in \mathbb{N}$ and $g \in C([a,b])$ and of bounded variation, $g(a) \neq g(b)$. The rest are as in this chapter. Then by Theorem 2 ([32])

we obtain the *representation formula*

$$\Gamma_f(\mu_1,\mu_2) = \frac{1}{(g(b)-g(a))} \int_a^b f(s_1)dg(s_1)$$

$$+ \frac{1}{(g(b)-g(a))} \sum_{k=0}^{n-2} \left(\int_a^b f^{(k+1)}(s_1)dg(s_1) \right)$$

$$\cdot \left(\int_X q \left(\int_a^b \cdots \int_a^b P\left(g\left(\frac{p}{q}\right), g(s_1) \right) \right. \right.$$

$$\left. \left. \cdot \prod_{i=1}^k P(g(s_i),g(s_{i+1}))ds_1 ds_2 \cdots ds_{k+1} d\lambda \right) + \mathcal{R}_7, \quad (7.86)$$

where

$$\mathcal{R}_7 = \int_X q \left(\int_a^b \cdots \int_a^b P\left(g\left(\frac{p}{q}\right), g(s_1) \right) \right.$$

$$\left. \cdot \prod_{i=1}^{n-1} P(g(s_i),g(s_{i+1}))f^{(n)}(s_n)ds_1 \cdots ds_n \right) d\lambda. \quad (7.87)$$

Clearly we get

$$|\mathcal{R}_7| \leq \|f^{(n)}\|_{\infty,[a,b]} \int_X q \left(\int_a^b \cdots \int_a^b \left| P\left(g\left(\frac{p}{q}\right), g(s_1) \right) \right| \right.$$

$$\left. \cdot \prod_{i=1}^{n-1} |P(g(s_i),g(s_{i+1}))|ds_1 \cdots ds_n \right) d\lambda. \quad (7.88)$$

Next we give some L_α results, $\alpha > 1$.

Remark 7.21. We repeat from Remark 7.5 the remainder(7.26),

$$\mathcal{R}_1 := \int_X q(x) \left(\int_a^b P^* \left(\frac{p(x)}{q(x)}, t \right) f'(t)dt \right) d\lambda(x),$$

where f is as in this chapter and $f \in C^1([a,b])$, P^* is given by (7.24). Let $\alpha, \beta > 1$: $\frac{1}{\alpha} + \frac{1}{\beta} = 1$. Then

$$|\mathcal{R}_1| \leq \|f'\|_{\alpha,[a,b]} \int_X q(x) \left\| P^* \left(\frac{p(x)}{q(x)}, \cdot \right) \right\|_{\beta,[a,b]} d\lambda(x) \quad (7.89)$$

and

$$|\mathcal{R}_1| \leq \|f'\|_{1,[a,b]} \int_X q(x) \left\| P^* \left(\frac{p(x)}{q(x)}, \cdot \right) \right\|_{\infty,[a,b]} d\lambda(x). \quad (7.90)$$

Similarly we obtain for $\mathcal{R}_2$ (see (7.30)) of Remark 7.6, here $f \in C^{(2)}([a,b])$ and as in this chapter that

$$|\mathcal{R}_2| \leq \|f''\|_{\alpha,[a,b]} \int_X q \left(\int_a^b \left| P^* \left(\frac{p}{q}, t \right) \right| \|P^*(t,\cdot)\|_{\beta,[a,b]}dt \right) d\lambda, \quad (7.91)$$

and

$$|\mathcal{R}_2| \le \|f''\|_{1,[a,b]} \int_X q\left(\int_a^b \left|P^*\left(\frac{p}{q},t\right)\right| \|P^*(t,\cdot)\|_{\infty,[a,b]} dt\right) d\lambda. \tag{7.92}$$

Similarly we get for $\mathcal{R}_3$ (see (7.38)) of Remark 7.8, here $f \in C^{(3)}([a,b])$ and as in this chapter, P as in (7.34), that

$$|\mathcal{R}_3| \le \|f'''\|_{\alpha,[a,b]} \int_X q\left(\int_a^b \int_a^b \left|P\left(\frac{p}{q},s_1\right)\right| |P(s_1,s_2)| \|P(s_2,\cdot)\|_{\beta,[a,b]} ds_1 ds_2\right) d\lambda \tag{7.93}$$

and

$$|\mathcal{R}_3| \le \|f'''\|_{1,[\alpha,\beta]} \int_X q\left(\int_a^b \int_a^b \left|P\left(\frac{p}{q},s_1\right)\right| |P(s_1,s_2)| \|P(s_2,\cdot)\|_{\infty,[a,b]} ds_1 ds_2\right) d\lambda. \tag{7.94}$$

Next let things be as in Remark 7.12 and $\mathcal{R}_4$ as in (7.48). Then

$$|\mathcal{R}_4| \le \|f'\|_{\alpha,[a,b]} \int_X q(x) \left\|P\left(g\left(\frac{p(x)}{q(x)}\right),g(\cdot)\right)\right\|_{\beta,[a,b]} d\lambda(x) \tag{7.95}$$

and

$$|\mathcal{R}_4| \le \|f'\|_{1,[a,b]} \int_X q(x) \left\|P\left(g\left(\frac{p(x)}{q(x)}\right),g(\cdot)\right)\right\|_{\infty,[a,b]} d\lambda(x). \tag{7.96}$$

Let things be as in Remark 7.16 and $\mathcal{R}_5$ as in (7.77). Then

$$|\mathcal{R}_5| \le \|f''\|_{\alpha,[a,b]} \int_X q\left(\int_a^b \left|P\left(g\left(\frac{p}{q}\right),g(t)\right)\right| \|P(g(t),g(\cdot))\|_{\beta,[a,b]} dt\right) d\lambda \tag{7.97}$$

and

$$|\mathcal{R}_5| \le \|f''\|_{1,[a,b]} \int_X q\left(\int_a^b \left|P\left(g\left(\frac{p}{q}\right),g(t)\right)\right| \|P(g(t),g(\cdot))\|_{\infty,[a,b]} dt\right) d\lambda. \tag{7.98}$$

Let things be as in Remark 7.18 and $\mathcal{R}_6$ as in (7.83). Then

$$|\mathcal{R}_6| \le \|f^{(n)}\|_{\alpha,[a,b]} \int_X q\left(\int_a^b \cdots \int_a^b \left|P\left(\frac{p}{q},s_1\right)\right|\right.$$
$$\left. \cdot \left(\prod_{i=1}^{n-2} |P(s_i,s_{i+1})|\right) \|P(s_{n-1},\cdot)\|_{\beta,[a,b]} ds_1 \cdots ds_{n-1}\right) d\lambda, \tag{7.99}$$

and

$$|\mathcal{R}_6| \le \|f^{(n)}\|_{1,[a,b]} \int_X q\left(\int_a^b \cdots \int_a^b \left|P\left(\frac{p}{q},s_1\right)\right|\right.$$
$$\left. \cdot \left(\prod_{i=1}^{n-2} |P(s_i,s_{i+1})|\right) \|P(s_{n-1},\cdot)\|_{\infty,[a,b]} ds_1 \cdots ds_{n-1}\right) d\lambda. \tag{7.100}$$

Finally let things be as in Remark 7.20 and $\mathcal{R}_7$ as in (7.87). Then

$$|\mathcal{R}_7| \le \|f^{(n)}\|_{\alpha,[a,b]} \int_X q(x) \left(\int_a^b \cdots \int_a^b \left| P\left(g\left(\frac{p(x)}{q(x)}\right), g(s_1)\right) \right| \right.$$

$$\cdot \left(\prod_{i=1}^{n-2} |P(g(s_i), g(s_{i+1}))| \right)$$

$$\left. \|P(g(s_{n-1}), g(\cdot))\|_{\beta,[a,b]} ds_1 \cdots ds_{n-1} \right) d\lambda(x), \tag{7.101}$$

and

$$|\mathcal{R}_7| \le \|f^{(n)}\|_{1,[a,b]} \int_X q(x) \left(\int_a^b \cdots \int_a^b \left| P\left(g\left(\frac{p(x)}{q(x)}\right), g(s_1)\right) \right| \right.$$

$$\cdot \left(\prod_{i=1}^{n-2} |P(g(s_i), g(s_{i+1}))| \right)$$

$$\left. \|P(g(s_{n-1}), g(\cdot))\|_{\infty,[a,b]} ds_1 \cdots ds_{n-1} \right) d\lambda(x). \tag{7.102}$$

At the end we simplify and expand results of Remark 7.21 in the following

Remark 7.22. Here $0 < a < b$ with $1 \in [a,b]$, and $\frac{p}{q} \in [a,b]$ a.e. on X.
i) We simplify (7.90). We notice that

$$\left| P^*\left(\frac{p}{q}, t\right) \right| = \left\{ \begin{array}{ll} t - a, & t \in \left[a, \frac{p}{q}\right] \\ \\ b - t, & t \in \left(\frac{p}{q}, b\right] \end{array} \right\}, \tag{7.103}$$

here P^* is as in (7.24). Then

$$\left\| P^*\left(\frac{p(x)}{q(x)}, \cdot\right) \right\|_{\infty,[a,b]} = \max\left(\frac{p(x)}{q(x)} - a, b - \frac{p(x)}{q(x)}\right). \tag{7.104}$$

Hence it holds

$$|\mathcal{R}_1| \le \|f'\|_{1,[a,b]} \int_X q(x) \max\left(\frac{p(x)}{q(x)} - a, b - \frac{p(x)}{q(x)}\right) d\lambda(x). \tag{7.105}$$

Here f is as in this chapter and $f \in C^1([a,b])$.

ii) Next we simplify (7.96). Take here g strictly increasing and continuous over $[a,b]$, e.g. $g(x) = e^x$, $\ln x$, $\sqrt{x}$, x^α with $\alpha > 1$. Also $f \in C^1([a,b])$ and as in this chapter. Then clearly we derive

$$|\mathcal{R}_4| \le \frac{\|f'\|_{1,[a,b]}}{(g(b) - g(a))} \int_X q(x) \max\left(g\left(\frac{p(x)}{q(x)}\right) - g(a), g(b) - g\left(\frac{p(x)}{q(x)}\right)\right) d\lambda(x). \tag{7.106}$$

In particular we obtain

1)

$$|\mathcal{R}_4| \le \frac{\|f'\|_{1,[a,b]}}{(e^b - e^a)} \int_X q(x) \max\left(e^{\frac{p(x)}{q(x)}} - e^a, e^b - e^{\frac{p(x)}{q(x)}} \right) d\lambda(x), \qquad (7.107)$$

2)

$$|\mathcal{R}_4| \le \frac{\|f'\|_{1,[a,b]}}{\ln b - \ln a} \int_X q(x) \max\left(\ln\left(\frac{p(x)}{q(x)}\right) - \ln a, \ln b - \ln\left(\frac{p(x)}{q(x)}\right) \right) d\lambda(x), \qquad (7.108)$$

3)

$$|\mathcal{R}_4| \le \frac{\|f'\|_{1,[a,b]}}{\sqrt{b} - \sqrt{a}} \int_X q(x) \max\left(\sqrt{\frac{p(x)}{q(x)}} - \sqrt{a}, \sqrt{b} - \sqrt{\frac{p(x)}{q(x)}} \right) d\lambda(x), \qquad (7.109)$$

and finally for $\alpha > 1$ we get

4)

$$|\mathcal{R}_4| \le \frac{\|f'\|_{1,[a,b]}}{b^\alpha - a^\alpha} \int_X q(x) \max\left(\frac{p^\alpha(x)}{q^\alpha(x)} - a^\alpha, b^\alpha - \frac{p^\alpha(x)}{q^\alpha(x)} \right) d\lambda(x). \qquad (7.110)$$

iii) Finally we simplify (7.89).

Let $\alpha, \beta > 1$ such that $\frac{1}{\alpha} + \frac{1}{\beta} = 1$. Also $f \in C^1([a,b])$ and as in this chapter. We obtain that

$$\left| P^*\left(\frac{p}{q}, t\right) \right|^\beta = \left\{ \begin{array}{ll} (t-a)^\beta, & t \in \left[a, \frac{p}{q}\right] \\[2mm] (b-t)^\beta, & t \in \left(\frac{p}{q}, b\right] \end{array} \right\}, \qquad (7.111)$$

and

$$\int_a^b \left| P^*\left(\frac{p}{q}, t\right) \right|^\beta dt = \frac{\left(\frac{p}{q} - a\right)^{\beta+1} + \left(b - \frac{p}{q}\right)^{\beta+1}}{\beta+1}. \qquad (7.112)$$

Thus

$$\left\| P^*\left(\frac{p(x)}{q(x)}, \cdot\right) \right\|_{\beta,[a,b]} = \sqrt[\beta]{\frac{(p(x) - aq(x))^{\beta+1} + (bq(x) - p(x))^{\beta+1}}{(\beta+1)q(x)^{\beta+1}}}. \qquad (7.113)$$

Here notice $q = 0$ a.e. on X. So we need only work with $q(x) > 0$. So all in (iii) make sense. Therefore

$$q(x) \left\| P^*\left(\frac{p(x)}{q(x)}, \cdot\right) \right\|_{\beta,[a,b]} = \sqrt[\beta]{\frac{(p(x) - aq(x))^{\beta+1} + (bq(x) - p(x))^{\beta+1}}{(\beta+1)q(x)}}, \qquad (7.114)$$

a.e. on X. So we have established that

$$|\mathcal{R}_1| \le \|f'\|_{\alpha,[a,b]} \int_X \left(\sqrt[\beta]{\frac{(p(x) - aq(x))^{\beta+1} + (bq(x) - p(x))^{\beta+1}}{(\beta+1)q(x)}} \right) d\lambda(x). \qquad (7.115)$$

Chapter 8

Sharp High Degree Estimation of Csiszar's f-Divergence

In this chapter are presented various sharp and nearly optimal probabilistic inequalities giving the high order approximation of Csiszar's f-divergence between two probability measures, which is the most essential and general tool for their comparison. The above are done through Taylor's formula, generalized Taylor-Widder's formula, an alternative recent expansion formula. Based on these we give many representation formulae of Csiszar's distance, then we estimate in all directions their remainders by using either the norms approach or the modulus of continuity way. Most of the last probabilistic estimates are sharp or nearly sharp, attained by basic simple functions. This treatment relies on [41].

8.1 Background

Throughout this chapter we use the following. Let f be a convex function from $(0, +\infty)$ into $\mathbb{R}$ which is strictly convex at 1 with $f(1) = 0$.

Let $(X, \mathcal{A}, \lambda)$ be a measure space, where λ is a finite or a σ-finite measure on $(X, \mathcal{A})$. And let μ_1, μ_2 be two probability measures on $(X, \mathcal{A})$ such that $\mu_1 \ll \lambda$, $\mu_2 \ll \lambda$ (absolutely continuous), e.g. $\lambda = \mu_1 + \mu_2$.

Denote by $p = \frac{d\mu_1}{d\lambda}$, $q = \frac{d\mu_2}{d\lambda}$ the (densities) Radon-Nikodym derivatives of μ_1, μ_2 with respect to λ. Here we suppose that

$$0 < a \leq \frac{p}{q} \leq b, \quad \text{a.e. on } X \text{ and } a \leq 1 \leq b.$$

The quantity

$$\Gamma_f(\mu_1, \mu_2) = \int_X q(x) f\left(\frac{p(x)}{q(x)}\right) d\lambda(x), \tag{8.1}$$

was introduced by I. Csiszar in 1967, see [140], and is called f-divergence of the probability measures μ_1 and μ_2.

By Lemma 1.1 of [140], the integral (8.1) is well-defined and $\Gamma_f(\mu_1, \mu_2) \geq 0$ with equality only when $\mu_1 = \mu_2$. Furthermore $\Gamma_f(\mu_1, \mu_2)$ does not depend on the choice of λ. The concept of f-divergence was introduced first in [139] as a generalization of Kullback's *information for discrimination* or *I-divergence* "*(generalized entropy)*"

73

[244], [243] and of Rényi's *information gain* (*I-divergence of order* α) [306]. In fact the *I-divergence of order* 1 equals

$$\Gamma_{u \log_2 u}(\mu_1, \mu_2).$$

The choice $f(u) = (u-1)^2$ produces again a known measure of difference of distributions that is called χ^2-*divergence*. Of course the *total variation* distance

$$|\mu_1 - \mu_2| = \int_X |p(x) - q(x)| \, d\lambda(x)$$

is equal to $\Gamma_{|u-1|}(\mu_1, \mu_2)$. Here by assuming $f(1) = 0$ we can consider $\Gamma_f(\mu_1, \mu_2)$, the f-divergence as a measure of the difference between the probability measures μ_1, μ_2.

The f-divergence is in general asymmetric in μ_1 and μ_2. But since f is convex and strictly convex at 1 so is

$$f^*(u) = uf\left(\frac{1}{u}\right) \tag{8.2}$$

and as in [140] we find

$$\Gamma_f(\mu_2, \mu_1) = \Gamma_{f^*}(\mu_1, \mu_2). \tag{8.3}$$

In Information Theory and Statistics many other divergences are used which are special cases of the above general Csiszar f-divergence, e.g. Hellinger distance D_H, α-divergence D_α, Bhattacharyya distance D_B, Harmonic distance D_{Ha}, Jeffrey's distance D_J, triangular discrimination D_Δ, for all these see, e.g. [102], [152].

The problem of finding and estimating the *proper distance* (*or difference or discrimination*) of two probability distributions is one of the major ones in Probability Theory. The above f-divergence measures in their various forms have been also applied to Anthropology, Genetics, Finance, Economics, Political Science, Biology, Approximation of Probability distributions, Signal Processing and Pattern Recognition.

A great motivation for this chapter has been the very important monograph on the topic by S. Dragomir [152].

8.2 Results Based on Taylor's Formula

We give the following

Theorem 8.1. *Let* $0 < a < 1 < b$, f *as in this chapter and* $f \in C^n([a,b])$, $n \geq 1$ *with* $|f^{(n)}(t) - f^{(n)}(1)|$ *be a convex function in* t. *Let* $0 < h < \min(1-a, b-1)$ *be fixed. Then*

$$\Gamma_f(\mu_1, \mu_2) \leq \sum_{k=2}^{n} \frac{|f^{(k)}(1)|}{k!} \left| \int_X q^{1-k}(p-q)^k d\lambda \right| + \frac{\omega_1(f^{(n)}, h)}{(n+1)!h} \int_X q^{-n}|p-q|^{n+1} d\lambda.$$

$$\tag{8.4}$$

Here ω_1 is the usual first modulus of continuity, $\omega_1(f^{(n)}, h) := \sup\{|f^{(n)}(x) - f^{(n)}(y)|; x, y \in [a, b], |x - y| \le h\}$. If $f^{(k)}(1) = 0$, $k = 2, \ldots, n$, then

$$\Gamma_f(\mu_1, \mu_2) \le \frac{\omega_1(f^{(n)}, h)}{(n+1)!h} \int_X q^{-n}|p - q|^{n+1} d\lambda. \tag{8.5}$$

Inequalities (8.4) and (8.5) when n is even are attained by

$$\tilde{f}(t) := \frac{|t - 1|^{n+1}}{(n+1)!}, \quad a \le t \le b. \tag{8.6}$$

Example 8.2. The function

$$g(x) := |x - 1|^{n+\alpha}, \quad n \in \mathbb{N}, \ \alpha > 1 \tag{8.7}$$

is convex and strictly convex at 1, also $g^{(k)}(1) = 0$, $k = 0, 1, 2, \ldots, n$. Furthermore

$$(g(x))^{(n)} = (n + \alpha)(n + \alpha - 1) \cdots (\alpha + 1)|x - 1|^\alpha (\text{sign}(x - 1))^n$$

with

$$|g(x)^{(n)}| = (n + \alpha)(n + \alpha - 1) \cdots (\alpha + 1)|x - 1|^\alpha$$

being convex. That is g fulfills all the assumptions of Theorem 8.1.

Proof. Here we have $(f(1) = 0)$

$$f(t) = \sum_{k=1}^{n} \frac{f^{(k)}(1)}{k!}(t - 1)^k + I_t, \tag{8.8}$$

where

$$I_t := \int_1^t \left(\int_1^{t_1} \cdots \left(\int_1^{t_{n-1}} (f^{(n)}(t_n) - f^{(n)}(1)) dt_n \cdots dt_1 \right) \right). \tag{8.9}$$

By Lemma 8.1.1 of [20], p. 243 we have that

$$|f^{(n)}(t) - f^{(n)}(1)| \le \frac{\omega_1(f^{(n)}, h)}{h}|t - 1|, \tag{8.10}$$

and

$$|I_t| \le \frac{\omega_1(f^{(n)}, h)}{h} \frac{|t - 1|^{n+1}}{(n+1)!}. \tag{8.11}$$

We observe by (8.8) that

$$qf\left(\frac{p}{q}\right) = f'(1)(p - q) + \sum_{k=2}^{n} \frac{f^{(k)}(1)}{k!} q^{1-k}(p - q)^k + qI_{p/q}, \tag{8.12}$$

true a.e. on X.

Here $q \ne 0$ a.e. on X. Integrating (8.12) against λ and using (8.11) we derive (8.4). Notice that for n even $\tilde{f}^{(n)}(t) = |t - 1|$ and $\tilde{f}^{(k)}(1) = 0$ for all $k = 0, 1, \ldots, n$, with $\|\tilde{f}^{(n+1)}\|_\infty = 1$. Furthermore,

$$\omega_1(\tilde{f}^{(n)}, h) = h.$$

In both cases in (8.4) and (8.5) their two sides equal

$$\frac{1}{(n+1)!}\int_X q^{-n}|p-q|^{n+1}d\lambda, \tag{8.13}$$

proving attainability and sharpness of (8.4) and (8.5). $\qquad\square$

We continue with the general

Theorem 8.3. *Let $0 < a \leq 1 \leq b$, f as in this chapter and $f \in C^n([a,b])$, $n \geq 1$. Suppose that $\omega_1(f^{(n)}, \delta) \leq w$, where $0 < \delta \leq b - a$, $w > 0$. Let $x \in \mathbb{R}$ and denote by*

$$\phi_n(x) := \int_0^{|x|} \left\lceil \frac{t}{\delta} \right\rceil \frac{(|x|-t)^{n-1}}{(n-1)!}dt, \tag{8.14}$$

where $\lceil \cdot \rceil$ is the ceiling of the number. Then

$$\Gamma_f(\mu_1,\mu_2) \leq \sum_{k=2}^n \frac{|f^{(k)}(1)|}{k!}\left|\int_X q^{1-k}(p-q)^k d\lambda\right| + w\int_X q\phi_n\left(\frac{p-q}{q}\right)d\lambda. \tag{8.15}$$

Inequality (8.15) is sharp, namely it is attained by the function

$$\tilde{f}_n(t) := w\phi_n(t-1), \quad a \leq t \leq b, \tag{8.16}$$

when n is even.

Proof. One has that

$$f(x) = \sum_{k=1}^n \frac{f^{(k)}(1)}{k!}(x-1)^k + \mathcal{R}(x), \tag{8.17}$$

for all $x \in [a,b]$, where

$$\mathcal{R}(x) := \int_1^x (f^{(n)}(t) - f^{(n)}(1))\frac{(x-t)^{n-1}}{(n-1)!}dt. \tag{8.18}$$

From p. 3, see (8) of [29] we obtain

$$|\mathcal{R}(x)| \leq \omega_1(f^{(n)}, \delta)\phi_n(x-1) \leq w\phi_n(x-1). \tag{8.19}$$

By (8.17) we get

$$qf\left(\frac{p}{q}\right) = \sum_{k=1}^n \frac{f^{(k)}(1)}{k!}q\left(\frac{p}{q}-1\right)^k + q\mathcal{R}\left(\frac{p}{q}\right), \tag{8.20}$$

a.e. on X. Furthermore we find

$$\Gamma_f(\mu_1,\mu_2) = \sum_{k=2}^n \frac{f^{(k)}(1)}{k!}\int_X q^{1-k}(p-q)^k d\lambda + \int_X q\mathcal{R}\left(\frac{p}{q}\right)d\lambda. \tag{8.21}$$

Using now (8.19) and taking absolute value over (8.21) we derive (8.15).

Next we prove the sharpness of (8.15). According to Remark 7.1.3 of [20], pp. 210–211 we have that

$$\phi_n(x) = \int_0^x \phi_{n-1}(t)dt, \quad x \in \mathbb{R}_+, \ n \geq 1, \tag{8.22}$$

where

$$\phi_0(t) := \left\lceil \frac{|t|}{\delta} \right\rceil, \quad t \in \mathbb{R}. \tag{8.23}$$

Clearly

$$\phi_n^{(k)}(x) = (\text{sign}(x))^k \phi_{n-k}(x), \quad k = 1, 2, \ldots, n, \ x \in \mathbb{R}. \tag{8.24}$$

That is

$$\tilde{f}_n^{(k)}(x) = (\text{sign}(x-1))^k \tilde{f}_{n-k}(x), \quad x \in \mathbb{R}, \ k = 1, \ldots, n. \tag{8.25}$$

Furthermore

$$\tilde{f}_n^{(k)}(1) = 0, \quad k = 0, 1, \ldots, n.$$

Since the function ϕ_n is even and convex over $\mathbb{R}$, then we have that $\tilde{f}_n$ is convex over $\mathbb{R}$, $n \geq 1$. Also since ϕ_n is strictly increasing over $\mathbb{R}_+$, then $\tilde{f}_n$ is strictly increasing over $[1, \infty)$ and strictly decreasing over $(-\infty, 1]$. Thus, $\tilde{f}_n$ is strictly convex at 1, $n \geq 1$. Here since n is even

$$\tilde{f}_n^{(n)}(x) = w \left\lceil \frac{|x-1|}{\delta} \right\rceil$$

and by Lemma 7.1.1, see p. 208 of [20] we find that

$$\omega_1(\tilde{f}_n^{(n)}, \delta) \leq w.$$

So $\tilde{f}_n$ fulfills basically all the assumptions of Theorem 8.3. Clearly then both sides of (8.15) equal to

$$w \int_X q\phi_n \left(\frac{p-q}{q} \right) d\lambda,$$

establishing that (8.15) is attained.

Next we see that $w \left\lceil \frac{|t-1|}{\delta} \right\rceil$ can be arbitrarily close approximated by a sequence of continuous functions that fulfill the assumptions of Theorem 8.3. More precisely $\left\lceil \frac{|t-1|}{\delta} \right\rceil$ is a step function over $\mathbb{R}$ which takes the values $j = 1, 2, \ldots$, respectively, over the intervals $[1 - (j+1)\delta, 1 - j\delta)$ and $(1 + (j-1)\delta, 1 + j\delta]$ with upward jumps of size 1, therefore

$$\tilde{f}_n^{(n)}(t) = w \left\lceil \frac{|t-1|}{\delta} \right\rceil$$

has finitely many discontinuities over $[a, b]$ and is approximated by continuous functions as follows.

Here for $N \in \mathbb{N}$ we define the continuous functions over $\mathbb{R}$,

$$f_{0N}(t) := \begin{cases} \dfrac{Nw|t|}{2\delta} + kw\left(1 - \dfrac{N}{2}\right), & \text{if } k\delta \leq |t| \leq \left(k + \dfrac{2}{N}\right)\delta, \\[4mm] (k+1)w, & \text{if } \left(k + \dfrac{2}{N}\right)\delta < |t| \leq (k+1)\delta, \end{cases} \tag{8.26}$$

where $k \in \mathbb{Z}_+$. Clearly $f_{0N}(|t|) > 0$ with $f_{0N}(0) = 0$ and f_{0N} is increasing over $\mathbb{R}_+$ and decreasing over $\mathbb{R}_-$. Also we define

$$f_{nN}(t) := w \int_0^{|t|} \left(\int_0^{t_1} \cdots \left(\int_0^{t_{n-1}} f_{0N}(t_n) dt_n \right) \cdots \right) dt_1, \qquad (8.27)$$

for $t \in \mathbb{R}$. Clearly $f_{nN} \in C^n(\mathbb{R})$, is even and convex over $\mathbb{R}$ and is strictly increasing on $\mathbb{R}_+$.

According to [20], pp. 218–219 we obtain

$$\lim_{N \to +\infty} f_{0N}(|t - 1|) = w \left\lceil \frac{|t - 1|}{\delta} \right\rceil, \quad \text{for } t \in [a, b],\ N \in \mathbb{N}, \qquad (8.28)$$

with

$$\omega_1(f_{0N}(|\cdot - 1|), \delta) = \omega_1(f_{0N}(|\cdot|), \delta) \leq w. \qquad (8.29)$$

Furthermore it invalid that

$$\lim_{N \to +\infty} f_{nN}(|t - 1|) = \tilde{f}_n(t), \quad \text{for } t > 0. \qquad (8.30)$$

Let $g(x) := f_{nN}(x - 1)$, $x \in \mathbb{R}$, then $g^{(k)}(x) = (\text{sign}(x - 1))^k f_{(n-k)N}(x - 1)$, $x \in \mathbb{R}$, $k = 1, 2, \ldots, n$, $g^{(k)}(1) = 0$ for $k = 0, 1, \ldots, n$, and for all $x \in \mathbb{R}$ it holds $g^{(n)}(x) = f_{0N}(x - 1)$ since n is even. Furthermore $g \in C^n(R)$, g convex on $\mathbb{R}$, g strictly increasing on $[1, +\infty)$ and g strictly decreasing on $(-\infty, 1]$, therefore g is strictly convex at 1. That is g with all the above properties and (8.29), is consistent with the assumptions of Theorem 8.3. That is, $\tilde{f}_n(t)$ is arbitrarily close approximated consistently with the assumptions of Theorem 8.3.

$\square$

Next we have

Corollary 8.4. (to Theorem 8.3). *It holds* $(0 < \delta \leq b - a)$

$$\Gamma_f(\mu_1, \mu_2) \leq \sum_{k=2}^{n} \frac{|f^{(k)}(1)|}{k!} \left| \int_X q^{1-k}(p - q)^k d\lambda \right|$$

$$+ \omega_1(f^{(n)}, \delta) \left\{ \frac{1}{(n+1)!\delta} \int_X q^{-n}|p - q|^{n+1} d\lambda + \frac{1}{2n!} \int_X q^{1-n}|p - q|^n d\lambda \right.$$

$$\left. + \frac{\delta}{8(n-1)!} \int_X q^{2-n}|p - q|^{n-1} d\lambda \right\}, \qquad (8.31)$$

and

$$\Gamma_f(\mu_1, \mu_2) \leq \sum_{k=2}^{n} \frac{|f^{(k)}(1)|}{k!} \left| \int_X q^{1-k}(p - q)^k d\lambda \right| \qquad (8.32)$$

$$+ \omega_1(f^{(n)}, \delta) \left\{ \frac{1}{n!} \int_X q^{1-n}|p - q|^n d\lambda + \frac{1}{n!(n+1)\delta} \int_X q^{-n}|p - q|^{n+1} d\lambda \right\}.$$

In particular we have

$$\Gamma_f(\mu_1, \mu_2) \le \omega_1(f', \delta)\left\{ \frac{1}{2\delta} \int_X q^{-1}(p - q)^2 d\lambda + \frac{1}{2} \int_X |p - q| d\lambda + \frac{\delta}{8} \right\}, \qquad (8.33)$$

and

$$\Gamma_f(\mu_1, \mu_2) \le \omega_1(f', \delta)\left\{ \int_X |p - q| d\lambda + \frac{1}{2\delta} \int_X q^{-1}(p - q)^2 d\lambda \right\}. \qquad (8.34)$$

Proof. From [20], p. 210, inequality (7.1.18) we have

$$\phi_n(x) \le \left(\frac{|x|^{n+1}}{(n+1)!\delta} + \frac{|x|^n}{2n!} + \frac{\delta|x|^{n-1}}{8(n-1)!} \right), \qquad x \in \mathbb{R}, \qquad (8.35)$$

with equality only at $x = 0$. We obtain that (see (8.18) and (8.19))

$$|\mathcal{R}(x)| \le \omega_1(f^{(n)}, \delta)\left(\frac{|x-1|^{n+1}}{(n+1)!\delta} + \frac{|x-1|^n}{2n!} + \frac{\delta|x-1|^{n-1}}{8(n-1)!} \right) \le w(\cdots), \qquad (8.36)$$

with equality at $x = 1$.

Also from [20], p. 217, inequality (7.2.9) we get

$$|\mathcal{R}(x)| \le \omega_1(f^{(n)}, \delta)\frac{|x-1|^n}{n!}\left(1 + \frac{|x-1|}{(n+1)\delta} \right),$$
$$\le w(\cdots), \qquad (8.37)$$

with equality at $x = 1$. Hence by (8.17), (8.18) and (8.36) we find

$$\Gamma_f(\mu_1, \mu_2) \le \sum_{k=2}^{n} \frac{|f^{(k)}(1)|}{k!}\left| \int_X q^{1-k}(p-q)^k d\lambda \right| + \omega_1(f^{(n)}, \delta)\left(\int_X \left(\frac{q^{-n}|p-q|^{n+1}}{(n+1)!\delta} \right.\right.$$
$$\left.\left. + \frac{q^{1-n}|p-q|^n}{2n!} + \frac{\delta q^{2-n}|p-q|^{n-1}}{8(n-1)!} \right) d\lambda \right), \qquad (8.38)$$

etc. Also by (8.17), (8.18) and (8.37) we derive

$$\Gamma_f(\mu_1, \mu_2) \le \sum_{k=2}^{n} \frac{|f^{(k)}(1)|}{k!}\left| \int_X q^{1-k}(p-q)^k d\lambda \right|$$
$$+ \omega_1(f^{(n)}, \delta)\left(\int_X \frac{|p-q|^n}{n!}\left(q^{1-n} + \frac{q^{-n}|p-q|}{(n+1)\delta} \right) d\lambda \right), \qquad (8.39)$$

etc. $\qquad\qquad\square$

We give

Corollary 8.5. (to Theorem 8.3). *Suppose here that $f^{(n)}$ is of Lipschitz type of order α, $0 < \alpha \le 1$, i.e.*

$$\omega_1(f^{(n)}, \delta) \le K\delta^\alpha, \qquad K > 0, \qquad (8.40)$$

for any $0 < \delta \le b - a$. Then

$$\Gamma_f(\mu_1, \mu_2) \le \sum_{k=2}^{n} \frac{|f^{(k)}(1)|}{k!}\left| \int_X q^{1-k}(p-q)^k d\lambda \right| + \frac{K}{\prod_{i=1}^{n}(\alpha + i)} \int_X q^{1-n-\alpha}|p-q|^{n+\alpha} d\lambda.$$
$$(8.41)$$

When n is even (8.41) is attained by $f^*(x) = c|x - 1|^{n+\alpha}$, where $c := K/\left(\prod_{i=1}^{n}(\alpha + i)\right) > 0$.

Proof. We have again that

$$f(x) = \sum_{k=1}^{n} \frac{f^{(k)}(1)}{k!}(x - 1)^k + \mathcal{R}(x),$$

where

$$\mathcal{R}(x) := \int_1^x \int_1^{x_1} \cdots \int_1^{x_{n-1}} (f^{(n)}(x_n) - f^{(n)}(1))dx_n \cdots dx_1. \tag{8.42}$$

From [29], p. 6 see (29)*, we get

$$|\mathcal{R}(x)| \leq K \cdot \frac{|x - 1|^{n+\alpha}}{\prod_{i=1}^{n}(\alpha + i)}. \tag{8.43}$$

(E.g. if $f \in C^{n+1}([a, b])$, then $K = \|f^{(n+1)}\|_\infty$ and $\alpha = 1$). Then

$$q\left|\mathcal{R}\left(\frac{p}{q}\right)\right| \leq \frac{Kq\left|\frac{p}{q} - 1\right|^{n+\alpha}}{\prod_{i=1}^{n}(\alpha + i)} = \frac{K}{\prod_{i=1}^{n}(\alpha + i)} q^{1-n-\alpha}|p - q|^{n+\alpha}, \tag{8.44}$$

a.e. on X, and

$$\int_X q\left|\mathcal{R}\left(\frac{p}{q}\right)\right| d\lambda \leq \frac{K}{\prod_{i=1}^{n}(\alpha + i)} \int_X q^{1-n-\alpha}|p - q|^{n+\alpha}d\lambda. \tag{8.45}$$

Thus we have established (8.41).

Clearly (8.41) is attained by f^* which is acceptable to Theorem 8.3 requirements. Notice that $f^{*(i)}(1) = 0$, $i = 0, 1, \ldots, n$ and $f^{*(n)}(x) = K|x - 1|^\alpha$ fulfills (8.40) in here by (105) of [29], p. 28. See also from [29], p. 28 at the bottom that $f^*(x) = |\mathcal{R}(x)|$, etc. $\qquad\square$

Corollary 8.6. (to Theorem 8.3). *Suppose that*

$$b - a \geq \int_X q^{-1}p^2 d\lambda - 1 > 0. \tag{8.46}$$

Then

$$\Gamma_f(\mu_1, \mu_2) \leq \omega_1\left(f', \left(\int_X q^{-1}p^2 d\lambda - 1\right)\right)\left\{\int_X |p - q|d\lambda + \frac{1}{2}\right\}. \tag{8.47}$$

Proof. From (8.34) by choosing

$$\delta := \int_X q^{-1}(p - q)^2 d\lambda = \int_X q^{-1}p^2 d\lambda - 1. \qquad\square$$

Corollary 8.7. (to Theorem 8.3). *Suppose (1.46), that is*

$$b - a \geq \int_X q^{-1} p^2 d\lambda - 1 > 0,$$

$$|p - q| \leq 1 \quad \text{a.e. on } X, \tag{8.48}$$

and

$$\lambda(X) < \infty. \tag{8.49}$$

Then

$$\Gamma_f(\mu_1, \mu_2) \leq \left(\lambda(X) + \frac{1}{2} \right) \omega_1 \left(f', \left(\int_X q^{-1} p^2 d\lambda - 1 \right) \right). \tag{8.50}$$

Proof. From (8.47) and (8.48). $\qquad\square$

Note. In our setting observe that always

$$\int_X q^{-1} p^2 d\lambda \geq 1, \tag{8.51}$$

and of course

$$a \leq \int_X q^{-1} p^2 d\lambda \leq b. \tag{8.52}$$

We present

Corollary 8.8. (to Theorem 8.3). *Suppose that*

$$\int_X |p - q| d\lambda > 0. \tag{8.53}$$

Then

$$\Gamma_f(\mu_1, \mu_2) \leq \omega_1 \left(f', \int_X |p - q| d\lambda \right) \left\{ \int_X |p - q| d\lambda + \frac{b - a}{2} \right\}$$

$$\leq \frac{3}{2}(b - a) \omega_1 \left(f', \int_X |p - q| d\lambda \right). \tag{8.54}$$

Proof. Notice that

$$q^{-1}(p - q)^2 = q^{-1} |p - q|^2 = |p - q| \left| \frac{p}{q} - 1 \right| \leq (b - a)|p - q|, \quad \text{a.e. on } X.$$

Hence

$$\int_X q^{-1}(p - q)^2 d\lambda \leq (b - a) \int_X |p - q| d\lambda. \tag{8.55}$$

We use (8.34), (8.53), (8.55) and we choose

$$\delta := \int_X |p - q| d\lambda \leq b - a, \tag{8.56}$$

etc. $\qquad\square$

Notice that (8.54) is good only for small $b - a > 0$.

Next we derive

Corollary 8.9. (to Theorem 8.3). *Assume that $r > 0$ and (8.46)- (8.49), respectively,*

$$b - a \geq r \left(\int_X q^{-1} p^2 d\lambda - 1 \right) > 0,$$

$$|p - q| \leq 1 \quad a.e. \ on \ X,$$

and

$$\lambda(X) < \infty.$$

Then

$$\Gamma_f(\mu_1, \mu_2) \leq \left(\lambda(X) + \frac{1}{2r} \right) \omega_1 \left(f', r \left(\int_X q^{-1} p^2 d\lambda - 1 \right) \right). \tag{8.57}$$

Proof. Here we choose in (8.34) that

$$\delta := r \int_X q^{-1} (p - q)^2 d\lambda, \tag{8.58}$$

etc. $\qquad\qquad\qquad\qquad\qquad\qquad\qquad\qquad\qquad\qquad\qquad\qquad\qquad\square$

Remark 8.10. (on Theorem 8.1). From (8.5) for $n = 1$ we obtain

$$\Gamma_f(\mu_1, \mu_2) \leq \frac{\omega_1(f', h)}{2h} \int_X q^{-1} (p - q)^2 d\lambda, \tag{8.59}$$

which is attained by $\frac{(t-1)^2}{2}$. Assuming that

$$0 < h := \int_X q^{-1} (p - q)^2 d\lambda < \min(1 - a, b - 1), \tag{8.60}$$

we derive the inequality

$$\Gamma_f(\mu_1, \mu_2) \leq \frac{1}{2} \omega_1 \left(f', \left(\int_X q^{-1} p^2 d\lambda - 1 \right) \right). \tag{8.61}$$

The counterpart of the above is

Proposition 8.11. *Let f and the rest as in this chapter, also assume $f \in C([a, b])$.*
i) *Suppose (8.53),*

$$\int_X |p - q| d\lambda > 0.$$

Then

$$\Gamma_f(\mu_1, \mu_2) \leq 2\omega_1 \left(f, \int_X |p - q| d\lambda \right). \tag{8.62}$$

ii) *Let $r > 0$ and*

$$b - a \geq r \int_X |p - q| d\lambda > 0. \tag{8.63}$$

Then

$$\Gamma_f(\mu_1, \mu_2) \le \left(1 + \frac{1}{r}\right) \cdot \omega_1\left(f, r \int_X |p - q| d\lambda\right). \tag{8.64}$$

Proof. We observe that

$$\Gamma_f(\mu_1, \mu_2) = \int_X qf\left(\frac{p}{q}\right) d\lambda = \int_X qf\left(\frac{p}{q}\right) d\lambda - f(1)$$

$$= \int_X q\left(f\left(\frac{p}{q}\right) - f(1)\right) d\lambda \le \int_X q\left|f\left(\frac{p}{q}\right) - f(1)\right| d\lambda$$

$$\le \text{ (by Lemma 7.1.1 of [20], p. 208)} \int_X q\omega_1(f, h)\left\lceil \frac{\left|\frac{p}{q} - 1\right|}{h}\right\rceil d\lambda$$

$$\le \omega_1(f, h)\left(\int_X q\left(1 + \frac{\left|\frac{p}{q} - 1\right|}{h}\right) d\lambda\right) = \omega_1(f, h)\left(1 + \frac{1}{h}\int_X |p - q| d\lambda\right).$$

That is, we obtain

$$\Gamma_f(\mu_1, \mu_2) \le \omega_1(f, h)\left(1 + \frac{1}{h}\int_X |p - q| d\lambda\right). \tag{8.65}$$

Setting

$$h := \int_X |p - q| d\lambda \tag{8.66}$$

we find

$$\Gamma_f(\mu_1, \mu_2) \le 2\omega_1(f, h),$$

proving (8.62).

By setting

$$h := r \int_X |p - q| d\lambda, \quad r > 0 \tag{8.67}$$

we derive

$$\Gamma_f(\mu_1, \mu_2) \le \omega_1(f, h)\left(1 + \frac{1}{r}\right). \tag{8.68}$$

$\square$

Also we give

Proposition 8.12. *Let f as in this setting and f is a Lipschitz function of order α, $0 < \alpha \le 1$, i.e. there exists $K > 0$ such that*

$$|f(x) - f(y)| \le K|x - y|^\alpha, \quad \text{all } x, y \in [a, b]. \tag{8.69}$$

Then

$$\Gamma_f(\mu_1, \mu_2) \le K \int_X q^{1-\alpha}|p - q|^\alpha d\lambda. \tag{8.70}$$

Proof. As in the proof of Proposition 8.11 we have

$$\Gamma_f(\mu_1, \mu_2) \le \int_X q\left|f\left(\frac{p}{q}\right) - f(1)\right| d\lambda \le \int_X qK\left|\frac{p}{q} - 1\right|^\alpha d\lambda = K \int_X q^{1-\alpha}|p - q|^\alpha d\lambda.$$

$\square$

8.3   Results Based on Generalized Taylor–Widder Formula

(see also [339], [29])

Let $f, u_0, u_1, \ldots, u_n \in C^{n+1}([a, b])$, $n \in \mathbb{Z}_+$, and let the Wronskians
$$W_i(x) := W[u_0(x), u_1(x), \ldots, u_i(x)] :$$

$$= \left| \begin{pmatrix} u_0(x) & u_1(x) & \cdots & u_i(x) \\ u_0'(x) & u_1'(x) & \cdots & u_i'(x) \\ \vdots & & & \\ u_0^{(i)}(x) & u_1^{(i)}(x) & \cdots & u_i^{(i)}(x) \end{pmatrix} \right|, \quad i = 0, 1, \ldots, n, \quad (8.71)$$

and suppose that $W_i(x)$ are positive over $[a, b]$. Clearly the functions

$$\phi_0(x) := W_0(x) := u_0(x), \quad \phi_1(x) := \frac{W_1(x)}{(W_0(x))^2}, \ldots$$

$$\phi_i(x) := \frac{W_i(x) W_{i-2}(x)}{(W_{i-1}(x))^2}, \quad i = 2, 3, \ldots, n, \quad (8.72)$$

are positive everywhere on $[a, b]$. Consider the linear differential operator of order $i \geq 0$

$$L_i f(x) := \frac{W[u_0(x), u_1(x), \ldots, u_{i-1}(x), f(x)]}{W_{i-1}(x)} \quad (8.73)$$

for $i = 1, \ldots, n+1$, where $L_0 f(x) := f(x)$ all $x \in [a, b]$.
Here $W[u_0(x), u_1(x), \ldots, u_{i-1}(x), f(x)]$ denotes the Wronskian of $u_0, u_1, \ldots, u_{i-1}$, f. Notice that for $i = 1, \ldots, n+1$, we obtain

$$L_i f(x) = \phi_0(x) \phi_1(x) \cdots \phi_{i-1}(x) \frac{d}{dx} \frac{1}{\phi_{i-1}(x)} \frac{d}{dx} \frac{1}{\phi_{i-2}(x)} \frac{d}{dx} \cdots \frac{d}{dx} \frac{1}{\phi_1(x)} \frac{d}{dx} \frac{f(x)}{\phi_0(x)}.$$
$$(8.74)$$

Consider also the functions

$$g_i(x, t) := \frac{1}{W_i(x)} \cdot \left| \begin{pmatrix} u_0(t) & u_1(t) & \cdots & u_i(t) \\ u_0'(t) & u_1'(t) & \cdots & u_i'(t) \\ \vdots & & & \\ u_0^{(i-1)}(t) & u_1^{(i-1)}(t) & \cdots & u_i^{(i-1)}(t) \\ u_0(x) & u_1(x) & \cdots & u_i(x) \end{pmatrix} \right| \quad (8.75)$$

for $i = 1, 2, \ldots, n$, where $g_0(x, t) := \frac{u_0(x)}{u_0(t)}$ all $x, t \in [a, b]$. Notice that $g_i(x, t)$, as a function of x, is a linear combination of $u_0(x), u_1(x), \ldots, u_i(x)$ and, furthermore, it holds

$$g_i(x, t) = \frac{\phi_0(x)}{\phi_0(t) \cdots \phi_i(t)} \int_t^x \phi_1(x_1) \int_t^{x_1}$$

$$\cdots \int_t^{x_{i-2}} \phi_{i-1}(x_{i-1}) \int_t^{x_{i-1}} \phi_i(x_i) dx_i dx_{i-1} \cdots dx_1$$

$$= \frac{1}{\phi_0(t) \cdots \phi_i(t)} \int_t^x \phi_0(s) \cdots \phi_i(s) g_{i-1}(x, s) ds \quad (8.76)$$

all $i = 1, 2, \ldots, n$. Put

$$N_n(x, t) := \int_t^x g_n(x, s)ds. \tag{8.77}$$

Example 8.13. Let $u_i(x) = x^i$, $i = 0, 1, \ldots, n$, be defined on $[a, b]$. Then $L_i f(t) = f^{(i)}(t)$, $i = 1, \ldots, n+1$ and $g_i(x, t) = \frac{(x-t)^i}{i!}$ for all $x, t \in [a, b]$ and $i = 0, 1, \ldots, n$. From [339, Theorem II, p. 138], we obtain

$$f(x) = \sum_{i=0}^{n} L_i f(t) \cdot g_i(x, t) + \int_t^x g_n(x, s) \cdot L_{n+1} f(s)ds \tag{8.78}$$

all $x, t \in [a, b]$, $n \in \mathbb{Z}_+$. This is the *Taylor–Widder formula*. Thus we have

$$f(x) = \sum_{i=0}^{n} L_i f(t)g_i(x, t) + L_{n+1} f(t) \cdot N_n(x, t) + R_n^*(x, t), \tag{8.79}$$

where

$$R_n^*(x, t) := \int_t^x g_n(x, s)(L_{n+1} f(s) - L_{n+1} f(t))ds, \quad n \in \mathbb{Z}_+. \tag{8.80}$$

Here we use a lot the above concepts. From the above definitions, we observe the following

$$\begin{cases} g_n(x, t) > 0, & x > t, n \geq 1, \\ g_n(x, t) < 0, & x < t, n \text{ odd}, \\ g_n(x, t) > 0, & x < t, n \text{ even}, \\ g_n(t, t) = 0, & n \geq 1, \ g_0(x, t) > 0 \end{cases} \tag{8.81}$$

for all $x, t \in [a, b]$.

Since $f(1) = 0$ we have

$$f(x) = \sum_{i=1}^{n} L_i f(1)g_i(x, 1) + L_{n+1} f(1)N_n(x, 1) + \mathcal{R}_n^*(x, 1), \tag{8.82}$$

where

$$\mathcal{R}_n^*(x, 1) := \int_1^x g_n(x, s)(L_{n+1} f(s) - L_{n+1} f(1))ds, \quad n \in \mathbb{Z}_+, \tag{8.83}$$

and

$$N_n(x, 1) := \int_1^x g_n(x, s)ds, \quad x \in [a, b]. \tag{8.84}$$

Let

$$\omega_1(L_{n+1} f, h) \leq w, \quad 0 < h \leq b - a, \quad w > 0. \tag{8.85}$$

Then by Theorem 23, p. 18 of [29] we get that

$$|\mathcal{R}_n^*(x, 1)| \leq w \left\lceil \frac{|x-1|}{h} \right\rceil |N_n(x, 1)| \leq w \left(1 + \frac{|x-1|}{h} \right) |N_n(x, 1)|. \tag{8.86}$$

Working in (8.82) we observe that

$$qf\left(\frac{p}{q}\right) = \sum_{i=1}^{n} L_i f(1)qg_i\left(\frac{p}{q},1\right) + L_{n+1}f(1)qN_n\left(\frac{p}{q},1\right) + q\mathcal{R}_n^*\left(\frac{p}{q},1\right), \text{a.e. on } X.$$

(8.87)

That is we have established the representation formula

$$\Gamma_f(\mu_1,\mu_2) = \sum_{i=1}^{n} L_i f(1)\int_X qg_i\left(\frac{p}{q},1\right)d\lambda + L_{n+1}f(1)\int_X qN_n\left(\frac{p}{q},1\right)d\lambda + T_1,$$

(8.88)

where

$$T_1 := \int_X q\mathcal{R}_n^*\left(\frac{p}{q},1\right)d\lambda.$$

(8.89)

From (8.86) we get

$$|T_1| \leq \int_X q\left|\mathcal{R}_n^*\left(\frac{p}{q},1\right)\right|d\lambda \leq w\int_X q\left\lceil\frac{\left|\frac{p}{q}-1\right|}{h}\right\rceil\left|N_n\left(\frac{p}{q},1\right)\right|d\lambda$$

$$\leq w\int_X \left(q+\frac{|p-q|}{h}\right)\left|N_n\left(\frac{p}{q},1\right)\right|d\lambda.$$

(8.90)

Based on the above we have now established

Theorem 8.14. *It holds*

$$\Gamma_f(\mu_1,\mu_2) \leq \sum_{i=1}^{n} |L_i f(1)|\left|\int_X qg_i\left(\frac{p}{q},1\right)d\lambda\right| + |L_{n+1}f(1)|\left|\int_X qN_n\left(\frac{p}{q},1\right)d\lambda\right|$$

$$+ w\int_X \left(q+\frac{|p-q|}{h}\right)\left|N_n\left(\frac{p}{q},1\right)\right|d\lambda.$$

(8.91)

Next let

$$\omega_1(L_{n+1}f,\delta) \leq A\delta^\alpha,$$

(8.92)

$0 < \delta \leq b-a$, $A > 0$, $0 \leq \alpha \leq 1$. Then from Theorem 25, p. 19 of [29] we obtain

$$|\mathcal{R}_n^*(x,t)| \leq A|x-t|^\alpha|N_n(x,t)|,$$

(8.93)

for all $x,t \in [a,b]$ and $n \geq 0$. Hence

$$q\left|\mathcal{R}_n^*\left(\frac{p}{q},1\right)\right| \leq Aq^{1-\alpha}|p-q|^\alpha\left|N_n\left(\frac{p}{q},1\right)\right|,$$

(8.94)

a.e. on X, and

$$|T_1| \leq \int_X q\left|\mathcal{R}_n^*\left(\frac{p}{q},1\right)\right|d\lambda \leq A\int_X q^{1-\alpha}|p-q|^\alpha\left|N_n\left(\frac{p}{q},1\right)\right|d\lambda.$$

(8.95)

Thus we have proved

Theorem 8.15. *It holds*

$$\Gamma_f(\mu_1,\mu_2) \leq \sum_{i=1}^{n} |L_i f(1)| \left| \int_X q g_i \left(\frac{p}{q},1\right) d\lambda \right| + |L_{n+1}f(1)| \left| \int_X q N_n \left(\frac{p}{q},1\right) d\lambda \right|$$

$$+ A \int_X q^{1-\alpha} |p-q|^\alpha \left| N_n \left(\frac{p}{q},1\right) \right| d\lambda. \tag{8.96}$$

From Theorem 26, p. 19 of [29] we have: let $1 \in (a,b)$ and choose $0 < h \leq \min(1-a, b-1)$, assuming that $|L_{n+1}f(x) - L_{n+1}f(1)|$ is convex in $x \in [a,b]$ and calling $w := \omega_1(L_{n+1}f, h)$ we get

$$|\mathcal{R}_n^*(x,1)| \leq \frac{w}{h}|x-1|\,|N_n(x,1)|, \tag{8.97}$$

for all $x \in [a,b]$ and $n \geq 0$. Therefore

$$q \left| \mathcal{R}_n^* \left(\frac{p}{q},1\right) \right| \leq \frac{w}{h}|p-q| \left| N_n \left(\frac{p}{q},1\right) \right|, \quad \text{a.e. on } X. \tag{8.98}$$

So that

$$|T_1| \leq \frac{w}{h} \int_X |p-q| \left| N_n \left(\frac{p}{q},1\right) \right| d\lambda. \tag{8.99}$$

We have established

Theorem 8.16. *It holds*

i)

$$\Gamma_f(\mu_1,\mu_2) \leq \sum_{i=1}^{n} |L_i f(1)| \left| \int_X q g_i \left(\frac{p}{q},1\right) d\lambda \right| + |L_{n+1}f(1)| \left| \int_X q N_n \left(\frac{p}{q},1\right) d\lambda \right|$$

$$+ \frac{w}{h} \int_X |p-q| \left| N_n \left(\frac{p}{q},1\right) \right| d\lambda, \quad n \geq 0. \tag{8.100}$$

ii) *Choosing and supposing*

$$0 < h := \int_X |p-q| \left| N_n \left(\frac{p}{q},1\right) \right| d\lambda \leq \min(1-a, b-1), \tag{8.101}$$

we derive

$$\Gamma_f(\mu_1,\mu_2) \leq \sum_{i=1}^{n} |L_i f(1)| \left| \int_X q g_i \left(\frac{p}{q},1\right) d\lambda \right| + |L_{n+1}f(1)| \left| \int_X q N_n \left(\frac{p}{q},1\right) d\lambda \right|$$

$$+ \omega_1 \left(L_{n+1}f, \int_X |p-q| \left| N_n \left(\frac{p}{q},1\right) \right| d\lambda \right). \tag{8.102}$$

Next by Lemmas 11.1.5, 11.1.6 of [20], p. 345 we have:

Let $a < 1 < b$, $u_0(x) := c > 0$, $u_1(x)$ concave for $x \leq 1$, and $u_1(x)$ convex for $x \geq 1$. Let

$$\tilde{G}_n(x,1) := |G_n(x,1)|, \tag{8.103}$$

where

$$G_n(x,1) := \int_1^x g_n(x,s) \left\lceil \frac{|s-1|}{h} \right\rceil ds, \tag{8.104}$$

all $x \in [a,b]$, $0 < h \le b - a$, $n \ge 0$. Then

$$\tilde{G}_n(x,1) \le \max\left\{ \frac{\tilde{G}_n(b,1)}{b-1}, \frac{\tilde{G}_n(a,1)}{1-a} \right\} |x-1| =: \theta|x-1|, \tag{8.105}$$

all $x \in [a,b]$, $n \ge 1$.

But from Theorem 21, p. 16 of [29] we get

$$|\mathcal{R}_n^*(x,1)| \le w\tilde{G}_n(x,1), \tag{8.106}$$

all $x \in [a,b]$, $n \in \mathbb{Z}_+$, with $\omega_1(L_{n+1}f,h) \le w$, $0 < h \le b - a$, $w > 0$. Consequently we obtain

$$|\mathcal{R}_n^*(x,1)| \le w\theta|x-1|, \tag{8.107}$$

all $x \in [a,b]$, $n \ge 1$.

Furthermore it holds

$$|T_1| \le \int_X q \left| \mathcal{R}_n^* \left(\frac{p}{q}, 1 \right) \right| d\lambda \le w\theta \int_X |p-q| d\lambda, \quad n \ge 1. \tag{8.108}$$

We have established

Theorem 8.17. *It holds*

$$\Gamma_f(\mu_1,\mu_2) \le \sum_{i=1}^n |L_i f(1)| \left| \int_X qg_i \left(\frac{p}{q}, 1 \right) d\lambda \right|$$

$$+ |L_{n+1}f(1)| \left| \int_X qN \left(\frac{p}{q}, 1 \right) d\lambda \right| + w\theta \int_X |p-q| d\lambda, \, n \ge 1, \tag{8.109}$$

where

$$\theta := \max\left\{ \frac{\tilde{G}_n(b,1)}{b-1}, \frac{\tilde{G}_n(a,1)}{1-a} \right\}, \tag{8.110}$$

with $1 \in (a,b)$.

It follows

Corollary 8.18. (to Theorem 8.17). *It holds*

$$\Gamma_f(\mu_1,\mu_2) \le |L_1 f(1)| \left| \int_X qg_1 \left(\frac{p}{q}, 1 \right) d\lambda \right| + |L_2 f(1)| \left| \int_X qN_1 \left(\frac{p}{q}, 1 \right) d\lambda \right|$$

$$+ \omega_1 \left(L_2 f, \int_X |p-q| d\lambda \right) \cdot \left(\int_X |p-q| d\lambda \right)$$

$$\cdot \left(\max\left\{ \frac{\tilde{G}_1(b,1)}{b-1}, \frac{\tilde{G}_1(a,1)}{1-a} \right\} \right), \tag{8.111}$$

under the assumption that

$$\int_X |p - q| d\lambda > 0.$$

From (8.78) we have

$$f(x) = \sum_{i=1}^{n} L_i f(1) g_i(x, 1) + \mathcal{R}_w(x), \tag{8.112}$$

where

$$\mathcal{R}_w(x) := \int_1^x g_n(x, s) L_{n+1} f(s) ds, \tag{8.113}$$

all $x \in [a, b]$, $n \in \mathbb{Z}_+$. Set also

$$N_n^*(x, 1) := \int_1^x |g_n(x, s)| ds. \tag{8.114}$$

Then we observe that

$$|\mathcal{R}_w(x)| \le \|L_{n+1} f\|_{\infty, [a,b]} |N_n^*(x, 1)|. \tag{8.115}$$

Hence

$$q \left| \mathcal{R}_w \left(\frac{p}{q} \right) \right| \le \|L_{n+1} f\|_{\infty, [a,b]} q \left| N_n^* \left(\frac{p}{q}, 1 \right) \right|$$

and

$$\int_X q \left| \mathcal{R}_w \left(\frac{p}{q} \right) \right| d\lambda \le \|L_{n+1} f\|_{\infty, [a,b]} \int_X q \left| N_n^* \left(\frac{p}{q}, 1 \right) \right| d\lambda. \tag{8.116}$$

We have proved

Theorem 8.19. *It holds*

$$\Gamma_f(\mu_1, \mu_2) \le \sum_{i=1}^{n} |L_i f(1)| \left| \int_X q g_i \left(\frac{p}{q}, 1 \right) d\lambda \right|$$

$$+ \|L_{n+1} f\|_{\infty, [a,b]} \int_X q \left| N_n^* \left(\frac{p}{q}, 1 \right) \right| d\lambda. \tag{8.117}$$

Next from (8.113) we obtain

$$|\mathcal{R}_w(x)| \le \left| \int_1^x |(L_{n+1} f)(s)| ds \right| \sup_{s \in [a,b]} |g_n(x, \cdot)|. \tag{8.118}$$

Furthermore it is true that

$$|\mathcal{R}_w(x)| \le \|L_{n+1} f\|_{1, [a,b]} \|g_n(x, \cdot)\|_{\infty, [a,b]}. \tag{8.119}$$

Thus

$$q \left| \mathcal{R}_w \left(\frac{p}{q} \right) \right| \le \|L_{n+1} f\|_{1, [a,b]} q \left\| g_n \left(\frac{p}{q}, \cdot \right) \right\|_{\infty, [a,b]}, \tag{8.120}$$

and

$$\int_X q\left|\mathcal{R}_w\left(\frac{p}{q}\right)\right|d\lambda \le \|L_{n+1}f\|_{1,[a,b]}\int_X q\left\|g_n\left(\frac{p}{q},\cdot\right)\right\|_{\infty,[a,b]}d\lambda. \tag{8.121}$$

We have established

Theorem 8.20. *It holds*

$$\Gamma_f(\mu_1,\mu_2) \le \sum_{i=1}^{n}|L_if(1)|\left|\int_X qg_i\left(\frac{p}{q},1\right)d\lambda\right|$$

$$+ \|L_{n+1}f\|_{1,[a,b]}\int_X q\left\|g_n\left(\frac{p}{q},\cdot\right)\right\|_{\infty,[a,b]}d\lambda. \tag{8.122}$$

Next we observe that for $\alpha,\beta > 1$ such that $\frac{1}{\alpha} + \frac{1}{\beta} = 1$ we have

$$|\mathcal{R}_w(x)| \le \|L_{n+1}f\|_{\alpha,[a,b]}\|g_n(x,\cdot)\|_{\beta,[a,b]}. \tag{8.123}$$

We derive

Theorem 8.21. *It holds*

$$\Gamma_f(\mu_1,\mu_2) \le \sum_{i=1}^{n}|L_if(1)|\left|\int_X qg_i\left(\frac{p}{q},1\right)d\lambda\right|$$

$$+ \|L_{n+1}f\|_{\alpha,[a,b]}\left(\int_X q\left\|g_n\left(\frac{p}{q},\cdot\right)\right\|_{\beta,[a,b]}d\lambda\right). \tag{8.124}$$

In the following we use Lemmas 12.1.4 and 12.1.5, p. 368 of [20].
Here $u_0(x) = c > 0$, $u_1(x)$ concave for $x \le 1$ and convex for $x \ge 1$. We put

$$\tilde{N}_n(x,1) := |N_n(x,1)|, \tag{8.125}$$

where

$$N_n(x,1) := \int_1^x g_n(x,s)ds, \tag{8.126}$$

all $x \in [a,b]$, $n \ge 0$. Let $1 \in (a,b)$ then

$$\tilde{N}_n(x,1) \le \psi|x-1|, \tag{8.127}$$

all $x \in [a,b]$, $n \ge 1$ where

$$\psi := \max\left\{\frac{\tilde{N}_n(b,1)}{b-1}, \frac{\tilde{N}_n(a,1)}{1-a}\right\}. \tag{8.128}$$

Here we take n even then $g_n(x,s) \ge 0$, all $x,s \in [a,b]$. Clearly then we have

$$|\mathcal{R}_w(x)| \le \|L_{n+1}f\|_{\infty,[a,b]}\left|\int_1^x g_n(x,s)ds\right| = \|L_{n+1}f\|_{\infty,[a,b]}\tilde{N}_n(x,1)$$

$$\le \psi\|L_{n+1}f\|_{\infty,[a,b]}|x-1|, \quad \text{all } x \in [a,b]. \tag{8.129}$$

We have proved

Theorem 8.22. *Here* $1 \in (a, b)$, n *even. It holds*

$$\Gamma_f(\mu_1, \mu_2) \leq \sum_{i=1}^{n} |L_i f(1)| \left| \int_X q g_i \left(\frac{p}{q}, 1\right) d\lambda \right|$$

$$+ \psi \|L_{n+1} f\|_{\infty,[a,b]} \left(\int_X |p - q| d\lambda \right). \tag{8.130}$$

Remark 8.23. Here again we use (8.78). Let $x \geq t$ then by (8.81) $g_n(x, t) \geq 0$, $n \in \mathbb{Z}_+$. Let $L_{n+1} f \geq 0$ (≤ 0) then

$$\int_t^x g_n(x, s) L_{n+1} f(s) ds \geq 0 \quad (\leq 0) \tag{8.131}$$

and

$$f(x) \geq (\leq) \sum_{i=0}^{n} L_i f(t) g_i(x, t), \quad x \geq t, \quad n \in \mathbb{Z}_+. \tag{8.132}$$

Let n be even then $g_n(x, s) \geq 0$ for any $x, s \in [a, b]$. If $L_{n+1} f \geq 0$ (≤ 0) and $x \leq t$, then

$$\int_t^x g_n(x, s) L_{n+1} f(s) ds \leq 0 \quad (\geq 0), \tag{8.133}$$

and

$$f(x) \leq (\geq) \sum_{i=0}^{n} L_i f(t) g_i(x, t). \tag{8.134}$$

If n is odd, then for $x \leq t$ by (8.81) we get $g_n(x, t) \leq 0$. If $L_{n+1} f \geq 0$ (≤ 0) and $x \leq t$, then (8.131) and (8.132) are again true.

Conclusion. If n is odd and $L_{n+1} f \geq 0$ (≤ 0), then

$$f(x) \geq (\leq) \sum_{i=0}^{n} L_i f(t) g_i(x, t), \tag{8.135}$$

for any $x, t \in [a, b]$, $n \in \mathbb{Z}_+$.

In particular, by $f(1) = 0$, $a \leq 1 \leq b$, $a \leq \frac{p}{q} \leq b$, a.e. on X we obtain:
If n is odd and $L_{n+1} f \geq 0$ (≤ 0), then

$$f(x) \geq (\leq) \sum_{i=1}^{n} L_i f(1) g_i(x, 1), \quad \text{any } x \in [a, b]. \tag{8.136}$$

Hence we have

$$q f\left(\frac{p}{q}\right) \geq (\leq) \sum_{i=1}^{n} L_i f(1) q g_i \left(\frac{p}{q}, 1\right), \quad \text{a.e. on } X.$$

Conclusion. We have established that

$$\Gamma_f(\mu_1, \mu_2) \geq (\leq) \sum_{i=1}^{n} L_i f(1) \int_X q g_i \left(\frac{p}{q}, 1\right) d\lambda, \tag{8.137}$$

when n is odd and $L_{n+1} f \geq 0$ (≤ 0).

8.4   Results Based on an Alternative Expansion Formula

We need the following basic general result (see also (3.53), p. 89 of [125] and [36]). We give our own proof.

Theorem 8.24. *Let $f \in C^n([a,b])$, $[a,b] \subset \mathbb{R}$, $n \in \mathbb{N}$, $x, z \in [a,b]$. Then*

$$\int_z^x f(t)dt = \sum_{k=0}^{n-1} (-1)^k f^{(k)}(x) \frac{(x-z)^{k+1}}{(k+1)!} + \frac{(-1)^n}{n!} \int_z^x (t-z)^n f^{(n)}(t)dt. \qquad (8.138)$$

Proof. From [327], p. 98 we use the generalized integration by parts formula

$$\int uv^{(n)}dt = uv^{(n-1)} - u'v^{(n-2)} + u''v^{(n-3)} - \cdots + (-1)^n \int u^{(n)}vdt, \qquad (8.139)$$

where here

$$u(t) = f(t), \quad v(t) = \frac{(t-z)^n}{n!}. \qquad (8.140)$$

Notice $u^{(k)}(t) = \frac{(t-z)^{n-k}}{(n-k)!}$, $k = 1, 2, \ldots, n-1$ and $v^{(n)}(t) = 1$, and $u^{(k)}(z) = 0$, $k = 1, 2, \ldots, n-1$. Using (8.139) we have

$$\int f(t)dt = f(t)(t-z) - f'(t)\frac{(t-z)^2}{2} + f''(t)\frac{(t-z)^3}{3!}$$
$$- f^{(3)}(t)\frac{(t-z)^4}{4!} \cdots \frac{(-1)^n}{n!} \int f^{(n)}(t)(t-z)^n dt. \qquad (8.141)$$

Thus

$$\int_z^x f(t)dt = f(x)(x-z) - f'(x)\frac{(x-z)^2}{2} + f''(x)\frac{(x-z)^3}{3!} - f^{(3)}(x)\frac{(x-z)^4}{4!} \cdots$$
$$+ \frac{(-1)^n}{n!} \int_z^x (t-z)^n f^{(n)}(t)dt$$
$$= \sum_{k=0}^{n-1} (-1)^k f^{(k)}(x) \frac{(x-z)^{k+1}}{(k+1)!} + \frac{(-1)^n}{n!} \int_z^x (t-z)^n f^{(n)}(t)dt, \qquad (8.142)$$

proving (8.138).   □

Here again $0 < a \leq 1 \leq b$, $f(1) = 0$, $a \leq \frac{p}{q} \leq b$, a.e. on X and all the rest as in this chapter and $f \in C^{n+1}([a,b])$, $n \in \mathbb{N}$. We plug into (8.138) instead of f, f' and $z = 1$, we obtain

$$f(x) = \sum_{k=0}^{n-1} (-1)^k f^{(k+1)}(x) \frac{(x-1)^{k+1}}{(k+1)!} + \frac{(-1)^n}{n!} \int_1^x (t-1)^n f^{(n+1)}(t)dt, \qquad (8.143)$$

all $x \in [a,b]$, see also (2.13), p. 5 of [101]. We call the remainder of (8.143)

$$\psi_1(x) := \frac{(-1)^n}{n!} \int_1^x (t-1)^n f^{(n+1)}(t)dt. \qquad (8.144)$$

Consequently we have

$$qf\left(\frac{p}{q}\right) = \sum_{k=0}^{n-1}(-1)^k f^{(k+1)}\left(\frac{p}{q}\right)\frac{q\left(\frac{p}{q}-1\right)^{k+1}}{(k+1)!} + \frac{(-1)^n}{n!}q\int_1^{p/q}(t-1)^n f^{(n+1)}(t)dt,$$

$$(8.145)$$

a.e. on X.

We have proved our way the following representation formula (see also (2.10), p. 4, of [101]).

Theorem 8.25. *All elements involved here as in this chapter and $f \in C^{n+1}([a,b])$, $n \in \mathbb{N}$, $0 < a \le 1 \le b$. Then*

$$\Gamma_f(\mu_1,\mu_2) = \sum_{k=0}^{n-1}\frac{(-1)^k}{(k+1)!}\int_X f^{(k+1)}\left(\frac{p}{q}\right)q^{-k}(p-q)^{k+1}d\lambda + \psi_2, \qquad (8.146)$$

where

$$\psi_2 := \frac{(-1)^n}{n!}\int_x q\left(\int_1^{p/q}(t-1)^n f^{(n+1)}(t)dt\right)d\lambda. \qquad (8.147)$$

Next we estimate ψ_2.

We see that

$$|\psi_1(x)| \le \frac{\|f^{(n+1)}\|_{\infty,[a,b]}}{(n+1)!}|x-1|^{n+1}, \quad \text{all } x \in [a,b]. \qquad (8.148)$$

Also

$$|\psi_1(x)| \le \frac{\|f^{(n+1)}\|_{1,[a,b]}}{n!}|x-1|^n, \quad \text{all } x \in [a,b]. \qquad (8.149)$$

Furthermore for $\alpha,\beta > 1$: $\frac{1}{\alpha} + \frac{1}{\beta} = 1$ we derive

$$|\psi_1(x)| \le \frac{\|f^{(n+1)}\|_{\alpha,[a,b]}}{n!(n\beta+1)^{1/\beta}}|x-1|^{n+\frac{1}{\beta}}, \quad \text{all } x \in [a,b]. \qquad (8.150)$$

Then

$$q\left|\psi_1\left(\frac{p}{q}\right)\right| \le \min \text{ of } \begin{cases} \dfrac{\|f^{(n+1)}\|_{\infty,[a,b]}}{(n+1)!}q^{-n}|p-q|^{n+1}, \\[2ex] \dfrac{\|f^{(n+1)}\|_{1,[a,b]}}{n!}q^{1-n}|p-q|^n, \\[2ex] \dfrac{\|f^{(n+1)}\|_{\alpha,[a,b]}}{n!(n\beta+1)^{1/\beta}}q^{1-n-\frac{1}{\beta}}|p-q|^{n+\frac{1}{\beta}}, \ \alpha,\beta>1: \dfrac{1}{\alpha}+\dfrac{1}{\beta}=1. \end{cases}$$

$$(8.151)$$

All the above are true a.e. on X.

We have proved (see also the similar Corollary 1, p. 10, of [101]).

Theorem 8.26. *All elements are as in this chapter and $f \in C^{n+1}([a,b])$, $n \in \mathbb{N}$. Here ψ_2 is as in (8.147). Then*

$$
|\psi_2| \leq \min \text{ of } \begin{cases} B_1 := \dfrac{\|f^{(n+1)}\|_{\infty,[a,b]}}{(n+1)!} \displaystyle\int_X q^{-n}|p-q|^{n+1} d\lambda, \\[3mm] B_2 := \dfrac{\|f^{(n+1)}\|_{1,[a,b]}}{n!} \displaystyle\int_X q^{1-n}|p-q|^n d\lambda, \text{ for } \alpha, \beta > 1: \dfrac{1}{\alpha} + \dfrac{1}{\beta} = 1, \\[3mm] B_3 := \dfrac{\|f^{(n+1)}\|_{\alpha,[a,b]}}{n!(n\beta+1)^{1/\beta}} \displaystyle\int_X q^{1-n-\frac{1}{\beta}}|p-q|^{n+\frac{1}{\beta}} d\lambda. \end{cases}
$$

$$(8.152)$$

Also it holds

$$
\Gamma_f(\mu_1, \mu_2) \leq \sum_{k=0}^{n-1} \frac{1}{(k+1)!} \left| \int_X f^{(k+1)}\left(\frac{p}{q}\right) q^{-k}(p-q)^{k+1} d\lambda \right| + \min(B_1, B_2, B_3).
$$

$$(8.153)$$

Corollary 8.27. (to Theorem 8.26).

Case of $n = 1$. It holds

$$
|\psi_2| \leq \min \text{ of } \begin{cases} B_{1,1} := \dfrac{\|f''\|_{\infty,[a,b]}}{2} \displaystyle\int_X q^{-1}(p-q)^2 d\lambda, \\[3mm] B_{2,1} := \|f''\|_{1,[a,b]} \displaystyle\int_X |p-q| d\lambda, \text{ for } \alpha, \beta > 1: \dfrac{1}{\alpha} + \dfrac{1}{\beta} = 1, \\[3mm] B_{3,1} := \dfrac{\|f''\|_{\alpha,[a,b]}}{(\beta+1)^{1/\beta}} \displaystyle\int_X q^{-\frac{1}{\beta}}|p-q|^{1+\frac{1}{\beta}} d\lambda. \end{cases}
$$

$$(8.154)$$

Also it holds

$$
\Gamma_f(\mu_1, \mu_2) \leq \left| \int_X f'\left(\frac{p}{q}\right)(p-q) d\lambda \right| + \min(B_{1,1}, B_{2,1}, B_{3,1}).
$$

$$(8.155)$$

Remark 8.28. (to Theorem 8.25). Notice that

$$
\frac{(-1)^n}{n!} \int_1^x (t-1)^n f^{(n+1)}(t) dt = \frac{(-1)^n}{n!} \int_1^x (t-1)^n [f^{(n+1)}(t) - f^{(n+1)}(1)] dt
$$
$$
+ (-1)^n f^{(n+1)}(1) \frac{(x-1)^{n+1}}{(n+1)!}. \qquad (8.156)
$$

Therefore from (8.143) we obtain

$$
f(x) = \sum_{k=0}^{n-1} (-1)^k f^{(k+1)}(x) \frac{(x-1)^{k+1}}{(k+1)!} + \frac{(-1)^n}{(n+1)!} f^{(n+1)}(1)(x-1)^{n+1} + \psi_3(x),
$$

$$(8.157)$$

where

$$\psi_3(x) := \frac{(-1)^n}{n!} \int_1^x (t-1)^n (f^{(n+1)}(t) - f^{(n+1)}(1))dt. \tag{8.158}$$

Then for $0 < h \le b - a$ we obtain

$$|\psi_3(x)| \le \frac{1}{n!} \left| \int_1^x |t-1|^n |f^{(n+1)}(t) - f^{(n+1)}(1)|dt \right|$$

$$\text{(by Lemma 7.1.1, p. 208, of [20])}$$

$$\le \frac{\omega_1(f^{(n+1)}, h)}{n!} \left| \int_1^x |t-1|^n \left\lceil \frac{|t-1|}{h} \right\rceil dt \right|$$

$$\le \frac{\omega_1(f^{(n+1)}, h)}{n!} \left| \int_1^x |t-1|^n \left(1 + \frac{|t-1|}{h}\right) dt \right|$$

$$= \frac{\omega_1(f^{(n+1)}, h)}{n!} \left| \int_1^x |t-1|^n dt + \frac{1}{h} \int_1^x |t-1|^{n+1} dt \right|$$

$$\le \frac{\omega_1(f^{(n+1)}, h)}{n!} \left\{ \left| \int_1^x |t-1|^n dt \right| + \frac{1}{h} \left| \int_1^x |t-1|^{n+1} dt \right| \right\}.$$

We have proved for $0 < h \le b - a$ that

$$|\psi_3(x)| \le \frac{\omega_1(f^{(n+1)}, h)}{n!} \left\{ \left| \int_1^x |t-1|^n dt \right| + \frac{1}{h} \left| \int_1^x |t-1|^{n+1} dt \right| \right\}, \tag{8.159}$$

all $x \in [a, b]$.

Next suppose that $|f^{(n+1)}(t) - f^{(n+1)}(1)|$ is convex in t and $0 < h \le \min(1 - a, b - 1)$ then

$$|\psi_3(x)| \le \frac{1}{n!} \left| \int_1^x |t-1|^n |f^{(n+1)}(t) - f^{(n+1)}(1)|dt \right|$$

$$\text{(by Lemma 8.1.1, p. 243, of [20])}$$

$$\le \frac{\omega_1(f^{(n+1)}, h)}{n!h} \left| \int_1^x |t-1|^{n+1} dt \right|. \tag{8.160}$$

By

$$\left| \int_1^x |t-1|^m dt \right| = \frac{|x-1|^{m+1}}{m+1}, \quad m \ne -1,$$

we get the following

For $0 < h \le b - a$ we have

$$|\psi_3(x)| \le \frac{\omega_1(f^{(n+1)}, h)}{n!} \left\{ \frac{|x-1|^{n+1}}{n+1} + \frac{1}{h} \frac{|x-1|^{n+2}}{(n+2)} \right\}, \tag{8.161}$$

and for $0 < h \le \min(1 - a, b - 1)$ and $|f^{(n+1)}(t) - f^{(n+1)}(1)|$ convex in t we obtain

$$|\psi_3(x)| \le \frac{\omega_1(f^{(n+1)}, h)}{n!(n+2)h} |x-1|^{n+2}, \tag{8.162}$$

$n \in \mathbb{N}$, all $x \in [a, b]$. The last inequality (8.162) is attained when n is even by

$$\hat{f}(t) := \frac{(t-1)^{n+2}}{(n+2)!}, a \le t \le b. \tag{8.163}$$

Really $\hat{f} \in C^{n+1}([a, b])$, it is convex and strictly convex at 1. We get $\hat{f}^{(k)}(1) = 0$, $k = 0, 1, \ldots, n+1$ where $\hat{f}^{(n+1)}(t) = t - 1$ with obviously $|\hat{f}(t)^{(n+1)}|$ convex and $\omega_1(\hat{f}^{(n+1)}, h) = h$.

For $\hat{f}$ the corresponding

$$\hat{\psi}_3(x) = \frac{(x-1)^{n+2}}{n!(n+2)}, \quad x \in [a, b], \tag{8.164}$$

proving (8.162) attained. Here again

$$qf\left(\frac{p}{q}\right) = \sum_{k=0}^{n-1} \frac{(-1)^k}{(k+1)!} f^{(k+1)}\left(\frac{p}{q}\right) q^{-k}(p-q)^{k+1}$$

$$+ \frac{(-1)^n}{(n+1)!} f^{(n+1)}(1)q^{-n}(p-q)^{n+1} + q\psi_3\left(\frac{p}{q}\right), \tag{8.165}$$

a.e. on X.

We have established

Theorem 8.29. *All elements involved here as in this chapter and* $f \in C^{n+1}([a, b])$, $n \in \mathbb{N}$, $0 < a \le 1 \le b$. *Then*

$$\Gamma_f(\mu_1, \mu_2) = \sum_{k=0}^{n-1} \frac{(-1)^k}{(k+1)!} \int_X f^{(k+1)}\left(\frac{p}{q}\right) q^{-k}(p-q)^{k+1} d\lambda$$

$$+ \frac{(-1)^n}{(n+1)!} f^{(n+1)}(1) \int_X q^{-n}(p-q)^{n+1} d\lambda + \psi_4, \tag{8.166}$$

where

$$\psi_4 := \int_X q\psi_3\left(\frac{p}{q}\right) d\lambda. \tag{8.167}$$

Next we estimate ψ_4.

Let $0 < h \le b - a$, then from (8.161) we have

$$q\left|\psi_3\left(\frac{p}{q}\right)\right| \le \frac{\omega_1(f^{(n+1)}, h)}{n!} \left\{ \frac{q^{-n}|p-q|^{n+1}}{n+1} + \frac{1}{h}\frac{q^{-n-1}|p-q|^{n+2}}{(n+2)} \right\}, \tag{8.168}$$

therefore

$$|\psi_4| \le \frac{\omega_1(f^{(n+1)}, h)}{n!} \left\{ \frac{1}{n+1} \int_X q^{-n}|p-q|^{n+1} d\lambda \right.$$

$$\left. + \frac{1}{(n+2)h} \int_X q^{-n-1}|p-q|^{n+2} d\lambda \right\}, n \in \mathbb{N}. \tag{8.169}$$

When $0 < h \le \min(1-a, b-1)$ and $|f^{(n+1)}(t) - f^{(n+1)}(1)|$ convex in t we obtain

$$\left| \psi_3 \left(\frac{p}{q} \right) \right| \le \frac{\omega_1(f^{(n+1)}, h)}{n!(n+2)h} q^{-n-1} |p - q|^{n+2}, \tag{8.170}$$

and

$$|\psi_4| \le \frac{\omega_1(f^{(n+1)}, h)}{n!(n+2)h} \int_X q^{-n-1} |p - q|^{n+2} d\lambda, \quad n \in \mathbb{N}. \tag{8.171}$$

For n even the last inequality (8.171) is attained by $\hat{f}$ by (8.163).

We have proved

Theorem 8.30. *All elements involved here as in this chapter and* $f \in C^{n+1}([a, b])$, $n \in \mathbb{N}$, $0 < a \le 1 \le b$ *and*

$$0 < \frac{1}{(n+2)} \int_X q^{-n-1} |p - q|^{n+2} d\lambda \le b - a. \tag{8.172}$$

Then (8.166) is again valid and

$$|\psi_4| \le$$

$$\frac{1}{n!} \omega_1 \left(f^{(n+1)}, \frac{1}{(n+2)} \int_X q^{-n-1} |p - q|^{n+2} d\lambda \right)$$

$$\left(1 + \frac{1}{n+1} \int_X q^{-n} |p - q|^{n+1} d\lambda \right). \tag{8.173}$$

Also we have

Theorem 8.31. *All elements involved here as in this chapter and* $f \in C^{n+1}([a, b])$, $n \in \mathbb{N}$, $0 < a \le 1 \le b$. *Furthermore we suppose that* $|f^{(n+1)}(t) - f^{(n+1)}(1)|$ *is convex in* t *and*

$$0 < \frac{1}{n!(n+2)} \int_X q^{-n-1} |p - q|^{n+2} d\lambda \le \min(1 - a, b - 1). \tag{8.174}$$

Then (8.166) is again valid and

$$|\psi_4| \le \omega_1 \left(f^{(n+1)}, \frac{1}{n!(n+2)} \int_X q^{-n-1} |p - q|^{n+2} d\lambda \right). \tag{8.175}$$

The last (8.175) is attained by $\hat{f}$ *of (8.163) when* n *is even.*

When $n = 1$ we derive

Corollary 8.32. (to Theorem 8.30). *All elements involved here as in this chapter and* $f \in C^2([a, b])$, $0 < a \le 1 \le b$ *and*

$$0 < \frac{1}{3} \int_X q^{-2} |p - q|^3 d\lambda \le b - a. \tag{8.176}$$

Then

$$\Gamma_f(\mu_1, \mu_2) = \int_X f' \left(\frac{p}{q} \right) (p - q) d\lambda - \frac{f''(1)}{2} \int_X q^{-1} (p - q)^2 d\lambda + \psi_{4,1}, \tag{8.177}$$

where

$$\psi_{4,1} := \int_X q\psi_{3,1}\left(\frac{p}{q}\right) d\lambda, \tag{8.178}$$

with

$$\psi_{3,1}(x) := -\int_1^x (t-1)(f''(t) - f''(1))dt, \quad x \in [a,b]. \tag{8.179}$$

Furthermore it holds

$$|\psi_{4,1}| \le \omega_1\left(f'', \frac{1}{3}\int_X q^{-2}|p-q|^3 d\lambda\right) \cdot \frac{1}{2}\left(\int_X q^{-1}p^2 d\lambda + 1\right). \tag{8.180}$$

Corollary 8.33. (to Theorem 8.31). *All elements involved here as in this chapter and $f \in C^2([a,b])$, $0 < a \le 1 \le b$. Furthermore we suppose that $|f''(t) - f''(1)|$ is convex in t and*

$$0 < \frac{1}{3}\int_X q^{-2}|p-q|^3 d\lambda \le \min(1-a, b-1). \tag{8.181}$$

Then again (8.177) is valid and

$$|\psi_{4,1}| \le \omega_1\left(f'', \frac{1}{3}\int_X q^{-2}|p-q|^3 d\lambda\right). \tag{8.182}$$

Remark 8.34 (to Theorem 8.29). Here all as in Theorem 8.29 are valid especially the conclusions (8.166) and (8.167). Additionally assume that

$$|f^{(n+1)}(x) - f^{(n+1)}(y)| \le K|x-y|^\alpha, \tag{8.183}$$

$K > 0$, $0 < \alpha \le 1$, all $x, y \in [a,b]$ with $(x-1)(y-1) \ge 0$. Then

$$|\psi_3| \le \frac{K}{n!}\left|\int_1^x |t-1|^{n+\alpha}dt\right| = \frac{K}{n!(n+\alpha+1)}|x-1|^{n+\alpha+1}, \tag{8.184}$$

all $x \in [a,b]$, $n \in \mathbb{N}$. Therefore it holds

$$q\left|\psi_3\left(\frac{p}{q}\right)\right| \le \frac{K}{n!(n+\alpha+1)}q^{-n-\alpha}|p-q|^{n+\alpha+1}, \quad \text{a.e. on } X. \tag{8.185}$$

We have established

Theorem 8.35. *All elements involved here as in this chapter and $f \in C^{n+1}([a,b])$, $n \in \mathbb{N}$, $0 < a \le 1 \le b$. Additionally assume (8.183),*

$$|f^{(n+1)}(x) - f^{(n+1)}(y)| \le K|x-y|^\alpha,$$

$K > 0$, $0 < \alpha \le 1$, *all $x, y \in [a,b]$ with $(x-1)(y-1) \ge 0$. Then we get (8.166),*

$$\Gamma_f(\mu_1, \mu_2) = \sum_{k=0}^{n-1} \frac{(-1)^k}{(k+1)!}\int_X f^{(k+1)}\left(\frac{p}{q}\right) q^{-k}(p-q)^{k+1} d\lambda$$

$$+ \frac{(-1)^n}{(n+1)!} f^{(n+1)}(1)\int_X q^{-n}(p-q)^{n+1}d\lambda + \psi_4.$$

Here we obtain that

$$|\psi_4| \le \frac{K}{n!(n+\alpha+1)} \int_X q^{-n-\alpha}|p-q|^{n+\alpha+1}d\lambda. \tag{8.186}$$

Inequality (8.186) is attained when n is even by

$$f^*(x) = \tilde{c}|x-1|^{n+\alpha+1}, \quad x \in [a,b] \tag{8.187}$$

where

$$\tilde{c} := \frac{K}{\prod\limits_{j=0}^{n}(n+\alpha+1-j)}. \tag{8.188}$$

Proof of sharpness of (8.186). Here $f^* \in C^{n+1}([a,b])$, is convex and strictly convex at $x = 1$, $f^*(1) = 0$.

We find that

$$f^{*(n+1)}(x) = K|x-1|^\alpha \text{ sign } (x-1) \tag{8.189}$$

with $f^{*(k)}(1) = 0$, $k = 1,2,\ldots,n+1$. Furthermore $f^{*(n+1)}$ clearly fulfills (8.183), see in p. 28 of [29] inequality (105).

Additionally we obtain the corresponding

$$\psi_3(x) = \frac{K|x-1|^{n+\alpha+1}}{n!(n+\alpha+1)}, \quad x \in [a,b]. \tag{8.190}$$

Then the corresponding

$$\psi_4 = \int_X q\left|\psi_3\left(\frac{p}{q}\right)\right|d\lambda = \frac{K}{n!(n+\alpha+1)} \int_X q^{-n-\alpha}|p-q|^{n+\alpha+1}d\lambda, \tag{8.191}$$

proving that (8.186) is attained. $\qquad\square$

Finally we present

Corollary 8.36 (to Theorem 8.35). *All elements involved here as in this chapter and $f \in C^2([a,b])$, $0 < a \le 1 \le b$. Additionally assume that*

$$|f''(x) - f''(y)| \le K|x-y|^\alpha, \tag{8.192}$$

$K > 0$, $0 < \alpha \le 1$, *all $x,y \in [a,b]$ with $(x-1)(y-1) \ge 0$. Then we get (8.177),*

$$\Gamma_f(\mu_1,\mu_2) = \int_X f'\left(\frac{p}{q}\right)(p-q)d\lambda - \frac{f''(1)}{2}\int_X q^{-1}(p-q)^2 d\lambda + \psi_{4,1},$$

where

$$\psi_{4,1} := \int_X q\psi_{3,1}\left(\frac{p}{q}\right)d\lambda,$$

with

$$\psi_{3,1}(x) := -\int_1^x (t-1)(f''(t) - f''(1))dt, \quad x \in [a,b].$$

It holds that

$$|\psi_{4,1}| \le \frac{K}{(\alpha+2)} \int_X q^{-1-\alpha}|p-q|^{\alpha+2}d\lambda. \tag{8.193}$$

Chapter 9

Csiszar's f-Divergence as a Measure of Dependence

In this chapter the Csiszar's f-Divergence or Csiszar's Discrimination, the most general distance among probability measures, is established as the most general measure of Dependence between two random variables which is a very important aspect of stochastics. Many estimates of it are given, leading to optimal or nearly optimal probabilistic inequalities. That is we are comparing and connecting this general measure of dependence to known other specific entities by involving basic parameters of our setting. This treatment relies on [39].

9.1 Background

Throughout this chapter we use the following. Let f by a convex function from $(0, +\infty)$ into $\mathbb{R}$ which is strictly convex at 1 with $f(1) = 0$. Let $(\mathbb{R}^2, \mathcal{B}^2, \lambda)$ be the measure space, where λ is the product Lebesgue measure on $(\mathbb{R}^2, \mathcal{B}^2)$ with $\mathcal{B}$ being the Borel σ-field. And let $X, Y : \Omega \to \mathbb{R}$ be random variables on the probability space (Ω, P). Consider the probability distributions μ_{XY} and $\mu_X \times \mu_Y$ on $\mathbb{R}^2$, where μ_{XY}, μ_X, μ_Y stand for the joint distribution of X and Y and their marginal distributions, respectively.

Here we assume as existing the following probability density functions, the joint pdf of μ_{XY} to be $t(x, y)$, $x, y \in \mathbb{R}$, the pdf of μ_X to be $p(x)$ and the pdf of μ_Y to be $q(y)$. Clearly $\mu_X \times \mu_Y$ has pdf $p(x)q(y)$. Here we further assume that $0 < a \leq \frac{t}{pq} \leq b$, a.e. on $\mathbb{R}^2$ and $a \leq 1 \leq b$.

The quantity

$$\Gamma_f(\mu_{XY}, \mu_X \times \mu_Y) = \int_{\mathbb{R}^2} p(x)q(y)f\left(\frac{t(x, y)}{p(x)q(y)}\right) d\lambda(x, y), \qquad (9.1)$$

is the *Csiszar's distance or f-divergence between μ_{XY} and $\mu_X \times \mu_Y$.*

Γ_f is the most general measure for comparing probability measures and it was first introduced by I. Csiszar in 1963, [139], see also [140], [152], [244], [243], [305], [306], and [101], [102], [125], [126], [37], [42], [43], [41].

In Information Theory and Statistics many various divergences are used which are special cases of the above Γ_f divergence. Γ_f has many applications also in

101

Anthropology, Genetics, Finance, Economics, Political Science, Biology, Approximation of Probability Distributions, Signal Processing and Pattern Recognition. Here X, Y are less dependent the closer the distributions μ_{XY} and $\mu_X \times \mu_Y$ are, thus $\Gamma_f(\mu_{XY}, \mu_X \times \mu_Y)$ can be considered *as a measure of dependence* of X and Y. For $f(u) = u \log_2 u$ we obtain the *mutual information* of X, and Y,

$$I(X, Y) = I(\mu_{XY} \parallel \mu_X \times \mu_Y) = \Gamma_{u \log_2 u}(\mu_{XY}, \mu_X \times \mu_Y),$$

see [140]. For $f(u) = (u - 1)^2$ we get the *mean square contingency* :

$$\varphi^2(X, Y) = \Gamma_{(u-1)^2}(\mu_{XY}, \mu_X \times \mu_Y),$$

see [305]. In the last we need $\mu_{XY} \ll \mu_X \times \mu_Y$, where $\ll$ denotes absolute continuity, but to cover the case of $\mu_{XY} \not\ll \mu_X \times \mu_Y$ we set $\varphi^2(X, Y) = +\infty$, then the last formula is always valid.

Clearly here $\mu_{XY}, \mu_X \times \mu_Y \ll \lambda$, also $\Gamma_f(\mu_{XY}, \mu_X \times \mu_Y) \geq 0$ with equality only when $\mu_{XY} = \mu_X \times \mu_Y$, i.e. when X, Y are independent r.v.'s. In this chapter we present applications of author's articles ([37]), ([42]), ([43]), ([41]).

9.2   Results

Part I. Here we apply results from [37]. We present the following.

Theorem 9.1. *Let $f \in C^1([a, b])$. Then*

$$\Gamma_f(\mu_{XY}, \mu_X \times \mu_Y) \leq \|f'\|_{\infty,[a,b]} \int_{\mathbb{R}^2} |t(x, y) - p(x)q(y)| \, d\lambda(x, y). \qquad (9.2)$$

Inequality (9.2) is sharp. The optimal function is $f^*(y) = |y - 1|^\alpha$, $\alpha > 1$ under the condition, either

i)

$$\max(b - 1, 1 - a) < 1 \qquad and \qquad C := \mathrm{esssup}(pq) < +\infty, \qquad (9.3)$$

     or

ii)

$$|t - pq| \leq pq \leq 1, \quad a.e. \ on \ \mathbb{R}^2. \qquad (9.4)$$

Proof. Based on Theorem 1 of [37]. $\square$

Next we give the more general

Theorem 9.2. *Let $f \in C^{n+1}([a, b])$, $n \in \mathbb{N}$, such that $f^{(k)}(1) = 0$, $k = 0, 2, 3, \ldots, n$. Then*

$$\Gamma_f(\mu_{XY}, \mu_X \times \mu_Y)$$
$$\leq \frac{\|f^{(n+1)}\|_{\infty,[a,b]}}{(n+1)!} \int_{\mathbb{R}^2} (p(x))^{-n} (q(y))^{-n} |t(x, y) - p(x)q(y)|^{n+1} d\lambda(x, y). \qquad (9.5)$$

Inequality (9.5) is sharp. The optimal function is $\tilde{f}(y) := |y - 1|^{n+\alpha}$, $\alpha > 1$ under the condition (i) or (ii) of Theorem 9.1.

Proof. As in Theorem 2 of [37]. $\qquad\square$

As a special case we have

Corollary 9.3 (to Theorem 9.2). *Let $f \in C^2([a,b])$, such that $f(1) = 0$. Then*

$$\Gamma_f(\mu_{XY}, \mu_X \times \mu_Y) \leq \frac{\|f^{(2)}\|_{\infty,[a,b]}}{2} \int_{\mathbb{R}^2} (p(x))^{-1}(q(y))^{-1}(t(x,y) - p(x)q(y))^2 \, d\lambda(x,y).$$

(9.6)

Inequality (9.6) is sharp. The optimal function is $\tilde{f}(y) := |y - 1|^{1+\alpha}$, $\alpha > 1$ under the condition (i) or (ii) of Theorem 9.1.

Proof. By Corollary 1 of [37]. $\qquad\square$

Next we connect Csiszar's f-divergence to the usual first modulus of continuity ω_1.

Theorem 9.4. *Assume that*

$$0 < h := \int_{\mathbb{R}^2} |t(x,y) - p(x)q(y)| \, d\lambda(x,y) \leq \min(1 - a, b - 1).$$

(9.7)

Then

$$\Gamma_f(\mu_{XY}, \mu_X \times \mu_Y) \leq \omega_1(f, h).$$

(9.8)

Inequality (9.8) is sharp, namely it is attained by

$$f^*(x) = |x - 1|.$$

Proof. Based on Theorem 3 of [37]. $\qquad\square$

Part II. Here we apply results from [42]. But first we need some basic background from [26], [122]. Let $\nu > 0$, $n := [\nu]$ and $\alpha = \nu - n$ $(0 < \alpha < 1)$. Let $x, x_0 \in [a,b] \subseteq \mathbb{R}$ such that $x \geq x_0$, where x_0 is fixed. Let $f \in C([a,b])$ and define

$$(\mathcal{J}_\nu^{x_0} f)(x) := \frac{1}{\Gamma(\nu)} \int_{x_0}^x (x - s)^{\nu-1} f(s) \, ds, \quad x_0 \leq x \leq b,$$

(9.9)

the *generalized Riemann-Liouville integral*, where Γ stands for the gamma function.

We define the subspace

$$C_{x_0}^\nu([a,b]) := \{f \in C^n([a,b]) : J_{1-\alpha}^{x_0} D^n f \in C^1([x_0, b])\}.$$

(9.10)

For $f \in C_{x_0}^\nu([a,b])$ we define the *generalized ν-fractional derivative off over $[x_0, b]$* as

$$D_{x_0}^\nu f := D J_{1-\alpha}^{x_0} f^{(n)} \quad \left(D := \frac{d}{dx}\right).$$

(9.11)

We present the following.

Theorem 9.5. *Let $a < b$, $1 \leq \nu < 2$, $f \in C_a^\nu([a,b])$, and $0 < a \leq \frac{t}{pq} \leq b$, a.e. on $\mathbb{R}^2$. Then*

$$\Gamma_f(\mu_{XY}, \mu_X \times \mu_Y)$$
$$\leq \frac{\|D_a^\nu f\|_{\infty,[a,b]}}{\Gamma(\nu+1)} \int_{\mathbb{R}^2} (p(x))^{1-\nu}(q(y))^{1-\nu}(t(x,y) - ap(x)q(y))^\nu \, d\lambda(x,y). \quad (9.12)$$

Proof. It follows from Theorem 2 of [42]. $\square$

The counterpart of the previous result comes next.

Theorem 9.6. *Let $a < b$, $\nu \geq 2$, $n := [\nu]$, $f \in C_a^\nu([a,b])$, $f^{(i)}(a) = 0$, $i = 0, 1, \ldots, n-1$, and $a \leq \frac{t(x,y)}{p(x)q(y)} \leq b$, a.e. on $\mathbb{R}^2$. Then*

$$\Gamma_f(\mu_{XY}, \mu_X \times \mu_Y)$$
$$\leq \frac{\|D_a^\nu f\|_{\infty,[a,b]}}{\Gamma(\nu+1)} \int_{\mathbb{R}^2} (p(x))^{1-\nu}(q(y))^{1-\nu}(t(x,y) - ap(x)q(y))^\nu d\lambda(x,y). \quad (9.13)$$

Proof. Based on Theorem 3 of [42]. $\square$

Next we give an $L_{\tilde{\alpha}}$ estimate.

Theorem 9.7. *Let $a < b$, $\nu \geq 1$, $n := [\nu]$, $f \in C_a^\nu([a,b])$, $f^{(i)}(a) = 0$, $i = 0, 1, \ldots, n-1$, and $a \leq \frac{t}{pq} \leq b$, a.e. on $\mathbb{R}^2$. Let $\tilde{\alpha}, \beta > 1$: $\frac{1}{\tilde{\alpha}} + \frac{1}{\beta} = 1$. Then*

$$\Gamma_f(\mu_{XY}, \mu_X \times \mu_Y) \leq \frac{\|D_a^\nu f\|_{\tilde{\alpha},[a,b]}}{\Gamma(\nu)(\beta(\nu-1)+1)^{1/\beta}}$$
$$\cdot \int_{\mathbb{R}^2} (p(x)q(y))^{2-\nu-\frac{1}{\beta}}(t(x,y) - ap(x)q(y))^{\nu-1+\frac{1}{\beta}} d\lambda(x,y). \quad (9.14)$$

Proof. Based on Theorem 4 of [42]. $\square$

It follows an L_∞ estimate.

Theorem 9.8. *Let $a < b$, $\nu \geq 1$, $n := [\nu]$, $f \in C_a^\nu([a,b])$, $f^{(i)}(a) = 0$, $i = 0, 1, \ldots, n-1$, and $a \leq \frac{t}{pq} \leq b$, a.e. on $\mathbb{R}^2$. Then*

$$\Gamma_f(\mu_{XY}, \mu_X \times \mu_Y) \leq \frac{\|D_a^\nu f\|_{1,[a,b]}}{\Gamma(\nu)} \int_{\mathbb{R}^2} (p(x)q(y))^{2-\nu}(t(x,y) - ap(x)q(y))^{\nu-1} d\lambda(x,y).$$
$$(9.15)$$

Proof. Based on Theorem 5 of [42]. $\square$

We continue with

Theorem 9.9. *Let $f \in C^1([a,b])$, $a \neq b$. Then*

$$\Gamma_f(\mu_{XY}, \mu_X \times \mu_Y) \leq \frac{1}{b-a} \int_a^b f(x)dx$$
$$- \int_{\mathbb{R}^2} f'\left(\frac{t(x,y)}{p(x)q(y)}\right)\left(\frac{p(x)q(y)(a+b)}{2} - t(x,y)\right) d\lambda(x,y). \quad (9.16)$$

Proof. Based on Theorem 6 of [42]. $\qquad\square$

Furthermore we obtain

Theorem 9.10. *Let n odd and $f \in C^{n+1}([a,b])$, such that $f^{(n+1)} \geq 0$ (≤ 0), $0 < a \leq \frac{t}{pq} \leq b$, a.e. on $\mathbb{R}^2$. Then*

$$\Gamma_f(\mu_{XY}, \mu_X \times \mu_Y) \leq (\geq) \frac{1}{b-a} \int_a^b f(x)dx \tag{9.17}$$

$$- \sum_{i=1}^n \frac{1}{(i+1)!} \left(\sum_{k=0}^i \int_{\mathbb{R}^2} f^{(i)} \left(\frac{t}{pq} \right) (pq)^{1-i} (bpq - t)^{(i-k)} (apq - t)^k d\lambda \right).$$

Proof. Based on Theorem 7 of [42]. $\qquad\square$

Part III. Here we apply results from [43].

We start with

Theorem 9.11. *Let $f, g \in C^1([a,b])$ where f as in this chapter, $g' \neq 0$ over $[a,b]$. Then*

$$\Gamma_f(\mu_{XY}, \mu_X \times \mu_Y) \leq \left\| \frac{f'}{g'} \right\|_{\infty,[a,b]} \int_{\mathbb{R}^2} p(x)q(y) \left| g \left(\frac{t(x,y)}{p(x)q(y)} \right) - g(1) \right| d\lambda(x,y).$$

$$\tag{9.18}$$

Proof. Based on Theorem 1 of [43]. $\qquad\square$

We give

Examples 9.12 (to Theorem 9.11).
1) Let $g(x) = \frac{1}{x}$. Then

$$\Gamma_f(\mu_{XY}, \mu_X \times \mu_Y) \leq \|x^2 f'\|_{\infty,[a,b]} \int_{\mathbb{R}^2} \frac{p(x)q(y)}{t(x,y)} |t(x,y) - p(x)q(y)| d\lambda(x,y). \tag{9.19}$$

2. Let $g(x) = e^x$. Then

$$\Gamma_f(\mu_{XY}, \mu_X \times \mu_Y) \leq \|e^{-x} f'\|_{\infty,[a,b]} \int_{\mathbb{R}^2} p(x)q(y) \left| e^{\frac{t(x,y)}{p(x)q(y)}} - e \right| d\lambda(x,y). \tag{9.20}$$

3) Let $g(x) = e^{-x}$. Then

$$\Gamma_f(\mu_{XY}, \mu_X \times \mu_Y) \leq \|e^x f'\|_{\infty,[a,b]} \int_{\mathbb{R}^2} p(x)q(y) \left| e^{-\frac{t(x,y)}{p(x)q(y)}} - e^{-1} \right| d\lambda(x,y). \tag{9.21}$$

4) Let $g(x) = \ell n x$. Then

$$\Gamma_f(\mu_{XY}, \mu_X \times \mu_Y) \leq \|x f'\|_{\infty,[a,b]} \int_{\mathbb{R}^2} p(x)q(y) \left| \ell n \left(\frac{t(x,y)}{p(x)q(y)} \right) \right| d\lambda(x,y). \tag{9.22}$$

5) Let $g(x) = x \ell n\, x$, with $a > e^{-1}$. Then

$$\Gamma_f(\mu_{XY}, \mu_X \times \mu_Y) \leq \left\| \frac{f'}{1 + \ell n\, x} \right\|_{\infty,[a,b]} \int_{\mathbb{R}^2} t(x,y) \left| \ell n \left(\frac{t(x,y)}{p(x)q(y)} \right) \right| d\lambda(x,y). \tag{9.23}$$

6) Let $g(x) = \sqrt{x}$. Then

$$\Gamma_f(\mu_{XY}, \mu_X \times \mu_Y) \le 2\|\sqrt{x}f'\|_{\infty,[a,b]} \int_{\mathbb{R}^2} \sqrt{p(x)q(y)}\,|\sqrt{t(x,y)} - \sqrt{p(x)q(y)}|d\lambda(x,y).$$
(9.24)

7) Let $g(x) = x^\alpha$, $\alpha > 1$. Then

$$\Gamma_f(\mu_{XY}, \mu_X \times \mu_Y)$$
$$\le \frac{1}{\alpha}\left\|\frac{f'}{x^{\alpha-1}}\right\|_{\infty,[a,b]} \int_{\mathbb{R}^2} (p(x)q(y))^{1-\alpha}|(t(x,y))^\alpha - (p(x)q(y))^\alpha|d\lambda(x,y). \quad (9.25)$$

Next we give

Theorem 9.13. *Let* $f \in C^1([a,b])$, $a \ne b$. *Then*

$$\left|\Gamma_f(\mu_{XY}, \mu_X \times \mu_Y) - \frac{1}{b-a}\int_a^b f(y)dy\right| \le \frac{\|f'\|_{\infty,[a,b]}}{(b-a)}\left[\left(\frac{a^2+b^2}{2}\right) - (a+b)\right.$$
$$\left. + \int_{\mathbb{R}^2} (p(x)q(y))^{-1}t^2(x,y)d\lambda(x,y)\right].$$
(9.26)

Proof. Based on Theorem 2 of [43]. $\qquad\square$

It follows

Theorem 9.14. *Let* $f \in C^{(2)}([a,b])$, $a \ne b$. *Then*

$$\left|\Gamma_f(\mu_{XY}, \mu_X \times \mu_Y) - \frac{1}{b-a}\int_a^b f(s)ds - \left(\frac{f(b)-f(a)}{b-a}\right)\left(1 - \frac{a+b}{2}\right)\right| \quad (9.27)$$
$$\le \frac{(b-a)^2}{2}\|f''\|_{\infty,[a,b]}\left\{\int_{\mathbb{R}^2} p(x)q(y)\left[\frac{\left(\frac{t(x,y)}{p(x)q(y)} - \left(\frac{a+b}{2}\right)\right)^2}{(b-a)^2} + \frac{1}{4}\right]^2 d\lambda(x,y) + \frac{1}{12}\right\}.$$

Proof. Based on (33) of [43]. $\qquad\square$

Working more generally we have

Theorem 9.15. *Let* $f \in C^1([a,b])$ *and* $g \in C([a,b])$ *of bounded variation,* $g(a) \ne g(b)$. *Then*

$$\left|\Gamma_f(\mu_{XY}, \mu_X \times \mu_Y) - \frac{1}{g(b)-g(a)}\int_a^b f(s)dg(s)\right|$$
$$\le \|f''\|_{\infty,[a,b]}\left\{\int_{\mathbb{R}^2} pq\left(\int_a^b \left|P\left(g\left(\frac{t}{pq}\right), g(s)\right)\right|ds\right)d\lambda\right\}, \quad (9.28)$$

where

$$P(g(z), g(s)) := \begin{cases} \dfrac{g(s)-g(a)}{g(b)-g(a)}, & a \le s \le z, \\[2mm] \dfrac{g(s)-g(b)}{g(b)-g(a)}, & z < s \le b. \end{cases} \quad (9.29)$$

Proof. By Theorem 5 of [43]. $\square$

We give

Examples 9.16 (to Theorem 9.15). Let $f \in C^1([a,b])$ and $g(x) = e^x$, $\ell n\, x$, $\sqrt{x}$, x^α, $\alpha > 1$; $x > 0$. Then by (71)–(74) of [43] we obtain

1)

$$\left| \Gamma_f(\mu_{XY}, \mu_X \times \mu_Y) - \frac{1}{e^b - e^a} \int_a^b f(s)e^s ds \right|$$

$$\leq \frac{\|f'\|_{\infty,[a,b]}}{(e^b - e^a)} \left\{ 2 \int_{\mathbb{R}^2} pq e^{\frac{t}{pq}} d\lambda - e^a(2 - a) + e^b(b - 2) \right\}, \qquad (9.30)$$

2)

$$\left| \Gamma_f(\mu_{XY}, \mu_X \times \mu_Y) - \left(\ell n \left(\frac{b}{a} \right) \right)^{-1} \int_a^b \frac{f(s)}{s} ds \right|$$

$$\leq \|f'\|_{\infty,[a,b]} \left(\ell n \left(\frac{b}{a} \right) \right)^{-1} \left\{ 2 \int_{\mathbb{R}^2} t\, \ell n\, \frac{t}{pq} d\lambda - \ell n(ab) + (a + b - 2) \right\}, \qquad (9.31)$$

3)

$$\left| \Gamma_f(\mu_{XY}, \mu_X \times \mu_Y) - \frac{1}{2(\sqrt{b} - \sqrt{a})} \int_a^b \frac{f(s)}{\sqrt{s}} ds \right|$$

$$\leq \frac{\|f'\|_{\infty,[a,b]}}{(\sqrt{b} - \sqrt{a})} \left\{ \frac{4}{3} \int_{\mathbb{R}^2} (pq)^{-1/2} t^{3/2} d\lambda + \frac{(a^{3/2} + b^{3/2})}{3} - (\sqrt{a} + \sqrt{b}) \right\}, \qquad (9.32)$$

4) $(\alpha > 1)$

$$\left| \Gamma_f(\mu_{XY}, \mu_X \times \mu_Y) - \frac{\alpha}{(b^\alpha - a^\alpha)} \int_a^b f(s)s^{\alpha-1} ds \right|$$

$$\leq \frac{\|f'\|_{\infty,[a,b]}}{(b^\alpha - a^\alpha)} \left\{ \frac{2}{(\alpha + 1)} \left(\int_{\mathbb{R}^2} (pq)^{-\alpha} t^{\alpha+1} d\lambda \right) \right.$$

$$\left. + \left(\frac{\alpha}{\alpha + 1} \right) (a^{\alpha+1} + b^{\alpha+1}) - (a^\alpha + b^\alpha) \right\}. \qquad (9.33)$$

We continue with

Theorem 9.17. *Let $f \in C^{(2)}([a,b])$, $g \in C([a,b])$ and of bounded variation, $g(a) \neq g(b)$. Then*

$$\left| \Gamma_f(\mu_{XY}, \mu_X \times \mu_Y) - \frac{1}{(g(b) - g(a))} \int_a^b f(s)dg(s) \right.$$

$$\left. - \frac{1}{(g(b) - g(a))} \left(\int_a^b f'(s_1)dg(s_1) \right) \cdot \left(\int_{\mathbb{R}^2} \left(pq \int_a^b P\left(g\left(\frac{t}{pq} \right), g(s) \right) ds \right) d\lambda \right) \right|$$

$$\leq \|f''\|_{\infty,[a,b]} \left(\int_{\mathbb{R}^2} \left(pq \left(\int_a^b \int_a^b \left| P\left(g\left(\frac{t}{pq} \right), g(s) \right) \right| P(g(s), g(s_1)) ds_1 ds \right) \right) d\lambda \right).$$

$$(9.34)$$

Proof. Apply (79) of [43].   □

Remark 9.18. Next we define

$$P^*(z, s) := \begin{cases} s - a, & s \in [a, z] \\ s - b, & s \in (z, b]. \end{cases} \tag{9.35}$$

Let $f \in C^1([a, b])$. Then by (25) of [43] we get the representation

$$\Gamma_f(\mu_{XY}, \mu_X \times \mu_Y) = \frac{1}{b - a} \left(\int_a^b f(s)ds + \mathcal{R}_1 \right),$$

where according to (26) of [43] we have

$$\mathcal{R}_1 = \int_{\mathbb{R}^2} p(x)q(y) \left(\int_a^b P^* \left(\frac{t(x, y)}{p(x)q(y)}, s \right) f'(s)ds \right) d\lambda(x, y). \tag{9.36}$$

Let again $f \in C^1([a, b])$, $g \in C([a, b])$ and of bounded variation, $g(a) \neq g(b)$. Then by (47) of [43] we get the representation

$$\Gamma_f(\mu_{XY}, \mu_X \times \mu_Y) = \frac{1}{(g(b) - g(a))} \int_a^b f(s)dg(s) + \mathcal{R}_4, \tag{9.37}$$

where by (48) of [43] we obtain

$$\mathcal{R}_4 = \int_{\mathbb{R}^2} pq \left(\int_a^b P \left(g \left(\frac{t}{pq} \right), g(s) \right) f'(s)ds \right) d\lambda. \tag{9.38}$$

Let $\tilde{\alpha}, \beta > 1$: $\frac{1}{\tilde{\alpha}} + \frac{1}{\beta} = 1$, then by (89) of [43] we have

$$|\mathcal{R}_1| \leq \|f'\|_{\tilde{\alpha}, [a,b]} \int_{\mathbb{R}^2} p(x)q(y) \left\| P^* \left(\frac{t(x, y)}{p(x)q(y)}, \cdot \right) \right\|_{\beta, [a,b]} d\lambda(x, y), \tag{9.39}$$

and by (90) of [43] we get

$$|\mathcal{R}_1| \leq \|f'\|_{1, [a,b]} \int_{\mathbb{R}^2} p(x)q(y) \left\| P^* \left(\frac{t(x, y)}{p(x)q(y)}, \cdot \right) \right\|_{\infty, [a,b]} d\lambda(x, y). \tag{9.40}$$

Furthermore by (95) and (96) of [43] we find

$$|\mathcal{R}_4| \leq \|f'\|_{\tilde{\alpha}, [a,b]} \int_{\mathbb{R}^2} p(x)q(y) \left\| P \left(g \left(\frac{t(x, y)}{p(x)q(y)} \right), g(\cdot) \right) \right\|_{\beta, [a,b]} d\lambda(x, y), \tag{9.41}$$

and

$$|\mathcal{R}_4| \leq \|f'\|_{1, [a,b]} \int_{\mathbb{R}^2} p(x)q(y) \left\| P \left(g \left(\frac{t(x, y)}{p(x)q(y)} \right), g(\cdot) \right) \right\|_{\infty, [a,b]} d\lambda(x, y). \tag{9.42}$$

Remark 9.19. Here $0 < a < b$ with $1 \in [a, b]$, and $\frac{t}{pq} \in [a, b]$ a.e. on $\mathbb{R}^2$.

(i) By (105) of [43] for $f \in C^1([a, b])$ we obtain

$$|\mathcal{R}_1| \leq \|f'\|_{1, [a,b]} \int_{\mathbb{R}^2} p(x)q(y) \max \left(\frac{t(x, y)}{p(x)q(y)} - a, b - \frac{t(x, y)}{p(x)q(y)} \right) d\lambda(x, y). \tag{9.43}$$

(ii) Let g be strictly increasing and continuous over $[a, b]$, e.g., $g(x) = e^x$, $\ell n\, x$, $\sqrt{x}$, x^α with $\alpha > 1$, and $x > 0$ whenever is needed. Also $f \in C^1([a, b])$ and as in this chapter. Then by (106) of [43] we find

$$|\mathcal{R}_4| \leq \frac{\|f'\|_{1,[a,b]}}{(g(b) - g(a))} \int_{\mathbb{R}^2} p(x)q(y) \max\left(g\left(\frac{t(x, y)}{p(x)q(y)} \right) \right.$$
$$\left. - g(a), g(b) - g\left(\frac{t(x, y)}{p(x)q(y)} \right) \right) d\lambda(x, y). \tag{9.44}$$

In particular via (107)–(110) of [43] we obtain
1)

$$|\mathcal{R}_4| \leq \frac{\|f'\|_{1,[a,b]}}{(e^b - e^a)} \int_{\mathbb{R}^2} p(x)q(y) \max\left(e^{\frac{t(x,y)}{p(x)q(y)}} - e^a, e^b - e^{\frac{t(x,y)}{p(x)q(y)}} \right) d\lambda(x, y), \tag{9.45}$$

2)

$$|\mathcal{R}_4| \leq \frac{\|f'\|_{1,[a,b]}}{(\ell n\, b - \ell n\, a)} \int_{\mathbb{R}^2} p(x)q(y) \max\left(\ell n\left(\frac{t(x, y)}{p(x)q(y)} \right) \right.$$
$$\left. - \ell n\, a, \ell n\, b - \ell n\left(\frac{t(x, y)}{p(x)q(y)} \right) \right) d\lambda(x, y), \tag{9.46}$$

3)

$$|\mathcal{R}_4| \leq \frac{\|f'\|_{1,[a,b]}}{\sqrt{b} - \sqrt{a}} \int_{\mathbb{R}^2} p(x)q(y) \max\left(\sqrt{\frac{t(x, y)}{p(x)q(y)}} \right.$$
$$\left. - \sqrt{a}, \sqrt{b} - \sqrt{\frac{t(x, y)}{p(x)q(y)}} \right) d\lambda(x, y), \tag{9.47}$$

and finally for $\alpha > 1$ we obtain
4)

$$|\mathcal{R}_4| \leq \frac{\|f'\|_{1,[a,b]}}{b^\alpha - a^\alpha} \int_{\mathbb{R}^2} p(x)q(y) \max\left(\frac{t^\alpha(x, y)}{(p(x)q(y))^\alpha} - a^\alpha, b^\alpha - \frac{t^\alpha(x, y)}{(p(x)q(y))^\alpha} \right) d\lambda(x, y). \tag{9.48}$$

At last

iii) Let $\tilde{\alpha}, \beta > 1$ such that $\frac{1}{\tilde{\alpha}} + \frac{1}{\beta} = 1$, then by (115) of [43] we have

$$|\mathcal{R}_1| \leq \|f'\|_{\tilde{\alpha},[a,b]}$$
$$\cdot \int_{\mathbb{R}^2} \left(\sqrt[\beta]{\frac{(t(x, y) - ap(x)q(y))^{\beta+1} + (bp(x)q(y) - t(x, y))^{\beta+1}}{(\beta + 1)p(x)q(y)}} \right) d\lambda(x, y). \tag{9.49}$$

Part IV. Here we apply results from [41].

We start with

Theorem 9.20. *Let* $0 < a < 1 < b$, f *as in this chapter and* $f \in C^n([a,b])$, $n \geq 1$ *with* $|f^{(n)}(s) - f^{(n)}(1)|$ *be a convex function in* s. *Let* $0 < h < \min(1-a, b-1)$ *be fixed. Then*

$$\Gamma_f(\mu_{XY}, \mu_X \times \mu_Y) \leq \sum_{k=2}^{n} \frac{|f^{(k)}(1)|}{k!} \left| \int_{\mathbb{R}^2} (pq)^{1-k}(t-pq)^k d\lambda \right|$$

$$+ \frac{\omega_1(f^{(n)}, h)}{(n+1)!h} \int_{\mathbb{R}^2} (pq)^{-n}|t-pq|^{n+1} d\lambda. \tag{9.50}$$

Here ω_1 *is the usual first modulus of continuity. If* $f^{(k)}(1) = 0$, $k = 2, \ldots, n$, *then*

$$\Gamma_f(\mu_{XY}, \mu_X \times \mu_Y) \leq \frac{\omega_1(f^{(n)}, h)}{(n+1)!h} \int_{\mathbb{R}^2} (pq)^{-n}|t-pq|^{n+1} d\lambda. \tag{9.51}$$

Inequalities (9.50) and (9.51) when n *is even are attained by*

$$\tilde{f}(s) = \frac{|s-1|^{n+1}}{(n+1)!}, \quad a \leq s \leq b. \tag{9.52}$$

Proof. Apply Theorem 1 of [41]. $\square$

We continue with the general

Theorem 9.21. *Let* $0 < a \leq 1 \leq b$, f *as in this chapter and* $f \in C^n([a,b])$, $n \geq 1$. *Assume that* $\omega_1(f^{(n)}, \delta) \leq w$, *where* $0 < \delta \leq b-a$, $w > 0$. *Let* $x \in \mathbb{R}$ *and denote by*

$$\phi_n(x) := \int_0^{|x|} \left\lceil \frac{s}{\delta} \right\rceil \frac{(|x|-s)^{n-1}}{(n-1)!} \, ds, \tag{9.53}$$

where $\lceil \cdot \rceil$ *is the ceiling of the number. Then*

$$\Gamma_f(\mu_{XY}, \mu_X \times \mu_Y) \leq \sum_{k=2}^{n} \frac{|f^{(k)}(1)|}{k!} \left| \int_{\mathbb{R}^2} (pq)^{1-k}(t-pq)^k d\lambda \right|$$

$$+ w \int_{\mathbb{R}^2} pq\phi_n \left(\frac{t-pq}{pq} \right) d\lambda. \tag{9.54}$$

Inequality (9.54) is sharp, namely it is attained by the function

$$\tilde{f}_n(s) := w\phi_n(s-1), \quad a \leq s \leq b, \tag{9.55}$$

when n *is even.*

Proof. Apply Theorem 2 of [41]. $\square$

It follows

Corollary 9.22 (to Theorem 9.21). *It holds* $(0 < \delta \leq b-a)$

$$\Gamma_f(\mu_{XY}, \mu_X \times \mu_Y) \leq \sum_{k=2}^{n} \frac{|f^{(k)}(1)|}{k!} \left| \int_{\mathbb{R}^2} (pq)^{1-k}(t-pq)^k d\lambda \right|$$

$$+ \omega_1(f^{(n)}, \delta) \left\{ \frac{1}{(n+1)!\delta} \int_{\mathbb{R}^2} (pq)^{-n}|t-pq|^{n+1} d\lambda \right.$$

$$+ \frac{1}{2n!} \int_{\mathbb{R}^2} (pq)^{1-n}|t-pq|^n d\lambda$$

$$+ \left. \frac{\delta}{8(n-1)!} \int_{\mathbb{R}^2} (pq)^{2-n}|t-pq|^{n-1} d\lambda \right\}, \tag{9.56}$$

and

$$\Gamma_f(\mu_{XY}, \mu_X \times \mu_Y) \leq \sum_{k=2}^{n} \frac{|f^{(k)}(1)|}{k!} \left| \int_{\mathbb{R}^2} (pq)^{1-k}(t-pq)^k d\lambda \right|$$

$$+ \omega_1(f^{(n)}, \delta) \left\{ \frac{1}{n!} \int_{\mathbb{R}^2} (pq)^{1-n}|t-pq|^n d\lambda \right.$$

$$+ \left. \frac{1}{n!(n+1)\delta} \int_{\mathbb{R}^2} (pq)^{-n}|t-pq|^{n+1} d\lambda \right\}. \qquad (9.57)$$

In particular we have

$$\Gamma_f(\mu_{XY}, \mu_X \times \mu_Y) \leq \omega_1(f', \delta) \left\{ \frac{1}{2\delta} \int_{\mathbb{R}^2} (pq)^{-1}(t-pq)^2 d\lambda \right.$$

$$+ \left. \frac{1}{2} \int_{\mathbb{R}^2} |t-pq| d\lambda + \frac{\delta}{8} \right\}, \qquad (9.58)$$

and

$$\Gamma_f(\mu_{XY}, \mu_X \times \mu_Y) \leq \omega_1(f', \delta) \left\{ \int_{\mathbb{R}^2} |t-pq| d\lambda + \frac{1}{2\delta} \int_{\mathbb{R}^2} (pq)^{-1}(t-pq)^2 d\lambda \right\}. \quad (9.59)$$

Proof. By Corollary 1 of [41]. $\qquad\qquad\square$

Also we have

Corollary 9.23 (to Theorem 9.21). *Assume here that $f^{(n)}$ is of Lipschitz type of order α, $0 < \alpha \leq 1$, i.e.*

$$\omega_1(f^{(n)}, \delta) \leq K\delta^\alpha, \qquad K > 0, \qquad (9.60)$$

for any $0 < \delta \leq b - a$. Then

$$\Gamma_f(\mu_{XY}, \mu_X \times \mu_Y) \leq \sum_{k=2}^{n} \frac{|f^{(k)}(1)|}{k!} \left| \int_{\mathbb{R}^2} (pq)^{1-k}(t-pq)^k d\lambda \right|$$

$$+ \frac{K}{\prod\limits_{i=1}^{n}(\alpha+i)} \int_{\mathbb{R}^2} (pq)^{1-n-\alpha}|t-pq|^{n+\alpha} d\lambda. \qquad (9.61)$$

When n is even (9.61) is attained by $f^*(x) = c|x-1|^{n+\alpha}$, where $c := K/\left(\prod\limits_{i=1}^{n}(\alpha+i) \right) > 0$.

Proof. Application of Corollary 2 of [41]. $\qquad\qquad\square$

Next comes

Corollary 9.24 (to Theorem 9.21). *Assume that*

$$b - a \geq \int_{\mathbb{R}^2} (pq)^{-1}t^2 d\lambda - 1 > 0. \qquad (9.62)$$

then

$$\Gamma_f(\mu_{XY}, \mu_X \times \mu_Y) \leq \omega_1 \left(f', \left(\int_{\mathbb{R}^2} (pq)^{-1} t^2 d\lambda - 1 \right) \right) \left\{ \int_{\mathbb{R}^2} |t - pq| d\lambda + \frac{1}{2} \right\}.$$
(9.63)

Proof. Application of Corollary 3 of [41]. $\square$

Corollary 9.25 (to Theorem 9.21). *Assume that*

$$\int_{\mathbb{R}^2} |t - pq| d\lambda > 0.$$
(9.64)

Then

$$\Gamma_f(\mu_{XY}, \mu_X \times \mu_Y) \leq \omega_1 \left(f', \int_{\mathbb{R}^2} |t - pq| d\lambda \right) \left\{ \int_{\mathbb{R}^2} |t - pq| d\lambda + \frac{b-a}{2} \right\}$$
$$\leq \frac{3}{2}(b-a)\omega_1 \left(f', \int_{\mathbb{R}^2} |t - pq| d\lambda \right).$$
(9.65)

Proof. Application of Corollary 5 of [41]. $\square$

We present

Proposition 9.26. *Let f as in this chapter and $f \in C([a,b])$.*
i) *Assume that*

$$\int_{\mathbb{R}^2} |t - pq| d\lambda > 0.$$
(9.66)

Then

$$\Gamma_f(\mu_{XY}, \mu_X \times \mu_Y) \leq 2\omega_1 \left(f, \int_{\mathbb{R}^2} |t - pq| d\lambda \right).$$
(9.67)

ii) *Let $r > 0$ and*

$$b - a \geq r \int_{\mathbb{R}^2} |t - pq| d\lambda > 0.$$
(9.68)

Then

$$\Gamma_f(\mu_{XY}, \mu_X \times \mu_Y) \leq \left(1 + \frac{1}{r} \right) \omega_1 \left(f, r \int_{\mathbb{R}^2} |t - pq| d\lambda \right).$$
(9.69)

Proof. Application of Proposition 1 of [41]. $\square$

Also we give

Proposition 9.27. *Let f as in this setting and f is a Lipschitz function of order α, $0 < \alpha \leq 1$, i.e., there exists $K > 0$ such that*

$$|f(x) - f(y)| \leq K|x - y|^\alpha, \quad all \ x, y \in [a, b].$$
(9.70)

Then

$$\Gamma_f(\mu_{XY}, \mu_X \times \mu_Y) \leq K \int_{\mathbb{R}^2} (pq)^{1-\alpha} |t - pq|^\alpha d\lambda.$$
(9.71)

Proof. By Proposition 2 of [41]. $\qquad\square$

Next we present some alternative type of results. We start with

Theorem 9.28. *Let* $f \in C^{n+1}([a,b])$, $n \in \mathbb{N}$, *then we get the representation*

$$\Gamma_f(\mu_{XY}, \mu_X \times \mu_Y) = \sum_{k=0}^{n-1} \frac{(-1)^k}{(k+1)!} \int_{\mathbb{R}^2} f^{(k+1)}\left(\frac{t}{pq}\right)(pq)^{-k}(t-pq)^{k+1}d\lambda + \psi_2, \quad (9.72)$$

where

$$\psi_2 = \frac{(-1)^n}{n!} \int_{\mathbb{R}^2}(pq)\left(\int_1^{\frac{t}{pq}}(s-1)^n f^{(n+1)}(s)ds\right)d\lambda. \quad (9.73)$$

Proof. By Theorem 12 of [41]. $\qquad\square$

Next we estimate ψ_2. We give

Theorem 9.29. *Let* $f \in C^{n+1}([a,b])$, $n \in \mathbb{N}$. *Then*

$$|\psi_2| \le \min \ of \begin{cases} B_1 := \dfrac{\|f^{(n+1)}\|_{\infty,[a,b]}}{(n+1)!} \displaystyle\int_{\mathbb{R}^2}(pq)^{-n}|t-pq|^{n+1}d\lambda, \\[2ex] B_2 := \dfrac{\|f^{(n+1)}\|_{1,[a,b]}}{n!} \displaystyle\int_{\mathbb{R}^2}(pq)^{1-n}|t-pq|^{n}d\lambda, \\[2ex] \qquad and\ for\ \tilde\alpha, \beta > 1: \dfrac{1}{\tilde\alpha} + \dfrac{1}{\beta} = 1, \\[2ex] B_3 := \dfrac{\|f^{(n+1)}\|_{\tilde\alpha,[a,b]}}{n!(n\beta+1)^{1/\beta}} \displaystyle\int_{\mathbb{R}^2}(pq)^{1-n-\frac{1}{\beta}}|t-pq|^{n+\frac{1}{\beta}}d\lambda. \end{cases} \quad (9.74)$$

Also it holds

$$\Gamma_f(\mu_{XY}, \mu_X \times \mu_Y) \le \sum_{k=0}^{n-1} \frac{1}{(k+1)!}\left|\int_{\mathbb{R}^2} f^{(k+1)}\left(\frac{t}{pq}\right)(pq)^{-k}(t-pq)^{k+1}d\lambda\right|$$
$$+ \min(B_1, B_2, B_3). \quad (9.75)$$

Proof. See Theorem 13 of [41]. $\qquad\square$

We have

Corollary 9.30 (to Theorem 9.29). *Case of* $n = 1$. *It holds*

$$|\psi_2| \le \min \ of \begin{cases} B_{1,1} := \dfrac{\|f''\|_{\infty,[a,b]}}{2} \displaystyle\int_{\mathbb{R}^2}(pq)^{-1}(t-pq)^2 d\lambda, \\[2ex] B_{2,1} := \|f''\|_{1,[a,b]} \displaystyle\int_{\mathbb{R}^2}|t-pq|d\lambda, \\[2ex] \qquad and\ for\ \tilde\alpha, \beta > 1, \dfrac{1}{\tilde\alpha} + \dfrac{1}{\beta} = 1, \\[2ex] B_{3,1} := \dfrac{\|f''\|_{\tilde\alpha,[a,b]}}{(\beta+1)^{1/\beta}} \displaystyle\int_{\mathbb{R}^2}(pq)^{-1/\beta}|t-pq|^{1+\frac{1}{\beta}}d\lambda. \end{cases} \quad (9.76)$$

Also it holds

$$\Gamma_f(\mu_{XY}, \mu_X \times \mu_Y) \leq \left| \int_{\mathbb{R}^2} f'\left(\frac{t}{pq}\right)(t - pq)d\lambda \right| + \min(B_{1,1}, B_{2,1}, B_{3,1}). \qquad (9.77)$$

Proof. See Corollary 8 of [41]. $\square$

We further present

Theorem 9.31. *Let $f \in C^{n+1}([a,b])$, $n \in \mathbb{N}$, then we get the representation*

$$\Gamma_f(\mu_{XY}, \mu_X \times \mu_Y) = \sum_{k=0}^{n-1} \frac{(-1)^k}{(k+1)!} \int_{\mathbb{R}^2} f^{(k+1)}\left(\frac{t}{pq}\right)(pq)^{-k}(t-pq)^{k+1}d\lambda$$

$$+ \frac{(-1)^n}{(n+1)!} f^{(n+1)}(1) \int_{\mathbb{R}^2} (pq)^{-n}(t-pq)^{n+1}d\lambda + \psi_4, \qquad (9.78)$$

where

$$\psi_4 = \int_{\mathbb{R}^2} pq\psi_3\left(\frac{t}{pq}\right)d\lambda, \qquad (9.79)$$

with

$$\psi_3(x) := \frac{(-1)^n}{n!} \int_1^x (s-1)^n (f^{(n+1)}(s) - f^{(n+1)}(1))ds. \qquad (9.80)$$

Proof. Based on Theorem 14 of [41]. $\square$

Next we present estimations of ψ_4.

Theorem 9.32. *Let $f \in C^{n+1}([a,b])$, $n \in \mathbb{N}$, and*

$$0 < \frac{1}{(n+2)} \int_{\mathbb{R}^2} (pq)^{-n-1}|t-pq|^{n+2}d\lambda \leq b - a. \qquad (9.81)$$

Then (9.78) is again valid and

$$|\psi_4| \leq \frac{1}{n!}\omega_1\left(f^{(n+1)}, \frac{1}{n+2}\int_{\mathbb{R}^2}(pq)^{-n-1}|t-pq|^{n+2}d\lambda\right)$$

$$\cdot \left(1 + \frac{1}{n+1}\int_{\mathbb{R}^2}(pq)^{-n}|t-pq|^{n+1}d\lambda\right). \qquad (9.82)$$

Proof. By Theorem 15 of [41]. $\square$

Also we have

Theorem 9.33. *Let $f \in C^{n+1}([a,b])$, $n \in \mathbb{N}$, and $|f^{(n+1)}(s) - f^{(n+1)}(1)|$ is convex in s and*

$$0 < \frac{1}{n!(n+2)} \int_{\mathbb{R}^2}(pq)^{-n-1}|t-pq|^{n+2}d\lambda \leq \min(1-a, b-1). \qquad (9.83)$$

Then (9.78) is again valid and

$$|\psi_4| \leq \omega_1\left(f^{(n+1)}, \frac{1}{n!(n+2)}\int_{\mathbb{R}^2}(pq)^{-n-1}|t-pq|^{n+2}d\lambda\right). \qquad (9.84)$$

The last (9.84) is attained by

$$\hat{f}(s) := \frac{(s-1)^{n+2}}{(n+2)!}, \quad a \le s \le b \tag{9.85}$$

when n is even.

Proof. By Theorem 16 of [41]. $\qquad\square$

When $n = 1$ we obtain

Corollary 9.34 (to Theorem 9.32). *Let $f \in C^2([a,b])$ and*

$$0 < \frac{1}{3} \int_{\mathbb{R}^2} (pq)^{-2}|t - pq|^3 d\lambda \le b - a. \tag{9.86}$$

Then

$$\Gamma_f(\mu_{XY}, \mu_X \times \mu_Y) = \int_{\mathbb{R}^2} f'\left(\frac{t}{pq}\right)(t - pq)d\lambda - \frac{f''(1)}{2} \int_{\mathbb{R}^2} (pq)^{-1}(t - pq)^2 d\lambda + \psi_{4,1}, \tag{9.87}$$

where

$$\psi_{4,1} := \int_{\mathbb{R}^2} pq\psi_{3,1}\left(\frac{t}{pq}\right) d\lambda, \tag{9.88}$$

with

$$\psi_{3,1}(x) := -\int_1^x (s - 1)(f''(s) - f''(1))ds, \quad x \in [a,b]. \tag{9.89}$$

Furthermore it holds

$$|\psi_{4,1}| \le \omega_1\left(f'', \frac{1}{3} \int_{\mathbb{R}^2} (pq)^{-2}|t - pq|^3 d\lambda\right) \frac{1}{2}\left(\int_{\mathbb{R}^2} (pq)^{-1}t^2 d\lambda + 1\right). \tag{9.90}$$

Proof. See Corollary 9 of [41]. $\qquad\square$

Also we have

Corollary 9.35 (to Theorem 9.33). *Let $f \in C^2([a,b])$, and $|f^{(2)}(s) - f^{(2)}(1)|$ is convex in s and*

$$0 < \frac{1}{3} \int_{\mathbb{R}^2} (pq)^{-2}|t - pq|^3 d\lambda \le \min(1 - a, b - 1). \tag{9.91}$$

Then again (9.87) is true and

$$|\psi_{4,1}| \le \omega_1\left(f'', \frac{1}{3} \int_{\mathbb{R}^2} (pq)^{-2}|t - pq|^3 d\lambda\right). \tag{9.92}$$

Proof. See Corollary 10 of [41]. $\qquad\square$

The last main result of the chapter follows.

Theorem 9.36. *Let $f \in C^{n+1}([a,b])$, $n \in \mathbb{N}$. Additionally assume that*

$$|f^{(n+1)}(x) - f^{(n+1)}(y)| \le K|x - y|^\alpha, \tag{9.93}$$

$K > 0$, $0 < \alpha \leq 1$, all $x, y \in [a, b]$ with $(x - 1)(y - 1) \geq 0$. Then

$$\Gamma_f(\mu_{XY}, \mu_X \times \mu_Y) = \sum_{k=0}^{n-1} \frac{(-1)^k}{(k+1)!} \int_{\mathbb{R}^2} f^{(k+1)}\left(\frac{t}{pq}\right)(pq)^{-k}(t - pq)^{k+1}d\lambda$$

$$+ \frac{(-1)^n}{(n+1)!} f^{(n+1)}(1) \int_{\mathbb{R}^2} (pq)^{-n}(t - pq)^{n+1}d\lambda + \psi_4. \quad (9.94)$$

Here we find that

$$|\psi_4| \leq \frac{K}{n!(n + \alpha + 1)} \int_{\mathbb{R}^2} (pq)^{-n-\alpha}|t - pq|^{n+\alpha+1}d\lambda. \quad (9.95)$$

Inequality (9.95) is attained when n is even by

$$f^*(x) = \tilde{c}|x - 1|^{n+\alpha+1}, \quad x \in [a, b] \quad (9.96)$$

where

$$\tilde{c} := \frac{K}{\displaystyle\prod_{j=0}^{n}(n + \alpha + 1 - j)}. \quad (9.97)$$

Proof. By Theorem 17 of [41]. $\square$

    Finally we give

Corollary 9.37 (to Theorem 9.36). *Let $f \in C^2([a, b])$. Additionally assume that*

$$|f''(x) - f''(y)| \leq K|x - y|^\alpha, \quad K > 0, \ 0 < \alpha \leq 1, \ \text{all } x, y \in [a, b] \quad (9.98)$$

with $(x - 1)(y - 1) \geq 0$. Then

$$\Gamma_f(\mu_{XY}, \mu_X \times \mu_Y) = \int_{\mathbb{R}^2} f'\left(\frac{t}{pq}\right)(t - pq)d\lambda - \frac{f''(1)}{2} \int_{\mathbb{R}^2} (pq)^{-1}(t - pq)^2 d\lambda + \psi_{4,1}.$$

$$(9.99)$$

Here

$$\psi_{4,1} := \int_{\mathbb{R}^2} pq\psi_{3,1}\left(\frac{t}{pq}\right)d\lambda \quad (9.100)$$

with

$$\psi_{3,1}(x) := -\int_1^x (s - 1)(f''(s) - f''(1))ds, \quad x \in [a, b]. \quad (9.101)$$

It holds that

$$|\psi_{4,1}| \leq \frac{K}{(\alpha + 2)} \int_{\mathbb{R}^2} (pq)^{-1-\alpha}|t - pq|^{\alpha+2}d\lambda. \quad (9.102)$$

Proof. See Corollary 11 of [41]. $\square$

Chapter 10

Optimal Estimation of Discrete Csiszar f-Divergence

In this chapter are established a number of sharp and nearly optimal discrete probabilistic inequalities giving various types of order of approximation of Discrete Csiszar f-divergence between two discrete probability measures. This distance is the most essential and general tool for the comparison of these measures. We give also various types of representation of this discrete Csiszar distance and then we estimate tight their remainders, that is leading to very close discrete probabilistic inequalities involving various norms. We give also plenty of interesting applications. This treatment follows [45].

10.1 Background

The basic general background here has as follows.

Let f by a convex function from $(0, +\infty)$ into $\mathbb{R}$ which is strictly convex at 1 with $f(1) = 0$.

Let (X, A, λ) be a measure space, where λ is a finite or a σ-finite measure on (X, A). Let μ_1, μ_2 be two probability measures on (X, A) such that $\mu_1 \ll \lambda$, $\mu_2 \ll \lambda$ (absolutely continuous), e.g. $\lambda = \mu_1 + \mu_2$.

Denote by $p = \frac{d\mu_1}{d\lambda}$, $q = \frac{d\mu_2}{d\lambda}$ the (densities) Radon–Nikodym derivatives of μ_1, μ_2 with respect to λ Typically we assume that

$$0 < a \le \frac{p}{q} \le b, \quad \text{a.e. on } X \text{ and } a \le 1 \le b.$$

The quantity

$$\Gamma_f(\mu_1, \mu_2) = \int_X q(x) f\left(\frac{p(x)}{q(x)}\right) d\lambda(x), \tag{10.1}$$

was introduced by I. Csiszar in 1967, see [140], and is called f-*divergence* of the probability measures μ_1 and μ_2.

By Lemma 1.1 of [140], the integral (10.1) is well-defined and $\Gamma_f(\mu_1, \mu_2) \ge 0$ with equality only when $\mu_1 = \mu_2$. Furthermore by [140], [42] $\Gamma_f(\mu_1, \mu_2)$ does not depend on the choice of λ. The concept of f-divergence was introduced first in [139] as a generalization of Kullback's "*information for discrimination*" or I-*divergence* (*generalized entropy*) [244], [243], and of Rényi's "*information gain*" (*I-divergence*) *of order* α [306]. In fact the *I-divergence of order* 1 equals

$$\Gamma_{u \log_2 u}(\mu_1, \mu_2).$$

The choice $f(u) = (u-1)^2$ produces again a known measure of difference of distributions that is called χ^2-*divergence*. Of course the *total variation* distance

$$|\mu_1 - \mu_2| = \int_X |p(x) - q(x)| \, d\lambda(x)$$

is equal to $\Gamma_{|u-1|}(\mu_1, \mu_2)$. Here by assuming $f(1) = 0$ we can consider $\Gamma_f(\mu_1, \mu_2)$ the f-divergence as a measure of the difference between the probability measures μ_1, μ_2.

This is a chapter where we apply to the discrete case a great variety of interesting results of the above continuous case coming from the author's earlier but recent articles [37], [42], [43], [41].

Throughout this chapter we use the following. Let again f be a convex function from $(0, +\infty)$ into $\mathbb{R}$ which is strictly convex at 1 with $f(1) = 0$. Let $X = \{x_1, x_2, \ldots\}$ be a countable arbitrary set and A be the set of all subsets of X. Then any probability distribution μ on (X, A) is uniquely determined by the sequence $P = \{p_1, p_2, \ldots\}$, $p_i = \mu(\{x_i\})$, $i = 1, \ldots$. We call M the set of all distributions on (X, A) such that all p_i, $i = 1, \ldots$, are positive.

Denote here by $\mu_1, \mu_2 \in M$ the distributions with corresponding sequences $P = \{p_1, \ldots\}$, $Q = \{q_1, q_2, \ldots\}$ ($p_i > 0$, $q_i > 0$, $i = 1, 2, \ldots$).

We consider here only the set $M^* \subset M$ of distributions μ_1, μ_2 such that

$$0 < a \le \frac{p_i}{q_i} \le b, \quad i = 1, \ldots, \quad \text{on } X \text{ and } a \le 1 \le b, \tag{10.2}$$

where a, b are fixed.

In our discrete case we choose λ such that $\lambda(\{x_i\}) = 1$, all $i = 1, 2, \ldots$. Clearly here $\mu_1 \ll \lambda$, $\mu_2 \ll \lambda$. Therefore by (10.1) we obtain the special case of

$$\Gamma_f(\mu_1, \mu_2) = \sum_{i=1}^{\infty} q_i f\left(\frac{p_i}{q_i}\right). \tag{10.3}$$

So (10.3) is the discrete f-*divergence* or *Csiszar distance* between two discrete distributions from M^*, see also [139], [141]. The quantity Γ_f as in (10.3) is also called *generalized measure of information*.

The above discrete general concept of $\Gamma_f(\mu_1, \mu_2)$ given by (10.3) incorporates a large variety of discrete specific Γ_f distances such of Kullback–Leibler so called information divergence, Hellinger, Rényi α-order entropy, χ^2, Variational, Triangular discrimination, Bhattacharyya, Jeffreys distance. For all these and a whole of very interesting general theory please see S. Dragomir's important book [152], especially his particular article [153]. In there S. Dragomir deals with finite sums similar to (10.3), while we deal with finite or infinite ones.

The problem of finding and estimating the proper distance (or difference or discrimination) of two discrete probability distributions is one of the major ones in Probability Theory. The above discrete f-divergence measure (10.3) for various specific f's, of finite or infinite sums have been applied a lot to Anthropology, Genetics, Finance, Economics, Political Science, Biology, Approximation of Probability distributions, Signal Processing and Pattern Recognition.

In the following the proofs of all formulated and presented results are based a lot on this Section 10.1, especially on the transfer from the continuous to the discrete case.

10.2　Results

Part I. Here we apply results from [37].

We present

Theorem 10.1. *Let $f \in C^1([a,b])$ and $\mu_1, \mu_2 \in M^*$. Then*

$$\Gamma_f(\mu_1, \mu_2) \le \|f'\|_{\infty,[a,b]} \left(\sum_{i=1}^{\infty} |p_i - q_i| \right). \tag{10.4}$$

Inequality (10.4) is sharp. The optimal function is $f^*(y) = |y - 1|^{\alpha}$, $\alpha > 1$ under the condition, either
　　i) $\max(b - 1, 1 - a) < 1$ *and*

$$C := \sup(Q) < +\infty, \tag{10.5}$$

or
　ii)

$$|p_i - q_i| \le q_i \le 1, \quad i = 1, 2, \dots . \tag{10.6}$$

Proof. Based on Theorem 1 of [37]. □

Next we give the more general

Theorem 10.2. *Let $f \in C^{n+1}([a,b])$, $n \in \mathbb{N}$, such that $f^{(k)}(1) = 0$, $k = 0, 2, \dots, n$, and $\mu_1, \mu_2 \in M^*$. Then*

$$\Gamma_f(\mu_1, \mu_2) \le \frac{\|f^{(n+1)}\|_{\infty,[a,b]}}{(n+1)!} \left(\sum_{i=1}^{\infty} q_i^{-n} |p_i - q_i|^{n+1} \right). \tag{10.7}$$

Inequality (10.7) is sharp. The optimal function is $\tilde{f}(y) := |y - 1|^{n+\alpha}$, $\alpha > 1$ under the condition (i) or (ii) of Theorem 10.1.

Proof. Based on Theorem 2 of [37]. □

As a special case we have

Corollary 10.3 (to Theorem 10.2). *Let* $f \in C^2([a,b])$ *and* $\mu_1, \mu_2 \in M^*$. *Then*

$$\Gamma_f(\mu_1, \mu_2) \leq \frac{\|f^{(2)}\|_{\infty,[a,b]}}{2} \left(\sum_{i=1}^{\infty} q_i^{-1}(p_i - q_i)^2 \right). \tag{10.8}$$

Inequality (10.8) is sharp. The optimal function is $\tilde{f}(y) := |y - 1|^{1+\alpha}$, $\alpha > 1$ *under the condition* (i) *or* (ii) *of Theorem 10.1.*

Proof. By Corollary 1 of [37]. $\qquad\square$

Next we connect discrete Csiszar f-divergence to the usual first modulus of continuity ω_1.

Theorem 10.4. *Assume that*

$$0 < h := \sum_{i=1}^{\infty} |p_i - q_i| \leq \min(1 - a, b - 1). \tag{10.9}$$

Let $\mu_1, \mu_2 \in M^*$. *Then*

$$\Gamma_f(\mu_1, \mu_2) \leq \omega_1(f, h). \tag{10.10}$$

Inequality (10.10) is sharp, namely it is attained by $f^*(x) := |x - 1|$.

Proof. By Theorem 3 of [37]. $\qquad\square$

Part II. Here we apply results from [42]. But first we need some basic concepts from [122], [26].

Let $\nu > 0$, $n := [\nu]$ and $\alpha = \nu - n$ $(0 < \alpha < 1)$. Let $x, x_0 \in [a, b] \subseteq \mathbb{R}$ such that $x \geq x_0$, where x_0 is fixed. Let $f \in C([a, b])$ and define

$$(J_\nu^{x_0} f)(x) := \frac{1}{\Gamma(\nu)} \int_{x_0}^{x} (x - s)^{\nu-1} f(s) ds, \quad x_0 \leq x \leq b, \tag{10.11}$$

the *generalized Riemann–Liouville integral*, where Γ stands for the gamma function. We define the subspace

$$C_{x_0}^\nu([a, b]) := \left\{ f \in C^n([a, b]) \colon J_{1-\alpha}^{x_0} D^n f \in C^1([x_0, b]) \right\}. \tag{10.12}$$

For $f \in C_{x_0}^\nu([a, b])$ we define the *generalized ν-fractional derivative of f over $[x_0, b]$* as

$$D_{x_0}^\nu f := D J_{1-\alpha}^{x_0} f^{(n)} \quad \left(D := \frac{d}{dx} \right). \tag{10.13}$$

We present the following.

Theorem 10.5. *Let* $a < b$, $1 \leq \nu < 2$, $f \in C_a^\nu([a, b])$, $\mu_1, \mu_2 \in M^*$. *Then*

$$\Gamma_f(\mu_1, \mu_2) \leq \frac{\|D_a^\nu f\|_{\infty,[a,b]}}{\Gamma(\nu + 1)} \left(\sum_{i=1}^{\infty} q_i^{1-\nu}(p_i - aq_i)^\nu \right). \tag{10.14}$$

Proof. By Theorem 2 of [42]. $\qquad\square$

The counterpart of the previous result follows.

Theorem 10.6. *Let* $a < b$, $\nu \geq 2$, $n := [\nu]$, $f \in C_a^\nu([a,b])$, $f^{(i)}(a) = 0$, $i = 0, 1, \ldots, n-1$, $\mu_1, \mu_2 \in M^*$. *Then*

$$\Gamma_f(\mu_1, \mu_2) \leq \frac{\|D_a^\nu f\|_{\infty, [a,b]}}{\Gamma(\nu+1)} \left(\sum_{i=1}^{\infty} q_i^{1-\nu}(p_i - aq_i)^\nu \right). \tag{10.15}$$

Proof. By Theorem 3 of [42]. $\qquad\square$

Next we give an $L_{\tilde{\alpha}}$ estimate.

Theorem 10.7. *Let* $a < b$, $\nu \geq 1$, $n := [\nu]$, $f \in C_a^\nu([a,b])$, $f^{(i)}(a) = 0$, $i = 0, 1, \ldots, n-1$, $\mu_1, \mu_2 \in M^*$. *Let* $\tilde{\alpha}, \beta > 1$: $\frac{1}{\tilde{\alpha}} + \frac{1}{\beta} = 1$. *Then*

$$\Gamma_f(\mu_1, \mu_2) \leq \frac{\|D_a^\nu f\|_{\tilde{\alpha}, [a,b]}}{\Gamma(\nu)(\beta(\nu-1)+1)^{1/\beta}} \left(\sum_{i=1}^{\infty} q_i^{2-\nu-\frac{1}{\beta}}(p_i - aq_i)^{\nu-1+\frac{1}{\beta}} \right). \tag{10.16}$$

Proof. By Theorem 4 of [42]. $\qquad\square$

It follows an L_∞ estimate.

Theorem 10.8. *Let* $a < b$, $\nu \geq 1$, $n := [\nu]$, $f \in C_a^\nu([a,b])$, $f^{(i)}(a) = 0$, $i = 0, 1, \ldots, n-1$, $\mu_1, \mu_2 \in M^*$. *Then*

$$\Gamma_f(\mu_1, \mu_2) \leq \frac{\|D_a^\nu f\|_{1, [a,b]}}{\Gamma(\nu)} \left(\sum_{i=1}^{\infty} q_i^{2-\nu}(p_i - aq_i)^{\nu-1} \right). \tag{10.17}$$

Proof. Based on Theorem 5 of [42]. $\qquad\square$

We continue with

Theorem 10.9. *Let* $f \in C^1([a,b])$, $a \neq b$, $\mu_1, \mu_2 \in M^*$. *Then*

$$\Gamma_f(\mu_1, \mu_2) \leq \frac{1}{b-a} \int_a^b f(x)dx - \sum_{i=1}^{\infty} f'\left(\frac{p_i}{q_i}\right) \left(\frac{q_i(a+b)}{2} - p_i\right). \tag{10.18}$$

Proof. Based on Theorem 6 of [42]. $\qquad\square$

Furthermore we obtain

Theorem 10.10. *Let* n *odd and* $f \in C^{n+1}([a,b])$, *such that* $f^{(n+1)} \geq 0 \ (\leq 0)$, $\mu_1, \mu_2 \in M^*$. *Then*

$$\Gamma_f(\mu_1, \mu_2) \leq (\geq) \frac{1}{b-a} \int_a^b f(x)dx - \sum_{i=1}^{n} \frac{1}{(i+1)!} \left(\sum_{k=0}^{i} \left(\sum_{j=1}^{\infty} f^{(i)}\left(\frac{p_j}{q_j}\right) q_j^{1-i} \right. \right.$$

$$\left. \left. \cdot (bq_j - p_j)^{(i-k)}(aq_j - p_j)^k \right) \right). \tag{10.19}$$

Proof. Based on Theorem 7 of [42]. $\qquad\square$

Part III. Here we apply results from [43]. We begin with

Theorem 10.11. *Let $f, g \in C^1([a,b])$ where f as in this chapter, $g' \neq 0$ over $[a,b]$, $\mu_1, \mu_2 \in M^*$. Then*

$$\Gamma_f(\mu_1, \mu_2) \leq \left\| \frac{f'}{g'} \right\|_{\infty,[a,b]} \left(\sum_{i=1}^{\infty} q_i \left| g\left(\frac{p_i}{q_i}\right) - g(1) \right| \right). \tag{10.20}$$

Proof. Based on Theorem 1 of [43]. $\square$

Examples 10.12 (to Theorem 10.11).

1) Let $g(x) = \frac{1}{x}$. Then

$$\Gamma_f(\mu_1, \mu_2) \leq \|x^2 f'\|_{\infty,[a,b]} \left(\sum_{i=1}^{\infty} \frac{q_i}{p_i} |p_i - q_i| \right). \tag{10.21}$$

2) Let $g(x) = e^x$. Then

$$\Gamma_f(\mu_1, \mu_2) \leq \|e^{-x} f'\|_{\infty,[a,b]} \left(\sum_{i=1}^{\infty} q_i \left| e^{\frac{p_i}{q_i}} - e \right| \right). \tag{10.22}$$

3) Let $g(x) = e^{-x}$. Then

$$\Gamma_f(\mu_1, \mu_2) \leq \|e^x f'\|_{\infty,[a,b]} \left(\sum_{i=1}^{\infty} q_i \left| e^{-\frac{p_i}{q_i}} - e^{-1} \right| \right). \tag{10.23}$$

4) Let $g(x) = \ln x$. Then

$$\Gamma_f(\mu_1, \mu_2) \leq \|x f'\|_{\infty,[a,b]} \left(\sum_{i=1}^{\infty} q_i \left| \ln\left(\frac{p_i}{q_i}\right) \right| \right). \tag{10.24}$$

5) Let $g(x) = x \ln x$, with $a > e^{-1}$. Then

$$\Gamma_f(\mu_1, \mu_2) \leq \left\| \frac{f'}{1 + \ln x} \right\|_{\infty,[a,b]} \left(\sum_{i=1}^{\infty} p_i \left| \ln\left(\frac{p_i}{q_i}\right) \right| \right). \tag{10.25}$$

6) Let $g(x) = \sqrt{x}$. Then

$$\Gamma_f(\mu_1, \mu_2) \leq 2\|\sqrt{x} f'\|_{\infty,[a,b]} \left(\sum_{i=1}^{\infty} \sqrt{q_i} \left| \sqrt{p_i} - \sqrt{q_i} \right| \right). \tag{10.26}$$

7) Let $g(x) = x^\alpha$, $\alpha > 1$. Then

$$\Gamma_f(\mu_1, \mu_2) \leq \frac{1}{\alpha} \left\| \frac{f'}{x^{\alpha-1}} \right\|_{\infty,[a,b]} \left(\sum_{i=1}^{\infty} q_i^{1-\alpha} |p_i^\alpha - q_i^\alpha| \right). \tag{10.27}$$

Next we present

Theorem 10.13. *Let $f \in C^1([a,b])$, $a \neq b$, $\mu_1, \mu_2 \in M^*$. Then*

$$\left| \Gamma_f(\mu_1, \mu_2) - \frac{1}{b-a} \int_a^b f(x)dx \right| \leq \frac{\|f'\|_{\infty,[a,b]}}{(b-a)} \left[\left(\frac{a^2 + b^2}{2} \right) \right.$$
$$\left. - (a+b) + \sum_{i=1}^{\infty} q_i^{-1} p_i^2 \right]. \tag{10.28}$$

Proof. Based on Theorem 2 of [43].   □

It follows

Theorem 10.14. *Let $f \in C^{(2)}([a,b])$, $a \neq b$, $\mu_1, \mu_2 \in M^*$. Then*

$$\left| \Gamma_f(\mu_1, \mu_2) - \frac{1}{b-a} \int_a^b f(x)dx - \left(\frac{f(b) - f(a)}{b-a} \right) \left(1 - \frac{a+b}{2} \right) \right|$$

$$\leq \frac{(b-a)^2}{2} \|f''\|_{\infty,[a,b]} \left\{ \sum_{i=1}^{\infty} q_i \left[\frac{\left(\frac{p_i}{q_i} - \left(\frac{a+b}{2} \right) \right)^2}{(b-a)^2} + \frac{1}{4} \right]^2 + \frac{1}{12} \right\}. \qquad (10.29)$$

Proof. Based on (33) of [43].   □

Working more generally we have

Theorem 10.15. *Let $f \in C^1([a,b])$ and $g \in C([a,b])$ of bounded variation, $g(a) \neq g(b)$, $\mu_1, \mu_2 \in M^*$. Then*

$$\left| \Gamma_f(\mu_1, \mu_2) - \frac{1}{g(b) - g(a)} \int_a^b f(x)dg(x) \right|$$

$$\leq \|f''\|_{\infty,[a,b]} \left\{ \sum_{i=1}^{\infty} q_i \left(\int_a^b \left| P\left(g\left(\frac{p_i}{q_i} \right), q(t) \right) \right| dt \right) \right\}, \qquad (10.30)$$

where,

$$P(g(z), g(t)) := \begin{cases} \dfrac{g(t) - g(a)}{g(b) - g(a)}, & a \leq t \leq z, \\[2mm] \dfrac{g(t) - g(b)}{g(b) - g(a)}, & z < t \leq b. \end{cases} \qquad (10.31)$$

Proof. By Theorem 5 of [43].   □

Example 10.16 (to Theorem 10.15). Let $f \in C^1([a,b])$, $a \neq b$, and $g(x) = e^x$, $\ln x$, $\sqrt{x}$, x^α, $\alpha > 1$; $x > 0$. Then by (71)-(74) of [43] we derive

1)

$$\left| \Gamma_f(\mu_1, \mu_2) - \frac{1}{e^b - e^a} \int_a^b f(x) - e^x dx \right|$$

$$\leq \frac{\|f'\|_{\infty,[a,b]}}{(e^b - e^a)} \left\{ 2 \sum_{i=1}^{\infty} q_i e^{\frac{p_i}{q_i}} - e^a(2 - a) + e^b(b - 2) \right\}, \qquad (10.32)$$

2)

$$\left| \Gamma_f(\mu_1, \mu_2) - \left(\ln\left(\frac{b}{a} \right) \right)^{-1} \int_a^b \frac{f(x)}{x} dx \right|$$

$$\leq \|f'\|_{\infty,[a,b]} \left(\ln\left(\frac{b}{a} \right) \right)^{-1} \left\{ 2 \sum_{i=1}^{\infty} p_i \ln \frac{p_i}{q_i} - \ln(ab) + (a + b - 2) \right\}, \qquad (10.33)$$

3)

$$\left| \Gamma_f(\mu_1, \mu_2) - \frac{1}{2(\sqrt{b} - \sqrt{a})} \int_a^b \frac{f(x)}{\sqrt{x}} dx \right|$$

$$\leq \frac{\|f'\|_{\infty,[a,b]}}{(\sqrt{b} - \sqrt{a})} \left\{ \frac{4}{3} \sum_{i=1}^{\infty} q_i^{-1/2} p_i^{3/2} + \frac{(a^{3/2} + b^{3/2})}{3} - (\sqrt{a} + \sqrt{b}) \right\}, \quad (10.34)$$

4) $(\alpha > 1)$

$$\left| \Gamma_f(\mu_1, \mu_2) - \frac{\alpha}{(b^\alpha - a^\alpha)} \int_a^b f(x) x^{\alpha-1} dx \right|$$

$$\leq \frac{\|f'\|_{\infty,[a,b]}}{(b^\alpha - a^\alpha)} \left\{ \frac{2}{(\alpha+1)} \left(\sum_{i=1}^{\infty} q_i^{-\alpha} p_i^{\alpha+1} \right) \right.$$

$$\left. + \left(\frac{\alpha}{\alpha+1} (a^{\alpha+1} + b^{\alpha+1}) - (a^\alpha + b^\alpha) \right) \right\}. \quad (10.35)$$

We continue with

Theorem 10.17. *Let* $f \in C^{(2)}([a,b])$, $g \in C([a,b])$ *and of bounded variation,* $g(a) \neq g(b)$, $\mu_1, \mu_2 \in M^*$. *Then*

$$\left| \Gamma_f(\mu_1, \mu_2) - \frac{1}{(g(b) - g(a))} \int_a^b f(x) dg(x) \right.$$

$$\left. - \frac{1}{(g(b) - g(a))} \left(\int_a^b f'(x) dg(x) \right) \left[\sum_{i=1}^{\infty} \left(q_i \left(\int_a^b P\left(g\left(\frac{p_i}{q_i} \right), g(x) \right) dx \right) \right) \right] \right|$$

$$\leq \|f''\|_{\infty,[a,b]} \left(\sum_{i=1}^{\infty} \left(q_i \left(\int_a^b \int_a^b \left| P\left(g\left(\frac{p_i}{q_i} \right), g(x) \right) \right| |P(g(x), g(x_1))| dx_1 dx \right) \right) \right).$$

$$(10.36)$$

Proof. Apply (79) of [43]. $\square$

Remark 10.18. Next we define

$$P^*(z, s) := \begin{cases} s - a, & s \in [a, z] \\ s - b, & s \in (z, b]. \end{cases} \quad (10.37)$$

Let $f \in C^1([a,b])$, $a \neq b$, $\mu_1, \mu_2 \in M^*$. Then by (25) of [43] we get the representation

$$\Gamma_f(\mu_1, \mu_2) = \frac{1}{b-a} \left(\int_a^b f(x) dx + \mathcal{R}_1 \right), \quad (10.38)$$

where by (26) of [43] we have

$$\mathcal{R}_1 := \sum_{i=1}^{\infty} q_i \left(\int_a^b P^*\left(\frac{p_i}{q_i}, x \right) f'(x) dx \right). \quad (10.39)$$

Let again $f \in C^1([a,b])$, $g \in C([a,b])$ and of bounded variation, $g(a) \neq g(b)$, $\mu_1, \mu_2 \in M^*$.

Then by (47) of [43] we get the representation formula

$$\Gamma_f(\mu_1, \mu_2) = \frac{1}{g(b) - g(a)} \int_a^b f(x) dg(x) + \mathcal{R}_4,$$ (10.40)

where by (48) of [43] we have

$$\mathcal{R}_4 := \sum_{i=1}^\infty q_i \left(\int_a^b P\left(g\left(\frac{p_i}{q_i}\right), g(x)\right) f'(x) dx \right).$$ (10.41)

Let $\tilde{\alpha}, \beta > 1$: $\frac{1}{\tilde{\alpha}} + \frac{1}{\beta} = 1$, then by (89) of [43] we have

$$|\mathcal{R}_1| \leq \|f'\|_{\tilde{\alpha},[a,b]} \left(\sum_{i=1}^\infty q_i \left\| P^*\left(\frac{p_i}{q_i}, \cdot\right) \right\|_{\beta,[a,b]} \right)$$ (10.42)

and by (90) of [43] we obtain

$$|\mathcal{R}_1| \leq \|f'\|_{1,[a,b]} \left(\sum_{i=1}^\infty q_i \left\| P^*\left(\frac{p_i}{q_i}, \cdot\right) \right\|_{\infty,[a,b]} \right).$$ (10.43)

Furthermore by (95) and (96) of [43] we derive

$$|\mathcal{R}_4| \leq \|f'\|_{\tilde{\alpha},[a,b]} \left(\sum_{i=1}^\infty q_i \left\| P\left(g\left(\frac{p_i}{q_i}\right), g(\cdot)\right) \right\|_{\beta,[a,b]} \right)$$

and

$$|\mathcal{R}_4| \leq \|f'\|_{1,[a,b]} \left(\sum_{i=1}^\infty q_i \left\| P\left(g\left(\frac{p_i}{q_i}\right), g(\cdot)\right) \right\|_{\infty,[a,b]} \right).$$ (10.44)

Remark 10.19. Let $a < b$.

(i) By (105) of [43] we get

$$|\mathcal{R}_1| \leq \|f'\|_{1,[a,b]} \left(\sum_{i=1}^\infty q_i \max\left(\frac{p_i}{q_i} - a, b - \frac{p_i}{q_i}\right) \right), \quad f \in C^1([a,b]).$$ (10.45)

(ii) Let g be strictly increasing and continuous over $[a,b]$, e.g. $g(x) = e^x$, $\ln x$, $\sqrt{x}$, x^α with $\alpha > 1$, and $x > 0$ whenever is needed. Also let $f \in C^1([a,b])$ and as in this chapter always. Then by (106) of [43] we find

$$|\mathcal{R}_4| \leq \frac{\|f'\|_{1,[a,b]}}{(g(b) - g(a))} \left(\sum_{i=1}^\infty q_i \max\left(g\left(\frac{p_i}{q_i}\right) - g(a), g(b) - g\left(\frac{p_i}{q_i}\right)\right) \right).$$ (10.46)

In particular via (107)-(110) of [43] we obtain

1)

$$|\mathcal{R}_4| \leq \frac{\|f'\|_{1,[a,b]}}{(e^b - e^a)} \left(\sum_{i=1}^\infty q_i \max\left(e^{\frac{p_i}{q_i}} - e^a, e^b - e^{\frac{p_i}{q_i}}\right) \right),$$ (10.47)

2)

$$|\mathcal{R}_4| \leq \frac{\|f'\|_{1,[a,b]}}{\ln b - \ln a} \left(\sum_{i=1}^{\infty} q_i \max\left(\ln\left(\frac{p_i}{q_i}\right) - \ln a, \ln b - \ln\left(\frac{p_i}{q_i}\right) \right) \right), \tag{10.48}$$

3)

$$|\mathcal{R}_4| \leq \frac{\|f'\|_{1,[a,b]}}{\sqrt{b} - \sqrt{a}} \left(\sum_{i=1}^{\infty} q_i \max\left(\sqrt{\frac{p_i}{q_i}} - \sqrt{a}, \sqrt{b} - \sqrt{\frac{p_i}{q_i}} \right) \right), \tag{10.49}$$

and finally for $\alpha > 1$ we derive

4)

$$|\mathcal{R}_4| \leq \frac{\|f'\|_{1,[a,b]}}{b^\alpha - a^\alpha} \left(\sum_{i=1}^{\infty} q_i \max\left(\frac{p_i^\alpha}{q_i^\alpha} - a^\alpha, b^\alpha - \frac{p_i^\alpha}{q_i^\alpha} \right) \right). \tag{10.50}$$

iii) At last let $\tilde{\alpha}, \beta > 1$ such that $\frac{1}{\tilde{\alpha}} + \frac{1}{\beta} = 1$, then by (115) of [43] we have

$$|\mathcal{R}_1| \leq \|f'\|_{\tilde{\alpha},[a,b]} \left(\sum_{i=1}^{\infty} \left(\sqrt[\beta]{\frac{(p_i - aq_i)^{\beta+1} + (bq_i - p_i)^{\beta+1}}{(\beta+1)q_i}} \right) \right). \tag{10.51}$$

Part IV. Here we apply results from [41].

We start with

Theorem 10.20. *Let $0 < a < 1 < b$, f as in this chapter and $f \in C^n([a,b])$, $n \geq 1$ with $|f^{(n)}(t) - f^{(n)}(1)|$ be a convex function in t. Let $0 < h < \min(1 - a, b - 1)$ be fixed, $\mu_1, \mu_2 \in M^*$. Then*

$$\Gamma_f(\mu_1, \mu_2) \leq \sum_{k=2}^{n} \frac{|f^{(k)}(1)|}{k!} \left| \sum_{i=1}^{\infty} q_i^{1-k}(p_i - q_i)^k \right|$$

$$+ \frac{\omega_1(f^{(n)}, h)}{(n+1)!h} \left(\sum_{i=1}^{\infty} q_i^{-n}|p_i - q_i|^{n+1} \right). \tag{10.52}$$

Here ω_1 is the usual first modulus of continuity. If $f^{(k)}(1) = 0$, $k = 2, \ldots, n$, then

$$\Gamma_f(\mu_1, \mu_2) \leq \frac{\omega_1(f^{(n)}, h)}{(n+1)!h} \left(\sum_{i=1}^{\infty} q_i^{-n}|p_i - q_i|^{n+1} \right). \tag{10.53}$$

Inequalities (10.52) and (10.53) when n is even are attained by

$$\tilde{f}(t) := \frac{|t - 1|^{n+1}}{(n+1)!}, \quad a \leq t \leq b. \tag{10.54}$$

Proof. By Theorem 1 of [41]. $\qquad\qquad\square$

We continue with the general

Theorem 10.21. *Let $f \in C^n([a,b])$, $n \geq 1$, $\mu_1, \mu_2 \in M^*$. Assume that $\omega_1(f^{(n)}, \delta) \leq w$, where $0 < \delta \leq b - a$, $w > 0$. Let $x \in \mathbb{R}$ and denote by*

$$\phi_n(x) := \int_0^{|x|} \left\lceil \frac{t}{\delta} \right\rceil \frac{(|x| - t)^{n-1}}{(n-1)!} dt, \tag{10.55}$$

where $\lceil \cdot \rceil$ is the ceiling of the number. Then

$$\Gamma_f(\mu_1, \mu_2) \le \sum_{k=2}^{n} \frac{|f^{(k)}(1)|}{k!} \left| \sum_{i=1}^{\infty} q_i^{1-k}(p_i - q_i)^k \right| + w\left(\sum_{i=1}^{\infty} q_i \phi_n \left(\frac{p_i - q_i}{q_i} \right) \right). \tag{10.56}$$

Inequality (10.56) is sharp, namely it is attained by the function

$$\tilde{f}_n(t) := w\phi_n(t-1), \quad a \le t \le b, \tag{10.57}$$

when n is even.

Proof. By Theorem 2 of [41]. $\qquad\square$

It follows

Corollary 10.22 (to Theorem 10.21). *Let $\mu_1, \mu_2 \in M^*$. It holds $(0 < \delta \le b - a)$*

$$\begin{aligned}
\Gamma_f(\mu_1, \mu_2) \le{} & \sum_{k=2}^{n} \frac{|f^{(k)}(1)|}{k!} \left(\sum_{i=1}^{\infty} q_i^{1-k}(p_i - q_i)^k \right) \\
& + \omega_1(f^{(n)}, \delta) \left\{ \frac{1}{(n+1)!\delta} \left(\sum_{i=1}^{\infty} q_i^{-n}|p_i - q_i|^{n+1} \right) \right. \\
& + \frac{1}{2n!} \left(\sum_{i=1}^{\infty} q_i^{1-n}|p_i - q_i|^{n} \right) \\
& \left. + \frac{\delta}{8(n-1)!} \left(\sum_{i=1}^{\infty} q_i^{2-n}|p_i - q_i|^{n-1} \right) \right\},
\end{aligned} \tag{10.58}$$

and

$$\begin{aligned}
\Gamma_f(\mu_1, \mu_2) \le{} & \sum_{k=2}^{n} \frac{|f^{(k)}(1)|}{k!} \left| \sum_{i=1}^{\infty} q_i^{1-k}(p_i - q_i)^k \right| \\
& + \omega_1(f^{(n)}, \delta) \left\{ \frac{1}{n!} \left(\sum_{i=1}^{\infty} q_i^{1-n}|p_i - q_i|^{n} \right) \right. \\
& \left. + \frac{1}{n!(n+1)\delta} \left(\sum_{i=1}^{\infty} q_i^{-n}|p_i - q_i|^{n+1} \right) \right\}.
\end{aligned} \tag{10.59}$$

In particular we have

$$\Gamma_f(\mu_1, \mu_2) \le \omega_1(f', \delta) \left\{ \frac{1}{2\delta} \left(\sum_{i=1}^{\infty} q_i^{-1}(p_i - q_i)^2 \right) + \frac{1}{2} \left(\sum_{i=1}^{\infty} |p_i - q_i| \right) + \frac{\delta}{8} \right\}, \tag{10.60}$$

and

$$\Gamma_f(\mu_1, \mu_2) \le \omega_1(f', \delta) \left\{ \left(\sum_{i=1}^{\infty} |p_i - q_i| \right) + \frac{1}{2\delta} \left(\sum_{i=1}^{\infty} q_i^{-1}(p_i - q_i)^2 \right) \right\}. \tag{10.61}$$

Proof. Based on Corollary 1 of [41]. $\qquad\square$

Also we have

Corollary 10.23 (to Theorem 10.21). *Let $\mu_1, \mu_2 \in M^*$. Assume here that $f^{(n)}$ is of Lipschitz type of order α, $0 < \alpha \le 1$, i.e.*

$$\omega_1(f^{(n)}, \delta) \le K\delta^\alpha, \quad K > 0, \tag{10.62}$$

for any $0 < \delta \le b - a$. Then

$$\Gamma_f(\mu_1, \mu_2) \le \sum_{k=2}^{n} \frac{|f^{(k)}(1)|}{k!} \left| \sum_{i=1}^{\infty} q_i^{1-k}(p_i - q_i)^k \right| + \frac{K}{\prod_{i=1}^{n}(\alpha + i)} \left(\sum_{i=1}^{\infty} q_i^{1-n-\alpha}|p_i - q_i|^{n+\alpha} \right). \tag{10.63}$$

When n is even (10.63) is attained by $f^*(x) = c|x - 1|^{n+\alpha}$, where $c := K/\left(\prod_{i=1}^{n}(\alpha + i)\right) > 0$.

Proof. Based on Corollary 2 of [41]. □

Corollary 10.24 (to Theorem 10.21). *Let $\mu_1, \mu_2 \in M^*$. Suppose that*

$$b - a \ge \sum_{i=1}^{\infty} q_i^{-1} p_i^2 - 1 > 0. \tag{10.64}$$

Then

$$\Gamma_f(\mu_1, \mu_2) \le \omega_1\left(f', \left(\sum_{i=1}^{\infty} q_i^{-1} p_i^2 - 1\right)\right) \left\{ \sum_{i=1}^{\infty} |p_i - q_i| + \frac{1}{2} \right\}. \tag{10.65}$$

Proof. Based on Corollary 3 of [41]. □

Corollary 10.25 (to Theorem 10.21). *Let $\mu_1, \mu_2 \in M^*$. Suppose that*

$$\sum_{i=1}^{\infty} |p_i - q_i| > 0. \tag{10.66}$$

Then

$$\Gamma_f(\mu_1, \mu_2) \le \omega_1\left(f', \left(\sum_{i=1}^{\infty} |p_i - q_i|\right)\right) \left\{ \left(\sum_{i=1}^{\infty} |p_i - q_i|\right) + \frac{b - a}{2} \right\}$$

$$\le \frac{3}{2}(b - a)\omega_1\left(f', \left(\sum_{i=1}^{\infty} |p_i - q_i|\right)\right). \tag{10.67}$$

Proof. Based on Corollary 5 of [41]. □

We give

Proposition 10.26. *Let $f \in C([a, b])$, $\mu_1, \mu_2 \in M^*$.*
  i) *Assume that*

$$\sum_{i=1}^{\infty} |p_i - q_i| > 0. \tag{10.68}$$

Then

$$\Gamma_f(\mu_1,\mu_2) \le 2\omega_1\left(f,\left(\sum_{i=1}^{\infty}|p_i-q_i|\right)\right). \tag{10.69}$$

ii) *Let $r>0$ and*

$$b-a \ge r\left(\sum_{i=1}^{\infty}|p_i-q_i|\right) > 0. \tag{10.70}$$

Then

$$\Gamma_f(\mu_1,\mu_2) \le \left(1+\frac{1}{r}\right)\omega_1\left(f,r\left(\sum_{i=1}^{\infty}|p_i-q_i|\right)\right). \tag{10.71}$$

Proof. Based on Proposition 1 of [41]. $\qquad\square$

Also we give

Proposition 10.27. *Let f as in this setting and f is a Lipschitz function of order α, $0 < \alpha \le 1$, i.e. there exists $K>0$ such that*

$$|f(x)-f(y)| \le K|x-y|^{\alpha}, \quad all\ x,y\in[a,b]. \tag{10.72}$$

Let $\mu_1,\mu_2 \in M^*$. Then

$$\Gamma_f(\mu_1,\mu_2) \le K\left(\sum_{i=1}^{\infty}q_i^{1-\alpha}|p_i-q_i|^{\alpha}\right). \tag{10.73}$$

Proof. Based on Proposition 2 of [41]. $\qquad\square$

Next we present some alternative type of results.
We start with

Theorem 10.28. *All elements involved here as in this chapter and $f \in C^{n+1}([a,b])$, $n \in \mathbb{N}$. Let $\mu_1,\mu_2 \in M^*$. Then*

$$\Gamma_f(\mu_1,\mu_2) = \sum_{k=0}^{n-1}\frac{(-1)^k}{(k+1)!}\left(\sum_{i=1}^{\infty}f^{(k+1)}\left(\frac{p_i}{q_i}\right)q_i^{-k}(p_i-q_i)^{k+1}\right) + \psi_2, \tag{10.74}$$

where

$$\psi_2 := \frac{(-1)^n}{n!}\left(\sum_{i=1}^{\infty}q_i\left(\int_1^{\frac{p_i}{q_i}}(t-1)^n f^{(n+1)}(t)dt\right)\right). \tag{10.75}$$

Proof. By Theorem 12 of [41]. $\qquad\square$

Next we estimate ψ_2. We give

Theorem 10.29. *All here as in Theorem 10.28. Then*

$$
|\psi_2| \leq \min \; of \begin{cases} B_1 := \dfrac{\|f^{(n+1)}\|_{\infty,[a,b]}}{(n+1)!} \left(\displaystyle\sum_{i=1}^{\infty} q_i^{-n} |p_i - q_i|^{n+1} \right), \\[4mm] B_2 := \dfrac{\|f^{(n+1)}\|_{1,[a,b]}}{n!} \left(\displaystyle\sum_{i=1}^{\infty} q_i^{1-n} |p_i - q_i|^{n} \right), \\[4mm] \qquad and \; for \; \tilde{\alpha}, \beta > 1 : \dfrac{1}{\tilde{\alpha}} + \dfrac{1}{\beta} = 1, \\[4mm] B_3 := \dfrac{\|f^{(n+1)}\|_{\tilde{\alpha},[a,b]}}{n!(n\beta+1)^{1/\beta}} \left(\displaystyle\sum_{i=1}^{\infty} q_i^{1-n-\frac{1}{\beta}} |p_i - q_i|^{n+\frac{1}{\beta}} \right). \end{cases}
$$

(10.76)

Also it holds

$$
\Gamma_f(\mu_1, \mu_2) \leq \sum_{k=0}^{n-1} \frac{1}{(k+1)!} \left| \sum_{i=1}^{\infty} f^{(k+1)}\left(\frac{p_i}{q_i}\right) q_i^{-k}(p_i - q_i)^{k+1} \right| + \min(B_1, B_2, B_3).
$$

(10.77)

Proof. By Theorem 13 of [41]. $\qquad\qquad\square$

Corollary 10.30 (to Theorem 10.29). *Case of $n = 1$. It holds*

$$
|\psi_2| \leq \min \; of \begin{cases} B_{1,1} := \dfrac{\|f''\|_{\infty,[a,b]}}{2} \left(\displaystyle\sum_{i=1}^{\infty} q_i^{-1}(p_i - q_i)^2 \right), \\[4mm] B_{2,1} := \|f''\|_{1,[a,b]} \left(\displaystyle\sum_{i=1}^{\infty} |p_i - q_i| \right), \\[4mm] \qquad and \; for \; \tilde{\alpha}, \beta > 1 : \dfrac{1}{\tilde{\alpha}} + \dfrac{1}{\beta} = 1, \\[4mm] B_{3,1} := \dfrac{\|f''\|_{\tilde{\alpha},[a,b]}}{(\beta+1)^{1/\beta}} \left(\displaystyle\sum_{i=1}^{\infty} q_i^{-\frac{1}{\beta}} |p_i - q_i|^{1+\frac{1}{\beta}} \right). \end{cases}
$$

(10.78)

Also it holds

$$
\Gamma_f(\mu_1, \mu_2) \leq \left| \sum_{i=1}^{\infty} f'\left(\frac{p_i}{q_i}\right)(p_i - q_i) \right| + \min(B_{1,1}, B_{2,1}, B_{3,1}).
$$

(10.79)

Proof. By Corollary 8 of [41]. $\qquad\qquad\square$

We further give

Theorem 10.31. *Let all elements involved as in this chapter, $f \in C^{n+1}([a,b])$, $n \in \mathbb{N}$ and $\mu_1, \mu_2 \in M^*$. Then*

$$\Gamma_f(\mu_1, \mu_2) = \sum_{k=0}^{n-1} \frac{(-1)^k}{(k+1)!} \left(\sum_{i=1}^{\infty} f^{(k+1)}\left(\frac{p_i}{q_i}\right) q_i^{-k}(p_i - q_i)^{k+1} \right)$$
$$+ \frac{(-1)^n}{(n+1)!} f^{(n+1)}(1) \left(\sum_{i=1}^{\infty} q_i^{-n}(p_i - q_i)^{n+1} \right) + \psi_4, \qquad (10.80)$$

where

$$\psi_4 := \sum_{i=1}^{\infty} q_i \psi_3 \left(\frac{p_i}{q_i} \right), \qquad (10.81)$$

with

$$\psi_3(x) := \frac{(-1)^n}{n!} \int_1^x (s-1)^n (f^{(n+1)}(s) - f^{(n+1)}(1)) ds. \qquad (10.82)$$

Proof. Based on Theorem 14 of [41]. $\qquad\square$

Next we present estimations of ψ_4.

Theorem 10.32. *Let all elements involved here as in this chapter and $f \in C^{n+1}([a,b])$, $n \in \mathbb{N}$, $\mu_1, \mu_2 \in M^*$ and*

$$0 < \frac{1}{(n+2)} \left(\sum_{i=1}^{\infty} q_i^{-n-1} |p_i - q_i|^{n+2} \right) \leq b - a. \qquad (10.83)$$

Then (10.80) is again valid and

$$|\psi_4| \leq \frac{1}{n!} \omega_1 \left(f^{(n+1)}, \frac{1}{(n+2)} \left(\sum_{i=1}^{\infty} q_i^{-n-1} |p_i - q_i|^{n+2} \right) \right)$$
$$\cdot \left(1 + \frac{1}{(n+1)} \left(\sum_{i=1}^{\infty} q_i^{-n} |p_i - q_i|^{n+1} \right) \right). \qquad (10.84)$$

Proof. Based on Theorem 15 of [41]. $\qquad\square$

Also we have

Theorem 10.33. *All elements involved here as in this chapter and $f \in C^{n+1}([a,b])$, $n \in \mathbb{N}$, $\mu_1, \mu_2 \in M^*$. Furthermore we assume that $|f^{(n+1)}(t) - f^{(n+1)}(1)|$ is convex in t and*

$$0 < \frac{1}{n!(n+2)} \left(\sum_{i=1}^{\infty} q_i^{-n-1} |p_i - q_i|^{n+2} \right) \leq \min(1 - a, b - 1). \qquad (10.85)$$

Then (10.80) is again valid and

$$|\psi_4| \leq \omega_1 \left(f^{(n+1)}, \frac{1}{n!(n+2)} \left(\sum_{i=1}^{\infty} q_i^{-n-1} |p_i - q_i|^{n+2} \right) \right). \qquad (10.86)$$

The last (10.86) is attained by

$$\hat{f}(s) := \frac{(s-1)^{n+2}}{(n+2)!}, \quad a \le s \le b, \tag{10.87}$$

when n is even.

Proof. By Theorem 16 of [41].   □

When $n = 1$ we obtain

Corollary 10.34 (to Theorem 10.32). *Let $f \in C^2([a,b])$, $\mu_1, \mu_2 \in M^*$ and*

$$0 < \frac{1}{3}\left(\sum_{i=1}^{\infty} q_i^{-2}|p_i - q_i|^3\right) \le b - a. \tag{10.88}$$

Then

$$\Gamma_f(\mu_1, \mu_2) = \sum_{i=1}^{\infty} f'\left(\frac{p_i}{q_i}\right)(p_i - q_i) - \frac{f''(1)}{2}\left(\sum_{i=1}^{\infty} q_i^{-1}(p_i - q_i)^2\right) + \psi_{4,1}, \tag{10.89}$$

where

$$\psi_{4,1} := \sum_{i=1}^{\infty} q_i \psi_{3,1}\left(\frac{p_i}{q_i}\right), \tag{10.90}$$

with

$$\psi_{3,1}(x) := -\int_1^x (t-1)(f''(t) - f''(1))dt, \quad x \in [a,b]. \tag{10.91}$$

Furthermore it holds

$$|\psi_{4,1}| \le \omega_1\left(f'', \frac{1}{3}\left(\sum_{i=1}^{\infty} q_i^{-2}|p_i - q_i|^3\right)\right)\frac{1}{2}\left(\left(\sum_{i=1}^{\infty} q_i^{-1}p_i^2\right) + 1\right). \tag{10.92}$$

Proof. Based on Corollary 9 of [41].   □

Also we have

Corollary 10.35 (to Theorem 10.33). *Let $f \in C^2([a,b])$ and $|f''(t) - f''(1)|$ is convex in t. Here $\mu_1, \mu_2 \in M^*$ such that*

$$0 < \frac{1}{3}\left(\sum_{i=1}^{\infty} q_i^{-2}|p_i - q_i|^3\right) \le \min(1 - a, b - 1). \tag{10.93}$$

Then again (10.89) is valid and

$$|\psi_{4,1}| \le \omega_1\left(f'', \frac{1}{3}\left(\sum_{i=1}^{\infty} q_i^{-2}|p_i - q_i|^3\right)\right). \tag{10.94}$$

Proof. By Corollary 10 of [41].   □

The last main result of the chapter follows.

Theorem 10.36. *All elements involved here in this chapter and* $f \in C^{n+1}([a,b])$, $n \in \mathbb{N}$, $\mu_1, \mu_2 \in M^*$. *Additionally we assume that*

$$|f^{(n+1)}(x) - f^{(n+1)}(y)| \leq K|x - y|^\alpha, \tag{10.95}$$

$K > 0$, $0 < \alpha \leq 1$, *all* $x, y \in [a,b]$ *with* $(x-1)(y-1) \geq 0$. *Then*

$$\Gamma_f(\mu_1, \mu_2) = \sum_{k=0}^{n-1} \frac{(-1)^k}{(k+1)!} \left(\sum_{i=1}^{\infty} f^{(k+1)} \left(\frac{p_i}{q_i} \right) q_i^{-k} (p_i - q_i)^{k+1} \right)$$
$$+ \frac{(-1)^n}{(n+1)!} f^{(n+1)}(1) \left(\sum_{i=1}^{\infty} q_i^{-n} (p_i - q_i)^{n+1} \right) + \psi_4. \tag{10.96}$$

Here it holds

$$|\psi_4| \leq \frac{K}{n!(n+\alpha+1)} \left(\sum_{i=1}^{\infty} q_i^{-n-\alpha} |p_i - q_i|^{n+\alpha+1} \right). \tag{10.97}$$

Inequality (10.97) is attained when n *is even by*

$$f^*(x) = \tilde{c}|x - 1|^{n+\alpha+1}, \quad x \in [a,b] \tag{10.98}$$

where

$$\tilde{c} := \frac{K}{\prod\limits_{j=0}^{n} (n + \alpha + 1 - j)}. \tag{10.99}$$

Proof. By Theorem 17 of [41]. $\qquad\qquad\square$

Next we give

Corollary 10.37 (to Theorem 10.36). *Let* $f \in C^2([a,b])$, $\mu_1, \mu_2 \in M^*$. *Additionally we suppose that*

$$|f''(x) - f''(y)| \leq K|x - y|^\alpha, \tag{10.100}$$

$K > 0$, $0 < \alpha \leq 1$, *all* $x, y \in [a,b]$ *with* $(x-1)(y-1) \geq 0$. *Then*

$$\Gamma_f(\mu_1, \mu_2) = \sum_{i=1}^{\infty} f'\left(\frac{p_i}{q_i} \right)(p_i - q_i) - \frac{f''(1)}{2} \left(\sum_{i=1}^{\infty} q_i^{-1}(p_i - q_i)^2 \right) + \psi_{4,1}. \tag{10.101}$$

Here

$$\psi_{4,1} := \sum_{i=1}^{\infty} q_i \psi_{3,1} \left(\frac{p_i}{q_i} \right) \tag{10.102}$$

with

$$\psi_{3,1}(x) := - \int_1^x (t-1)(f''(t) - f''(1))dt, \quad x \in [a,b]. \tag{10.103}$$

It holds that

$$|\psi_{4,1}| \leq \frac{K}{(\alpha+2)} \left(\sum_{i=1}^{\infty} q_i^{-1-\alpha} |p_i - q_i|^{\alpha+2} \right). \tag{10.104}$$

Proof. By Corollary 11 of [41]. $\qquad\qquad\square$

Chapter 11

About a General Discrete Measure of Dependence

In this chapter the discrete Csiszar's f-Divergence or discrete Csiszar's Discrimination, the most general distance among discrete probability measures, is established as the most general measure of Dependence between two discrete random variables which is a very important matter of stochastics. Many discrete estimates of it are given, leading to optimal or nearly optimal discrete probabilistic inequalities. That is we are comparing and connecting this general discrete measure of dependence to known other specific discrete entities by involving basic parameters of our setting. This treatment is based on [38].

11.1 Background

The basic general background here has as follows.

Let f by a convex function from $(0, +\infty)$ into $\mathbb{R}$ which is strictly convex at 1 with $f(1) = 0$.

Let (Ω, A, λ) be a measure space, where λ is a finite or a σ-finite measure on (Ω, A). Let μ_1, μ_2 be two probability measures on (Ω, A) such that $\mu_1 \ll \lambda$, $\mu_2 \ll \lambda$ (absolutely continuous), e.g. $\lambda = \mu_1 + \mu_2$.

Denote by $p = \frac{d\mu_1}{d\lambda}$, $q = \frac{d\mu_2}{d\lambda}$ the (densities) Radon–Nikodym derivatives of μ_1, μ_2 with respect to λ. Typically we assume that

$$0 < a \leq \frac{p}{q} \leq b, \quad \text{a.e. on } \Omega \text{ and } a \leq 1 \leq b.$$

The quantity

$$\Gamma_f(\mu_1, \mu_2) = \int_\Omega q(x) f\left(\frac{p(x)}{q(x)}\right) d\lambda(x), \tag{11.1}$$

was introduced by I. Csiszar in 1967, see [140], and is called *f-divergence* of the probability measures μ_1 and μ_2.

By Lemma 1.1 of [140], the integral (11.1) is well-defined and $\Gamma_f(\mu_1, \mu_2) \geq 0$ with equality only when $\mu_1 = \mu_2$. Furthermore by [140], [42] $\Gamma_f(\mu_1, \mu_2)$ does not depend on the choice of λ. The concept of f-divergence was introduced first in [139] as a generalization of Kullback's *"information for discrimination"* or *I-divergence* (*generalized entropy*) [244], [243], and of Rényi's *"information gain"* (*I-divergence*) *of order* α [306]. In fact the *I-divergence of order* 1 equals

$$\Gamma_{u\log_2 u}(\mu_1,\mu_2).$$

The choice $f(u) = (u-1)^2$ produces again a known measure of difference of distributions that is called χ^2-*divergence*. Of course the *total variation* distance

$$|\mu_1 - \mu_2| = \int_\Omega |p(x) - q(x)|\, d\lambda(x)$$

is equal to $\Gamma_{|u-1|}(\mu_1,\mu_2)$. Here by assuming $f(1) = 0$ we can consider $\Gamma_f(\mu_1,\mu_2)$ the f-divergence as a measure of the difference between the probability measures μ_1, μ_2.

This is an expository, survey and applications chapter where we apply to the discrete case a large variety of interesting results of the above continuous case coming from the author's papers [37], [42], [43], [41], [39]. Namely here we present a complete study of discrete Csiszar's *f-divergence* as a *measure of dependence* of two discrete random variables.

Throughout this chapter we use the following.

Let again f be a convex function from $(0, +\infty)$ into $\mathbb{R}$ which is strictly convex at 1 with $f(1) = 0$.

Let the probability space (Ω, P) and X, Y be discrete random variables from Ω into $\mathbb{R}$ such that their ranges $\mathcal{R}_X = \{x_1, x_2, \ldots\}$, $\mathcal{R}_Y = \{y_1, y_2, \ldots\}$, respectively, are finite or countable subsets of $\mathbb{R}$.

Let the probability functions of X and Y be

$$\begin{aligned}
p_i &= p(x_i) = P(X = x_i), \quad i = 1, 2, \ldots, \\
q_i &= q(y_j) = P(Y = y_j), \quad j = 1, 2, \ldots,
\end{aligned} \tag{11.2}$$

respectively. Without loss of generality we assume that all $p_i, q_j > 0$, $i, j = 1, 2, \ldots$.

Consider the probability distributions μ_{XY} and $\mu_X \times \mu_Y$ on $\mathcal{R}_X \times \mathcal{R}_Y$, where μ_{XY}, μ_X, μ_Y stand for the joint probability distribution of X and Y and their marginal distributions, respectively, while $\mu_X \times \mu_Y$ is the product probability distribution of X and Y. Here μ_{XY} is uniquely determined by the double sequence

$$t_{ij} = t(x_i, y_j) = P(X = x_i, Y = y_j) = \mu_{XY}(\{(x_i, y_j)\}), \quad i, j = 1, 2, \ldots, \tag{11.3}$$

which t is the probability function of (X, Y).

Again without loss of generality we assume that $t_{ij} > 0$ for all $i, j = 1, 2, \ldots$. Clearly also here μ_X, μ_Y are uniquely determined by $p_i = \mu_X(\{x_i\})$, $q_j = \mu_Y(\{y_j\})$, respectively, $i, j = 1, 2, \ldots$. It is obvious that the product probability distribution $\mu_X \times \mu_Y$ has associated probability function pq.

In this article we assume that

$$0 < a \le \frac{t_{ij}}{p_i q_j} \le b, \quad \text{all } i, j = 1, 2, \ldots \text{ and } a \le 1 \le b, \tag{11.4}$$

where a, b are fixed.

Let $\mathcal{P}$ be the set of all subsets of $\mathcal{R}_X \times \mathcal{R}_Y$. We define λ to be a finite or σ-finite measure on $(\mathcal{R}_X \times \mathcal{R}_Y, \mathcal{P})$ such that $\lambda(\{(x_i, y_i)\}) = 1$, for all $i, j = 1, 2, \ldots$ We notice that μ_{XY} and $\mu_X \times \mu_Y$ are discrete probability measures on $(\mathcal{R}_X \times \mathcal{R}_Y, \mathcal{P})$ and are absolutely continuous ($\ll$) with respect to λ.

By applying (11.1) we obtain the special case

$$\Gamma_f(\mu_{XY}, \mu_X \times \mu_Y) = \sum_{i=1}^{\infty} \sum_{j=1}^{\infty} p_i q_j f\left(\frac{t_{ij}}{p_i q_j}\right). \tag{11.5}$$

This is the *discrete Csiszar's distance* or *f-divergence between* μ_{XY} *and* $\mu_X \times \mu_Y$. The random variables X, Y are the less dependent the closer the distributions μ_{XY} and $\mu_X \times \mu_Y$ are, thus $\Gamma_f(\mu_{XY}, \mu_X \times \mu_Y)$ can be considered as a *measure of dependence* of X and Y, see also [305], [140], [39].

From the above derives that $\Gamma_f(\mu_{XY}, \mu_X \times \mu_Y) \geq 0$ with equality only when $\mu_{XY} = \mu_X \times \mu_Y$, i.e. when X, Y are independent discrete random variables. As related references see here [139], [140], [141], [152], [244], [243], [305], [306], [37], [42], [43], [41], [39].

For $f(u) = u \log_2 u$ we obtain the *mutual information* of X and Y,

$$I(X, Y) = I(\mu_{XY} \| \mu_X \times \mu_Y) = \Gamma_{u \log_2 u}(\mu_{XY}, \mu_X \times \mu_Y),$$

see [140]. For $f(u) = (u - 1)^2$ we get the *mean square contingency*:

$$\varphi^2(X, Y) = \Gamma_{(u-1)^2}(\mu_{XY}, \mu_X \times \mu_Y),$$

see [140].

In Information Theory and Statistics many various divergences are used which are special cases of the above Γ_f divergence. Γ_f has many applications also in Anthropology, Genetics, Finance, Economics, Political Science, Biology, Approximation of Probability Distributions, Signal Processing and Pattern Recognition.

In the next the proofs of all formulated and presented results are based a lot on this Section 11.1 especially on the transfer from the continuous to the discrete case.

11.2 Results

Part I. Here we apply results from [37], [39]. We present the following

Theorem 11.1. *Let* $f \in C^1([a, b])$. *Then*

$$\Gamma_f(\mu_{XY}, \mu_X \times \mu_Y) \leq \|f'\|_{\infty, [a,b]} \left(\sum_{i=1}^{\infty} \sum_{j=1}^{\infty} |t_{ij} - p_i q_j|\right). \tag{11.6}$$

Inequality (11.6) is sharp. The optimal function is $f^*(y) = |y - 1|^\alpha$, $\alpha > 1$ *under the condition, either*

i)

$$\max(b-1, 1-a) < 1 \quad and \quad C := \sup_{i,j \in \mathbb{N}} (p_i q_j) < +\infty, \qquad (11.7)$$

or

ii)

$$|t_{ij} - p_i q_j| \le p_i q_j \le 1, \quad all \; i,j = 1, 2, \ldots. \qquad (11.8)$$

Proof. Based on Theorem 1 of [37] and Theorem 1 of [39]. $\square$

Next we give the more general

Theorem 11.2. *Let* $f \in C^{n+1}([a,b])$, $n \in \mathbb{N}$, *such that* $f^{(k)}(1) = 0$, $k = 0, 2, 3, \ldots, n$. *Then*

$$\Gamma_f(\mu_{XY}, \mu_X \times \mu_Y) \le \frac{\|f^{(n+1)}\|_{\infty,[a,b]}}{(n+1)!} \left(\sum_{i=1}^{\infty} \sum_{j=1}^{\infty} p_i^{-n} q_j^{-n} |t_{ij} - p_i q_j|^{n+1} \right). \qquad (11.9)$$

Inequality (11.9) is sharp. The optimal function is $\tilde{f}(y) := |y-1|^{n+\alpha}$, $\alpha > 1$ *under the condition (i) or (ii) of Theorem 11.1.*

Proof. As in Theorems 2 of [37], [39], respectively. $\square$

As a special case we have

Corollary 11.3 (to Theorem 11.2). *Let* $f \in C^2([a,b])$. *Then*

$$\Gamma_f(\mu_{XY}, \mu_X \times \mu_Y) \le \frac{\|f^{(2)}\|_{\infty,[a,b]}}{2} \left(\sum_{i=1}^{\infty} \sum_{j=1}^{\infty} p_i^{-1} q_j^{-1} (t_{ij} - p_i q_j)^2 \right). \qquad (11.10)$$

Inequality (11.10) is sharp. The optimal function is $\tilde{f}(y) := |y-1|^{1+\alpha}$, $\alpha > 1$ *under the condition (i) or (ii) of Theorem 11.1.*

Proof. By Corollaries 1 of [37] and [39]. $\square$

Next we connect Csiszar's f-divergence to the usual first modulus of continuity ω_1.

Theorem 11.4. *Suppose that*

$$0 < h := \left(\sum_{i=1}^{\infty} \sum_{j=1}^{\infty} |t_{ij} - p_i q_j| \right) \le \min(1-a, b-1). \qquad (11.11)$$

Then

$$\Gamma_f(\mu_{XY}, \mu_X \times \mu_Y) \le \omega_1(f, h). \qquad (11.12)$$

Inequality (11.12) is sharp, i.e. attained by

$$f^*(x) = |x-1|.$$

Proof. By Theorems 3 of [37] and [39]. $\square$

Part II. Here we apply results from [42] and [39]. But first we need to give some basic background from [122], [26]. Let $\nu > 0$, $n := [\nu]$ and $\alpha = \nu - n$ $(0 < \alpha < 1)$. Let $x, x_0 \in [a, b] \subseteq \mathbb{R}$ such that $x \geq x_0$, where x_0 is fixed and $f \in C([a, b])$ and define

$$(J_\nu^{x_0} f)(x) := \frac{1}{\Gamma(\nu)} \int_{x_0}^{x} (x - s)^{\nu-1} f(s)\,ds, \quad x_0 \leq x \leq b, \qquad (11.13)$$

the *generalized Riemann–Liouville integral*, where Γ stands for the gamma function.

We define the subspace

$$C_{x_0}^\nu([a, b]) := \left\{ f \in C^n([a, b]) : J_{1-\alpha}^{x_0} f^{(n)} \in C^1([x_0, b]) \right\}. \qquad (11.14)$$

For $f \in C_{x_0}^\nu([a, b])$ we define the *generalized ν-fractional derivative of f* over $[x_0, b]$ as

$$D_{x_0}^\nu f = D J_{1-\alpha}^{x_0} f^{(n)} \left(D := \frac{d}{dx} \right). \qquad (11.15)$$

We present the following

Theorem 11.5. *Let* $a < b$, $1 \leq \nu < 2$, $f \in C_a^\nu([a, b])$. *Then*

$$\Gamma_f(\mu_{XY}, \mu_X \times \mu_Y) \leq \frac{\|D_a^\nu f\|_{\infty, [a,b]}}{\Gamma(\nu + 1)} \left(\sum_{i,j=1}^{\infty} (p_i q_j)^{1-\nu} (t_{ij} - a p_i q_j)^\nu \right). \qquad (11.16)$$

Proof. It follows from Theorem 2 of [42] and Theorem 4 of [39]. $\square$

The counterpart of the previous result comes next.

Theorem 11.6. *Let* $a < b$, $\nu \geq 2$, $n := [\nu]$, $f \in C_a^\nu([a, b])$, $f^{(i)}(a) = 0$, $i = 0, 1, \ldots, n - 1$. *Then*

$$\Gamma_f(\mu_{XY}, \mu_X \times \mu_Y) \leq \frac{\|D_a^\nu f\|_{\infty, [a,b]}}{\Gamma(\nu + 1)} \left(\sum_{i,j=1}^{\infty} (p_i q_j)^{1-\nu} (t_{ij} - a p_i q_j)^\nu \right). \qquad (11.17)$$

Proof. By Theorem 3 of [42] and Theorem 5 of [39]. $\square$

Next we give an $L_{\tilde{\alpha}}$ estimate.

Theorem 11.7. *Let* $a < b$, $\nu \geq 1$, $n := [\nu]$, $f \in C_a^\nu([a, b])$, $f^{(i)}(a) = 0$, $i = 0, 1, \ldots, n - 1$ *and let* $\tilde{\alpha}, \beta > 1$: $\frac{1}{\tilde{\alpha}} + \frac{1}{\beta} = 1$. *Then*

$$\Gamma_f(\mu_{XY}, \mu_X \times \mu_Y) \leq \frac{\|D_a^\nu f\|_{\tilde{\alpha}, [a,b]}}{\Gamma(\nu)(\beta(\nu - 1) + 1)^{1/\beta}} \left(\sum_{i,j=1}^{\infty} (p_i q_j)^{2-\nu-\frac{1}{\beta}} (t_{ij} - a p_i q_j)^{\nu-1+\frac{1}{\beta}} \right).$$
$$(11.18)$$

Proof. Based on Theorem 4 of [42] and Theorem 6 of [39]. $\square$

It follows an L_∞ estimate.

Theorem 11.8. *Let $a < b$, $\nu \geq 1$, $n := [\nu]$, $f \in C_a^\nu([a,b])$, $f^{(i)}(a) = 0$, $i = 0, 1, \ldots, n-1$. Then*

$$\Gamma_f(\mu_{XY}, \mu_X \times \mu_Y) \leq \frac{\|D_a^\nu f\|_{1,[a,b]}}{\Gamma(\nu)} \left(\sum_{i,j=1}^{\infty} (p_i q_j)^{2-\nu} (t_{ij} - a p_i q_j)^{\nu-1} \right). \qquad (11.19)$$

Proof. Based on Theorem 5 of [42] and Theorem 7 of [39]. $\qquad\qquad \square$

     We continue with

Theorem 11.9. *Let $f \in C^1([a,b])$, $a < b$. Then*

$$\Gamma_f(\mu_{XY}, \mu_X \times \mu_Y) \leq \frac{1}{b-a} \int_a^b f(x)dx - \left(\sum_{i,j=1}^{\infty} f'\left(\frac{t_{ij}}{p_i q_j}\right) \left(\frac{p_i q_j (a+b)}{2} - t_{ij}\right) \right).$$
$$(11.20)$$

Proof. Based on Theorem 6 of [42] and Theorem 8 of [39]. $\qquad\qquad \square$

     Furthermore we get

Theorem 11.10. *Let n odd and $f \in C^{n+1}([a,b])$, $a < b$, such that $f^{(n+1)} \geq 0$ (≤ 0). Then*

$$\Gamma_f(\mu_{XY}, \mu_X \times \mu_Y) \leq (\geq) \frac{1}{b-a} \int_a^b f(x)dx$$
$$- \sum_{i=1}^n \frac{1}{(i+1)!} \left[\sum_{k=0}^i \left(\sum_{\ell=1}^{\infty} \sum_{j=1}^{\infty} f^{(i)}\left(\frac{t_{\ell j}}{p_\ell q_j}\right) \right. \right.$$
$$\left. \left. \cdot (p_\ell q_j)^{1-i} (b p_\ell q_j - t_{\ell j})^{(i-k)} (a p_\ell q_j - t_{\ell j})^k \right) \right]. \qquad (11.21)$$

Proof. Based on Theorem 7 of [42] and Theorem 9 of [39]. $\qquad\qquad \square$

Part III. Here we apply results from [43] and [39].

     We start with

Theorem 11.11. *Let $f, g \in C^1([a,b])$ where f as in this chapter, $g' \neq 0$ over $[a,b]$. Then*

$$\Gamma_f(\mu_{XY}, \mu_X \times \mu_Y) \leq \left\| \frac{f'}{g'} \right\|_{\infty,[a,b]} \left(\sum_{i,j=1}^{\infty} p_i q_j \left| g\left(\frac{t_{ij}}{p_i q_j}\right) - g(1) \right| \right). \qquad (11.22)$$

Proof. Based on Theorem 1 of [43] and Theorem 10 of [39]. $\qquad\qquad \square$

     We give

Examples 11.12 (to Theorem 11.11).

1) Let $g(x) = \frac{1}{x}$. Then

$$\Gamma_f(\mu_{XY}, \mu_X \times \mu_Y) \le \|x^2 f'\|_{\infty,[a,b]} \left(\sum_{i,j=1}^{\infty} \frac{p_i q_j}{t_{ij}} |t_{ij} - p_i q_j| \right). \tag{11.23}$$

2) Let $g(x) = e^x$. Then

$$\Gamma_f(\mu_{XY}, \mu_X \times \mu_Y) \le \|e^{-x} f'\|_{\infty,[a,b]} \left(\sum_{i,j=1}^{\infty} p_i q_j |e^{t_{ij}/p_i q_j} - e| \right). \tag{11.24}$$

3) Let $g(x) = e^{-x}$. Then

$$\Gamma_f(\mu_{XY}, \mu_X \times \mu_Y) \le \|e^x f'\|_{\infty[a,b]} \left(\sum_{i,j=1}^{\infty} p_i q_j |e^{-t_{ij}/p_i q_j} - e^{-1}| \right). \tag{11.25}$$

4) Let $g(x) = \ln x$. Then

$$\Gamma_f(\mu_{XY}, \mu_X \times \mu_Y) \le \|x f'\|_{\infty,[a,b]} \left(\sum_{i,j=1}^{\infty} p_i q_j \left| \ln\left(\frac{t_{ij}}{p_i q_j} \right) \right| \right). \tag{11.26}$$

5) Let $g(x) = x \ln x$, with $a > e^{-1}$. Then

$$\Gamma_f(\mu_{XY}, \mu_X \times \mu_Y) \le \left\| \frac{f'}{1 + \ln x} \right\|_{\infty,[a,b]} \left(\sum_{i,j=1}^{\infty} t_{ij} \left| \ln\left(\frac{t_{ij}}{p_i q_j} \right) \right| \right). \tag{11.27}$$

6) Let $g(x) = \sqrt{x}$. Then

$$\Gamma_f(\mu_{XY}, \mu_X \times \mu_Y) \le 2\|\sqrt{x} f'\|_{\infty,[a,b]} \left(\sum_{i,j=1}^{\infty} \sqrt{p_i q_j} |\sqrt{t_{ij}} - \sqrt{p_i q_j}| \right). \tag{11.28}$$

7) Let $g(x) = x^\alpha$, $\alpha > 1$. Then

$$\Gamma_f(\mu_{XY}, \mu_X \times \mu_Y) \le \frac{1}{\alpha} \left\| \frac{f'}{x^{\alpha-1}} \right\|_{\infty,[a,b]} \left(\sum_{i,j=1}^{\infty} (p_i q_j)^{1-\alpha} |t_{ij}^\alpha - (p_i q_j)^\alpha| \right). \tag{11.29}$$

Next we give

Theorem 11.13. *Let $f \in C^1([a,b])$, $a < b$. Then*

$$\left| \Gamma_f(\mu_{XY}, \mu_X \times \mu_Y) - \frac{1}{b-a} \int_a^b f(x)dx \right|$$

$$\le \frac{\|f'\|_{\infty,[a,b]}}{(b-a)} \left[\left(\frac{a^2 + b^2}{2} \right) - (a + b) + \left(\sum_{i,j=1}^{\infty} (p_i q_j)^{-1} t_{ij}^2 \right) \right]. \tag{11.30}$$

Proof. By Theorem 2 of [43] and Theorem 11 of [39]. $\qquad\square$

It follows

Theorem 11.14. *Let $f \in C^{(2)}([a,b])$, $a < b$. Then*

$$\left| \Gamma_f(\mu_{XY}, \mu_X \times \mu_Y) - \frac{1}{b-a} \int_a^b f(s)ds - \left(\frac{f(b) - f(a)}{b-a} \right) \left(1 - \frac{a+b}{2} \right) \right|$$

$$\leq \frac{(b-a)^2}{2} \|f''\|_{\infty,[a,b]} \left\{ \left\{ \sum_{i,j=1}^\infty p_i q_j \left[\frac{\left(\frac{t_{ij}}{p_i q_j} - \left(\frac{a+b}{2} \right) \right)^2}{(b-a)^2} + \frac{1}{4} \right]^2 \right\} + \frac{1}{12} \right\}.$$

$$(11.31)$$

Proof. Based on (33) of [43] and Theorem 12 of [39]. $\qquad\square$

Working more generally we have

Theorem 11.15. *Let $f \in C^1([a,b])$ and $g \in C([a,b])$ of bounded variation, $g(a) \neq g(b)$. Then*

$$\left| \Gamma_f(\mu_{XY}, \mu_X \times \mu_Y) - \frac{1}{g(b) - g(a)} \int_a^b f(s)dg(s) \right|$$

$$\leq \|f''\|_{\infty,[a,b]} \left\{ \sum_{i,j=1}^\infty p_i q_j \left(\int_a^b \left| P\left(g\left(\frac{t_{ij}}{p_i q_j} \right), g(s) \right) \right| ds \right) \right\}, \qquad (11.32)$$

where

$$P(g(z), g(s)) := \begin{cases} \dfrac{g(s) - g(a)}{g(b) - g(a)}, & a \leq s \leq z, \\[3mm] \dfrac{g(s) - g(b)}{g(b) - g(a)}, & z < s \leq b. \end{cases} \qquad (11.33)$$

Proof. By Theorem 5 of [43] and Theorem 13 of [39]. $\qquad\square$

We give

Examples 11.16 (to Theorem 11.15). Let $f \in C^1([a,b])$ and $g(x) = e^x$, $\ln x$, $\sqrt{x}$, x^α, $\alpha > 1$; $x > 0$. Then by (71)–(74) of [43] and (30)–(33) of [39] we obtain

1)

$$\left| \Gamma_f(\mu_{XY}, \mu_X \times \mu_Y) - \frac{1}{e^b - e^a} \int_a^b f(s)e^s ds \right|$$

$$\leq \frac{\|f'\|_{\infty,[a,b]}}{(e^b - e^a)} \left\{ 2 \sum_{i,j=1}^\infty p_i q_j e^{t_{ij}/p_i q_j} - e^a(2-a) + e^b(b-2) \right\}, \qquad (11.34)$$

2)

$$\left| \Gamma_f(\mu_{XY}, \mu_X \times \mu_Y) - \left(\ln\left(\frac{b}{a} \right) \right)^{-1} \int_a^b \frac{f(s)}{s} ds \right|$$

$$\leq \|f'\|_{\infty,[a,b]} \left(\ln\left(\frac{b}{a} \right) \right)^{-1} \left\{ 2 \left(\sum_{i,j=1}^\infty t_{ij} \ln\left(\frac{t_{ij}}{p_i q_j} \right) \right) - \ln(ab) + (a+b-2) \right\},$$

$$(11.35)$$

3)

$$\left| \Gamma_f(\mu_{XY}, \mu_X \times \mu_Y) - \frac{1}{2(\sqrt{b} - \sqrt{a})} \int_a^b \frac{f(s)}{\sqrt{s}} ds \right|$$

$$\leq \frac{\|f'\|_{\infty[a,b]}}{(\sqrt{b} - \sqrt{a})} \left\{ \frac{4}{3} \left(\sum_{i,j=1}^{\infty} (p_i q_j)^{-1/2} t_{ij}^{3/2} \right) + \frac{(a^{3/2} + b^{3/2})}{3} - (\sqrt{a} + \sqrt{b}) \right\},$$

$$(11.36)$$

4) $(\alpha > 1)$

$$\left| \Gamma_f(\mu_{XY}, \mu_X \times \mu_Y) - \frac{\alpha}{(b^\alpha - a^\alpha)} \int_a^b f(s) s^{\alpha-1} ds \right|$$

$$\leq \frac{\|f'\|_{\infty,[a,b]}}{(b^\alpha - a^\alpha)} \left\{ \frac{2}{(\alpha + 1)} \left(\sum_{i,j=1}^{\infty} (p_i q_j)^{-\alpha} t_{ij}^{\alpha+1} \right) \right.$$

$$\left. + \left(\frac{\alpha}{\alpha + 1} \right) (a^{\alpha+1} + b^{\alpha+1}) - (a^\alpha + b^\alpha) \right\}. \qquad (11.37)$$

We continue with

Theorem 11.17. *Let $f \in C^{(2)}([a,b])$, $g \in C([a,b])$ and of bounded variation, $g(a) \neq g(b)$. Then*

$$\left| \Gamma_f(\mu_{XY}, \mu_X \times \mu_Y) \right.$$

$$- \frac{1}{(g(b) - g(a))} \int_a^b f(s) dg(s) - \frac{1}{(g(b) - g(a))} \left(\int_a^b f'(s_1) dg(s_1) \right)$$

$$\left. \cdot \left\{ \sum_{i,j=1}^{\infty} \left\{ p_i q_j \left(\int_a^b P \left(g \left(\frac{t_{ij}}{p_i q_j} \right), g(s) \right) ds \right) \right\} \right\} \right|$$

$$\leq \|f''\|_{\infty,[a,b]} \left\{ \sum_{i,j=1}^{\infty} \left(p_i q_j \left(\int_a^b \int_a^b \left| P \left(g \left(\frac{t_{ij}}{p_i q_j} \right), g(s) \right) \right| |P(g(s), g(s_1))| ds_1 ds \right) \right) \right\}.$$

$$(11.38)$$

Proof. Apply (79) of [43] and Theorem 14 of [39]. $\qquad \square$

Remark 11.18. Next we define

$$P^*(z, s) := \begin{cases} s - a, & s \in [a, z], \\ s - b, & s \in (z, b]. \end{cases} \qquad (11.39)$$

Let $f \in C^1([a,b])$. Then by (25) of [43] we get the representation

$$\Gamma_f(\mu_{XY}, \mu_X \times \mu_Y) = \frac{1}{b - a} \left(\int_a^b f(s) ds + \mathcal{R}_1 \right),$$

where according to (26) of [43] and (36) of [39] we have

$$\mathcal{R}_1 = \sum_{i,j=1}^{\infty} p_i q_j \left(\int_a^b P^* \left(\frac{t_{ij}}{p_i q_j}, s \right) f'(s) ds \right). \tag{11.40}$$

Let again $f \in C^1([a,b])$, $g \in C([a,b])$ and of bounded variation, $g(a) \neq g(b)$. Then by (47) of [43] we get the representation

$$\Gamma_f(\mu_{XY}, \mu_X \times \mu_Y) = \frac{1}{(g(b) - g(a))} \int_a^b f(s) dg(s) + \mathcal{R}_4, \tag{11.41}$$

where by (48) of [43] and (38) of [39] we obtain

$$\mathcal{R}_4 = \sum_{i,j=1}^{\infty} p_i q_j \left(\int_a^b P \left(g \left(\frac{t_{ij}}{p_i q_j} \right), g(s) \right) f'(s) ds \right). \tag{11.42}$$

Let $\tilde{\alpha}, \beta > 1 : \frac{1}{\tilde{\alpha}} + \frac{1}{\beta} = 1$, then by (89) of [43] and (39) of [39] we have

$$|\mathcal{R}_1| \leq \|f'\|_{\tilde{\alpha},[a,b]} \left(\sum_{i,j=1}^{\infty} p_i q_j \left\| P^* \left(\frac{t_{ij}}{p_i q_j}, \cdot \right) \right\|_{\beta,[a,b]} \right), \tag{11.43}$$

and by (90) of [43] and (40) of [39] we get

$$|\mathcal{R}_1| \leq \|f'\|_{1,[a,b]} \left(\sum_{i,j=1}^{\infty} p_i q_j \left\| P^* \left(\frac{t_{ij}}{p_i q_j}, \cdot \right) \right\|_{\infty,[a,b]} \right). \tag{11.44}$$

Furthermore by (95), (96) of [43] and (41), (42) of [39] we find

$$|\mathcal{R}_4| \leq \|f'\|_{\tilde{\alpha},[a,b]} \left(\sum_{i,j=1}^{\infty} p_i q_j \left\| P \left(g \left(\frac{t_{ij}}{p_i q_j} \right), g(\cdot) \right) \right\|_{\beta,[a,b]} \right), \tag{11.45}$$

and

$$|\mathcal{R}_4| \leq \|f'\|_{1,[a,b]} \left(\sum_{i,j=1}^{\infty} p_i q_j \left\| P \left(g \left(\frac{t_{ij}}{p_i q_j} \right), g(\cdot) \right) \right\|_{\infty,[a,b]} \right), \tag{11.46}$$

Remark 11.19. Let $a < b$.

(i) Here by (105) of [43] and (43) of [39] we obtain

$$|\mathcal{R}_1| \leq \|f'\|_{1,[a,b]} \left(\sum_{i,j=1}^{\infty} p_i q_j \max \left(\frac{t_{ij}}{p_i q_j} - a, b - \frac{t_{ij}}{p_i q_j} \right) \right), \quad f \in C^1([a,b]). \tag{11.47}$$

(ii) Let g be strictly increasing and continuous over $[a,b]$, e.g., $g(x) = e^x$, $\ln x$, $\sqrt{x}$, x^α with $\alpha > 1$, and $x > 0$ whenever is needed. Also $f \in C^1([a,b])$. Then by (106) of [43] and (44) of [39] we find

$$|\mathcal{R}_4| \leq \frac{\|f'\|_{1,[a,b]}}{(g(b) - g(a))} \left(\sum_{i,j=1}^{\infty} p_i q_j \max \left(g \left(\frac{t_{ij}}{p_i q_j} \right) - g(a), g(b) - g \left(\frac{t_{ij}}{p_i q_j} \right) \right) \right). \tag{11.48}$$

In particular via (107)–(110) of [43] and (45)–(48) of [39] we obtain

1)

$$|\mathcal{R}_4| \le \frac{\|f'\|_{1,[a,b]}}{(e^b - e^a)} \left(\sum_{i,j=1}^{\infty} p_i q_j \max\left(e^{t_{ij}/p_i q_j} - e^a, e^b - e^{t_{ij}/p_i q_j}\right) \right),$$

(11.49)

2)

$$|\mathcal{R}_4| \le \frac{\|f'\|_{1,[a,b]}}{(\ln b - \ln a)} \left(\sum_{i,j=1}^{\infty} p_i q_j \max\left(\ln\left(\frac{t_{ij}}{p_i q_j}\right) - \ln a, \ln b - \ln\left(\frac{t_{ij}}{p_i q_j}\right) \right) \right),$$

(11.50)

3)

$$|\mathcal{R}_4| \le \frac{\|f'\|_{1,[a,b]}}{\sqrt{b} - \sqrt{a}} \left(\sum_{i,j=1}^{\infty} p_i q_j \max\left(\sqrt{\frac{t_{ij}}{p_i q_j}} - \sqrt{a}, \sqrt{b} - \sqrt{\frac{t_{ij}}{p_i q_j}} \right) \right),$$

(11.51)

and finally for $\alpha > 1$ we obtain

4)

$$|\mathcal{R}_4| \le \frac{\|f'\|_{1,[a,b]}}{b^\alpha - a^\alpha} \left(\sum_{i,j=1}^{\infty} p_i q_j \max\left(\frac{t_{ij}^\alpha}{(p_i q_j)^\alpha} - a^\alpha, b^\alpha - \frac{t_{ij}^\alpha}{(p_i q_j)^\alpha} \right) \right).$$

(11.52)

At last

iii) Let $\tilde{\alpha}, \beta > 1$ such that $\frac{1}{\tilde{\alpha}} + \frac{1}{\beta} = 1$, then by (115) of [43] and (49) of [39] we have

$$|\mathcal{R}_1| \le \|f'\|_{\tilde{\alpha},[a,b]} \left(\sum_{i,j=1}^{\infty} \left(\sqrt[\beta]{\frac{(t_{ij} - a p_i q_j)^{\beta+1} + (b p_i q_j - t_{ij})^{\beta+1}}{(\beta + 1) p_i q_j}} \right) \right).$$

(11.53)

Part IV. Here we apply results from [41] and [39].

We start with

Theorem 11.20. *Let* $a < b$, $f \in C^n([a,b])$, $n \ge 1$ *with* $|f^{(n)}(s) - f^{(n)}(1)|$ *be a convex function in* s. *Let* $0 < h < \min(1 - a, b - 1)$ *be fixed. Then*

$$\Gamma_f(\mu_{XY}, \mu_X \times \mu_Y) \le \sum_{k=2}^{n} \frac{|f^{(k)}(1)|}{k!} \left| \sum_{i,j=1}^{\infty} (p_i q_j)^{1-k} (t_{ij} - p_i q_j)^k \right|$$

$$+ \frac{\omega_1(f^{(n)}, h)}{(n+1)! h} \left(\sum_{i,j=1}^{\infty} (p_i q_j)^{-n} |t_{ij} - p_i q_j|^{n+1} \right).$$

(11.54)

Here ω_1 *is the usual first modulus of continuity. If* $f^{(k)}(1) = 0$, $k = 2, \ldots, n$, *then*

$$\Gamma_f(\mu_{XY}, \mu_X \times \mu_Y) \le \frac{\omega_1(f^{(n)}, h)}{(n+1)! h} \left(\sum_{i,j=1}^{\infty} (p_i q_j)^{-n} |t_{ij} - p_i q_j|^{n+1} \right).$$

(11.55)

Inequalities (11.54) and (11.55) when n is even are attained by

$$\tilde{f}(s) = \frac{|s-1|^{n+1}}{(n+1)!}, \quad a \le s \le b. \tag{11.56}$$

Proof. Apply Theorem 1 of [41] and Theorem 15 of [39].   □

We continue with the general

Theorem 11.21. *Let $f \in C^n([a,b])$, $n \ge 1$. Assume that $\omega_1(f^{(n)}, \delta) \le w$, where $0 < \delta \le b - a$, $w > 0$. Let $x \in \mathbb{R}$ and denote by*

$$\phi_n(x) := \int_0^{|x|} \left\lceil \frac{s}{\delta} \right\rceil \frac{(|x|-s)^{n-1}}{(n-1)!} ds, \tag{11.57}$$

where $\lceil \cdot \rceil$ is the ceiling of the number. Then

$$\Gamma_f(\mu_{XY}, \mu_X \times \mu_Y) \le \sum_{k=2}^{n} \frac{|f^{(k)}(1)|}{k!} \left| \sum_{i,j=1}^{\infty} (p_i q_j)^{1-k} (t_{ij} - p_i q_j)^k \right|$$

$$+ w \left(\sum_{i,j=1}^{\infty} p_i q_j \phi_n \left(\frac{t_{ij} - p_i q_j}{p_i q_j} \right) \right). \tag{11.58}$$

Inequality (11.58) is sharp, namely it is attained by the function

$$\tilde{f}_n(s) := w \phi_n(s-1), \quad a \le s \le b, \tag{11.59}$$

when n is even.

Proof. Based on Theorem 2 of [41] and Theorem 16 of [39].   □

It follows

Corollary 11.22 (to Theorem 11.21). *It holds $(0 < \delta \le b - a)$*

$$\Gamma_f(\mu_{XY}, \mu_X \times \mu_Y) \le \sum_{k=2}^{n} \frac{|f^{(k)}(1)|}{k!} \left| \sum_{i,j=1}^{\infty} (p_i q_j)^{1-k} (t_{ij} - p_i q_j)^k \right|$$

$$+ \omega_1(f^{(n)}, \delta) \left\{ \frac{1}{(n+1)!\delta} \left(\sum_{i,j=1}^{\infty} (p_i q_j)^{-n} |t_{ij} - p_i q_j|^{n+1} \right) \right.$$

$$+ \frac{1}{2n!} \left(\sum_{i,j=1}^{\infty} (p_i q_j)^{1-n} |t_{ij} - p_i q_j|^n \right)$$

$$\left. + \frac{\delta}{8(n-1)!} \left(\sum_{i,j=1}^{\infty} (p_i q_j)^{2-n} |t_{ij} - p_i q_j|^{n-1} \right) \right\}, \tag{11.60}$$

and

$$\Gamma_f(\mu_{XY}, \mu_X \times \mu_Y) \leq \sum_{k=2}^{n} \frac{|f^{(k)}(1)|}{k!} \left| \sum_{i,j=1}^{\infty} (p_i q_j)^{1-k} (t_{ij} - p_i q_j)^k \right|$$

$$+ \omega_1(f^{(n)}, \delta) \left\{ \frac{1}{n!} \left(\sum_{i,j=1}^{\infty} (p_i q_j)^{1-n} |t_{ij} - p_i q_j|^n \right) \right.$$

$$\left. + \frac{1}{n!(n+1)\delta} \left(\sum_{i,j=1}^{\infty} (p_i q_j)^{-n} |t_{ij} - p_i q_j|^{n+1} \right) \right\}. \quad (11.61)$$

In particular we have

$$\Gamma_f(\mu_{XY}, \mu_X \times \mu_Y) \leq \omega_1(f', \delta) \left\{ \frac{1}{2\delta} \left(\sum_{i,j=1}^{\infty} (p_i q_j)^{-1} (t_{ij} - p_i q_j)^2 \right) \right.$$

$$\left. + \frac{1}{2} \left(\sum_{i,j=1}^{\infty} |t_{ij} - p_i q_j| \right) + \frac{\delta}{8} \right\}, \quad (11.62)$$

and

$$\Gamma_f(\mu_{XY}, \mu_X \times \mu_Y) \leq \omega_1(f', \delta) \left\{ \left(\sum_{i,j=1}^{\infty} |t_{ij} - p_i q_j| \right) \right.$$

$$\left. + \frac{1}{2\delta} \left(\sum_{i,j=1}^{\infty} (p_i q_j)^{-1} (t_{ij} - p_i q_j)^2 \right) \right\}. \quad (11.63)$$

Proof. By Corollary 1 of [41] and Corollary 2 of [39]. $\square$

Also we have

Corollary 11.23 (to Theorem 11.21). *Assume here that $f^{(n)}$ is of Lipschitz type of order α, $0 < \alpha \leq 1$, i.e.*

$$\omega_1(f^{(n)}, \delta) \leq K\delta^\alpha, \quad K > 0, \quad (11.64)$$

for any $0 < \delta \leq b - a$. Then

$$\Gamma_f(\mu_{XY}, \mu_X \times \mu_Y) \leq \sum_{k=2}^{n} \frac{|f^{(k)}(1)|}{k!} \left| \sum_{i,j=1}^{\infty} (p_i q_j)^{1-k} (t_{ij} - p_i q_j)^k \right|$$

$$+ \frac{K}{\prod\limits_{i=1}^{n}(\alpha + i)} \left(\sum_{i,j=1}^{\infty} (p_i q_j)^{1-n-\alpha} |t_{ij} - p_i q_j|^{n+\alpha} \right). \quad (11.65)$$

When n is even (11.65) is attained by $f^*(x) = c|x - 1|^{n+\alpha}$, where $c := K/\left(\prod\limits_{i=1}^{n}(\alpha + i) \right) > 0$.

Proof. Application of Corollary 2 of [41] and Corollary 3 of [39]. $\qquad\square$

Next comes

Corollary 11.24 (to Theorem 11.21). *Assume that*

$$b - a \geq \left(\sum_{i,j=1}^{\infty} (p_i q_j)^{-1} t_{ij}^2 \right) - 1 > 0. \tag{11.66}$$

Then

$$\Gamma_f(\mu_{XY}, \mu_X \times \mu_Y) \leq \omega_1 \left(f', \left(\sum_{i,j=1}^{\infty} (p_i q_j)^{-1} t_{ij}^2 - 1 \right) \right) \left\{ \left(\sum_{i,j=1}^{\infty} |t_{ij} - p_i q_j| \right) + \frac{1}{2} \right\}. \tag{11.67}$$

Proof. Application of Corollary 3 of [41] and Corollary 4 of [39]. $\qquad\square$

Corollary 11.25 (to Theorem 11.21). *Assume that*

$$\sum_{i,j=1}^{\infty} |t_{ij} - p_i q_j| > 0. \tag{11.68}$$

Then

$$\Gamma_f(\mu_{XY}, \mu_X \times \mu_Y) \leq \omega_1 \left(f', \left(\sum_{i,j=1}^{\infty} |t_{ij} - p_i q_j| \right) \right) \left\{ \left(\sum_{i,j=1}^{\infty} |t_{ij} - p_i q_j| \right) + \frac{b-a}{2} \right\}$$

$$\leq \frac{3}{2}(b-a)\omega_1 \left(f', \left(\sum_{i,j=1}^{\infty} |t_{ij} - p_i q_j| \right) \right). \tag{11.69}$$

Proof. Application of Corollary 5 of [41] and Corollary 5 of [39]. $\qquad\square$

We present

Proposition 11.26. *Let* $f \in C([a,b])$.

i) *Suppose that*

$$\left(\sum_{i,j=1}^{\infty} |t_{ij} - p_i q_j| \right) > 0. \tag{11.70}$$

Then

$$\Gamma_f(\mu_{XY}, \mu_X \times \mu_Y) \leq 2\omega_1 \left(f, \left(\sum_{i,j=1}^{\infty} |t_{ij} - p_i q_j| \right) \right). \tag{11.71}$$

ii) *Let* $r > 0$ *and*

$$b - a \geq r \left(\sum_{i,j=1}^{\infty} |t_{ij} - p_i q_j| \right) > 0. \tag{11.72}$$

Then

$$\Gamma_f(\mu_{XY}, \mu_X \times \mu_Y) \le \left(1 + \frac{1}{r}\right) \omega_1 \left(f, r\left(\sum_{i,j=1}^{\infty} |t_{ij} - p_i q_j|\right)\right). \tag{11.73}$$

Proof. By Propositions 1 of [41] and [39]. $\qquad \square$

Also we give

Proposition 11.27. *Let f be a Lipschitz function of order α, $0 < \alpha \le 1$, i.e. there exists $K > 0$ such that*

$$|f(x) - f(y)| \le K|x - y|^{\alpha}, \quad all\ x, y \in [a, b]. \tag{11.74}$$

Then

$$\Gamma_f(\mu_{XY}, \mu_X \times \mu_Y) \le K \left(\sum_{i,j=1}^{\infty} (p_i q_j)^{1-\alpha} |t_{ij} - p_i q_j|^{\alpha}\right). \tag{11.75}$$

Proof. By Propositions 2 of [41] and [39]. $\qquad \square$

Next we present some alternative type of results. We start with

Theorem 11.28 *Let $f \in C^{n+1}([a, b])$, $n \in \mathbb{N}$, then we get the representation*

$$\Gamma_f(\mu_{XY}, \mu_X \times \mu_Y) = \sum_{k=0}^{n-1} \frac{(-1)^k}{(k+1)!} \left(\sum_{i,j=1}^{\infty} f^{(k+1)}\left(\frac{t_{ij}}{p_i q_j}\right)(p_i q_j)^{-k}(t_{ij} - p_i q_j)^{k+1}\right) + \psi_2, \tag{11.76}$$

where

$$\psi_2 = \frac{(-1)^n}{n!} \left(\sum_{i,j=1}^{\infty} (p_i q_j) \left(\int_1^{\left(\frac{t_{ij}}{p_i q_j}\right)} (s-1)^n f^{(n+1)}(s)\,ds\right)\right). \tag{11.77}$$

Proof. By Theorem 12 of [41] and Theorem 17 of [39]. $\qquad \square$

Next we estimate ψ_2. We present

Theorem 11.29. *Let $f \in C^{n+1}([a, b])$, $n \in \mathbb{N}$. Then*

$$|\psi_2| \le \min\ of
\begin{cases}
B_1 := \dfrac{\|f^{(n+1)}\|_{\infty,[a,b]}}{(n+1)!} \left(\displaystyle\sum_{i,j=1}^{\infty} (p_i q_j)^{-n} |t_{ij} - p_i q_j|^{n+1}\right), \\[2.5em]
B_2 := \dfrac{\|f^{(n+1)}\|_{1,[a,b]}}{n!} \left(\displaystyle\sum_{i,j=1}^{\infty} (p_i q_j)^{1-n} |t_{ij} - p_i q_j|^{n}\right), \\[2.5em]
\qquad\quad and\ for\ \tilde{\alpha}, \beta > 1: \dfrac{1}{\tilde{\alpha}} + \dfrac{1}{\beta} = 1, \\[2.5em]
B_3 := \dfrac{\|f^{(n+1)}\|_{\tilde{\alpha},[a,b]}}{n!(n\beta+1)^{1/\beta}} \left(\displaystyle\sum_{i,j=1}^{\infty} (p_i q_j)^{1-n-1/\beta} |t_{ij} - p_i q_j|^{n+1/\beta}\right).
\end{cases} \tag{11.78}$$

Also it holds

$$\Gamma_f(\mu_{XY}, \mu_X \times \mu_Y) \leq \sum_{k=0}^{n-1} \frac{1}{(k+1)!} \left| \sum_{i,j=1}^{\infty} f^{(k+1)}\left(\frac{t_{ij}}{p_i q_j}\right) (p_i q_j)^{-k} (t_{ij} - p_i q_j)^{k+1} \right|$$

$$+ \min(B_1, B_2, B_3). \tag{11.79}$$

Proof. See Theorem 13 of [41] and Theorem 18 of [39]. $\qquad\square$

We have

Corollary 11.30 (to Theorem 11.29). *Case of $n = 1$. It holds*

$$|\psi_2| \leq \min \ of \begin{cases} B_{1,1} := \dfrac{\|f''\|_{\infty,[a,b]}}{2} \left(\displaystyle\sum_{i,j=1}^{\infty} (p_i q_j)^{-1} |t_{ij} - p_i q_j|^2 \right), \\[2em] B_{2,1} := \|f''\|_{1,[a,b]} \left(\displaystyle\sum_{i,j=1}^{\infty} |t_{ij} - p_i q_j| \right), \\[2em] and \ for \ \tilde{\alpha}, \beta > 1, \dfrac{1}{\tilde{\alpha}} + \dfrac{1}{\beta} = 1, \\[2em] B_{3,1} := \dfrac{\|f''\|_{\tilde{\alpha},[a,b]}}{(\beta+1)^{1/\beta}} \left(\displaystyle\sum_{i,j=1}^{\infty} (p_i q_j)^{-1/\beta} |t_{ij} - p_i q_j|^{1+1/\beta} \right). \end{cases} \tag{11.80}$$

Also it holds

$$\Gamma_f(\mu_{XY}, \mu_X \times \mu_Y) \leq \left| \sum_{i,j=1}^{\infty} f'\left(\frac{t_{ij}}{p_i q_j}\right) (t_{ij} - p_i q_j) \right| + \min(B_{1,1}, B_{2,1}, B_{3,1}). \tag{11.81}$$

Proof. See Corollary 8 of [41] and Corollary 6 of [39]. $\qquad\square$

We further give

Theorem 11.31 *Let $f \in C^{n+1}([a,b])$, $n \in \mathbb{N}$, then we get the representation*

$$\Gamma_f(\mu_{XY}, \mu_X \times \mu_Y) = \sum_{k=0}^{n-1} \frac{(-1)^k}{(k+1)!} \left(\sum_{i,j=1}^{\infty} f^{(k+1)}\left(\frac{t_{ij}}{p_i q_j}\right) (p_i q_j)^{-k} (t_{ij} - p_i q_j)^{k+1} \right)$$

$$+ \frac{(-1)^n}{(n+1)!} f^{(n+1)}(1) \left(\sum_{i,j=1}^{\infty} (p_i q_j)^{-n} (t_{ij} - p_i q_j)^{n+1} \right) + \psi_4, \tag{11.82}$$

where

$$\psi_4 = \sum_{i,j=1}^{\infty} p_i q_j \psi_3 \left(\frac{t_{ij}}{p_i q_j}\right), \tag{11.83}$$

with

$$\psi_3(x) := \frac{(-1)^n}{n!} \int_1^x (s-1)^n \big(f^{(n+1)}(s) - f^{(n+1)}(1)\big)ds. \qquad (11.84)$$

Proof. Based on Theorem 14 of [41] and Theorem 19 of [39]. $\qquad\square$

Next we present estimations of ψ_4.

Theorem 11.32. *Let* $f \in C^{n+1}([a,b])$, $n \in \mathbb{N}$, *and*

$$0 < \frac{1}{(n+2)} \left(\sum_{i,j=1}^{\infty} (p_i q_j)^{-n-1} |t_{ij} - p_i q_j|^{n+2} \right) \leq b - a. \qquad (11.85)$$

Then (11.82) is again valid and

$$|\psi_4| \leq \frac{1}{n!} \omega_1 \left(f^{(n+1)}, \frac{1}{(n+2)} \left(\sum_{i,j=1}^{\infty} (p_i q_j)^{-n-1} |t_{ij} - p_i q_j|^{n+2} \right) \right)$$

$$\cdot \left(1 + \frac{1}{(n+1)} \left(\sum_{i,j=1}^{\infty} (p_i q_j)^{-n} |t_{ij} - p_i q_j|^{n+1} \right) \right). \qquad (11.86)$$

Proof. By Theorem 15 of [41] and Theorem 20 of [39]. $\qquad\square$

Also we have

Theorem 11.33. *Let* $f \in C^{n+1}([a,b])$, $n \in \mathbb{N}$, *and* $|f^{(n+1)}(s) - f^{(n+1)}(1)|$ *is convex in* s *and*

$$0 < \frac{1}{n!(n+2)} \left(\sum_{i,j=1}^{\infty} (p_i q_j)^{-n-1} |t_{ij} - p_i q_j|^{n+2} \right) \leq \min(1-a, b-1). \qquad (11.87)$$

Then (11.82) is again valid and

$$|\psi_4| \leq \omega_1 \left(f^{(n+1)}, \frac{1}{n!(n+2)} \left(\sum_{i,j=1}^{\infty} (p_i q_j)^{-n-1} |t_{ij} - p_i q_j|^{n+2} \right) \right). \qquad (11.88)$$

The last (11.88) is attained by

$$\hat{f}(s) := \frac{(s-1)^{n+2}}{(n+2)!}, \quad a \leq s \leq b \qquad (11.89)$$

when n *is even.*

Proof. By Theorem 16 of [41] and Theorem 21 of [39]. $\qquad\square$

When $n = 1$ we obtain

Corollary 11.34 (to Theorem 11.32). *Let* $f \in C^2([a,b])$ *and*

$$0 < \frac{1}{3} \left(\sum_{i,j=1}^{\infty} (p_i q_j)^{-2} |t_{ij} - p_i q_j|^3 \right) \leq b - a. \qquad (11.90)$$

Then

$$\Gamma_f(\mu_{XY}, \mu_X \times \mu_Y) = \sum_{i,j=1}^{\infty} f'\left(\frac{t_{ij}}{p_i q_j}\right)(t_{ij} - p_i q_j)$$

$$-\frac{f''(1)}{2}\left(\sum_{i,j=1}^{\infty}(p_i q_j)^{-1}(t_{ij} - p_i q_j)^2\right) + \psi_{4,1}, \quad (11.91)$$

where

$$\psi_{4,1} := \sum_{i,j=1}^{\infty} p_i q_j \psi_{3,1}\left(\frac{t_{ij}}{p_i q_j}\right), \quad (11.92)$$

with

$$\psi_{3,1}(x) := -\int_1^x (s-1)(f''(s) - f''(1))ds, \quad x \in [a,b]. \quad (11.93)$$

Furthermore it holds

$$|\psi_{4,1}| \leq \omega_1\left(f'', \frac{1}{3}\left(\sum_{i,j=1}^{\infty}(p_i q_j)^{-2}|t_{ij} - p_i q_j|^3\right)\right)\frac{1}{2}\left(\left(\sum_{i,j=1}^{\infty}(p_i q_j)^{-1}t_{ij}^2\right) + 1\right).$$

$$(11.94)$$

Proof. See Corollary 9 of [41] and Corollary 7 of [39]. $\square$

Also we have

Corollary 11.35 (to Theorem 11.33). *Let $f \in C^2([a,b])$, and $|f^{(2)}(s) - f^{(2)}(1)|$ is convex in s and*

$$0 < \frac{1}{3}\left(\sum_{i,j=1}^{\infty}(p_i q_j)^{-2}|t_{ij} - p_i q_j|^3\right) \leq \min(1-a, b-1). \quad (11.95)$$

Then again (11.91) is true and

$$|\psi_{4,1}| \leq \omega_1\left(f'', \frac{1}{3}\left(\sum_{i,j=1}^{\infty}(p_i q_j)^{-2}|t_{ij} - p_i q_j|^3\right)\right). \quad (11.96)$$

Proof. See Corollary 10 of [41] and Corollary 8 of [39]. $\square$

The last main result of the chapter follows.

Theorem 11.36. *Let $f \in C^{n+1}([a,b])$, $n \in \mathbb{N}$. Additionally assume that*

$$|f^{(n+1)}(x) - f^{(n+1)}(y)| \leq K|x-y|^{\alpha}, \quad (11.97)$$

$K > 0$, $0 < \alpha \leq 1$, all $x, y \in [a,b]$ with $(x-1)(y-1) \geq 0$. Then

$$\Gamma_f(\mu_{XY}, \mu_X \times \mu_Y) = \sum_{k=0}^{n-1} \frac{(-1)^k}{(k+1)!}\left(\sum_{i,j=1}^{\infty} f^{(k+1)}\left(\frac{t_{ij}}{p_i q_j}\right)(p_i q_j)^{-k}(t_{ij} - p_i q_j)^{k+1}\right)$$

$$+ \frac{(-1)^n}{(n+1)!}f^{(n+1)}(1)\left(\sum_{i,j=1}^{\infty}(p_i q_j)^{-n}(t_{ij} - p_i q_j)^{n+1}\right) + \psi_4.$$

$$(11.98)$$

Here we find that

$$|\psi_4| \le \frac{K}{n!(n+\alpha+1)}\left(\sum_{i,j=1}^{\infty}(p_iq_j)^{-n-\alpha}|t_{ij}-p_iq_j|^{n+\alpha+1}\right). \tag{11.99}$$

Inequality (11.99) is attained when n is even by

$$f^*(x) = \tilde{c}|x-1|^{n+\alpha+1}, \quad x \in [a,b] \tag{11.100}$$

where

$$\tilde{c} := \frac{K}{\prod\limits_{j=0}^{n}(n+\alpha+1-j)}. \tag{11.101}$$

Proof. By Theorem 17 of [41] and Theorem 22 of [39]. $\qquad\square$

Finally we give

Corollary 11.37 (to Theorem 11.36). *Let $f \in C^2([a,b])$. Additionally assume that*

$$|f''(x) - f''(y)| \le K|x-y|^{\alpha}, \quad K > 0,\ 0 < \alpha \le 1, \tag{11.102}$$

all $x, y \in [a,b]$ with $(x-1)(y-1) \ge 0$. Then

$$\Gamma_f(\mu_{XY}, \mu_X \times \mu_Y) = \sum_{i,j=1}^{\infty} f'\left(\frac{t_{ij}}{p_iq_j}\right)(t_{ij}-p_iq_j)$$

$$- \frac{f''(1)}{2}\left(\sum_{i,j=1}^{\infty}(p_iq_j)^{-1}(t_{ij}-p_iq_j)^2\right) + \psi_{4,1}. \tag{11.103}$$

Here

$$\psi_{4,1} := \sum_{i,j=1}^{\infty} p_iq_j\psi_{3,1}\left(\frac{t_{ij}}{p_iq_j}\right) \tag{11.104}$$

with

$$\psi_{3,1}(x) := -\int_1^x (s-1)(f''(s) - f''(1))ds, \quad x \in [a,b]. \tag{11.105}$$

It holds that

$$|\psi_{4,1}| \le \frac{K}{(\alpha+2)}\left(\sum_{i,j=1}^{\infty}(p_iq_j)^{-1-\alpha}|t_{ij}-p_iq_j|^{\alpha+2}\right). \tag{11.106}$$

Proof. See Corollary 11 of [41] and Corollary 9 of [39]. $\qquad\square$

Chapter 12

Hölder-Like Csiszar's f-Divergence Inequalities

In this chapter we present a set of a large variety of probabilistic inequalities regarding Csiszar's f-divergence of two probability measures. These are in parallel to Hölder's type inequalities and some of them are attained. This treatment is based on [40].

12.1 Background

Here we follow [140].

Let f be an arbitrary convex function from $(0, +\infty)$ into $\mathbb{R}$ which is strictly convex at 1. We agree with the following notational conventions:

$$f(0) = \lim_{u \to 0+} f(u),$$

$$0 \cdot f\left(\frac{0}{0}\right) = 0, \tag{12.1}$$

$$0 f\left(\frac{a}{0}\right) = \lim_{\varepsilon \to 0+} \varepsilon f\left(\frac{a}{\varepsilon}\right) = a \lim_{u \to +\infty} \frac{f(u)}{u}, \quad (0 < a < +\infty).$$

Let $(X, \mathcal{A}, \lambda)$ be an arbitrary measure space with λ being a finite or σ-finite measure. Let also μ_1, μ_2 probability measures on X such that $\mu_1, \mu_2 \ll \lambda$ (absolutely continuous).

Denote the Radon–Nikodym derivatives (densities) of μ_i with respect to λ by $p_i(x)$:

$$p_i(x) = \frac{\mu_i(dx)}{\lambda(dx)}, \quad i = 1, 2.$$

The quantity (see [140]).

$$\Gamma_f(\mu_1, \mu_2) = \int_X p_2(x) f\left(\frac{p_1(x)}{p_2(x)}\right) \lambda(dx) \tag{12.2}$$

is called the *f-divergence* of the probability measures μ_1 and μ_2. The function f is called the *base function*. By Lemma 1.1 of [140] $\Gamma_f(\mu_1, \mu_2)$ is always well-defined

155

and $\Gamma_f(\mu_1, \mu_2) \geq f(1)$ with equality only for $\mu_1 = \mu_2$. Also from [140] it derives that $\Gamma_f(\mu_1, \mu_2)$ does not depend on the choice of λ. When assuming $f(1) = 0$ then Γ_f can be considered as the most general measure of difference between probability measures.

The Csiszar's f-divergence Γ_f incorporated most of special cases of probability measure distances, e.g. the variation distance, χ^2-divergence, information for discrimination or generalized entropy, information gain, mutual information, mean square contingency, etc. Γ_f has tremendous applications to almost all applied sciences where stochastics enters. For more see [37], [42], [43], [41], [139], [140], [141], [152].

In this chapter all the base functions f appearing in the function Γ_f are assumed to have all the above properties of f. Notice that $\Gamma_f(\mu_1, \mu_2) \leq \Gamma_{|f|}(\mu_1, \mu_2)$. For arbitrary functions f, g we notice that

$$f^r \circ g = (f \circ g)^r, \quad r \in \mathbb{R} \quad \text{and} \quad |f \circ g| = |f| \circ g.$$

Clearly one can have widely products of (strictly) convex functions to be (strictly) convex.

12.2   Results

We start with

Theorem 12.1. *Let $p, q > 1$ such that $\frac{1}{p} + \frac{1}{q} = 1$. Then*

$$\Gamma_{|f_1 f_2|}(\mu_1, \mu_2) \leq \left(\Gamma_{|f_1|^p}(\mu_1, \mu_2)\right)^{1/p} \left(\Gamma_{|f_2|^q}(\mu_1, \mu_2)\right)^{1/q}. \tag{12.3}$$

Equality in (12.3) is true when $|f_1|^p = c|f_2|^q$ for $c > 0$. More precisely, inequality (12.3) holds as equality iff there are $a, b \in \mathbb{R}$ not both zero such that

$$p_2 \cdot \left((a|f_1|^p - b|f_2|^q)\left(\frac{p_1}{p_2}\right)\right) = 0, \quad \text{a.e. } [\lambda].$$

Proof. Here we use the Hölder's inequality. We see that

$$\Gamma_{|f_1 f_2|}(\mu_1, \mu_2) = \int_X p_2 |f_1|\left(\frac{p_1}{p_2}\right) \cdot |f_2|\left(\frac{p_1}{p_2}\right) d\lambda$$

$$= \int_X \left(p_2^{1/p}|f_1|\left(\frac{p_1}{p_2}\right)\right)\left(p_2^{1/q}|f_2|\left(\frac{p_1}{p_2}\right)\right) d\lambda$$

$$\leq \left(\int_X p_2 |f_1|^p \left(\frac{p_1}{p_2}\right) d\lambda\right)^{1/p} \left(\int_X p_2 |f_2|^q \left(\frac{p_1}{p_2}\right) d\lambda\right)^{1/q}$$

$$= \left(\Gamma_{|f_1|^p}(\mu_1, \mu_2)\right)^{1/p} \left(\Gamma_{|f_2|^q}(\mu_1, \mu_2)\right)^{1/q}. \qquad \square$$

Corollary 12.2. *It holds*

$$\Gamma_{|f_1 f_2|}(\mu_1, \mu_2) \leq \left(\Gamma_{f_1^2}(\mu_1, \mu_2)\right)^{1/2} \left(\Gamma_{f_2^2}(\mu_1, \mu_2)\right)^{1/2}. \tag{12.4}$$

Proof. By (12.3) setting $p = q = 2$. $\qquad\square$

We continue with a generalization of Theorem 12.1.

Theorem 12.3. *Let* $a_1, a_2, \ldots, a_n > 1$, $n \in \mathbb{N}$: $\sum_{i=1}^{n} \frac{1}{a_i} = 1$. *Then*

$$\Gamma_{\left|\prod_{i=1}^{n} f_i\right|}(\mu_1, \mu_2) \leq \prod_{i=1}^{n} \left(\Gamma_{|f_i|^{a_i}}(\mu_1, \mu_2)\right)^{1/a_i}. \tag{12.5}$$

Proof. Here we use the generalized Hölder's inequality, see p. 186, [235] and p. 200, [214]. We observe that

$$\Gamma_{\left|\prod_{i=1}^{n} f_i\right|}(\mu_1, \mu_2) = \int_X p_2 \left|\prod_{i=1}^{n} f_i\left(\frac{p_1}{p_2}\right)\right| d\lambda = \int_X \prod_{i=1}^{n} \left(p_2^{1/a_i} \left|f_i\left(\frac{p_1}{p_2}\right)\right|\right) d\lambda$$
$$\leq \prod_{i=1}^{n} \left(\int_X p_2 |f_i|^{\alpha_i}\left(\frac{p_1}{p_2}\right) d\lambda\right)^{1/a_i} = \prod_{i=1}^{n} \left(\Gamma_{|f_i|^{\alpha_i}}(\mu_1, \mu_2)\right)^{1/a_i}. \quad \square$$

It follows the counterpart of Theorem 12.1.

Theorem 12.4. *Let* $0 < p < 1$ *and* $q < 0$ *such that* $\frac{1}{p} + \frac{1}{q} = 1$. *We suppose that* $p_2 > 0$ *a.e.* $[\lambda]$. *Then we have*

$$\Gamma_{|f_1 f_2|}(\mu_1, \mu_2) \geq \left(\Gamma_{|f_1|^p}(\mu_1, \mu_2)\right)^{1/p} \Gamma_{|f_2|^q}(\mu_1, \mu_2)^{1/q} \tag{12.6}$$

unless $\Gamma_{|f_2|^q}(\mu_1, \mu_2) = 0$. *Inequality (12.6) holds as an equality iff there exist* $A, B \in \mathbb{R}_+$ $(A + B \neq 0)$ *such that*

$$A p_2^{1/p} \left|f_1\left(\frac{p_1}{p_2}\right)\right| = B \left(p_2^{1/q} \left|f_2\left(\frac{p_1}{p_2}\right)\right|\right)^{1/p-1}, \quad \text{a.e. } [\lambda].$$

Proof. It is based on Hölder's inequality for $0 < p < 1$, see p. 191 of [214]. Indeed we have

$$\Gamma_{|f_1 f_2|}(\mu_1, \mu_2) = \int_X p_2 \left|f_1\left(\frac{p_1}{p_2}\right)\right| \left|f_2\left(\frac{p_1}{p_2}\right)\right| d\lambda$$
$$= \int_X \left(p_2^{1/p} |f_1|\left(\frac{p_1}{p_2}\right)\right) \left(p_2^{1/q} |f_2|\left(\frac{p_1}{p_2}\right)\right) d\lambda$$
$$\geq \left(\int_X p_2 |f_1|^p \left(\frac{p_1}{p_2}\right) d\lambda\right)^{1/p} \left(\int_X p_2 |f_2|^q \left(\frac{p_1}{p_2}\right) d\lambda\right)^{1/q}$$
$$= \left(\Gamma_{|f_1|^p}(\mu_1, \mu_2)\right)^{1/p} \left(\Gamma_{|f_2|^q}(\mu_1, \mu_2)\right)^{1/q}. \qquad\square$$

Next we have

Proposition 12.5. *Let* $f_2\left(\frac{p_1}{p_2}\right) \in L_\infty(X)$. *Then*

$$\Gamma_{|f_1 f_2|}(\mu_1, \mu_2) \leq \Gamma_{|f_1|}(\mu_1, \mu_2) \cdot \left(\operatorname{esssup} |f_2|\left(\frac{p_1}{p_2}\right)\right). \tag{12.7}$$

Proof. We notice that

$$\Gamma_{|f_1 f_2|}(\mu_1,\mu_2) = \int_X p_2 \cdot |f_1 \cdot f_2| \left(\frac{p_1}{p_2}\right) d\lambda = \int_X \left(p_2|f_1|\left(\frac{p_1}{p_2}\right)\right) |f_2|\left(\frac{p_1}{p_2}\right) d\lambda$$

$$\leq \left(\int_X p_2|f_1|\left(\frac{p_1}{p_2}\right) d\lambda\right) \cdot \operatorname{esssup}\left(|f_2|\left(\frac{p_1}{p_2}\right)\right)$$

$$= \Gamma_{|f_1|}(\mu_1,\mu_2) \cdot \operatorname{esssup}\left(|f_2|\left(\frac{p_1}{p_2}\right)\right). \qquad \square$$

In the following we use from [185], p. 120, see also [309],

Theorem 12.6. *Let f and g be positive, measurable functions on X for a measure space $(X,\mathcal{F},\mu)$. Let $0 < t < r < m < \infty$. Then*
(i) *The Rogers inequality*

$$\left(\int_X fg^r \, d\mu\right)^{m-t} \leq \left(\int_X fg^t \, d\mu\right)^{m-r} \left(\int_X fg^m \, d\mu\right)^{r-t} \qquad (12.8)$$

provided that the integrals on the right are finite.
(ii) *For $m > 1$, the "historical Hölder's inequality"*

$$\left(\int_X fg \, d\mu\right)^{m} \leq \left(\int_X f \, d\mu\right)^{m-1} \left(\int_X fg^m \, d\mu\right) \qquad (12.9)$$

provided that the integrals on the right are finite.

Using the last results we present the final result of the chapter.

Theorem 12.7. *Let f_1, f_2 with $|f_1|, |f_2| > 0$ and $p_2(x) > 0$ a.e. $[\lambda]$. Let $0 < t < r < m < \infty$. Then it holds*
i)

$$\left(\Gamma_{|f_1||f_2|^r}(\mu_1,\mu_2)\right)^{m-t} \leq \left(\Gamma_{|f_1||f_2|^t}(\mu_1,\mu_2)\right)^{m-r} \left(\Gamma_{|f_1||f_2|^m}(\mu_1,\mu_2)\right)^{r-t}. \qquad (12.10)$$

ii) *For $m > 1$,*

$$\left(\Gamma_{|f_1||f_2|}(\mu_1,\mu_2)\right)^{m} \leq \left(\Gamma_{|f_1|}(\mu_1,\mu_2)\right)^{m-1}\Gamma_{|f_1||f_2|^m}(\mu_1,\mu_2). \qquad (12.11)$$

Proof. i) By applying (12.8) we obtain

$$\left(\Gamma_{|f_1||f_2|^r}(\mu_1,\mu_2)\right)^{m-t} = \left(\int_X p_2|f_1|\left(\frac{p_1}{p_2}\right)|f_2|^r\left(\frac{p_1}{p_2}\right) d\lambda\right)^{m-t}$$

$$\leq \left(\int_X p_2|f_1|\left(\frac{p_1}{p_2}\right)|f_2|^t\left(\frac{p_1}{p_2}\right) d\lambda\right)^{m-r}$$

$$\cdot \left(\int_X p_2|f_1|\left(\frac{p_1}{p_2}\right)|f_2|^m\left(\frac{p_1}{p_2}\right) d\lambda\right)^{r-t}$$

$$= \left(\Gamma_{|f_1||f_2|^t}(\mu_1,\mu_2)\right)^{m-r} \left(\Gamma_{|f_1||f_2|^m}(\mu_1,\mu_2)\right)^{r-t}.$$

ii) By applying (12.9) we derive

$$
\begin{aligned}
\left(\Gamma_{|f_1||f_2|}(\mu_1,\mu_2)\right)^m &= \left(\int_X p_2|f_1|\left(\frac{p_1}{p_2}\right)|f_2|\left(\frac{p_1}{p_2}\right)d\lambda\right)^m \\
&\leq \left(\int_X p_2|f_1|\left(\frac{p_1}{p_2}\right)d\lambda\right)^{m-1}\left(\int_X p_2|f_1|\left(\frac{p_1}{p_2}\right)|f_2|^m\left(\frac{p_1}{p_2}\right)d\lambda\right) \\
&= \left(\Gamma_{|f_1|}(\mu_1,\mu_2)\right)^{m-1}\Gamma_{|f_1||f_2|^m}(\mu_1,\mu_2). \qquad \square
\end{aligned}
$$

Note 12.8. Let X be a finite or countable discrete set, $\mathcal{A}$ be its power set $\mathcal{P}(X)$ and λ has mass 1 for each $x \in X$, then Γ_f in (12.2) becomes a finite or infinite sum, respectively. As a result all the above presented integral inequalities are discretized and become summation inequalities.

Chapter 13

Csiszar's Discrimination and Ostrowski Inequalities via Euler-type and Fink Identities

The Csiszar's discrimination is the most essential and general measure for the comparison between two probability measures. In this chapter we provide probabilistic representation formulae for it via a generalized Euler-type identity and Fink identity. Then we give a large variety of tight estimates for their remainders in various cases. We finish with some generalized Ostrowski type inequalities involving $\|\cdot\|_p$, $p > 1$, using the Euler-type identity. This treatment follows [46].

13.1 Background

(I) Throughout Part A we use the following.

Let f be a convex function from $(0, +\infty)$ into $\mathbb{R}$ which is strictly convex at 1 with $f(1) = 0$. Let $(X, \mathcal{A}, \lambda)$ be a measure space, where λ is a finite or a σ-finite measure on $(X, \mathcal{A})$. And let μ_1, μ_2 be two probability measures on $(X, \mathcal{A})$ such that $\mu_1 \ll \lambda$, $\mu_2 \ll \lambda$ (absolutely continuous), e.g. $\lambda = \mu_1 + \mu_2$.

Denote by $p = \frac{d\mu_1}{d\lambda}$, $q = \frac{d\mu_2}{d\lambda}$ the (densities) Radon–Nikodym derivatives of μ_1, μ_2 with respect to λ. Here we suppose that

$$0 < a \le \frac{p}{q} \le b, \quad \text{a.e. on } X \text{ and } a \le 1 \le b.$$

The quantity

$$\Gamma_f(\mu_1, \mu_2) = \int_X q(x) f\left(\frac{p(x)}{q(x)}\right) d\lambda(x), \tag{13.1}$$

was introduced by I. Csiszar in 1967, see [140], and is called *f-divergence* of the probability measures μ_1 and μ_2. By Lemma 1.1 of [140], the integral (13.1) is well-defined and $\Gamma_f(\mu_1, \mu_2) \ge 0$ with equality only when $\mu_1 = \mu_2$. Furthermore $\Gamma_f(\mu_1, \mu_2)$ does not depend on the choice of λ. The concept of f-divergence was introduced first in [139] as a generalization of Kullback's *"information for discrimination"* or *I-divergence (generalized entropy)* [244], [243] and of Rényi's *"information gain"* (*I-divergence of order α*) [306]. In fact the *I-divergence of order* 1 equals

$$\Gamma_{u\log_2 u}(\mu_1, \mu_2).$$

The choice $f(u) = (u - 1)^2$ produces again a known measure of difference of distributions that is called χ^2-*divergence*. Of course the *total variation* distance $|\mu_1 - \mu_2| = \int_X |p(x) - q(x)| d\lambda(x)$ is equal to $\Gamma_{|u-1|}(\mu_1, \mu_2)$.

Here by assuming $f(1) = 0$ we can consider $\Gamma_f(\mu_1, \mu_2)$, the f-divergence as a measure of the difference between the probability measures μ_1, μ_2. The f-divergence is in general asymmetric in μ_1 and μ_2. But since f is convex and strictly convex at 1 so is

$$f^*(u) = u f\left(\frac{1}{u}\right) \tag{13.2}$$

and as in [140] we find

$$\Gamma_f(\mu_2, \mu_1) = \Gamma_{f^*}(\mu_1, \mu_2). \tag{13.3}$$

In Information Theory and Statistics many other divergences are used which are special cases of the above general Csiszar f-divergence, e.g. Hellinger distance D_H, α-divergence D_α, Bhattacharyya distance D_B, Harmonic distance D_{Ha}, Jeffrey's distance D_J, triangular discrimination D_Δ, for all these see, e.g. [102], [152]. The problem of finding and estimating the *proper distance (or difference or discrimination)* of two probability distributions is one of the major ones in Probability Theory.

The above f-divergence measures in their various forms have been also applied to Anthropology, Genetics, Finance, Economics, Political Science, Biology, Approximation of Probability distributions, Signal Processing and Pattern Recognition. A great inspiration for this chapter has been the very important monograph on the topic by S. Dragomir [152].

In Part A we find representations and estimates for $\Gamma_f(\mu_1, \mu_2)$ via the generalized Euler type and Fink identities.

We use here the sequence $\{B_k(t), k \geq 0\}$ of Bernoulli polynomials which is uniquely determined by the following identities:

$$B_k'(t) = k B_{k-1}(t), \quad k \geq 1, \quad B_0(t) = 1$$

and

$$B_k(t + 1) - B_k(t) = k t^{k-1}, \quad k \geq 0.$$

The values $B_k = B_k(0)$, $k \geq 0$ are the known Bernoulli numbers. We need to mention

$$B_0(t) = 1, \quad B_1(t) = t - \frac{1}{2}, \quad B_2(t) = t^2 - t + \frac{1}{6},$$

$$B_3(t) = t^3 - \frac{3}{2}t^2 + \frac{1}{2}t, \quad B_4(t) = t^4 - 2t^3 + t^2 - \frac{1}{30},$$

$$B_5(t) = t^5 - \frac{5}{2}t^4 + \frac{5}{3}t^3 - \frac{1}{6}t, \quad \text{and} \quad B_6(t) = t^6 - 3t^5 + \frac{5}{2}t^4 - \frac{t^2}{2} + \frac{1}{42}.$$

Let $\{B_k^*(t), k \geq 0\}$ be a sequence of periodic functions of period 1, related to Bernoulli polynomials as

$$B_k^*(t) = B_k(t), \quad 0 \leq t < 1, \quad B_k^*(t + 1) = B_k^*(t), \quad t \in \mathbb{R}.$$

We have that $B_0^*(t) = 1$, B_1^* is a discontinuous function with a jump of -1 at each integer, while B_k^*, $k \geq 2$ are continuous functions. Notice that $B_k(0) = B_k(1) = B_k$, $k \geq 2$.

We use the following result, see [53], [143].

Theorem 13.1. *Let any $a, b \in \mathbb{R}$, $f \colon [a, b] \to \mathbb{R}$ be such that $f^{(n-1)}$, $n \geq 1$, is a continuous function and $f^{(n)}(x)$ exists and is finite for all but a countable set of x in (a, b) and that $f^{(n)} \in L_1([a, b])$. Then for every $x \in [a, b]$ we have*

$$f(x) = \frac{1}{b-a} \int_a^b f(t)dt + \sum_{k=1}^{n-1} \frac{(b-a)^{k-1}}{k!} B_k\left(\frac{x-a}{b-a}\right)[f^{(k-1)}(b) - f^{(k-1)}(a)]$$

$$+ \frac{(b-a)^{n-1}}{n!} \int_{[a,b]} \left[B_n\left(\frac{x-a}{b-a}\right) - B_n^*\left(\frac{x-t}{b-a}\right) \right] f^{(n)}(t)dt. \tag{13.4}$$

The sum in (13.4) when $n = 1$ is zero. If $f^{(n-1)}$ is absolutely continuous then (13.4) is valid again.

We need also

Theorem 13.2 (see [195]). *Let any $a, b \in \mathbb{R}$, $f \colon [a, b] \to \mathbb{R}$, $n \geq 1$, $f^{(n-1)}$ is absolutely continuous on $[a, b]$. Then*

$$f(x) = \frac{n}{b-a} \int_a^b f(t)dt - \sum_{k=1}^{n-1} F_k(x)$$

$$+ \frac{1}{(n-1)!(b-a)} \int_a^b (x-t)^{n-1} k(t, x) f^{(n)}(t)dt, \tag{13.5}$$

where

$$F_k(x) := \left(\frac{n-k}{k!}\right) \left(\frac{f^{(k-1)}(a)(x-a)^k - f^{(k-1)}(b)(x-b)^k}{b-a}\right), \tag{13.6}$$

and

$$k(t, x) := \begin{cases} t - a, & a \leq t \leq x \leq b, \\ t - b, & a \leq x < t \leq b. \end{cases} \tag{13.7}$$

(II) In Part B we give L_p form, $1 < p < \infty$, Ostrowski type inequalities via the generalized Euler type identity (13.4). We mention as inspiration writing this chapter the famous Ostrowski inequality [277]:

$$\left| f(x) - \frac{1}{b-a} \int_a^b f(t)dt \right| \leq \left[\frac{1}{4} + \frac{\left(x - \frac{a+b}{2}\right)^2}{(b-a)^2} \right] (b-a)\|f'\|_\infty,$$

where $f \in C'([a, b])$, $x \in [a, b]$, which is a sharp inequality, see [21].

Other motivations come from [53], [32], [21], [143]. Here our inequalities involve higher order derivatives of f.

13.2　Main Results

Part A

We give

Theorem 13.3. *Let $f\colon [a,b] \to \mathbb{R}$ as in Section 13.1 (I) and be such that $f^{(n-1)}$, $n \geq 1$, is a continuous function and $f^{(n)}(x)$ exists and is finite for all but a countable set of x in (a,b) and that $f^{(n)} \in L_{p_1}([a,b])$, for $1 \leq p_1 \leq \infty$. Then*

$$\Gamma_f(\mu_1,\mu_2) = \frac{1}{b-a}\int_a^b f(t)\,dt \tag{13.8}$$
$$+ \sum_{k=1}^{n-1} \frac{(b-a)^{k-1}}{k!}\left(f^{(k-1)}(b) - f^{(k-1)}(a)\right)\int_X qB_k\left(\frac{\frac{p}{q}-a}{b-a}\right)d\lambda + \mathcal{R}_n,$$

where

$$\mathcal{R}_n = \frac{(b-a)^{n-1}}{n!}\int_X q\left(\int_{[a,b]}\left(B_n\left(\frac{\frac{p}{q}-a}{b-a}\right) - B_n^*\left(\frac{\frac{p}{q}-t}{b-a}\right)\right)f^{(n)}(t)\,dt\right)d\lambda. \tag{13.9}$$

One has

$$|R_n| \leq \frac{(b-a)^{n-1}}{n!}\int_X q\left(\int_{[a,b]}\left|B_n\left(\frac{\frac{p}{q}-a}{b-a}\right) - B_n^*\left(\frac{\frac{p}{q}-t}{b-a}\right)\right||f^{(n)}(t)|\,dt\right)d\lambda. \tag{13.10}$$

Proof. In (13.4), set $x = \frac{p}{q}$, multiply it by q and integrate against λ. $\qquad\square$

Next we present related estimates to Γ_f.

Theorem 13.4. *All assumptions as in Theorem 13.3. Then*

(i)

$$|R_n| \leq \frac{(b-a)^n}{n!}\left(\int_X q\left(\int_0^1\left|B_n(t) - B_n\left(\frac{\frac{p}{q}-a}{b-a}\right)\right|dt\right)d\lambda\right)\|f^{(n)}\|_\infty, \tag{13.11}$$

if $f^{(n)} \in L_\infty([a,b])$,

(ii)

$$|R_n| \leq \frac{(b-a)^n}{n!}\left(\int_X q\left(\int_0^1\left|B_n(t) - B_n\left(\frac{\frac{p}{q}-a}{b-a}\right)\right|^{q_1}dt\right)^{1/q_1}d\lambda\right)\|f^{(n)}\|_{p_1},$$
$$\text{for } p_1, q_1 > 1 : \frac{1}{p_1} + \frac{1}{q_1} = 1, \tag{13.12}$$

if $f^{(n)} \in L_{p_1}([a,b])$,

(iii)

$$|R_n| \leq \frac{(b-a)^{n-1}}{n!}\left(\int_X q\left\|B_n\left(\frac{\frac{p}{q}-a}{b-a}\right) - B_n(t)\right\|_{\infty,[0,1]}d\lambda\right)\|f^{(n)}\|_1, \tag{13.13}$$

$$if\ f^{(n)} \in L_1([a,b]).$$

Proof. We are using (13.10). Part (iii) is obvious by [143], p. 347. For Part (i) we notice that

$$|R_n| \overset{(13.10)}{\leq} \frac{(b-a)^{n-1}}{n!} \left(\int_X q \left(\int_{[a,b]} \left| B_n\left(\frac{\frac{p}{q}-a}{b-a}\right) - B_n^*\left(\frac{\frac{p}{q}-t}{b-a}\right) \right| dt \right) d\lambda \right) \|f^{(n)}\|_\infty$$

and by the change of variable method the claim is proved.

For part (ii) we apply Hölder's inequality on (13.10) and we have

$$|R_n| \leq \frac{(b-a)^{n-1}}{n!} \left(\int_X q \left(\int_{[a,b]} \left| B_n\left(\frac{\frac{p}{q}-a}{b-a}\right) \right.\right.\right.$$
$$\left.\left.\left. - B_n^*\left(\frac{\frac{p}{q}-t}{b-a}\right) \right|^{q_1} dt \right)^{\frac{1}{q_1}} d\lambda \right) \|f^{(n)}\|_{p_1},$$

again by change of variable claim is established. $\square$

Next we derive

Theorem 13.5. *Let f as in Section 13.1, (I) such that $f^{(n-1)}$ is absolutely continuous function on $[a,b]$, $n \geq 1$ and that $f^{(n)} \in L_{p_1}([a,b])$, where $1 \leq p_1 \leq \infty$. Then*

$$\Gamma_f(\mu_1, \mu_2) = \frac{n}{b-a} \int_a^b f(t)dt - \sum_{k=1}^{n-1} \binom{n-k}{k!} \tag{13.14}$$

$$\times \left(\frac{f^{(k-1)}(a) \int_X q^{1-k}(p-aq)^k d\lambda - f^{(k-1)}(b) \int_X q^{1-k}(p-bq)^k d\lambda}{b-a} \right) + G_n,$$

where

$$G_n := \frac{1}{(n-1)!(b-a)} \int_X q^{2-n} \left(\int_a^b (p-tq)^{n-1} k\left(t, \frac{p}{q}\right) f^{(n)}(t)dt \right) d\lambda, \tag{13.15}$$

and

$$|G_n| \leq \frac{1}{(n-1)!(b-a)} \int_X q^{2-n} \left(\int_a^b |p-tq|^{n-1} \left| k\left(t, \frac{p}{q}\right) \right| |f^{(n)}(t)|dt \right) d\lambda. \tag{13.16}$$

Proof. In (13.5) set $x = \frac{p}{q}$, multiply it by q and integrate against λ. $\square$

We give the estimates

Theorem 13.6. *All suppositions as in Theorem 13.5. Then*

(i)

$$|G_n| \leq \frac{1}{(n+1)!(b-a)} \left(\int_X q^{-n}((p-aq)^{n+1} + (bq-p)^{n+1})d\lambda \right) \|f^{(n)}\|_\infty, \tag{13.17}$$

if $f^{(n)} \in L_\infty([a,b])$,

(ii)

$$|G_n| \le \frac{(B(q_1 + 1, q_1(n - 1) + 1))^{1/q_1}}{(n - 1)!(b - a)} \tag{13.18}$$

$$\times \left(\int_X q^{1 - n - \frac{1}{q_1}} \left\{ (p - aq)^{q_1 n + 1} + (bq - p)^{q_1 n + 1} \right\}^{1/q_1} d\lambda \right) \|f^{(n)}\|_{p_1},$$

for $p_1, q_1 > 1 : \frac{1}{p_1} + \frac{1}{q_1} = 1$, *if* $f^{(n)} \in L_{p_1}([a, b])$,
and (iii)

$$|G_n| \le \frac{1}{(n - 1)!(b - a)} \left(\int_X q \left(\max\left(\frac{p}{q} - a, b - \frac{p}{q} \right) \right)^n d\lambda \right) \|f^{(n)}\|_1, \tag{13.19}$$

if $f^{(n)} \in L_1([a, b])$.

Proof. (i) We have by (13.16) that

$$|G_n| \le \frac{1}{(n - 1)!(b - a)} \left(\int_X q^{2 - n} \left(\int_a^b |p - tq|^{n - 1} \left| K\left(t, \frac{p}{q} \right) \right| dt \right) d\lambda \right) \|f^{(n)}\|_\infty,$$

if $f^{(n)} \in L_\infty([a, b])$.
    Here

$$K\left(t, \frac{p}{q} \right) = \begin{cases} t - a, & a \le t \le \dfrac{p}{q} \le b, \\[2mm] t - b, & a \le \dfrac{p}{q} < t \le b. \end{cases}$$

Then by [338], p. 256 we obtain

$$\int_a^b |p - tq|^{n - 1} \left| K\left(t, \frac{p}{q} \right) \right| dt = \frac{q^{-2}}{n(n + 1)} \left((p - aq)^{n + 1} + (bq - p)^{n + 1} \right),$$

which proves the claim.
    (ii) Again by (13.16) we derive

$$|G_n| \le \frac{1}{(n - 1)!(b - a)} \left(\int_X q^{2 - n} \left(\int_a^b |p - tq|^{q_1(n - 1)} \right. \right.$$

$$\left. \left. \left| K\left(t, \frac{p}{q} \right) \right|^{q_1} dt \right)^{1/q_1} d\lambda \right) \|f^{(n)}\|_{p_1},$$

for $p_1, q_1 > 1 : \frac{1}{p_1} + \frac{1}{q_1} = 1$, *if* $f^{(n)} \in L_{p_1}([a, b])$. Again by [338], p. 256 we find

$$\left(\int_a^b |p - tq|^{q_1(n - 1)} \left| K\left(t, \frac{p}{q} \right) \right|^{q_1} dt \right)^{1/q_1}$$

$$= q^{-\left(1 + \frac{1}{q_1} \right)} \left(B(q_1 + 1, q_1(n - 1) + 1) \right)^{\frac{1}{q_1}} \left\{ (p - aq)^{q_1 n + 1} + (bq - p)^{q_1 n + 1} \right\}^{1/q_1},$$

proving the claim.

(iii) From (13.16) we obtain

$$|G_n| \leq \frac{1}{(n-1)!(b-a)} \left(\int_X q^{2-n} \|p - tq\|_{\infty,[a,b]}^{n-1} \left\| K\left(t, \frac{p}{q}\right) \right\|_{\infty,[a,b]} d\lambda \right) \|f^{(n)}\|_1,$$

if $f^{(n)} \in L_1([a,b])$. We notice that

$$\left\| K\left(t, \frac{p}{q}\right) \right\|_{\infty,[a,b]} = \left\| \frac{p}{q} - t \right\|_{\infty,[a,b]} = \max\left(\frac{p}{q} - a, b - \frac{p}{q} \right).$$

The last establishes the claim. $\qquad\qquad\square$

Next we study some important cases. For that we need

Theorem 13.7. *Let any* $a, b \in \mathbb{R}$, $a \leq x \leq b$, *call* $\tilde{\lambda} := \frac{x-a}{b-a}$. *It holds*
  (i)

$$I_1(\tilde{\lambda}) := \int_0^1 |B_1(t) - B_1(\tilde{\lambda})| dt = \frac{1}{4} + \frac{\left(x - \frac{a+b}{2}\right)^2}{(b-a)^2} \leq \frac{1}{2}, \ \forall x \in [a,b], \qquad (13.20)$$

with equality at $x = a, b$,
  (ii) ([143], p. 343)

$$I_2(\tilde{\lambda}) := \int_0^1 |B_2(t) - B_2(\tilde{\lambda})| dt = \frac{8}{3}\delta^3(x) - \delta^2(x) + \frac{1}{12} \leq \frac{1}{6}, \ \forall x \in [a,b], \quad (13.21)$$

with equality at $x = a, b$, *where*

$$\delta(x) := \frac{\left|x - \frac{a+b}{2}\right|}{b-a}, \quad x \in [a,b], \qquad (13.22)$$

  (iii) ([13])

$$I_3(\tilde{\lambda}) := \int_0^1 |B_3(t) - B_3(\tilde{\lambda})| dt =$$

$$\begin{cases}
-\dfrac{3}{2}t_1^4 + 2t_1^3 - \dfrac{t_1^2}{2} + \dfrac{3}{2}\tilde{\lambda}^4 - \tilde{\lambda}^3 - \tilde{\lambda}^2 + \dfrac{\tilde{\lambda}}{2}, & \tilde{\lambda} \in \left[0, \dfrac{3-\sqrt{3}}{6}\right], \\[2ex]
\dfrac{3}{2}t_1^4 - 2t_1^3 + \dfrac{t_1^2}{2} - \dfrac{3}{2}\tilde{\lambda}^4 + 3\tilde{\lambda}^3 - 2\tilde{\lambda}^2 + \dfrac{\tilde{\lambda}}{2}, & \tilde{\lambda} \in \left(\dfrac{3-\sqrt{3}}{6}, \dfrac{1}{2}\right], \\[2ex]
\dfrac{3}{2}t_2^4 - 2t_2^3 + \dfrac{t_2^2}{2} - \dfrac{3}{2}\tilde{\lambda}^4 + \tilde{\lambda}^3 + \tilde{\lambda}^2 - \dfrac{\tilde{\lambda}}{2}, & \tilde{\lambda} \in \left(\dfrac{1}{2}, \dfrac{3+\sqrt{3}}{6}\right], \\[2ex]
-\dfrac{3}{2}t_2^4 + 2t_2^3 - \dfrac{t_2^2}{2} + \dfrac{3}{2}\tilde{\lambda}^4 - 3\tilde{\lambda}^3 + 2\tilde{\lambda}^2 - \dfrac{\tilde{\lambda}}{2}, & \tilde{\lambda} \in \left(\dfrac{3+\sqrt{3}}{6}, 1\right],
\end{cases} \qquad (13.23)$$

where

$$t_1 := \frac{3}{4} - \frac{\tilde{\lambda}}{2} - \frac{1}{2}\sqrt{\frac{1}{4} + 3\tilde{\lambda} - 3\tilde{\lambda}^2},$$

$$t_2 := \frac{3}{4} - \frac{\tilde{\lambda}}{2} + \frac{1}{2}\sqrt{\frac{1}{4} + 3\tilde{\lambda} - 3\tilde{\lambda}^2}, \qquad (13.24)$$

and

$$I_3(\tilde{\lambda}) \leq \frac{\sqrt{3}}{36}, \tag{13.25}$$

with equality at $\tilde{\lambda} = \frac{3\pm\sqrt{3}}{6}$.

   (iv) ([53])

$$I_4(\tilde{\lambda}) := \int_0^1 |B_4(t) - B_4(\tilde{\lambda})|dt$$

$$= \begin{cases} \dfrac{16\tilde{\lambda}^5}{5} - 7\tilde{\lambda}^4 + \dfrac{14}{3}\tilde{\lambda}^3 - \tilde{\lambda}^2 + \dfrac{1}{30}, & 0 \leq \tilde{\lambda} \leq \frac{1}{2}, \\[3mm] -\dfrac{16\tilde{\lambda}^5}{5} + 9\tilde{\lambda}^4 - \dfrac{26\tilde{\lambda}^3}{3} + 3\tilde{\lambda}^2 - \dfrac{1}{10}, & \frac{1}{2} \leq \tilde{\lambda} \leq 1, \end{cases} \tag{13.26}$$

with

$$I_4(\tilde{\lambda}) \leq \frac{1}{30}, \tag{13.27}$$

attained at $\tilde{\lambda} = 0, 1$, i.e. when $x = a, b$, and

   (v) ([143]) *for $n \geq 1$ we have*

$$I_n(\tilde{\lambda}) := \int_0^1 |B_n(t) - B_n(\tilde{\lambda})|dt \leq \left(\int_0^1 (B_n(t) - B_n(\tilde{\lambda}))^2 dt\right)^{1/2}$$

$$= \sqrt{\frac{(n!)^2}{(2n)!}|B_{2n}| + B_n^2(\tilde{\lambda})}. \tag{13.28}$$

Consequently by (13.11) and Theorem 13.7 we derive the precise estimates.

Theorem 13.8. *All assumptions as in Theorem 13.3 with $f^{(n)} \in L_\infty([a,b])$, $n \geq 1$. Here*

$$\tilde{\lambda} := \tilde{\lambda}(x) := \frac{\frac{p(x)}{q(x)} - a}{b - a} = \frac{\frac{p}{q} - a}{b - a}, \quad x \in X, \tag{13.29}$$

i.e. $\tilde{\lambda} \in [0, 1]$, a.e. on X. Then

$$|R_n| \leq \frac{(b-a)^n}{n!}\left(\int_X q I_n(\tilde{\lambda})d\lambda\right)\|f^{(n)}\|_\infty, \tag{13.30}$$

and

$$|R_1| \leq \frac{(b-a)}{2}\|f'\|_\infty, \tag{13.31}$$

$$|R_2| \leq \frac{(b-a)^2}{12}\|f''\|_\infty, \tag{13.32}$$

$$|R_3| \leq \frac{\sqrt{3}(b-a)^3}{216}\|f'''\|_\infty, \tag{13.33}$$

and

$$|R_4| \leq \frac{(b-a)^4}{720}\|f^{(4)}\|_\infty. \tag{13.34}$$

We have also by (13.12) and (13.28) that

Corollary 13.9. *All assumptions as in Theorem 13.3 with $f^{(n)} \in L_2([a,b])$. Then*

$$|R_n| \le \frac{(b-a)^n}{n!} \left(\int_X q \left(\sqrt{\frac{(n!)^2}{(2n)!}|B_{2n}| + B_n^2 \left(\frac{\frac{p}{q}-a}{b-a} \right)} \right) d\lambda \right) \|f^{(n)}\|_2, \;\; n \ge 1.$$

(13.35)

We need

Note 13.10. From [143], p. 347 for $r \in \mathbb{N}$, $x \in [a,b]$ we have that

$$\max_{t \in [0,1]} \left| B_{2r}(t) - B_{2r}\left(\frac{x-a}{b-a} \right) \right| = (1 - 2^{-2r})|B_{2r}| + \left| 2^{-2r}B_{2r} - B_{2r}\left(\frac{x-a}{b-a} \right) \right|. \quad (13.36)$$

And from [143], p. 348 we obtain

$$\max_{t \in [0,1]} \left| B_{2r+1}(t) - B_{2r+1}\left(\frac{x-a}{b-a} \right) \right|$$
$$\le \frac{2(2r+1)!}{(2\pi)^{2r+1}(1-2^{-2r})} + \left| B_{2r+1}\left(\frac{x-a}{b-a} \right) \right|, \quad (13.37)$$

along with

$$\max_{t \in [0,1]} \left| B_1(t) - B_1\left(\frac{x-a}{b-a} \right) \right| = \frac{1}{2} + \left| x - \frac{a+b}{2} \right|. \quad (13.38)$$

Using (13.13) and Note 13.10 we find

Theorem 13.11. *All assumptions as in Theorem 13.3 with $f^{(n)} \in L_1([a,b])$, $r,n \ge 1$. Then*

$$|R_1| \le \left\{ \frac{1}{2} + \int_X \left| p - \frac{(a+b)}{2} \right| q \, d\lambda \right\} \|f'\|_1, \quad (13.39)$$

when $n = 2r$ we have

$$|R_{2r}| \le \frac{(b-a)^{2r-1}}{(2r)!} \left\{ (1 - 2^{-2r})|B_{2r}| \right.$$
$$\left. + \int_X q \left| 2^{-2r}B_{2r} - B_{2r}\left(\frac{\frac{p}{q}-a}{b-a} \right) \right| d\lambda \right\} \|f^{(2r)}\|_1, \quad (13.40)$$

and for $n = 2r+1$ we obtain

$$|R_{2r+1}| \le \frac{(b-a)^{2r}}{(2r+1)!} \left\{ \frac{2(2r+1)!}{(2\pi)^{2r+1}(1-2^{-2r})} \right.$$
$$\left. + \int_X q \left| B_{2r+1}\left(\frac{\frac{p}{q}-a}{b-a} \right) \right| d\lambda \right\} \|f^{(2r+1)}\|_1. \quad (13.41)$$

Part B

Remark 13.12. Let f be as in Theorem 13.1, we denote

$$\Delta_n(x) := f(x) - \frac{1}{b-a} \int_a^b f(t)dt \tag{13.42}$$

$$- \sum_{k=1}^{n-1} \frac{(b-a)^{k-1}}{k!} B_k\left(\frac{x-a}{b-a}\right)[f^{(k-1)}(b) - f^{(k-1)}(a)], \ x \in [a,b], \ n \geq 1.$$

We have by (13.4) that

$$\Delta_n(x) = \frac{(b-a)^{n-1}}{n!} \int_{[a,b]} \left[B_n\left(\frac{x-a}{b-a}\right) - B_n^*\left(\frac{x-t}{b-a}\right) \right] f^{(n)}(t)dt. \tag{13.43}$$

Let here $1 < p < \infty$, we see by Hölder's inequality that

$$|\Delta_n(x)| \leq \frac{(b-a)^{n-1}}{n!} \left(\int_a^b \left| B_n\left(\frac{x-a}{b-a}\right) - B_n^*\left(\frac{x-t}{b-a}\right) \right|^q dt \right)^{1/q} \|f^{(n)}\|_p =: (*), \tag{13.44}$$

where $q > 1$ such that $\frac{1}{p} + \frac{1}{q} = 1$, given that $f^{(n)} \in L_p([a,b])$. By change of variable we find that

$$(*) = \frac{(b-a)^{n-1+\frac{1}{q}}}{n!} \left(\int_0^1 \left| B_n(t) - B_n\left(\frac{x-a}{b-a}\right) \right|^q dt \right)^{1/q} \|f^{(n)}\|_p. \tag{13.45}$$

We have established the generalized Ostrowski type inequality.

Theorem 13.13. *Let any $a,b \in \mathbb{R}$, $f\colon [a,b] \to \mathbb{R}$ be such that $f^{(n-1)}$, $n \geq 1$, is a continuous function and $f^{(n)}(x)$ exists and is finite for all but a countable set of x in (a,b) and that $f^{(n)} \in L_p([a,b])$, $1 < p,q < \infty$: $\frac{1}{p} + \frac{1}{q} = 1$. Then for every $x \in [a,b]$ we have*

$$|\Delta_n(x)| \leq \frac{(b-a)^{n-1+\frac{1}{q}}}{n!} \left(\int_0^1 \left| B_n(t) - B_n\left(\frac{x-a}{b-a}\right) \right|^q dt \right)^{1/q} \|f^{(n)}\|_p. \tag{13.46}$$

The last theorem appeared first in [143] as Theorem 9 there, but with missing two of the assumptions only by assuming $f^{(n)} \in L_p([a,b])$.

We give

Corollary 13.14. *Let any $a,b \in \mathbb{R}$, $f\colon [a,b] \to \mathbb{R}$ be such that $f^{(n-1)}$, $n \geq 1$, is a continuous function and $f^{(n)}(x)$ exists and is finite for all but a countable set of x in (a,b) and that $f^{(n)} \in L_2([a,b])$. Then for any $x \in [a,b]$ we have*

$$|\Delta_n(x)| \leq \frac{(b-a)^{n-\frac{1}{2}}}{n!} \left(\sqrt{\frac{(n!)^2}{(2n)!}|B_{2n}| + B_n^2\left(\frac{x-a}{b-a}\right)} \right) \|f^{(n)}\|_2. \tag{13.47}$$

Proof. By Theorem 13.13, $p = 2$ and that

$$\int_0^1 \left(B_n(t) - B_n\left(\frac{x-a}{b-a}\right) \right)^2 dt = \frac{(n!)^2}{(2n)!}|B_{2n}| + B_n^2\left(\frac{x-a}{b-a}\right), \tag{13.48}$$

see [143], p. 352. $\qquad\qquad\square$

Here again, (13.47) appeared first in [143], p. 351, as Corollary 5 there, erroneously under the only assumption $f^{(n)} \in L_2([a,b])$.

Remark 13.15. We find the trapezoidal formula

$$\Delta_4(a) = \Delta_4(b) = \left(\frac{f(a) + f(b)}{2}\right) - \frac{(b-a)}{12}(f'(b) - f'(a)) - \frac{1}{b-a}\int_a^b f(t)dt, \quad (13.49)$$

and the midpoint formula

$$\Delta_4\left(\frac{a+b}{2}\right) = f\left(\frac{a+b}{2}\right) + \frac{(b-a)}{24}(f'(b) - f'(a)) - \frac{1}{b-a}\int_a^b f(t)dt. \quad (13.50)$$

Furthermore we obtain the trapezoidal formula

$$\Delta_5(a) = \Delta_5(b) = \Delta_6(a) = \Delta_6(b) = \frac{(f(a) + f(b))}{2} - \frac{(b-a)}{12}(f'(b) - f'(a))$$

$$+ \frac{(b-a)^3}{720}(f'''(b) - f'''(a)) - \frac{1}{b-a}\int_a^b f(t)dt. \quad (13.51)$$

We also derive the midpoint formula

$$\Delta_5\left(\frac{a+b}{2}\right) = \Delta_6\left(\frac{a+b}{2}\right) = f\left(\frac{a+b}{2}\right) + \frac{(b-a)}{24}(f'(b) - f'(a))$$

$$- \frac{7(b-a)^3}{5760}(f'''(b) - f'''(a)) - \frac{1}{b-a}\int_a^b f(t)dt. \quad (13.52)$$

Using Mathematica 4 and (13.47) we establish

Theorem 13.16. *Let any $a,b \in \mathbb{R}$, $f\colon [a,b] \to \mathbb{R}$ be such that $f^{(3)}$ is a continuous function and $f^{(4)}(x)$ exists and is finite for all but a countable set of x in (a,b) and that $f^{(4)} \in L_2([a,b])$. Then*

$$|\Delta_4(a)| \le \frac{(b-a)^{3.5}}{72\sqrt{70}}\|f^{(4)}\|_2, \quad (13.53)$$

$$\left|\Delta_4\left(\frac{a+b}{2}\right)\right| \le \frac{1}{1152}\sqrt{\frac{107}{35}}(b-a)^{3.5}\|f^{(4)}\|_2. \quad (13.54)$$

We continue with

Theorem 13.17. *Let any $a,b \in \mathbb{R}$, $f\colon [a,b] \to \mathbb{R}$ be such that $f^{(4)}$ is a continuous function and $f^{(5)}(x)$ exists and is finite for all but a countable set of x in (a,b) and that $f^{(5)} \in L_2([a,b])$. Then*

$$|\Delta_5(a)| \le \frac{(b-a)^{4.5}}{144\sqrt{2310}}\|f^{(5)}\|_2, \quad (13.55)$$

$$\left|\Delta_5\left(\frac{a+b}{2}\right)\right| \le \frac{(b-a)^{4.5}}{144\sqrt{2310}}\|f^{(5)}\|_2. \quad (13.56)$$

We finish with

Theorem 13.18. *Let any $a, b \in \mathbb{R}$, $f: [a, b] \to \mathbb{R}$ be such that $f^{(5)}$ is a continuous function and $f^{(6)}(x)$ exists and is finite for all but a countable set of x in (a, b) and that $f^{(6)} \in L_2([a, b])$. Then*

$$|\Delta_6(a)| \leq \frac{1}{1440} \sqrt{\frac{101}{30030}} (b - a)^{5.5} \|f^{(6)}\|_2, \tag{13.57}$$

and

$$\left|\Delta_6\left(\frac{a + b}{2}\right)\right| \leq \frac{1}{46080} \sqrt{\frac{7081}{2145}} (b - a)^{5.5} \|f^{(6)}\|_2. \tag{13.58}$$

Chapter 14

Taylor-Widder Representations and Grüss, Means, Ostrowski and Csiszar's Inequalities

In this chapter based on the very general Taylor–Widder formula several representation formulae are developed. Applying these are established very general inequalities of types: Grüss, comparison of means,Ostrowski, Csiszar f-divergence. The approximations involve L_p norms, any $1 \le p \le \infty$. This treatment follows [56].

14.1 Introduction

The article that motivates to most this chapter is of D. Widder (1928), see [339], where he generalizes the Taylor formula and series, see also Section 14.2. Based on that approach we give several representation formulae for the integral of a function f over $[a, b] \subseteq \mathbb{R}$, and the most important of these involve also $f(x)$, for any $x \in [a, b]$, see Theorems 14.6, 14.8, 14.9, 14.10 and Corollary 14.12. Then being motivated by the works of Ostrowski [277], Grüss [205], of the author [48] and Csiszar [140] we give our related generalized results. We would like to mention the following inspiring results.

Theorem 14.1 (Ostrowski, 1938, [277]). *Let $f\colon [a, b] \to \mathbb{R}$ be continuous on $[a, b]$ and differentiable on (a, b) whose derivative $f'\colon (a, b) \to \mathbb{R}$ is bounded on (a, b), i.e.,*
$$\|f'\|_\infty = \sup_{t \in (a,b)} |f'(t)| < +\infty. \text{ Then}$$

$$\left| \frac{1}{b-a} \int_a^b f(t)dt - f(x) \right| \le \left[\frac{1}{4} + \frac{\left(x - \frac{a+b}{2}\right)^2}{(b-a)^2} \right] (b-a)\|f'\|_\infty,$$

for any $x \in [a, b]$. The constant $\frac{1}{4}$ is the best possible.

Theorem 14.2 (Grüss, 1935, [205]). *Let f, g integrable functions from $[a, b]$ into $\mathbb{R}$, such that $m \le f(x) \le M$, $\rho \le g(x) \le \sigma$, for all $x \in [a, b]$, where $m, M, \rho, \sigma \in \mathbb{R}$. Then*

$$\left| \frac{1}{b-a} \int_a^b f(x)g(x)dx - \left(\frac{1}{b-a} \int_a^b f(x)dx \right) \left(\frac{1}{b-a} \int_a^b g(x)dx \right) \right| \le \frac{1}{4}(M-m)(\sigma-\rho).$$

173

So our generalizations here are over an *extended complete Tschebyshev system* $\{u_i\}_{i=0}^n \in C^{n+1}([a,b])$, $n \geq 0$, see Theorems 14.3, 14.4, and [223]. Thus our Ostrowski type general inequality is presented in Theorem 14.14 and is sharp.

Next we generalize Grüss inequality in Theorems 14.15, 14.16, 14.17 and Corollary 14.18. It follows the generalized comparison of integral averages in Theorem 14.19, where we compare the Riemann integral to arbitrary general integrals involving finite measures. We finish with generalized representation and estimates for the Csiszar f-divergence, see Theorems 14.21, 14.22. The last is the most general and best measure for the difference between probability measures.

14.2   Background

The following are taken from [339]. Let $f, u_0, u_1, \ldots, u_n \in C^{n+1}([a,b])$, $n \geq 0$, and the Wronskians

$$W_i(x) := W[u_0(x), u_1(x), \ldots, u_i(x)] := \begin{vmatrix} u_0(x) & u_1(x) & \cdots & u_i(x) \\ u_0'(x) & u_1'(x) & \cdots & u_i'(x) \\ \vdots & & & \\ u_0^{(i)}(x) & u_1^{(i)}(x) & \cdots & u_i^{(i)}(x) \end{vmatrix}$$

$i = 0, 1, \ldots, n$.

Suppose $W_i(x) > 0$ over $[a,b]$. Clearly then

$$\phi_0(x) := W_0(x) = u_0(x), \quad \phi_1(x) := \frac{W_1(x)}{(W_0(x))^2}, \ldots, \phi_i(x) := \frac{W_i(x)W_{i-2}(x)}{(W_{i-1}(x))^2},$$

$i = 2, 3, \ldots, n$ are positive on $[a,b]$.

For $i \geq 0$, the linear differentiable operator of order i:

$$L_i f(x) := \frac{W[u_0(x), u_1(x), \ldots, u_{i-1}(x), f(x)]}{W_{i-1}(x)}, \quad i = 1, \ldots, n+1,$$

$$L_0 f(x) := f(x), \quad \forall x \in [a,b].$$

Then for $i = 1, \ldots, n+1$ we have

$$L_i f(x) = \phi_0(x)\phi_1(x) \cdots \phi_{i-1}(x) \frac{d}{dx} \frac{1}{\phi_{i-1}(x)} \frac{d}{dx} \frac{1}{\phi_{i-2}(x)} \frac{d}{dx} \cdots \frac{d}{dx} \frac{1}{\phi_1(x)} \frac{d}{dx} \frac{f(x)}{\phi_0(x)}.$$

Consider also

$$g_i(x,t) := \frac{1}{W_i(t)} \begin{vmatrix} u_0(t) & u_1(t) & \cdots & u_i(t) \\ u_0'(t) & u_1'(t) & \cdots & u_i'(t) \\ \vdots & \vdots & & \vdots \\ u_0^{(i-1)}(t) & u_1^{(i-1)}(t) & \cdots & u_i^{(i-1)}(t) \\ u_0(x) & u_1(x) & \cdots & u_i(x) \end{vmatrix},$$

$i = 1, 2, \ldots, n$; $g_0(x,t) := \frac{u_0(x)}{u_0(t)}$, $\forall x, t \in [a,b]$.

Note that $g_i(x,t)$ as a function of x is a linear combination of $u_0(x)$, $u_1(x), \ldots, u_i(x)$ and it holds

$$g_i(x,t)$$
$$= \frac{\phi_0(x)}{\phi_0(t)\cdots\phi_i(t)} \int_t^x \phi_1(x_1) \int_t^{x_1} \cdots \int_t^{x_{i-2}} \phi_{i-1}(x_{i-1}) \int_t^{x_{i-1}} \phi_i(x_i)dx_i dx_{i-1} \cdots dx_1$$
$$= \frac{1}{\phi_0(t)\cdots\phi_i(t)} \int_t^x \phi_0(s)\cdots\phi_i(s)g_{i-1}(x,s)ds, \quad i = 1, 2, \ldots, n.$$

Example [339]. The sets

$$\{1, x, x^2, \ldots, x^n\}, \{1, \sin x, -\cos x, -\sin 2x, \cos 2x, \ldots, (-1)^{n-1}\sin nx, (-1)^n \cos nx\}$$

fulfill the above theory.

We mention

Theorem 14.3 (Karlin and Studden (1966), see p. 376, [223]). *Let* $u_0, u_1, \ldots, u_n \in C^n([a,b])$, $n \geq 0$. *Then* $\{u_i\}_{i=0}^n$ *is an extended complete Tschebyshev system on* $[a,b]$ *iff* $W_i(x) > 0$ *on* $[a,b]$, $i = 0, 1, \ldots, n$.

We also mention

Theorem 14.4 (D. Widder, p. 138, [339]). *Let the functions* $f(x)$, $u_0(x)$, $u_1(x), \ldots, u_n(x) \in C^{n+1}([a,b])$, *and the Wronskians* $W_0(x), W_1(x), \ldots, W_n(x) > 0$ *on* $[a,b]$, $x \in [a,b]$. *Then for* $t \in [a,b]$ *we have*

$$f(x) = f(t)\frac{u_0(x)}{u_0(t)} + L_1 f(t)g_1(x,t) + \cdots + L_n f(t)g_n(x,t) + R_n(x),$$

where

$$R_n(x) := \int_t^x g_n(x,s)L_{n+1}f(s)ds.$$

E.g. one could take $u_0(x) = c > 0$. *If* $u_i(x) = x^i$, $i = 0, 1, \ldots, n$, *defined on* $[a,b]$, *then*

$$L_i f(t) = f^{(i)}(t) \quad and \quad g_i(x,t) = \frac{(x-t)^i}{i!}, \quad t \in [a,b].$$

So under the assumptions of Theorem 14.4 it holds

$$f(x) = f(y)\frac{u_0(x)}{u_0(y)} + \sum_{i=1}^n L_i f(y)g_i(x,y) + \int_y^x g_n(x,t)L_{n+1}f(t)dt, \quad \forall x,y \in [a,b].$$

$$(14.1)$$

14.3   Main Results

Define the Kernel

$$K(t,x) := \begin{cases} t - a, & a \leq t \leq x \leq b; \\ t - b, & a \leq x < t \leq b. \end{cases} \tag{14.2}$$

Integrating (14.1) with respect to y we find

$$f(x)(b-a) = u_0(x) \int_a^b \frac{f(y)}{u_0(y)} dy + \sum_{i=1}^n \int_a^b L_i f(y) g_i(x,y) dy$$

$$+ \int_a^b \left(\int_y^x g_n(x,t) L_{n+1} f(t) dt \right) dy, \quad \forall x \in [a,b]. \tag{14.3}$$

We notice that (see also [195])

$$\int_a^b dy \int_y^x dt = \int_a^x dy \int_y^x dt + \int_x^b dy \int_y^x dt$$

$$= \int_a^x dt \int_a^t dy - \int_x^b dy \int_x^y dt = \int_a^x dt \int_a^t dy - \int_x^b dt \int_t^b dy. \tag{14.4}$$

By letting $* := g_n(x,t) L_{n+1} f(t)$ we better notice that

$$\int_a^b \left(\int_y^x * \, dt \right) dy = \int_a^x \left(\int_y^x * \, dt \right) dy + \int_x^b \left(\int_y^x * \, dt \right) dy$$

$$= \int_a^x \left(\int_a^t * \, dy \right) dt - \int_x^b \left(\int_x^y * \, dt \right) dy$$

$$= \int_a^x * \left(\int_a^t dy \right) dt - \int_x^b * \left(\int_t^b dy \right) dt$$

$$= \int_a^x * \, (t-a) dt + \int_x^b * \, (t-b) dt = \int_a^b * \, k(t,x) dt. \tag{14.5}$$

Above we have that

$$\begin{Bmatrix} a \leq y \leq x \\ y \leq t \leq x \end{Bmatrix} \Leftrightarrow \begin{Bmatrix} a \leq t \leq x \\ a \leq y \leq t \end{Bmatrix},$$

and

$$\begin{Bmatrix} x \leq y \leq b \\ x \leq t \leq y \end{Bmatrix} \Leftrightarrow \begin{Bmatrix} x \leq t \leq b \\ t \leq y \leq b \end{Bmatrix}. \tag{14.6}$$

So we obtain

Lemma 14.5. *It holds*

$$\int_a^b \left(\int_y^x g_n(x,t) L_{n+1} f(t) dt \right) dy = \int_a^b g_n(x,t) L_{n+1} f(t) K(t,x) dt. \tag{14.7}$$

We give the general representation result, see also [195], as a special case.

Theorem 14.6. *Let* $f(x)$, $u_0(x), u_1(x), \ldots, u_n(x) \in C^{n+1}([a,b])$, $n \geq 0$, *the Wronskians* $W_i(x) > 0$, $i = 0, 1, \ldots, n$, *over* $[a,b]$, $x \in [a,b]$. *Then*

$$f(x) = \frac{u_0(x)}{b-a} \int_a^b \frac{f(y)}{u_0(y)} dy + \frac{1}{b-a} \left(\sum_{i=1}^n \int_a^b L_i f(y) g_i(x,y) dy \right)$$

$$+ \frac{1}{b-a} \int_a^b g_n(x,t) L_{n+1} f(t) K(t,x) dt, \quad \forall x \in [a,b]. \tag{14.8}$$

When $u_0(x) = c > 0$, *then*

$$f(x) = \frac{1}{b-a} \int_a^b f(y) dy + \frac{1}{b-a} \left(\sum_{i=1}^n \int_a^b L_i f(y) g_i(x,y) dy \right)$$

$$+ \frac{1}{b-a} \int_a^b g_n(x,t) L_{n+1} f(t) K(t,x) dt, \quad \forall x \in [a,b]. \tag{14.9}$$

Proof. Based on (14.3) and Lemma 14.5. $\qquad\square$

We also need

Lemma 14.7. *Let* f, $\{u_i\}_{i=0}^n \in C^n([a,b])$, *and* $W_i(x) > 0$, $i = 0, 1, \ldots, n$, $n \geq 0$, *over* $[a,b]$. *Then*

$$\sum_{i=1}^n \int_a^b L_i f(y) g_i(x,y) dy = \sum_{i=0}^{n-1} (n-i) \Bigg\{ \left[L_i f(b) g_{i+1}(x,b) - L_i f(a) g_{i+1}(x,a) \right]$$

$$+ \int_a^b \left(\frac{L_i f(y) g_{i+1}(x,y)}{\phi_{i+1}(y)} \right) \phi'_{i+1}(y) dy \Bigg\} + n u_0(x) \int_a^b \frac{f(y)}{u_0(y)} dy, \ \forall x \in [a,b]. \tag{14.10}$$

In case of $u_0(x) = c > 0$ *then we have*

$$\sum_{i=1}^n \int_a^b L_i f(y) g_i(x,y) dy = \sum_{i=0}^{n-1} (n-i) \Bigg\{ \left[L_i f(b) g_{i+1}(x,b) - L_i f(a) g_{i+1}(x,a) \right]$$

$$+ \int_a^b \left(\frac{L_i f(y) g_{i+1}(x,y)}{\phi_{i+1}(y)} \right) \phi'_{i+1}(y) dy \Bigg\} + n \int_a^b f(y) dy, \ \forall x \in [a,b]. \tag{14.11}$$

Proof. By [339] for $i = 0, 1, \ldots, n-1$, we get

$$\left(\prod_{k=0}^{i+1} \phi_k(y) \right) g_{i+1}(x,y) = - \int_x^y \left(\prod_{k=0}^{i+1} \phi_k(s) \right) g_i(x,s) ds.$$

Thus

$$\left(\left(\prod_{k=0}^{i+1} \phi_k(y) \right) g_{i+1}(x,y) \right)' = - \left(\left(\prod_{k=0}^{i+1} \phi_k(y) \right) g_i(x,y) \right)$$

and

$$-d \left(\left(\prod_{k=0}^{i+1} \phi_k(y) \right) g_{i+1}(x,y) \right) = \left(\prod_{k=0}^{i+1} \phi_k(y) \right) g_i(x,y) dy. \tag{14.12}$$

Consequently for $i = 0, 1, \ldots, n - 1$ we have

$$J_i(x) := \int_a^b L_i f(y) g_i(x, y) dy = \int_a^b \frac{L_i f(y)}{\left(\prod_{k=0}^{i+1} \phi_k(y)\right)} \left(\prod_{k=0}^{i+1} \phi_k(y)\right) g_i(x, y) dy$$

$$\overset{(14.12)}{=} -\int_a^b \left(\frac{L_i f(y)}{\prod_{k=0}^{i+1} \phi_k(y)}\right) d\left(\left(\prod_{k=0}^{i+1} \phi_k(y)\right) g_{i+1}(x, y)\right)$$

$$= -L_i f(y) g_{i+1}(x, y) \Big|_a^b + \int_a^b \left(\prod_{k=0}^{i+1} \phi_k(y)\right) g_{i+1}(x, y) d\left(\frac{L_i f(y)}{\prod_{k=0}^{i+1} \phi_k(y)}\right).$$

$$(14.13)$$

So we find for $i = 0, 1, \ldots, n - 1$ that

$$J_i(x) = L_i f(a) g_{i+1}(x, a) - L_i f(b) g_{i+1}(x, b) + \theta_i(x), \tag{14.14}$$

where

$$\theta_i(x) := \int_a^b \left(\prod_{k=0}^{i+1} \phi_k(y)\right) g_{i+1}(x, y) d\left(\frac{L_i f(y)}{\prod_{k=0}^{i+1} \phi_k(y)}\right). \tag{14.15}$$

Again from [339], for $i = 0, 1, \ldots, n - 1$, we have

$$\frac{L_i f(y)}{\prod_{k=0}^{i-1} \phi_k(y)} = \frac{d}{dy} \frac{1}{\phi_{i-1}(y)} \frac{d}{dy} \frac{1}{\phi_{i-2}(y)} \frac{d}{dy} \cdots \frac{d}{dy} \frac{1}{\phi_1(y)} \frac{d}{dy} \frac{f(y)}{\phi_0(y)}, \tag{14.16}$$

and

$$\frac{L_{i+1} f(y)}{\prod_{k=0}^{i} \phi_k(y)} = \frac{d}{dy} \frac{1}{\phi_i(y)} \frac{d}{dy} \frac{1}{\phi_{i-1}(y)} \frac{d}{dy} \cdots \frac{d}{dy} \frac{1}{\phi_1(y)} \frac{d}{dy} \frac{f(y)}{\phi_0(y)}. \tag{14.17}$$

That is, for $i = 0, 1, \ldots, n - 1$ we derive

$$\frac{L_{i+1} f(y)}{\prod_{k=0}^{i} \phi_k(y)} = \frac{d}{dy} \left(\frac{L_i f(y) \phi_{i+1}(y)}{\prod_{k=0}^{i+1} \phi_k(y)}\right)$$

$$= \left(\frac{d}{dy}\left(\frac{L_i f(y)}{\prod_{k=0}^{i+1} \phi_k(y)}\right)\right) \phi_{i+1}(y) + \left(\frac{L_i f(y)}{\prod_{k=0}^{i+1} \phi_k(y)}\right) \phi'_{i+1}(y). \tag{14.18}$$

Hence we find

$$\left(\frac{d}{dy} \left(\frac{L_i f(y)}{\prod_{k=0}^{i+1} \phi_k(y)} \right) \right) = \frac{\frac{L_{i+1} f(y)}{\prod_{k=0}^{i} \phi_k(y)} - \left(\frac{L_i f(y)}{\prod_{k=0}^{i+1} \phi_k(y)} \right) \phi'_{i+1}(y)}{\phi_{i+1}(y)} , \qquad (14.19)$$

for $i = 0, 1, \ldots, n-1$. That is

$$d \left(\frac{L_i f(y)}{\prod_{k=0}^{i+1} \phi_k(y)} \right) = \left[\left(\frac{L_{i+1} f(y)}{\prod_{k=0}^{i+1} \phi_k(y)} \right) - \left(\frac{L_i f(y)}{\prod_{k=0}^{i+1} \phi_k(y)} \right) \frac{\phi'_{i+1}(y)}{\phi_{i+1}(y)} \right] dy, \quad i = 0, 1, \ldots, n-1.$$

$$(14.20)$$

Applying (14.20) into (14.15) we get

$$\theta_i(x) = \int_a^b g_{i+1}(x, y) \left[L_{i+1} f(y) - L_i f(y) \frac{\phi'_{i+1}(y)}{\phi_{i+1}(y)} \right] dy$$

$$= \int_a^b g_{i+1}(x, y) L_{i+1} f(y) dy - \int_a^b L_i f(y) g_{i+1}(x, y) \frac{d\phi_{i+1}(y)}{\phi_{i+1}(y)} .$$

That is, we proved that

$$\theta_i(x) = J_{i+1}(x) - \int_a^b L_i f(y) g_{i+1}(x, y) \frac{d\phi_{i+1}(y)}{\phi_{i+1}(y)} , \quad \text{for } i = 0, 1, \ldots, n-1. \quad (14.21)$$

Therefore it holds

$$J_i(x) = L_i f(a) g_{i+1}(x, a) - L_i f(b) g_{i+1}(x, b) + J_{i+1}(x) - \int_a^b L_i f(y) g_{i+1}(x, y) \frac{d\phi_{i+1}(y)}{\phi_{i+1}(y)} ,$$

and

$$J_{i+1}(x) = J_i(x) + \left[L_i f(b) g_{i+1}(x, b) - L_i f(a) g_{i+1}(x, a) \right]$$

$$+ \int_a^b \frac{L_i f(y) g_{i+1}(x, y)}{\phi_{i+1}(y)} d\phi_{i+1}(y), \qquad (14.22)$$

for $i = 0, 1, \ldots, n-1$, with

$$J_0(x) = u_0(x) \int_a^b \frac{f(y)}{u_0(y)} dy. \qquad (14.23)$$

Thus we obtain

$$\sum_{i=0}^{n-1} (n-i)(J_{i+1}(x) - J_i(x)) = \sum_{i=0}^{n-1} (n-i) \left[L_i f(b) g_{i+1}(x, b) - L_i f(a) g_{i+1}(x, a) \right]$$

$$+ \sum_{i=0}^{n-1} (n-i) \int_a^b \left(\frac{L_i f(y) g_{i+1}(x, y)}{\phi_{i+1}(y)} \right) d\phi_{i+1}(y).$$

$$(14.24)$$

Clearly, finally we have

$$\sum_{i=0}^{n-1}(n-i)(J_{i+1}(x) - J_i(x)) = \sum_{i=1}^{n} J_i(x) - nJ_0(x), \qquad (14.25)$$

proving the claim. $\qquad\qquad\qquad\qquad\qquad\qquad\qquad\qquad\qquad\qquad\qquad\quad\square$

We give

Theorem 14.8. *Let* $f, \{u_i\}_{i=0}^{n} \in C^n([a,b])$, *and* $W_i(x) > 0$, $i = 0, 1, \ldots, n$, $n \geq 0$, *over* $[a, b]$. *Then*

$$u_0(x) \int_a^b \frac{f(y)}{u_0(y)} dy = \frac{1}{n}\left(\sum_{i=1}^{n} \int_a^b L_i f(y) g_i(x, y) dy\right)$$

$$+ \sum_{i=0}^{n-1}\left(1 - \frac{i}{n}\right)\left[L_i f(a) g_{i+1}(x, a) - L_i f(b) g_{i+1}(x, b)\right]$$

$$- \sum_{i=0}^{n-1}\left(1 - \frac{i}{n}\right)\int_a^b \left(\frac{L_i f(y) g_{i+1}(x, y)}{\phi_{i+1}(y)}\right)\phi_{i+1}'(y) dy, \quad \forall x \in [a, b].$$

$$(14.26)$$

When $u_0(x) = c > 0$, *then we obtain the representation*

$$\int_a^b f(y) dy = \frac{1}{n}\left(\sum_{i=1}^{n} \int_a^b L_i f(y) g_i(x, y) dy\right)$$

$$+ \sum_{i=0}^{n-1}\left(1 - \frac{i}{n}\right)\left[L_i f(a) g_{i+1}(x, a) - L_i f(b) g_{i+1}(x, b)\right]$$

$$- \sum_{i=0}^{n-1}\left(1 - \frac{i}{n}\right)\int_a^b \left(\frac{L_i f(y) g_{i+1}(x, y)}{\phi_{i+1}(y)}\right)\phi_{i+1}'(y) dy, \quad \forall x \in [a, b].$$

$$(14.27)$$

Proof. By Lemma 14.7. $\qquad\qquad\qquad\qquad\qquad\qquad\qquad\qquad\qquad\qquad\quad\square$

We give also the alternative result

Theorem 14.9. *Let* $f, \{u_i\}_{i=0}^{n} \in C^n([a,b])$, $n \geq 0$, $x \in [a,b]$, *all* $W_i(x) > 0$ *on* $[a, b]$, $i = 0, 1, \ldots, n$. *Then*

$$u_0(x) \int_a^b \frac{f(y)}{u_0(y)} dy = \int_a^b L_n f(y) g_n(x, y) dy$$

$$+ \sum_{i=0}^{n-1}\left\{\left[L_i f(a) g_{i+1}(x, a) - L_i f(b) g_{i+1}(x, b)\right]\right.$$

$$\left. - \int_a^b \frac{L_i f(y) g_{i+1}(x, y)}{\phi_{i+1}(y)}\phi_{i+1}'(y) dy\right\}, \quad \forall x \in [a, b]. \qquad (14.28)$$

When $u_0(x) = c > 0$, then

$$\int_a^b f(y)dy = \int_a^b L_n f(y)g_n(x,y)dy$$
$$+ \sum_{i=0}^{n-1} \Bigg\{ \left[L_i f(a)g_{i+1}(x,a) - L_i f(b)g_{i+1}(x,b) \right]$$
$$- \int_a^b \frac{L_i f(y)g_{i+1}(x,y)}{\phi_{i+1}(y)} \phi'_{i+1}(y)dy \Bigg\}, \quad \forall x \in [a,b]. \qquad (14.29)$$

Proof. Set

$$A_i(x) := \Bigg[L_i f(b)g_{i+1}(x,b) - L_i f(a)g_{i+1}(x,a) \right]$$
$$+ \int_a^b \frac{L_i f(y)g_{i+1}(x,y)}{\phi_{i+1}(y)} \phi'_{i+1}(y)dy \Bigg], \quad i = 0,1,\ldots,n-1. \qquad (14.30)$$

From (14.22) we have

$$J_{i+1} = J_i + A_i, \quad i = 0,1,\ldots,n-1. \qquad (14.31)$$

That is

$$J_1 = J_0 + A_0,$$
$$J_2 = J_1 + A_1 = J_0 + A_0 + A_1$$
$$J_3 = J_2 + A_2 = J_0 + A_0 + A_1 + A_2$$
$$\vdots$$
$$J_n = J_{n-1} + A_{n-1} = J_0 + \sum_{i=0}^{n-1} A_i,$$

i.e.

$$J_n(x) = J_0(x) + \sum_{i=0}^{n-1} A_i(x), \quad \forall x \in [a,b]. \qquad (14.32)$$

Hence we established

$$\int_a^b L_n f(y)g_n(x,y)dy = u_0(x) \int_a^b \frac{f(y)}{u_0(y)}dy$$
$$+ \sum_{i=0}^{n-1} \Bigg\{ \left[L_i f(b)g_{i+1}(x,b) - L_i f(a)g_{i+1}(x,a) \right]$$
$$+ \int_a^b \frac{L_i f(y)g_{i+1}(x,y)}{\phi_{i+1}(y)} \phi'_{i+1}(y)dy \Bigg\}, \quad \forall x \in [a,b], \qquad (14.33)$$

proving (14.28). $\qquad \square$

At last we give the following general representation formulae.

Theorem 14.10. *Let* $f, \{u_i\}_{i=0}^n \in C^{n+1}([a,b])$, $n \geq 0$, $x \in [a,b]$, *all* $W_i(x) > 0$ *on* $[a,b]$, $i = 0, 1, \ldots, n$. *Then*

$$f(x) = (n+1)\frac{u_0(x)}{b-a} \int_a^b \frac{f(y)}{u_0(y)}dy + \sum_{i=0}^{n-1}(n-i)\left\{ \frac{[L_if(b)g_{i+1}(x,b) - L_if(a)g_{i+1}(x,a)]}{b-a} \right.$$

$$+ \frac{1}{b-a}\int_a^b \left(\frac{L_if(y)g_{i+1}(x,y)\phi'_{i+1}(y)}{\phi_{i+1}(y)} \right) dy \left. \right\}$$

$$+ \frac{1}{b-a}\int_a^b g_n(x,t)L_{n+1}f(t)K(t,x)dt, \quad \forall x \in [a,b]. \tag{14.34}$$

If $u_0(x) = c > 0$, *then*

$$f(x) = \frac{(n+1)}{b-a} \int_a^b f(y)dy + \sum_{i=0}^{n-1}(n-i)\left\{ \frac{[L_if(b)g_{i+1}(x,b) - L_if(a)g_{i+1}(x,a)]}{b-a} \right.$$

$$+ \frac{1}{b-a}\int_a^b \left(\frac{L_if(y)g_{i+1}(x,y)\phi'_{i+1}(y)}{\phi_{i+1}(y)} \right) dy \left. \right\}$$

$$+ \frac{1}{b-a}\int_a^b g_n(x,t)L_{n+1}f(t)K(t,x)dt, \quad \forall x \in [a,b]. \tag{14.35}$$

Proof. By Theorem 14.6 and Lemma 14.7. $\qquad\qquad\square$

Note 14.11. Clearly (14.8),(14.9),(14.34) and (14.35) generalize the Fink identity (see [195]).

We have

Corollary 14.12. *Let* $f, \{u_i\}_{i=0}^n \in C^{n+1}([a,b])$, $n \geq 0$, $x \in [a,b]$, *all* $W_i(x) > 0$ *on* $[a,b]$, $i = 0, 1, \ldots, n$. *Then*

$$\widehat{\mathcal{E}}_{n+1}(x) := \frac{f(x)}{(n+1)} - \frac{u_0(x)}{b-a}\int_a^b \frac{f(y)}{u_0(y)}dy$$

$$+ \frac{1}{(n+1)}\sum_{i=0}^{n-1}(n-i)\left\{ \frac{[L_if(a)g_{i+1}(x,a) - L_if(b)g_{i+1}(x,b)]}{b-a} \right.$$

$$- \frac{1}{b-a}\int_a^b \left(\frac{L_if(y)g_{i+1}(x,y)\phi'_{i+1}(y)}{\phi_{i+1}(y)} \right) dy \left. \right\}$$

$$= \frac{1}{(n+1)(b-a)}\int_a^b g_n(x,t)L_{n+1}f(t)K(t,x)dt, \quad \forall x \in [a,b]. \tag{14.36}$$

If $u_0(x) = c > 0$, then

$$\mathcal{E}_{n+1}(x) := \frac{f(x)}{(n+1)} - \frac{1}{b-a} \int_a^b f(y)dy$$

$$+ \frac{1}{(n+1)} \sum_{i=0}^{n-1} (n-i) \left\{ \frac{[L_i f(a)g_{i+1}(x,a) - L_i f(b)g_{i+1}(x,b)]}{b-a} \right.$$

$$\left. - \frac{1}{b-a} \int_a^b \left(\frac{L_i f(y)g_{i+1}(x,y)\phi'_{i+1}(y)}{\phi_{i+1}(y)} \right) dy \right\}$$

$$= \frac{1}{(n+1)(b-a)} \int_a^b g_n(x,t)L_{n+1}f(t)K(t,x)dt, \quad \forall x \in [a,b]. \quad (14.37)$$

Proof. By Theorem 14.10. $\qquad\square$

Note 14.13. If $\phi'_{i+1}(y) = 0$, $i = 0, 1, \ldots, n-1$, then $\widehat{\mathcal{E}}_{n+1}(x)$, $\mathcal{E}_{n+1}(x)$ simplify a lot, e.g. case of $u_i(x) = x^i$, $i = 0, 1, \ldots, n$, where $\phi_{i+1}(y) = i+1$.

Next we estimate $\widehat{\mathcal{E}}_{n+1}(x)$, $\mathcal{E}_{n+1}(x)$ via the following Ostrowski type inequality:

Theorem 14.14. *Let* f, $\{u_i\}_{i=0}^n \in C^{n+1}([a,b])$, $n \geq 0$, $x \in [a,b]$, *all* $W_i(x) > 0$ *on* $[a,b]$, $i = 0, 1, \ldots, n$. *Then*

$$\left\{ \begin{array}{l} |\widehat{\mathcal{E}}_{n+1}(x)|, \\ |\mathcal{E}_{n+1}(x)| \end{array} \right\} \leq \frac{1}{(n+1)(b-a)} \min \left\{ \left[\int_a^x |g_n(x,t)|(t-a)dt \right. \right.$$

$$\left. + \int_x^b |g_n(x,t)|(b-t)dt \right] \|L_{n+1}f\|_\infty, \|g_n(x,\cdot)\|_\infty$$

$$\max((x-a),(b-x))\|L_{n+1}f\|_1, \|g_n(x,\cdot)K(\cdot,x)\|_q \|L_{n+1}f\|_p \bigg\},$$

$$(14.38)$$

where $p, q > 1$: $\frac{1}{p} + \frac{1}{q} = 1$, *any* $x \in [a,b]$. *The last inequality (14.38) is sharp, namely it is attained by an* $\hat{f} \in C^{n+1}[a,b]$, *such that*

$$|L_{n+1}\hat{f}(t)| = A|g_n(x,t)K(t,x)|^{q/p},$$

when $[L_{n+1}\hat{f}(t)g_n(x,t)K(t,x)]$ *is of fixed sign, where* $A > 0$, $\forall t \in [a,b]$, *for a fixed* $x \in [a,b]$.

Proof. We have at first that

$$|\widehat{\mathcal{E}}_{n+1}(x)| \leq \frac{1}{(n+1)(b-a)} \int_a^b |g_n(x,t)| \, |L_{n+1}f(t)| \, |K(t,x)|dt$$

$$\leq \frac{\|L_{n+1}f\|_\infty}{(n+1)(b-a)} \int_a^b |g_n(x,t)| \, K(t,x)|dt$$

$$= \frac{\|L_{n+1}f\|_\infty}{(n+1)(b-a)} \left[\int_a^x |g_n(x,t)|(t-a)dt + \int_x^b |g_n(x,t)|(b-t)dt \right].$$

$$(14.39)$$

Next we see that

$$|\widehat{\mathcal{E}}_{n+1}(x)| \leq \left(\frac{\|L_{n+1}f\|_1}{(n+1)(b-a)} \right) \|g_n(x,\cdot)\|_\infty \max\big((x-a),(b-x)\big). \tag{14.40}$$

Finally we observe for $p, q > 1$ with $\frac{1}{p} + \frac{1}{q} = 1$, that

$$|\widehat{\mathcal{E}}_{n+1}(x)| \leq \frac{\|L_{n+1}f\|_p \|g_n(x,\cdot)k(\cdot,x)\|_q}{(n+1)(b-a)}. \tag{14.41}$$

The last is equality when

$$|L_{n+1}f(t)|^p = B|g_n(x,t)k(t,x)|^q, \tag{14.42}$$

where $B > 0$. The claim now is obvious. $\qquad\qquad\qquad\qquad\square$

Next we present a basic general Grüss type inequality.

Theorem 14.15. *Let* f, h, $\{u_i\}_{i=0}^n \in C^{n+1}([a,b])$, $n \geq 0$, *all* $W_i(x) > 0$ *on* $[a,b]$, $i = 0, 1, \ldots, n$. *Then*

$$D := \left| \frac{1}{b-a} \int_a^b f(x)h(x)dx - \left(\frac{1}{b-a} \int_a^b f(x)dx \right) \left(\frac{1}{b-a} \int_a^b h(x)dx \right) \right.$$

$$\left. - \frac{1}{2(b-a)^2} \left(\sum_{i=1}^n \int_a^b \int_a^b \big(h(x)L_i f(y) + f(x)L_i h(y)\big) g_i(x,y)dydx \right) \right|$$

$$\leq \frac{1}{2(b-a)^2} \left(\int_a^b \int_a^b \big(|h(x)| \, \|L_{n+1}f\|_\infty \right.$$

$$\left. + |f(x)| \, \|L_{n+1}h\|_\infty\big)|g_n(x,t)| \, |k(t,x)|dtdx \right) =: M_1. \tag{14.43}$$

Proof. We have by (14.9) that

$$f(x) = \frac{1}{b-a} \int_a^b f(y)dy + \frac{1}{b-a} \left(\sum_{i=1}^n \int_a^b L_i f(y)g_i(x,y)dy \right)$$

$$+ \frac{1}{b-a} \int_a^b g_n(x,t)L_{n+1}f(t)K(t,x)dt, \quad \forall x \in [a,b], \tag{14.44}$$

and

$$h(x) = \frac{1}{b-a} \int_a^b h(y)dy + \frac{1}{b-a} \left(\sum_{i=1}^n \int_a^b L_i h(y)g_i(x,y)dy \right)$$

$$+ \frac{1}{b-a} \int_a^b g_n(x,t)L_{n+1}h(t)K(t,x)dt, \quad \forall x \in [a,b]. \tag{14.45}$$

Therefore it holds

$$f(x)h(x) = \frac{h(x)}{b-a} \int_a^b f(x)dx + \frac{1}{b-a} \left(\sum_{i=1}^n \int_a^b h(x)L_i f(y)g_i(x,y)dy \right)$$

$$+ \frac{1}{b-a} \int_a^b h(x)L_{n+1}f(t)g_n(x,t)K(t,x)dt, \tag{14.46}$$

and

$$f(x)h(x) = \frac{f(x)}{b-a} \int_a^b h(x)dx + \frac{1}{b-a}\left(\sum_{i=1}^n \int_a^b f(x)L_ih(y)g_i(x,y)dy\right)$$

$$+ \frac{1}{b-a} \int_a^b f(x)L_{n+1}h(t)g_n(x,t)K(t,x)dt, \quad \forall x \in [a,b]. \quad (14.47)$$

Integrating (14.46) and (14.47) we have

$$\frac{1}{(b-a)} \int_a^b f(x)h(x)dx = \frac{\left(\int_a^b f(x)dx\right)\left(\int_a^b h(x)dx\right)}{(b-a)^2}$$

$$+ \frac{1}{(b-a)^2}\left(\sum_{i=1}^n \int_a^b \int_a^b h(x)L_if(y)g_i(x,y)dydx\right)$$

$$+ \frac{1}{(b-a)^2} \int_a^b \int_a^b h(x)L_{n+1}f(t)g_n(x,t)K(t,x)dtdx, \quad (14.48)$$

and

$$\frac{1}{(b-a)} \int_a^b f(x)h(x)dx = \frac{\left(\int_a^b f(x)dx\right)\left(\int_a^b h(x)dx\right)}{(b-a)^2}$$

$$+ \frac{1}{(b-a)^2}\left(\sum_{i=1}^n \int_a^b \int_a^b f(x)L_ih(y)g_i(x,y)dydx\right)$$

$$+ \frac{1}{(b-a)^2} \int_a^b \int_a^b f(x)L_{n+1}h(t)g_n(x,t)K(t,x)dtdx. \quad (14.49)$$

Consequently from (14.48) and (14.49) we derive

$$\frac{1}{(b-a)} \int_a^b f(x)h(x)dx - \frac{1}{(b-a)^2}\left(\int_a^b f(x)dx\right)\left(\int_a^b h(x)dx\right)$$

$$- \frac{1}{(b-a)^2}\left(\sum_{i=1}^n \int_a^b \int_a^b h(x)L_if(y)g_i(x,y)dydx\right)$$

$$= \frac{1}{(b-a)^2} \int_a^b \int_a^b h(x)L_{n+1}f(t)g_n(x,t)K(t,x)dtdx, \quad (14.50)$$

and

$$\frac{1}{(b-a)} \int_a^b f(x)h(x)dx - \frac{1}{(b-a)^2}\left(\int_a^b f(x)dx\right)\left(\int_a^b h(x)dx\right)$$

$$- \frac{1}{(b-a)^2}\left(\sum_{i=1}^n \int_a^b \int_a^b f(x)L_ih(y)g_i(x,y)dydx\right)$$

$$= \frac{1}{(b-a)^2} \int_a^b \int_a^b f(x)L_{n+1}h(t)g_n(x,t)K(t,x)dtdx. \quad (14.51)$$

By adding and dividing by 2 the equalities (14.50) and (14.51) we obtain

$$\Delta := \frac{1}{(b-a)} \int_a^b f(x)h(x)dx - \frac{1}{(b-a)^2}\left(\int_a^b f(x)dx\right)\left(\int_a^b h(x)dx\right)$$

$$- \frac{1}{2(b-a)^2}\left(\sum_{i=1}^n \int_a^b \int_a^b (h(x)L_i f(y) + f(x)L_i h(y))g_i(x,y)dydx\right)$$

$$= \frac{1}{2(b-a)^2}\left(\int_a^b \int_a^b (h(x)L_{n+1}f(t) + f(x)L_{n+1}h(t))g_n(x,t)K(t,x)dtdx\right).$$

$$(14.52)$$

Therefore we conclude that

$$|\Delta| \le \frac{1}{2(b-a)^2}\left(\int_a^b \int_a^b (|h(x)|\,\|L_{n+1}f\|_\infty\right.$$

$$\left. + |f(x)|\,\|L_{n+1}h\|_\infty)|g_n(x,t)|\,|K(t,x)|dtdx\right), \tag{14.53}$$

proving the claim. $\qquad\square$

The related Grüss type L_1 result follows.

Theorem 14.16. *Let f, h, $\{u_i\}_{i=0}^n \in C^{n+1}([a,b])$, $n \ge 0$, all $W_i(x) > 0$ on $[a,b]$, $i = 0, 1, \ldots, n$. Then*

$$D = \left| \frac{1}{b-a} \int_a^b f(x)h(x)dx - \left(\frac{1}{b-a}\int_a^b f(x)dx\right)\left(\frac{1}{b-a}\int_a^b h(x)dx\right) \right.$$

$$\left. - \frac{1}{2(b-a)^2}\left(\sum_{i=1}^n \int_a^b \int_a^b (h(x)L_i f(y) + f(x)L_i h(y))g_i(x,y)dydx\right) \right|$$

$$\le \frac{1}{2}\|g_n(x,t)\|_{\infty,[a,b]^2}\left(\|h\|_\infty\|L_{n+1}f\|_1 + \|f\|_\infty\|L_{n+1}h\|_1\right) =: M_2. \tag{14.54}$$

Proof. We obtain

$$|\Delta| = \frac{1}{2(b-a)^2}\left|\int_a^b\int_a^b h(x)L_{n+1}f(t)g_n(x,t)K(t,x)dtdx\right.$$

$$\left.+\int_a^b\int_a^b f(x)L_{n+1}h(t)g_n(x,t)K(t,x)dtdx\right|$$

$$\leq \frac{1}{2(b-a)}\left\{\int_a^b\int_a^b |h(x)|\,|L_{n+1}f(t)|\,|g_n(x,t)|dtdx\right.$$

$$\left.+\int_a^b\int_a^b |f(x)|\,|L_{n+1}h(t)|\,|g_n(x,t)|dtdx\right\}$$

$$\leq \frac{1}{2(b-a)}\left\{\left(\int_a^b\left(\int_a^b |L_{n+1}f(t)|dt\right)dx\right)\|h\|_\infty\|g_n(x,t)\|_{\infty,[a,b]^2}\right.$$

$$\left.+\left(\int_a^b\left(\int_a^b |L_{n+1}h(t)|dt\right)dx\right)\|f\|_\infty\|g_n(x,t)\|_{\infty,[a,b]^2}\right\}$$

$$= \frac{1}{2}\left\{\|L_{n+1}f\|_1\|h\|_\infty\|g_n(x,t)\|_{\infty,[a,b]^2}\right.$$

$$\left.+\|L_{n+1}h\|_1\|f\|_\infty\|g_n(x,t)\|_{\infty,[a,b]^2}\right\}$$

$$= \frac{1}{2}\|g_n(x,t)\|_{\infty,[a,b]^2}\left(\|h\|_\infty\|L_{n+1}f\|_1 + \|f\|_\infty\|L_{n+1}h\|_1\right),$$

establishing the claim. $\qquad\square$

The related Grüss type L_p result follows.

Theorem 14.17. *Let f, h, $\{u_i\}_{i=0}^n \in C^{n+1}([a,b])$, $n \geq 0$, all $W_i(x) > 0$ on $[a,b]$, $i = 0,1,\ldots,n$. Let also $p,q,r > 0$ such that $\frac{1}{p} + \frac{1}{q} + \frac{1}{r} = 1$. Then*

$$D = \left|\frac{1}{b-a}\int_a^b f(x)h(x)dx - \left(\frac{1}{b-a}\int_a^b f(x)dx\right)\left(\frac{1}{b-a}\int_a^b h(x)dx\right)\right.$$

$$\left.-\frac{1}{2(b-a)^2}\left(\sum_{i=1}^n\int_a^b\int_a^b (h(x)L_if(y) + f(x)L_ih(y))g_i(x,y)dydx\right)\right|$$

$$\leq 2^{-1}(b-a)^{-(1+\frac{1}{r})}\|g_n(x,t)K(t,x)\|_{r,[a,b]^2}\left[\|h\|_{p,[a,b]}\|L_{n+1}f\|_{q,[a,b]}\right.$$

$$\left.+\|f\|_{p,[a,b]}\|L_{n+1}h\|_{q,[a,b]}\right] =: M_3. \tag{14.55}$$

Proof. We derive by generalized Hölder's inequality that

$$|\Delta| \le \frac{1}{2(b-a)^2} \left(\int_a^b \int_a^b |h(x)| \, |L_{n+1}f(t)| \, |g_n(x,t)K(t,x)| dt dx \right.$$

$$\left. + \int_a^b \int_a^b |f(x)| \, |L_{n+1}h(t)| \, |g_n(x,t)K(t,x)| dt dx \right)$$

$$\le \frac{1}{2(b-a)^2} \left[\left(\int_a^b \int_a^b |h(x)|^p dt dx \right)^{1/p} \left(\int_a^b \int_a^b |L_{n+1}f(t)|^q dt dx \right)^{1/q} \right.$$

$$\cdot \, \|g_n(x,t)K(t,x)\|_{r,[a,b]^2}$$

$$\left. + \left(\int_a^b \int_a^b |f(x)|^p dt dx \right)^{1/p} \left(\int_a^b \int_a^b |L_{n+1}h(t)|^q dt dx \right)^{1/q} \|g_n(x,t)K(t,x)\|_{r,[a,b]^2} \right]$$

$$= \frac{1}{2(b-a)^2} \left[(b-a)^{1/p}\|h\|_{p,[a,b]}(b-a)^{1/q}\|L_{n+1}f\|_{q,[a,b]}\|g_n(x,t)K(t,x)\|_{r,[a,b]^2} \right.$$

$$\left. + (b-a)^{1/p}\|f\|_{p,[a,b]}(b-a)^{1/q}\|L_{n+1}h\|_{q,[a,b]}\|g_n(x,t)K(t,x)\|_{r,[a,b]^2} \right]$$

$$= \frac{(b-a)^{1-\frac{1}{r}}}{2(b-a)^2} \left[\|h\|_{p,[a,b]}\|L_{n+1}f\|_{q,[a,b]} \right.$$

$$\left. + \|f\|_{p,[a,b]}\|L_{n+1}h\|_{q,[a,b]} \right] \|g_n(x,t)K(t,x)\|_{r,[a,b]^2},$$

proving the claim. $\square$

We conclude the Grüss type results with

Corollary 14.18 (on Theorems 14.15, 14.16, 14.17). *Let f, h, $\{u_i\}_{i=0}^n \in C^{n+1}([a,b])$, $n \ge 0$, all $W_i(x) > 0$ on $[a,b]$, $i = 0,1,\ldots,n$. Let also $p,q,r > 0$ such that $\frac{1}{p} + \frac{1}{q} + \frac{1}{r} = 1$. Then*

$$D \le \min\{M_1, M_2, M_3\}. \tag{14.56}$$

The related comparison of integral means follows, see also [48].

Theorem 14.19. *Let f, $\{u_i\}_{i=0}^n \in C^{n+1}([a,b])$, with $u_0(x) = c > 0$, $n \ge 0$, all $W_i(x) > 0$ on $[a,b]$, $i = 0,1,\ldots,n$. Let μ be a finite measure of mass $m > 0$ on*

$([c,d], \mathcal{P}([c,d])), [c,d] \subseteq [a,b] \subseteq \mathbb{R}$, *where $\mathcal{P}$ stands for the power set. Then*

$$\left| \frac{1}{(n+1)m} \int_{[c,d]} f(x)d\mu(x) - \frac{1}{b-a} \int_a^b f(x)dx \right.$$

$$+ \frac{1}{(n+1)m} \sum_{i=0}^{n-1} (n-i)$$

$$\cdot \left\{ \frac{\left[L_i f(a) \int_{[c,d]} g_{i+1}(x,a)d\mu(x) - L_i f(b) \int_{[c,d]} g_{i+1}(x,b)d\mu(x) \right]}{b-a} \right.$$

$$\left. \left. - \frac{1}{b-a} \left\{ \int_{[c,d]} \left[\int_a^b \left(\frac{L_i f(y) g_{i+1}(x,y) \phi'_{i+1}(y)}{\phi_{i+1}(y)} \right) dy \right] d\mu(x) \right\} \right\} \right|$$

$$\leq \frac{1}{(n+1)(b-a)m} \min \left\{ \left(\int_{[c,d]} \left[\int_a^x |g_n(x,t)|(t-a)dt \right. \right. \right.$$

$$\left. \left. + \int_x^b |g_n(x,t)|(b-t)dt \right] d\mu(x) \right) \|L_{n+1}f\|_\infty,$$

$$\left(\int_{[c,d]} (\|g_n(x,\cdot)\|_\infty \max(x-a,b-x)) \, d\mu(x) \right) \|L_{n+1}f\|_1,$$

$$\left. \left(\int_{[c,d]} \|g_n(x,\cdot)K(\cdot,x)\|_q d\mu(x) \right) \|L_{n+1}f\|_p \right\}, \tag{14.57}$$

where $p, q > 1: \frac{1}{p} + \frac{1}{q} = 1$.

Proof. By Theorem 14.14. $\qquad\square$

Assumptions 14.20. Next we follow [140].

Let f be a convex function from $(0, +\infty)$ into $\mathbb{R}$ which is strictly convex at 1 with $f(1) = 0$. Let $(X, \mathcal{A}, \lambda)$ be a measure space, where λ is a finite or a σ-finite measure on $(X, \mathcal{A})$. And let μ_1, μ_2 be two probability measures on $(X, \mathcal{A})$ such that $\mu_1 \ll \lambda$, $\mu_2 \ll \lambda$ (absolutely continuous), e.g. $\lambda = \mu_1 + \mu_2$. Denote by $p = \frac{d\mu_1}{d\lambda}$, $q = \frac{d\mu_2}{d\lambda}$, the (densities) Radon–Nikodym derivatives of μ_1, μ_2 with respect to λ. Here we assume that

$$0 < a \leq \frac{p}{q} \leq b, \quad \text{a.e. on } X \text{ and } a \leq 1 \leq b.$$

The quantity

$$\Gamma_f(\mu_1, \mu_2) = \int_X q(x) f\left(\frac{p(x)}{q(x)} \right) d\lambda(x) \tag{14.58}$$

was introduced by I. Csiszar in 1967, see [140], and is called *f-divergence* of the probability measures μ_1 and μ_2. By Lemma 1.1 of [140], the integral (14.58) is well-defined and $\Gamma_f(\mu_1, \mu_2) \geq 0$ with equality only when $\mu_1 = \mu_2$. Furthermore $\Gamma_f(\mu_1, \mu_2)$ does not depend on the choice of λ. Here by assuming $f(1) = 0$ we can consider $\Gamma_f(\mu_1, \mu_2)$, the f-divergence, as a measure of the difference between the probability measures μ_1, μ_2.

Here we give a representation and estimates for $\Gamma_f(\mu_1, \mu_2)$ via formula (14.35).

Theorem 14.21. *Let f and Γ_f as in Assumptions 14.20. Additionally assume that f, $\{u_i\}_{i=0}^{n} \in C^{n+1}([a,b])$, with $u_0(x) = c > 0$, $n \geq 0$, all $W_i(t) > 0$ on $[a,b]$, $i = 0, 1, \ldots, n$. Then*

$$
\Gamma_f(\mu_1, \mu_2)
$$
$$
= \frac{(n+1)}{b-a} \int_a^b f(y)\,dy
$$
$$
+ \sum_{i=0}^{n-1} (n-i) \Bigg\{ \frac{\left[L_i f(b)\Gamma_{g_{i+1}(\cdot, b)}(\mu_1, \mu_2) - L_i f(a)\Gamma_{g_{i+1}(\cdot, a)}(\mu_1, \mu_2)\right]}{b-a}
$$
$$
+ \frac{1}{b-a}\left(\int_X q(x) \left(\int_a^b \left(\frac{L_i f(y) g_{i+1}\left(\frac{p(x)}{q(x)}, y\right)\phi'_{i+1}(y)}{\phi_{i+1}(y)} \right) dy \right) d\lambda(x) \right) \Bigg\} + \mathcal{R}_{n+1},
$$
$$
\tag{14.59}
$$

where

$$
\mathcal{R}_{n+1} := \frac{1}{b-a} \int_X q(x) \left(\int_a^b g_n\left(\frac{p(x)}{q(x)}, t\right) L_{n+1} f(t) K\left(t, \frac{p(x)}{q(x)}\right) dt \right) d\lambda(x). \tag{14.60}
$$

Proof. In (14.35), set $x = \frac{p}{q}$, multiply it by q and integrate against λ. $\qquad\square$

The estimate for $\Gamma_f(\mu_1, \mu_2)$ follows.

Theorem 14.22. *Let all as in Theorem 14.21. Then*

$$
|\mathcal{R}_{n+1}|
$$
$$
\leq \frac{1}{b-a} \min\Bigg\{ \left[\int_X q(x) \left(\int_a^b \left| g_n\left(\frac{p(x)}{q(x)}, t\right) \right| \left| K\left(t, \frac{p(x)}{q(x)}\right) \right| dt \right) d\lambda(x) \right] \|L_{n+1} f\|_{\infty, [a,b]},
$$
$$
\left(\int_X q(x) \left\| g_n\left(\frac{p(x)}{q(x)}, \cdot\right) K\left(\cdot, \frac{p(x)}{q(x)}\right) \right\|_{p_2, [a,b]} d\lambda(x) \right) \|L_{n+1} f\|_{p_1, [a,b]},
$$
$$
\left[\int_X q(x) \left\| g_n\left(\frac{p(x)}{q(x)}, \cdot\right) \right\|_{\infty, [a,b]} \max\left(\frac{p(x)}{q(x)} - a, b - \frac{p(x)}{q(x)} \right) d\lambda(x) \right] \|L_{n+1} f\|_{1, [a,b]} \Bigg\},
$$
$$
\tag{14.61}
$$

where $p_1, p_2 > 1$ such that $\frac{1}{p_1} + \frac{1}{p_2} = 1$.

Proof. From (14.60) we obtain

$$
|\mathcal{R}_{n+1}| \leq \frac{1}{b-a} \int_X q(x) \left(\int_a^b \left| g_n\left(\frac{p(x)}{q(x)}, t\right) \right| |L_{n+1} f(t)| \left| K\left(t, \frac{p(x)}{q(x)}\right) \right| dt \right) d\lambda(x)
$$
$$
=: \Lambda. \tag{14.62}
$$

We see that

$$\Lambda \leq \frac{1}{b-a}\left[\int_X q(x)\left(\int_a^b \left|g_n\left(\frac{p(x)}{q(x)},t\right)\right|\left|K\left(t,\frac{p(x)}{q(x)}\right)\right|dt\right)d\lambda(x)\right]\|L_{n+1}f\|_{\infty,[a,b]}.$$

(14.63)

Also we have true that

$$\Lambda \leq \frac{1}{b-a}\left(\int_X q(x)\left\|g_n\left(\frac{p(x)}{q(x)},\cdot\right)K\left(\cdot,\frac{p(x)}{q(x)}\right)\right\|_{p_2,[a,b]}d\lambda(x)\right)\|L_{n+1}f\|_{p_1,[a,b]}.$$

(14.64)

Finally we derive

$$\Lambda \leq \frac{1}{b-a}\left(\int_X q(x)\left\|g_n\left(\frac{p(x)}{q(x)},\cdot\right)\right\|_{\infty,[a,b]}\right.$$
$$\left.\max\left(\frac{p(x)}{q(x)}-a,b-\frac{p(x)}{q(x)}\right)d\lambda(x)\right)\|L_{n+1}f\|_{1,[a,b]}.$$

(14.65)

$\square$

Chapter 15

Representations of Functions and Csiszar's f-Divergence

In this chapter we present very general Montgomery type weighted representation identities. We give applications to Csiszar's f-Divergence. This treatment follows [62].

15.1 Introduction

This chapter is motivated by Montgomery's identity [265]:

Theorem A. *Let $f : [a,b] \to R$ be differentiable on $[a,b]$ and $f' : [a,b] \to R$ integrable on $[a,b]$.*

Then the Montgomery identity holds

$$f(x) = \frac{1}{b-a}\int_a^b f(t)dt + \frac{1}{b-a}\int_a^b P^*(x,t)f'(t)dt,$$

where $P^*(x,t)$ is the Peáno Kernel defined by

$$P^*(x,t) := \begin{cases} \frac{t-a}{b-a}, & a \le t \le x, \\ \frac{t-b}{b-a}, & x < t \le b. \end{cases}$$

We are also motivated by the following weighted Montgomery type identity:

Theorem B (see [294] Pecaric). *Let $w : [a,b] \to R_+$ is some probability density function, that is, an integrable function satisfying $\int_a^b w(t)dt = 1$, and $W(t) = \int_a^t w(x)dx$ for $t \in [a,b]$, $W(t) = 0$ for $t < a$, and $W(t) = 1$ for $t > b$.*

Then

$$f(x) = \int_a^b w(t)f(t)dt + \int_a^b P_W(x,t)f'(t)dt,$$

where the weighted Peáno Kernel is

$$P_W(x,t) := \begin{cases} W(t), & a \le t \le x, \\ W(t) - 1, & x < t \le b. \end{cases}$$

193

The aim of this chapter is to present a set of very general representation formulae to functions. They are 1- weighted and 2- weighted very diverse representations.

At the end we give several applications to representing Csiszar's f-Divergence.

15.2   Main Results

We need

Proposition 15.1 ([32]). *Let $f : [a, b] \to \mathbb{R}$ be differentiable on $[a, b]$. The derivative $f' : [a, b] \to R$ is integrable on $[a, b]$. Let $g : [a, b] \to R$ be of bounded variation and such that $g(a) \neq g(b)$. Let $x \in [a, b]$.*

We define

$$
P_W(g(x), g(t)) := \begin{cases} \frac{g(t) - g(a)}{g(b) - g(a)}, & a \leq t \leq x, \\[2mm] \frac{g(t) - g(b)}{g(b) - g(a)}, & x < t \leq b. \end{cases}
\tag{15.1}
$$

Then

$$
f(x) = \frac{1}{g(b) - g(a)} \int_a^b f(t) dg(t) + \int_a^b P(g(x), g(t)) f'(t) dt.
\tag{15.2}
$$

We also need

Theorem 15.2 ([241], p. 17). *Let $f \in C^n([a, b])$, $n \in \mathbb{N}$, $x \in [a, b]$.*
Then

$$
f(x) = \frac{1}{b - a} \int_a^b f(t) dt + \Phi_{n-1}(x) + R_n(x),
\tag{15.3}
$$

where for $m \in \mathbb{N}$ we call

$$
\Phi_m(x) := \sum_{k=1}^m \frac{(b - a)^{k-1}}{k!} B_k \left(\frac{x - a}{b - a} \right) \left[f^{(k-1)}(b) - f^{(k-1)}(a) \right]
\tag{15.4}
$$

with the convention $\Phi_0(x) = 0$, and

$$
R_n(x) := -\frac{(b - a)^{n-1}}{n!} \int_a^b \left[B_n^* \left(\frac{x - t}{b - a} \right) - B_n \left(\frac{x - a}{b - a} \right) \right] f^{(n)}(t) dt.
\tag{15.5}
$$

Here $B_k(x)$, $k \geq 0$ are the Bernoulli polynomials, $B_k = B_k(0)$, $k \geq 0$, the Bernoulli numbers, and B_k^*, $k \geq 0$, are periodic functions of period one related to the Bernoulli polynomials as

$$
B_k^*(x) = B_k(x), \quad 0 \leq x < 1,
\tag{15.6}
$$

and

$$
B_k^*(x + 1) = B_k^*(x), \quad x \in \mathbb{R}.
\tag{15.7}
$$

Some basic properties of Bernoulli polynomials follow (see [1], 23.1). We have

$$B_0(x) = 1, \ B_1(x) = x - \frac{1}{2}, \ B_2(x) = x^2 - x + \frac{1}{6}, \tag{15.8}$$

and

$$B_k'(x) = kB_{k-1}(x), \ k \in \mathbb{N}, \tag{15.9}$$

$$B_k(x+1) - B_k(x) = kx^{k-1}, \ k \geq 0. \tag{15.10}$$

Clearly $B_0^* = 1$, B_1^* is a discontinuous function with a jump of -1 at each integer, and B_k^*, $k \geq 2$, is a continuous function.

Notice that $B_k(0) = B_k(1) = B_k$, $k \geq 2$.

We need also the more general

Theorem 15.3 (see [53]). *Let $f : [a,b] \to \mathbb{R}$ be such that $f^{(n-1)}$, $n \geq 1$, is a continuous function and $f^{(n)}$ exists and is finite for all but a countable set of x in (a,b) and that $f^{(n)} \in L_1([a,b])$. Then for every $x \in [a,b]$ we have*

$$f(x) = \frac{1}{b-a} \int_a^b f(t)dt + \sum_{k=1}^{n-1} \frac{(b-a)^{k-1}}{k!} B_k\left(\frac{x-a}{b-a}\right) \left[f^{(k-1)}(b) - f^{(k-1)}(a)\right]$$

$$+ \frac{(b-a)^{n-1}}{n!} \int_{[a,b]} \left[B_n\left(\frac{x-a}{b-a}\right) - B_n^*\left(\frac{x-t}{b-a}\right)\right] f^{(n)}(t)dt. \tag{15.11}$$

The sum in (15.11) when $n = 1$ is zero.

If $f^{(n-1)}$ is just absolutely continuous then (15.11) is valid again.

Formula (15.11) is a generalized Euler type identity, see also [143].

We give

Remark 15.4. Let f, g be as in Proposition 15.1, then we have

$$f(z) = \frac{1}{g(b) - g(a)} \int_a^b f(x)dg(x) + \int_a^b P(g(z), g(x))f'(x)dx. \tag{15.12}$$

Further assume that f' be such that $f^{(n-1)}$, $n \geq 2$, is a continuous function and $f^{(n)}(x)$ exists and is finite for all but a countable set of x in (a,b) and that $f^{(n)} \in L_1([a,b])$.

Then, by Theorem 15.3, we have

$$f'(x) = \frac{f(b) - f(a)}{b - a} + T_{n-2}(x) + R_{n-1}(x), \tag{15.13}$$

where

$$T_{n-2}(x) := \sum_{k=1}^{n-2} \frac{(b-a)^{k-1}}{k!} B_k\left(\frac{x-a}{b-a}\right) \left[f^{(k)}(b) - f^{(k)}(a)\right], \tag{15.14}$$

and

$$R_{n-1}(x) := -\frac{(b-a)^{n-2}}{(n-1)!} \int_a^b \left[B_{n-1}^*\left(\frac{x-t}{b-a}\right) - B_{n-1}\left(\frac{x-a}{b-a}\right)\right] f^{(n)}(t)dt. \tag{15.15}$$

The sum in (15.14) when $n = 2$ is zero.

If $f^{(n-1)}$ is just absolutely continuous then (15.13) is valid again.

By plugging (15.13) into (15.12) we obtain our first result.

Theorem 15.5. *Let $f : [a,b] \to \mathbb{R}$ be such that $f^{(n-1)}$, $n \geq 2$, is a continuous function and $f^{(n)}$ exists and is finite for all but a countable set of x in (a,b) and that $f^{(n)} \in L_1([a,b])$. Let $g : [a,b] \to \mathbb{R}$ be of bounded variation and $g(a) \neq g(b)$. Let $z \in [a,b]$.*

Then

$$f(z) = \frac{1}{g(b) - g(a)} \int_a^b f(x)dg(x) + \frac{f(b) - f(a)}{b - a} \int_a^b P(g(z), g(x))dx$$

$$+ \sum_{k=1}^{n-2} \frac{(b-a)^{k-1}}{k!} \left(f^{(k)}(b) - f^{(k)}(a) \right) \int_a^b P(g(z), g(x)) B_k \left(\frac{x-a}{b-a} \right) dx$$

$$+ \frac{(b-a)^{n-2}}{(n-1)!} \int_a^b P(g(z), g(x)) \left(\int_a^b \left(B_{n-1} \left(\frac{x-a}{b-a} \right) \right) \right.$$

$$\left. - B_{n-1}^* \left(\frac{x-t}{b-a} \right) \right) f^{(n)}(t)dt \right) dx. \tag{15.16}$$

Here,

$$P(g(z), g(t)) := \begin{cases} \frac{g(t) - g(a)}{g(b) - g(a)}, & a \leq t \leq z, \\[2mm] \frac{g(t) - g(b)}{g(b) - g(a)}, & z < t \leq b. \end{cases} \tag{15.17}$$

The sum in (15.16) is zero when $n = 2$.

We need Fink's identity.

Theorem 15.6 (Fink, [195]). *Let any $a, b \in \mathbb{R}$, $a \neq b$, $f : [a,b] \to \mathbb{R}$, $n \geq 1$, $f^{(n-1)}$ is absolutely continuous on $[a,b]$.*

Then

$$f(x) = \frac{n}{b-a} \int_a^b f(t)dt - \sum_{k=1}^{n-1} \left(\frac{n-k}{k!} \right) \left(\frac{f^{(k-1)}(a)(x-a)^k - f^{(k-1)}(b)(x-b)^k}{b-a} \right)$$

$$+ \frac{1}{(n-1)!(b-a)} \int_a^b (x-t)^{n-1} K(t,x) f^{(n)}(t)dt, \tag{15.18}$$

where

$$K(t,x) := \begin{cases} t-a, & a \leq t \leq x \leq b, \\ t-b, & a \leq x < t \leq b. \end{cases} \tag{15.19}$$

When $n = 1$ the sum $\sum_{k=1}^{n-1}$ in (15.18) is zero.

We make

Remark 15.7. Let f, g be as in Proposition 15.1, then we have

$$f(z) = \frac{1}{g(b) - g(a)} \int_a^b f(x)dg(x) + \int_a^b P(g(z), g(x))f'(x)dx. \tag{15.20}$$

Further suppose that f' be such that $f^{(n-1)}, n \geq 2$, is absolutely continuous on $[a, b]$.

Then

$$f'(x) = \frac{(n-1)(f(b) - f(a))}{b - a}$$
$$+ \sum_{k=1}^{n-2} \left(\frac{n - k - 1}{k!} \right) \left(\frac{f^{(k)}(b)(x - b)^k - f^{(k)}(a)(x - a)^k}{b - a} \right)$$
$$+ \frac{1}{(n-2)! \, (b - a)} \int_a^b (x - t)^{n-2} K(t, x) f^{(n)}(t)dt. \tag{15.21}$$

When $n = 2$ the sum $\sum_{k=1}^{n-2}$ in (15.21) is zero.

By plugging (15.21) into (15.20) we obtain

Theorem 15.8. *Let* $f : [a, b] \to R$ *be such that* $f^{(n-1)}$, $n \geq 2$, *is absolutely continuous on* $[a, b]$. *Let* $g : [a, b] \to R$ *be of bounded variation and* $a \neq b$, $g(a) \neq g(b)$. *Let* $z \in [a, b]$.

Then

$$f(z) = \frac{1}{g(b) - g(a)} \int_a^b f(x)dg(x) + \frac{(n-1)(f(b) - f(a))}{b - a} + \int_a^b P(g(z), g(x))dx$$
$$+ \sum_{k=1}^{n-2} \left(\frac{n - k - 1}{(b - a) \, k!} \right) \int_a^b P(g(z), g(x)) \left(f^{(k)}(b)(x - b)^k - f^{(k)}(a)(x - a)^k \right) dx$$
$$+ \frac{1}{(n-2)! \, (b - a)} \int_a^b P(g(z), g(x)) \left(\int_a^b (x - t)^{n-2} K(t, x) f^{(n)}(t)dt \right) dx. \tag{15.22}$$

When $n = 2$ *the sum* $\sum_{k=1}^{n-2}$ *in (15.22) is zero.*

We need to mention

Theorem 15.9 (Aljinovic- Pecaric, [14]). *Let's suppose* $f^{(n-1)}$, $n > 1$, *is a continuous function of bounded variation on* $[a, b]$. *If* $w : [a, b] \to \mathbb{R}_+$ *is some probability density function, i.e. integrable function satisfying* $\int_a^b w(t)dt = 1$, *and*

$$W(t) = \int_a^t w(x)dx \ for \ t \in [a, b],$$

$W(t) = 0$ *for* $t < a$ *and* $W(t) = 1$ *for* $t > b$, *the weighted Peáno Kernel is*

$$P_W(x, t) := \begin{cases} W(t), \ a \leq t \leq x, \\ \\ W(t) - 1, \ x < t \leq b. \end{cases} \tag{15.23}$$

Then the following identity holds

$$f(x) = \int_a^b w(t)f(t)dt$$

$$+ \sum_{k=1}^{n-1} \frac{(b-a)^{k-2}}{(k-1)!} \left(\int_a^b P_W(x,t) B_{k-1}\left(\frac{t-a}{b-a}\right) dt \right) \left(f^{(k-1)}(b) - f^{(k-1)}(a) \right)$$

$$- \frac{(b-a)^{n-2}}{(n-1)!} \int_a^b \left(\int_a^b P_W(x,s) \left[B_{n-1}^*\left(\frac{s-t}{b-a}\right) - B_{n-1}\left(\frac{s-a}{b-a}\right) \right] ds \right) df^{(n-1)}(t).$$

$$(15.24)$$

We make

Assumption 15.10. Let $f : [a,b] \to R$ be such that $f^{(n-1)}$, $n > 1$, is a continuous function and $f^{(n)}(x)$ exists and is finite for all but a countable set of x in (a,b) and that $f^{(n)} \in L_1([a,b])$.

We give

Comment 15.11. Let f as in Assumption 15.10. Then $f^{(n-1)}(x)$ absolutely continuous and thus of bounded variation. Furthermore, we can write in the last integral of (15.24)

$$df^{(n-1)}(t) = f^{(n)}(t)dt. \qquad (15.25)$$

So, if we assume in Theorem 15.9 for f that $f^{(n-1)}$ is absolutely continuous we have that (15.24) and (15.25) are valid.

We make

Remark 15.12. Let f, g be as in Proposition 15.1 , then we have

$$f(z) = \frac{1}{g(b) - g(a)} \int_a^b f(x)dg(x) + \int_a^b P(g(z), g(x))f'(x)dx. \qquad (15.26)$$

Further assume that f be such that $f^{(n-1)}$, $n > 2$, is a continuous function and $f^{(n)}(x)$ exists and is finite for all but a countable set of x in (a,b) and that $f^{(n)} \in L_1([a,b])$.

Then using terms of Theorem 15.9 we find

$$f'(x) = \int_a^b w(t)f'(t)dt+$$

$$\sum_{k=1}^{n-2} \frac{(b-a)^{k-2}}{(k-1)!} \left(\int_a^b P_W(x,t) B_{k-1}\left(\frac{t-a}{b-a}\right) dt \right) \left(f^{(k)}(b) - f^{(k)}(a) \right)$$

$$- \frac{(b-a)^{n-3}}{(n-2)!} \int_a^b \left(\int_a^b P_W(x,s) \left[B_{n-2}^*\left(\frac{s-t}{b-a}\right) - B_{n-2}\left(\frac{s-a}{b-a}\right) \right] ds \right) f^{(n)}(t)dt.$$

$$(15.27)$$

Plugging (15.27) into (15.26) we get

Theorem 15.13. *Let $f : [a, b] \to \mathbb{R}$ be such that $f^{(n-1)}$, $n > 2$, is a continuous function and $f^{(n)}(x)$ exists and is finite for all but a countable set of x in (a, b) and that $f^{(n)} \in L_1([a, b])$.*

Let $g : [a, b] \to \mathbb{R}$ be of bounded variation and $g(a) \neq g(b)$.

Let w, W and P_W as in Theorem 15.9, and P as in (15.1). Let $z \in [a, b]$. Then

$$f(z) = \frac{1}{g(b) - g(a)} \int_a^b f(x) dg(x) + \left(\int_a^b P(g(z), g(x)) dx \right) \left(\int_a^b w(t) f'(t) dt \right)$$

$$+ \sum_{k=1}^{n-2} \frac{(b-a)^{k-2}}{(k-1)!} \left(f^{(k)}(b) - f^{(k)}(a) \right)$$

$$\cdot \int_a^b P(g(z), g(x)) \left(\int_a^b P_W(x, t) B_{k-1} \left(\frac{t-a}{b-a} \right) dt \right) dx - \frac{(b-a)^{n-3}}{(n-2)!}$$

$$\cdot \int_a^b P(g(z), g(x)) \left[\int_a^b \left(\int_a^b P_W(x, s) \left[B_{n-2}^* \left(\frac{s-t}{b-a} \right) \right.\right.\right.$$

$$\left.\left.\left. - B_{n-2} \left(\frac{s-a}{b-a} \right) \right] ds \right) f^{(n)}(t) dt \right] dx. \tag{15.28}$$

We need

Theorem 15.14 (see [16], Aljinovic et al) *Let $f : [a, b] \to R$ be such that $f^{(n-1)}$, is an absolutely continuous function on $[a, b]$ for some $n > 1$.*

If $w : [a, b] \to R_+$ is some probability density function, then the following identity holds:

$$f(x) = \int_a^b w(t) f(t) dt - \sum_{k=1}^{n-1} F_k(x) + \sum_{k=1}^{n-1} \int_a^b w(t) F_k(t) dt$$

$$+ \frac{1}{(n-1)! \, (b-a)} \int_a^b (x-y)^{n-1} K(y, x) f^{(n)}(y) dy$$

$$- \frac{1}{(n-1)! \, (b-a)} \int_a^b \left(\int_a^b w(t)(t-y)^{n-1} K(y, t) dt \right) f^{(n)}(y) dy. \tag{15.29}$$

Here

$$F_k(x) := \left(\frac{n-k}{k!} \right) \left(\frac{f^{(k-1)}(a)(x-a)^k - f^{(k-1)}(b)(x-b)^k}{b-a} \right), \tag{15.30}$$

$$K(t, x) := \begin{cases} t - a, \ a \leq t \leq x \leq b, \\ \\ t - b, \ a \leq x < t \leq b. \end{cases} \tag{15.31}$$

We make

Remark 15.15. Same assumptions as in Theorem 15.14 and $n > 2$.
Here put

$$\overline{F}_k(x) := \left(\frac{n-k-1}{k!}\right)\left(\frac{f^{(k)}(a)(x-a)^k - f^{(k)}(b)(x-b)^k}{b-a}\right). \tag{15.32}$$

Thus by (15.29) we obtain

$$f'(x) = \int_a^b w(t)f'(t)dt - \sum_{k=1}^{n-2}\overline{F}_k(x) + \sum_{k=1}^{n-2}\int_a^b w(t)\overline{F}_k(t)dt$$

$$+ \frac{1}{(n-2)!\,(b-a)}\int_a^b (x-y)^{n-2}K(y,x)f^{(n)}(y)dy$$

$$- \frac{1}{(n-2)!\,(b-a)}\int_a^b \left(\int_a^b w(t)(t-y)^{n-2}K(y,t)dt\right)f^{(n)}(y)dy. \tag{15.33}$$

Let $g : [a,b] \to \mathbb{R}$ be of bounded variation and $a \neq b$, $g(a) \neq g(b)$.
Then, by Proposition 15.1, it holds (15.26).
By plugging in (15.33) into (15.26) we obtain

Theorem 15.16. *Let $f : [a,b] \to R$ be such that $f^{(n-1)}$ is an absolutely continuous function on $[a,b]$ for $n > 2$. Here $w : [a,b] \to R_+$ is some probability density function, and $z \in [a,b]$.*
Let $g : [a,b] \to R$ be of bounded variation with $a \neq b$, $g(a) \neq g(b)$. Then

$$f(z) = \frac{1}{g(b)-g(a)}\int_a^b f(x)dg(x) + \left(\int_a^b P(g(z),g(x))dx\right)\left(\int_a^b w(x)f'(x)dx\right)$$

$$-\sum_{k=1}^{n-2}\int_a^b P(g(z),g(x))\overline{F}_k(x)dx + \left(\int_a^b P(g(z),g(x))dx\right)\left(\sum_{k=1}^{n-2}\int_a^b w(x)\overline{F}_k(x)dx\right)$$

$$+ \frac{1}{(n-2)!\,(b-a)}\int_a^b P(g(z),g(x))\left(\int_a^b (x-y)^{n-2}K(y,x)f^{(n)}(y)dy\right)dx$$

$$- \frac{1}{(n-2)!\,(b-a)}\left(\int_a^b P(g(z),g(x))dx\right)$$

$$\cdot \int_a^b \left(\int_a^b w(t)(t-y)^{n-2}K(y,t)dt\right)f^{(n)}(y)dy. \tag{15.34}$$

Here $\overline{F}_k(x)$ is as in (15.32).

We mention

Background 15.17 (see [144]). Harmonic representation formulae. Assume that $(P_k(t),\ k \geq 0)$ is a harmonic sequence of polynomials i.e. the sequence of polynomials satisfying

$$P_k'(t) = P_{k-1}(t), \quad k \geq 1; \quad P_0(t) = 1. \tag{15.35}$$

Define $P_k^*(t),\ k \geq 0$, to be a periodic function of period 1, related to $P_k(t),\ k \geq 0$, as

$$P_k^*(t) = P_k(t), \quad 0 \leq t < 1,$$

$$P_k^*(t+1) = P_k^*(t), \quad t \in \mathbb{R}. \tag{15.36}$$

Thus, $P_0^*(t) = 1$, while for $k \geq 1$, $P_k^*(t)$ is continuous on $R \setminus Z$ and has a jump of

$$\alpha_k = P_k(0) - P_k(1) \tag{15.37}$$

at every integer t, whenever $\alpha_k \neq 0$.

Note that $\alpha_1 = -1$, since $P_1(t) = t + c$, for some $c \in R$.

Also, note that from the definition it follows

$$P_k'^*(t) = P_{k-1}^*(t), \quad k \geq 1, \ t \in R \setminus Z. \tag{15.38}$$

Let $f : [a,b] \to \mathbb{R}$ be such that $f^{(n-1)}$ is a continuous function of bounded variation on $[a,b]$ for some $n \geq 1$.

In [144] the following two identities have been proved:

$$f(x) = \frac{1}{b-a} \int_a^b f(t)dt + \tilde{T}_n(x) + \tau_n(x) + \tilde{R}_n^1(x) \tag{15.39}$$

and

$$f(x) = \frac{1}{b-a} \int_a^b f(t)dt + \tilde{T}_{n-1}(x) + \tau_n(x) + \tilde{R}_n^2(x), \tag{15.40}$$

where

$$\tilde{T}_m(x) := \sum_{k=1}^m (b-a)^{k-1} P_k\left(\frac{x-a}{b-a}\right)\left[f^{(k-1)}(b) - f^{(k-1)}(a)\right], \tag{15.41}$$

for $1 \leq m \leq n$, and

$$\tau_m(x) := \sum_{k=2}^m (b-a)^{k-1}\alpha_k f^{(k-1)}(x), \tag{15.42}$$

with convention $\tilde{T}_0(x) = 0$, $\tau_1(x) = 0$, while

$$\tilde{R}_n^1(x) := -(b-a)^{n-1} \int_{[a,b]} P_n^*\left(\frac{x-t}{b-a}\right) df^{(n-1)}(t) \tag{15.43}$$

and

$$\tilde{R}_n^2(x) := -(b-a)^{n-1} \int_{[a,b]} \left[P_n^*\left(\frac{x-t}{b-a}\right) - P_n\left(\frac{x-a}{b-a}\right)\right] df^{(n-1)}(t). \tag{15.44}$$

The last two integrals are of Riemann- Stieltjes type.

We make

Remark 15.18. Let $f : [a, b] \to \mathbb{R}$ be such that $f^{(n-1)}$ is a continuous function of bounded variation on $[a, b]$ for some $n \geq 2$.

Then we apply (15.40) for f' and for $n - 1$ to obtain

$$f'(x) = \frac{f(b) - f(a)}{b - a} + \tilde{T}_{n-2}(x) + \tau_{n-1}(x) + \tilde{R}_{n-1}^2(x), \qquad (15.45)$$

where

$$\tilde{T}_{n-2}(x) := \sum_{k=1}^{n-2} (b - a)^{k-1} P_k \left(\frac{x - a}{b - a} \right) \left(f^{(k)}(b) - f^{(k)}(a) \right), \qquad (15.46)$$

$$\tau_{n-1}(x) := \sum_{k=2}^{n-1} (b - a)^{k-1} \alpha_k f^{(k)}(x) \qquad (15.47)$$

with convention

$$\tilde{T}_0(x) = 0, \quad \tau_1(x) = 0, \qquad (15.48)$$

while

$$\tilde{R}_{n-1}^2(x) := -(b - a)^{n-2} \int_{[a,b]} \left[P_{n-1}^* \left(\frac{x - t}{b - a} \right) - P_{n-1} \left(\frac{x - a}{b - a} \right) \right] df^{(n-1)}(t). \qquad (15.49)$$

Let $g : [a, b] \to \mathbb{R}$ be of bounded variation with $a \neq b$, $g(a) \neq g(b)$.

Then by plugging (15.49) into (15.26) we obtain

Theorem 15.19. *Let $f : [a, b] \to \mathbb{R}$ be such that $f^{(n-1)}$ is a continuous function of bounded variation on $[a, b]$ for some $n \geq 2$. Let $g : [a, b] \to \mathbb{R}$ be of bounded variation with $a \neq b$, $g(a) \neq g(b)$. The rest of the terms as in Background 15.17 and Remark 15.18. Let $z \in [a, b]$.*
Then

$$f(z) = \frac{1}{g(b) - g(a)} \int_a^b f(x) dg(x) + \left(\int_a^b P(g(z), g(x)) dx \right) \left(\frac{f(b) - f(a)}{b - a} \right)$$

$$+ \sum_{k=1}^{n-2} (b - a)^{k-1} \left(f^{(k)}(b) - f^{(k)}(a) \right) \int_a^b P(g(z), g(x)) P_k \left(\frac{x - a}{b - a} \right) dx$$

$$+ \sum_{k=2}^{n-1} (b - a)^{k-1} \alpha_k \int_a^b P(g(z), g(x)) f^{(k)}(x) dx + (b - a)^{n-2} \int_a^b P(g(z), g(x))$$

$$\left(\int_{[a,b]} \left[P_{n-1} \left(\frac{x - a}{b - a} \right) - P_{n-1}^* \left(\frac{x - t}{b - a} \right) \right] df^{(n-1)}(t) \right) dx. \qquad (15.50)$$

If $f^{(n)}$ exists and is integrable on $[a, b]$, then in the last integral of (15.50) one can write

$$df^{(n-1)}(t) = f^{(n)}(t)dt.$$

We need

Background 15.20 (see [12]). General Weighted Euler harmonic identities. For $a, b \in \mathbb{R}$, $a < b$, let $w : [a, b] \to [0, \infty)$ be a probability density function i.e. integrable function satisfying

$$\int_a^b w(t)dt = 1. \tag{15.51}$$

For $n \geq 1$ and $t \in [a, b]$ let

$$w_n(t) := \frac{1}{(n-1)!} \int_a^t (t-s)^{n-1} w(s)ds. \tag{15.52}$$

Also, let

$$w_0(t) = w(t), \quad t \in [a, b]. \tag{15.53}$$

It is well known that w_n is equal to the n- th indefinite integral of w, being equal to zero at a, i.e. $w_n^n(t) = w(t)$ and $w_n(a) = 0$, for every $n \geq 1$.

A sequence of functions $H_n : [a, b] \to \mathbb{R}$, $n \geq 0$, is called w- harmonic sequence of functions on $[a, b]$ if

$$H_n'(t) = H_{n-1}(t), \quad t \neq 1; \quad H_0(t) = w(t), \quad t \in [a, b]. \tag{15.54}$$

The sequence $(w_n(t), \ t \neq 0)$ is an example of w- harmonic sequence of functions on $[a, b]$.

Assume that $(H_n(t), \ n \geq 0)$ is a w- harmonic sequence of functions on $[a, b]$.

Define $H_n^*(t)$, for $n \geq 0$, to be a periodic function of period 1, related to $H_n(t)$ as

$$H_n^*(t) = \frac{H_n(a + (b-a)t)}{(b-a)^n}, \quad 0 \leq t < 1, \tag{15.55}$$

$$H_n^*(t+1) = H_n^*(t), \quad t \in R. \tag{15.56}$$

Thus, for $n \geq 1$, $H_n^*(t)$ is continuous on $R \setminus Z$ and has a jump of

$$\beta_n = \frac{H_n(a) - H_n(b)}{(b-a)^n} \tag{15.57}$$

at every $t \in Z$, whenever $\beta_n \neq 0$.

Note that

$$\beta_1 = -\frac{1}{b-a}, \tag{15.58}$$

since

$$H_1(t) = c + w_1(t) = c + \int_a^t w(s)ds, \tag{15.59}$$

for some $c \in R$.

Also, note that

$$H^{*\prime}_n(t) = H^*_{n-1}(t), \quad n \geq 1, \quad t \in R \setminus Z. \tag{15.60}$$

Let $f : [a,b] \to \mathbb{R}$ be such that $f^{(n-1)}$ exists on $[a,b]$ for some $n \geq 1$. For every $x \in [a,b]$ and $1 \leq m \leq n$ we introduce the following notation

$$S_m(x) := \sum_{k=1}^m H_k(x) \left[f^{(k-1)}(b) - f^{(k-1)}(a) \right] + \sum_{k=2}^m [H_k(a) - H_k(b)] \, f^{(k-1)}(x), \tag{15.61}$$

with convention $S_1(x) = H_1(x)\,[f(b) - f(a)]$.

Theorem 15.21 (see [12]). *Let $(H_k, \; k \geq 0)$ be w- harmonic sequence of functions on $[a,b]$ and $f : [a,b] \to R$ such that $f^{(n-1)}$ is a continuous function of bounded variation on $[a,b]$ for some $n \geq 1$.*

Then for every $x \in [a,b]$

$$f(x) = \int_a^b f(t)W_x(t)dt + S_n(x) + R_n^1(x), \tag{15.62}$$

where

$$R_n^1(x) := -(b-a)^n \int_{[a,b]} H_n^* \left(\frac{x-t}{b-a} \right) df^{(n-1)}(t), \tag{15.63}$$

with

$$W_x(t) := \begin{cases} w(a + x - t), & a \leq t \leq x, \\ w(b + x - t), & x < t \leq b. \end{cases} \tag{15.64}$$

Also we have

Theorem 15.22 (see [12]). *Let $(H_k, \; k \geq 0)$ be w- harmonic sequence of functions on $[a,b]$ and $f : [a,b] \to R$ such that $f^{(n-1)}$ is a continuous function of bounded variation on $[a,b]$ for some $n \geq 1$.*

Then for every $x \in [a,b]$ and $n \geq 2$

$$f(x) = \int_a^b f(t)W_x(t)dt + S_{n-1}(x) + [H_n(a) - H_n(b)] f^{(n-1)}(x) + R_n^2(x), \tag{15.65}$$

while for $n = 1$

$$f(x) = \int_a^b f(t)W_x(t)dt + R_1^2(x), \tag{15.66}$$

$$R_n^2(x) := -(b-a)^n \int_{[a,b]} \left[H_n^* \left(\frac{x-t}{b-a} \right) - \frac{H_n(x)}{(b-a)^n} \right] df^{(n-1)}(t), \tag{15.67}$$

for $n \geq 1$.

Comment 15.23 (see [12]). In the case when $\varphi : [a, b] \to \mathbb{R}$ is such that φ' exists and is integrable on $[a, b]$, then the Riemann- Stieltjes integral $\int_{[a,b]} g(t)d\varphi(t)$ is equal to the Riemann integral $\int_a^b g(t)\varphi'(t)dt$.

Therefore, if $f : [a, b] \to \mathbb{R}$ is such that $f^{(n)}$ exists and is integrable on $[a, b]$, for some $n \geq 1$, then Theorem 15.21 and Theorem 15.22 hold with

$$R_n^1(x) := -(b-a)^n \int_{[a,b]} H_n^* \left(\frac{x-t}{b-a} \right) f^{(n)}(t)dt, \tag{15.68}$$

and

$$R_n^2(x) := -(b-a)^n \int_{[a,b]} \left[H_n^* \left(\frac{x-t}{b-a} \right) - \frac{H_n(x)}{(b-a)^n} \right] f^{(n)}(t)dt. \tag{15.69}$$

We make

Remark 15.24. We use the assumptions of Theorem 15.22, $n \geq 2$. We apply first (15.65) for f' and $n-1$ to get for $n \geq 3$,

$$f'(x) = \int_a^b f'(t)W_x(t)dt + \tilde{S}_{n-2}(x) + [H_{n-1}(a) - H_{n-1}(b)] f^{(n-1)}(x) + \tilde{R}_{n-1}^2(x), \tag{15.70}$$

where

$$\tilde{S}_{n-2}(x) = \sum_{k=1}^{n-2} H_k(x) \left(f^{(k)}(b) - f^{(k)}(a) \right) + \sum_{k=2}^{n-2} (H_k(a) - H_k(b)) f^{(k)}(x), \tag{15.71}$$

and

$$\tilde{S}_1(x) = H_1(x)(f'(b) - f'(a)), \tag{15.72}$$

while for $n = 2$, by (15.66), we obtain

$$f'(x) = \int_a^b f'(t)W_x(t)dt + \tilde{R}_1^2(x), \tag{15.73}$$

here, by (15.67), we have

$$\tilde{R}_{n-1}^2(x) = -(b-a)^{n-1} \int_{[a,b]} \left[H_{n-1}^* \left(\frac{x-t}{b-a} \right) - \frac{H_{n-1}(x)}{(b-a)^{n-1}} \right] df^{(n-1)}(t), \tag{15.74}$$

for $n \geq 2$.

Plugging (15.70), (15.73) into (15.26) we derive

Theorem 15.25. *Let $(H_k, \ k \geq 0)$ be w- harmonic sequence of functions on $[a, b]$ and $f : [a, b] \to \mathbb{R}$ such that $f^{(n-1)}$ is a continuous function of bounded variation on $[a, b]$ for some $n \geq 2$.*

Let $g : [a, b] \to \mathbb{R}$ be of bounded variation on $[a, b]$ such that $g(a) \neq g(b)$. Here W_x is as in (15.64). Let $z \in [a, b]$.

If $n \geq 3$, then

$$f(z) = \frac{1}{g(b) - g(a)} \int_a^b f(x)dg(x) + \left[\int_a^b P(g(z), g(x)) \left(\int_a^b f'(t)W_x(t)dt \right) dx \right]$$

$$+ \sum_{k=1}^{n-2} \left(f^{(k)}(b) - f^{(k)}(a) \right) \left(\int_a^b P(g(z), g(x))H_k(x)dx \right)$$

$$+ \sum_{k=2}^{n-2} \left(H_k(a) - H_k(b) \right) \int_a^b P(g(z), g(x))f^{(k)}(x)dx$$

$$+ \left(H_{n-1}(a) - H_{n-1}(b) \right) \int_a^b P(g(z), g(x))f^{(n-1)}(x)dx$$

$$+(b-a)^{n-1} \int_a^b P(g(z), g(x)) \left(\int_{[a,b]} \left(\frac{H_{n-1}(x)}{(b-a)^{n-1}} - H_{n-1}^* \left(\frac{x-t}{b-a} \right) \right) df^{(n-1)}(t) \right) dx. \tag{15.75}$$

If $n = 2$, then

$$f(z) = \frac{1}{g(b) - g(a)} \int_a^b f(x)dg(x) + \left[\int_a^b P(g(z), g(x)) \left(\int_a^b f'(t)W_x(t)dt \right) dx \right]$$

$$+(b-a) \int_a^b P(g(z), g(x)) \left(\int_{[a,b]} \left(\frac{H_1(x)}{b-a} - H_1^* \left(\frac{x-t}{b-a} \right) \right) df'(t) \right) dx. \tag{15.76}$$

If $f^{(n)}$ exists and is integrable on $[a,b]$, then in the last integral of (15.75) one can write $df^{(n-1)}(t) = f^{(n)}(t)dt$.

Also if f'' exists and is integrable on $[a,b]$, then in the last integral of (15.76) one can write $df'(t) = f''(t)dt$.

We mention

Theorem 15.26 (see [36]). *Let $f : [a,b] \to \mathbb{R}$ be n- times differentiable on $[a,b]$, $n \in N$. The n-th derivative $f^{(n)} : [a,b] \to \mathbb{R}$ is integrable on $[a,b]$. Let $g : [a,b] \to \mathbb{R}$ be of bounded variation on $[a,b]$ and $g(a) \neq g(b)$. Let $t \in [a,b]$. The kernel P is defined as in Proposition 15.1.*

Then, we find

$$f(t) = \frac{1}{g(b) - g(a)} \int_a^b f(s_1)dg(s_1) + \frac{1}{g(b) - g(a)} \sum_{k=0}^{n-2} \left(\int_a^b f^{(k+1)}(s_1)dg(s_1) \right)$$

$$\cdot \left(\underbrace{\int_a^b \cdots \int_a^b}_{(k+1)th-integral} P(g(t), g(s_1)) \prod_{i=1}^{k} P(g(s_i), g(s_{i+1}))ds_1 ds_2 \cdots ds_{k+1} \right)$$

$$+ \int_a^b \cdots \int_a^b P(g(t), g(s_1)) \prod_{i=1}^{n-1} P(g(s_i), g(s_{i+1})) f^{(n)}(s_n) ds_1 \cdots ds_n. \tag{15.77}$$

We make

Remark 15.27. All assumptions as in Theorem 15.26. Let $w(t)$ continuous function on $[a, b]$ such that $\int_a^b w(t) g(t) dt = 1$.

Then, by (15.77) we obtain

$$f(t) w(t) = \frac{w(t)}{g(b) - g(a)} \int_a^b f(s_1) dg(s_1) + \frac{w(t)}{g(b) - g(a)} \sum_{k=0}^{n-2} \left(\int_a^b f^{(k+1)}(s_1) dg(s_1) \right)$$

$$\cdot \left(\underbrace{\int_a^b \cdots \int_a^b}_{(k+1)th-integral} P(g(t), g(s_1)) \prod_{i=1}^{k} P(g(s_i), g(s_{i+1})) ds_1 ds_2 \cdots ds_{k+1} \right)$$

$$+ w(t) \int_a^b \cdots \int_a^b P(g(t), g(s_1)) \prod_{i=1}^{n-1} P(g(s_i), g(s_{i+1})) f^{(n)}(s_n) ds_1 \cdots ds_n. \tag{15.78}$$

Hence

$$\int_a^b f(t) w(t) dg(t) = \frac{1}{g(b) - g(a)} \int_a^b f(s_1) dg(s_1) + \sum_{k=0}^{n-2} \frac{\int_a^b f^{(k+1)}(s_1) dg(s_1)}{g(b) - g(a)}$$

$$\cdot \int_a^b w(t) \left(\int_a^b \cdots \int_a^b P(g(t), g(s_1)) \prod_{i=1}^{k} P(g(s_i), g(s_{i+1})) ds_1 ds_2 \cdots ds_{k+1} \right) dg(t)$$

$$+ \int_a^b w(t) \left(\int_a^b \cdots \int_a^b P(g(t), g(s_1)) \prod_{i=1}^{n-1} P(g(s_i), g(s_{i+1})) f^{(n)}(s_n) ds_1 \cdots ds_n \right) dg(t). \tag{15.79}$$

Next we subtract (15.79) from, by Theorem 15.26 ,original formula

$$f(z) = \frac{1}{g(b) - g(a)} \int_a^b f(s_1) dg(s_1) + \frac{1}{g(b) - g(a)} \sum_{k=0}^{n-2} \left(\int_a^b f^{(k+1)}(s_1) dg(s_1) \right)$$

$$\cdot \left(\int_a^b \cdots \int_a^b P(g(z), g(s_1)) \prod_{i=1}^{k} P(g(s_i), g(s_{i+1})) ds_1 ds_2 \cdots ds_{k+1} \right)$$

$$+ \int_a^b \cdots \int_a^b P(g(z), g(s_1)) \prod_{i=1}^{n-1} P(g(s_i), g(s_{i+1})) f^{(n)}(s_n) ds_1 \cdots ds_n, \ \forall \ z \in [a, b], \tag{15.80}$$

where

$$P(g(z), g(t)) := \begin{cases} \frac{g(t)-g(a)}{g(b)-g(a)}, & a \leq t \leq z, \\[2mm] \frac{g(t)-g(b)}{g(b)-g(a)}, & z < t \leq b. \end{cases} \tag{15.81}$$

We derive

$$f(z) = \int_a^b f(t)w(t)dg(t) + \sum_{k=0}^{n-2} \frac{\int_a^b f^{(k+1)}(s_1)dg(s_1)}{g(b)-g(a)}.$$

$$\cdot \left[\int_a^b w(t) \left[\int_a^b \cdots \int_a^b P(g(z), g(s_1)) \prod_{i=1}^{k} P(g(s_i), g(s_{i+1}))ds_1 ds_2 \cdots ds_{k+1} \right] dg(t) \right.$$

$$- \int_a^b w(t) \left(\int_a^b \cdots \int_a^b P(g(t), g(s_1)) \prod_{i=1}^{k} P(g(s_i), g(s_{i+1}))ds_1 ds_2 \cdots ds_{k+1} \right) dg(t) \Bigg]$$

$$+ \int_a^b w(t) \left[\int_a^b \cdots \int_a^b P(g(z), g(s_1)) \prod_{i=1}^{n-1} P(g(s_i), g(s_{i+1}))f^{(n)}(s_n)ds_1 \cdots ds_n \right] dg(t)$$

$$- \int_a^b w(t) \left(\int_a^b \cdots \int_a^b P(g(t), g(s_1)) \prod_{i=1}^{n-1} P(g(s_i), g(s_{i+1}))f^{(n)}(s_n)ds_1 \cdots ds_n \right) dg(t). \tag{15.82}$$

We have proved the general 2- weighted Montgomery representation formula.

Theorem 15.28. *Let $f : [a,b] \to \mathbb{R}$ be n- times differentiable on $[a,b]$, $n \in \mathbb{N}$. The n-th derivative $f^{(n)} : [a,b] \to \mathbb{R}$ is integrable on $[a,b]$. Let $g : [a,b] \to \mathbb{R}$ be of bounded variation on $[a,b]$ and $g(a) \neq g(b)$. Let $z \in [a,b]$. The kernel P is defined as in (15.1). Let further $w \in C([a,b])$ such that $\int_a^b w(t)g(t) = 1$.*
Then

$$f(z) = \int_a^b f(t)w(t)dg(t) + \sum_{k=0}^{n-2} \frac{\int_a^b f^{(k+1)}(s_1)dg(s_1)}{g(b)-g(a)}$$

$$\cdot \left[\int_a^b \cdots \int_a^b w(t) \left[P(g(z), g(s_1)) - P(g(t), g(s_1)) \right] \right.$$

$$\cdot \prod_{i=1}^{k} P(g(s_i), g(s_{i+1}))ds_1 ds_2 \cdots ds_{k+1} dg(t) \Bigg]$$

$$+ \int_a^b \cdots \int_a^b w(t) \left[P(g(z), g(s_1)) - P(g(t), g(s_1)) \right]$$

$$\cdot \prod_{i=1}^{n-1} P(g(s_i), g(s_{i+1}))f^{(n)}(s_n)ds_1 \cdots ds_n dg(t). \tag{15.83}$$

Notice that the difference $[P(g(z), g(s_1)) - P(g(t), g(s_1))]$ takes on only the values $0, \pm 1$.

We apply above representations to Csiszar's f-Divergence.

For earlier related work see [43], [101], [103].

We need

Background 15.29. For the rest of the chapter we use the following. Let be a convex function from $(0, \infty)$ into R which is strictly convex at 1 with $f(1) = 0$. Let $(\mathcal{X}, \mathcal{A}, \lambda)$ be a measure space, where λ is a finite or a σ-finite measure on $(\mathcal{X}, \mathcal{A})$. And let μ_1, μ_2 be two probability measures on $(\mathcal{X}, \mathcal{A})$ such that $\mu_1 \ll \lambda$, $\mu_2 \ll \lambda$ (absolutely continuous), e.g. $\lambda = \mu_1 + \mu_2$.

Denote by $p = \frac{d\mu_1}{d\lambda}$, $q = \frac{d\mu_2}{d\lambda}$ the (densities) Radon-Nikodym derivatives of μ_1, μ_2 with respect to λ.

Here we assume that

$$0 < a \le \frac{p}{q} \le b, \ a.e. \ on \ \mathcal{X} \ and \ a \le 1 \le b. \tag{15.84}$$

The quantity

$$\Gamma_f(\mu_1, \mu_2) = \int_{\mathcal{X}} q(x) f\left(\frac{p(x)}{q(x)}\right) d\lambda(x), \tag{15.85}$$

was introduced by I. Csiszar in 1967, see [140], and is called f-divergence of the probability measures μ_1 and μ_2. By Lemma 1.1 of [140] the integral (15.85) is well-defined and $\Gamma_f(\mu_1, \mu_2) \ge 0$ with equality only when $\mu_1 = \mu_2$. Furthermore $\Gamma_f(\mu_1, \mu_2)$ does not depend on the choice of λ.

Here by assuming $f(1) = 0$ we can consider $\Gamma_f(\mu_1, \mu_2)$, the f-divergence as a measure of the difference between the probability measures μ_1, μ_2.

Furthermore we consider here $f \in C^n([a, b])$, $n \in \mathbb{N}$, $g \in C([a, b])$ of bounded variation, $a \ne b$, $g(a) \ne g(b)$.

We make

Remark 15.30. All as in Background 15.29 , $n \ge 2$. By (15.16) we find

$$qf\left(\frac{p}{q}\right) = \frac{q}{g(b) - g(a)} \int_a^b f(x) dg(x) + q \left(\int_a^b P\left(g\left(\frac{p}{q}\right), g(x)\right) dx\right) \left(\frac{f(b) - f(a)}{b - a}\right)$$

$$+ \sum_{k=1}^{n-2} \frac{(b-a)^{k-1}}{k!} \left(f^{(k)}(b) - f^{(k)}(a)\right) q \int_a^b P\left(g\left(\frac{p}{q}\right), g(x)\right) B_k\left(\frac{x-a}{b-a}\right) dx$$

$$+ \frac{(b-a)^{n-2}}{(n-1)!} q \int_a^b P\left(g\left(\frac{p}{q}\right), g(x)\right)$$

$$\cdot \left(\int_a^b \left[B_{n-1}\left(\frac{x-a}{b-a}\right) - B_{n-1}^*\left(\frac{x-t}{b-a}\right)\right] f^{(n)}(t) dt\right) dx. \tag{15.86}$$

By integrating (15.86) we obtain

Theorem 15.31. *All as in Background 15.29 , $n \geq 2$.*

Then

$$
\Gamma_f(\mu_1, \mu_2) = \frac{1}{g(b) - g(a)} \int_a^b f(x) dg(x)
$$

$$
+ \left(\frac{f(b) - f(a)}{b - a} \right) \left(\int_{\mathcal{X}} q \left(\int_a^b P\left(g\left(\frac{p}{q} \right), g(x) \right) dx \right) d\lambda \right)
$$

$$
+ \sum_{k=1}^{n-2} \frac{(b-a)^{k-1}}{k!} \left(f^{(k)}(b) - f^{(k)}(a) \right)
$$

$$
\cdot \left(\int_{\mathcal{X}} q \left(\int_a^b P\left(g\left(\frac{p}{q} \right), g(x) \right) B_k \left(\frac{x-a}{b-a} \right) dx \right) d\lambda \right) + \frac{(b-a)^{n-2}}{(n-1)!}
$$

$$
\cdot \left(\int_{\mathcal{X}} q \left(\int_a^b P\left(g\left(\frac{p}{q} \right), g(x) \right) \left(\int_a^b \left(B_{n-1} \left(\frac{x-a}{b-a} \right) \right. \right. \right. \right.
$$

$$
\left. \left. \left. \left. - B_{n-1}^* \left(\frac{x-t}{b-a} \right) \right) f^{(n)}(t) dt \right) dx \right) d\lambda \right). \tag{15.87}
$$

Using (15.22) we obtain

Theorem 15.32. *All as in Background 15.29, $n \geq 2$.*

Then

$$
\Gamma_f(\mu_1, \mu_2)
$$

$$
= \frac{1}{g(b) - g(a)} \int_a^b f(x) dg(x) + (n-1) \left(\frac{f(b) - f(a)}{b - a} \right)
$$

$$
\cdot \left(\int_{\mathcal{X}} q \left(\int_a^b P\left(g\left(\frac{p}{q} \right), g(x) \right) dx \right) d\lambda \right) + \sum_{k=1}^{n-2} \left(\frac{n-k-1}{(b-a)\, k!} \right)
$$

$$
\cdot \left(\int_{\mathcal{X}} q \left(\int_a^b P\left(g\left(\frac{p}{q} \right), g(x) \right) \left(f^{(k)}(b)(x-b)^k - f^{(k)}(a)(x-a)^k \right) dx \right) d\lambda \right)
$$

$$
+ \frac{1}{(n-2)!\,(b-a)} \left(\int_{\mathcal{X}} q \left(\int_a^b P\left(g\left(\frac{p}{q} \right), g(x) \right) \right. \right.
$$

$$
\left. \left. \cdot \left(\int_a^b (x-t)^{n-2} K(t,x) f^{(n)}(t) dt \right) dx \right) d\lambda \right). \tag{15.88}
$$

When $n = 2$ the sum $\sum_{k=1}^{n-2}$ in (15.88) is zero.

Using (15.28) we get

Theorem 15.33. *All as in Background 15.29, $n > 2$.*

Here P_W as in Theorem 15.9.

Then

$$
\Gamma_f(\mu_1, \mu_2) = \frac{1}{g(b) - g(a)} \int_a^b f(x) dg(x)
$$

$$+ \left(\int_a^b w(t) f'(t) dt \right) \left(\int_{\mathcal{X}} q \left(\int_a^b P \left(g \left(\frac{p}{q} \right), g(x) \right) dx \right) d\lambda \right)$$

$$+ \sum_{k=1}^{n-2} \frac{(b-a)^{k-2}}{(k-1)!} \left(f^{(k)}(b) - f^{(k)}(a) \right)$$

$$\cdot \left(\int_{\mathcal{X}} q \left(\int_a^b P \left(g \left(\frac{p}{q} \right), g(x) \right) \left(\int_a^b P_W(x,t) B_{k-1} \left(\frac{t-a}{b-a} \right) dt \right) dx \right) d\lambda \right)$$

$$- \frac{(b-a)^{n-3}}{(n-2)!} \cdot \left(\int_{\mathcal{X}} q \left(\int_a^b P \left(g \left(\frac{p}{q} \right), g(x) \right) \right. \right.$$

$$\cdot \left. \left(\int_a^b \left(\int_a^b P_W(x,s) [B_{n-2}^* \left(\frac{s-t}{b-a} \right) - B_{n-2} \left(\frac{s-a}{b-a} \right)] ds \right) f^{(n)}(t) dt \right) dx \right) d\lambda \right).$$

$$(15.89)$$

Using (15.34) we obtain

Theorem 15.34. *All as in Background 15.29, $n > 2$. Here $w : [a,b] \to \mathbb{R}_+$ is a probability density function, and $\overline{F}_k$ is as in (15.32).*
Then

$$\Gamma_f(\mu_1, \mu_2)$$

$$= \frac{1}{g(b) - g(a)} \int_a^b f(x) dg(x)$$

$$+ \left(\int_a^b w(x) f'(x) dx \right) \left(\int_{\mathcal{X}} q \left(\int_a^b P \left(g \left(\frac{p}{q} \right), g(x) \right) dx \right) d\lambda \right)$$

$$- \sum_{k=1}^{n-2} \left(\int_{\mathcal{X}} q \left(\int_a^b P \left(g \left(\frac{p}{q} \right), g(x) \right) \overline{F}_k(x) dx \right) d\lambda \right)$$

$$+ \left(\sum_{k=1}^{n-2} \int_a^b w(x) \overline{F}_k(x) dx \right) \left(\int_{\mathcal{X}} q \left(\int_a^b P \left(g \left(\frac{p}{q} \right), g(x) \right) dx \right) d\lambda \right)$$

$$+ \frac{1}{(n-2)!(b-a)} \cdot \left(\int_{\mathcal{X}} q \left(\int_a^b P \left(g \left(\frac{p}{q} \right), g(x) \right) \right. \right.$$

$$\cdot \left. \left(\int_a^b (x-y)^{n-2} K(y,x) f^{(n)}(y) dy \right) dx \right) d\lambda \right)$$

$$- \frac{1}{(n-2)!(b-a)} \cdot \left(\int_{\mathcal{X}} q \left(\int_a^b P \left(g \left(\frac{p}{q} \right), g(x) \right) dx \right) d\lambda \right)$$

$$\cdot \left(\int_a^b \left(\int_a^b w(t)(t-y)^{n-2} K(y,t) dt \right) f^{(n)}(y) dy \right).$$

$$(15.90)$$

Using (15.50) we find

Theorem 15.35. *All as in Background 15.29, $n \geq 2$. The other terms here are as in Background 15.17 and Remark 15.18.*
Then

$$\Gamma_f(\mu_1, \mu_2)$$

$$= \frac{1}{g(b) - g(a)} \int_a^b f(x)dg(x)$$

$$+ \left(\frac{f(b) - f(a)}{b - a} \right) \left(\int_{\mathcal{X}} q \left(\int_a^b P\left(g\left(\frac{p}{q}\right), g(x) \right) dx \right) d\lambda \right)$$

$$+ \sum_{k=1}^{n-2} (b-a)^{k-1} \left(f^{(k)}(b) - f^{(k)}(a) \right)$$

$$\cdot \left(\int_{\mathcal{X}} q \left(\int_a^b P\left(g\left(\frac{p}{q}\right), g(x) \right) P_k\left(\frac{x-a}{b-a} \right) dx \right) d\lambda \right)$$

$$+ \sum_{k=2}^{n-1} (b-a)^{k-1} \alpha_k \left(\int_{\mathcal{X}} q \left(\int_a^b P\left(g\left(\frac{p}{q}\right), g(x) \right) f^{(k)}(x) dx \right) d\lambda \right) + (b-a)^{n-2}$$

$$\cdot \left(\int_{\mathcal{X}} q \left(\int_a^b P\left(g\left(\frac{p}{q}\right), g(x) \right) \left(\int_{[a,b]} \left[P_{n-1}\left(\frac{x-a}{b-a} \right) \right. \right. \right. \right.$$

$$\left. \left. \left. \left. - P_{n-1}^*\left(\frac{x-t}{b-a} \right) \right] f^{(n)}(t) dt \right) dx \right) d\lambda \right). \tag{15.91}$$

Using (15.75) and (15.76) we obtain

Theorem 15.36. *All as in Background 15.29, $n \geq 2$. Let $(H_k, \; k \geq 0)$ be w-harmonic sequence of functions on $[a, b]$. Here W_x is as in (15.64).*
If $n \geq 3$, then

$$\Gamma_f(\mu_1, \mu_2)$$

$$= \frac{1}{g(b) - g(a)} \int_a^b f(x)dg(x)$$

$$+ \left(\int_{\mathcal{X}} q \left(\int_a^b P\left(g\left(\frac{p}{q}\right), g(x) \right) \left(\int_a^b f'(t)W_x(t)dt \right) dx \right) d\lambda \right)$$

$$+ \sum_{k=1}^{n-2} \left(f^{(k)}(b) - f^{(k)}(a) \right) \left(\int_{\mathcal{X}} q \left(\int_a^b P\left(g\left(\frac{p}{q}\right), g(x) \right) H_k(x) dx \right) d\lambda \right)$$

$$+ \sum_{k=2}^{n-2} \left(H_k(a) - H_k(b) \right) \left(\int_{\mathcal{X}} q \left(\int_a^b P\left(g\left(\frac{p}{q}\right), g(x) \right) f^{(k)}(x) dx \right) d\lambda \right)$$

$$+ \left(H_{n-1}(a) - H_{n-1}(b) \right)$$

$$\cdot \left(\int_{\mathcal{X}} q \left(\int_a^b P \left(g \left(\frac{p}{q} \right), g(x) \right) f^{(n-1)}(x)dx \right) d\lambda \right) + (b-a)^{n-1}$$

$$\cdot \left(\int_{\mathcal{X}} q \left(\int_a^b P \left(g \left(\frac{p}{q} \right), g(x) \right) \right.$$

$$\cdot \left. \left(\int_{[a,b]} \left(\frac{H_{n-1}(x)}{(b-a)^{n-1}} - H_{n-1}^* \left(\frac{x-t}{b-a} \right) \right) f^{(n)}(t)dt \right) dx \right) d\lambda \right). \qquad (15.92)$$

If $n = 2$, then

$$\Gamma_f(\mu_1, \mu_2)$$

$$= \frac{1}{g(b) - g(a)} \int_a^b f(x)dg(x)$$

$$+ \left(\int_{\mathcal{X}} q \left(\int_a^b P \left(g \left(\frac{p}{q} \right), g(x) \right) \left(\int_a^b f'(t)W_x(t)dt \right) dx \right) d\lambda \right) + (b-a)$$

$$\cdot \left(\int_{\mathcal{X}} q \left(\int_a^b P \left(g \left(\frac{p}{q} \right), g(x) \right) \right.$$

$$\cdot \left. \left(\int_{[a,b]} \left(\frac{H_1(x)}{b-a} - H_1^* \left(\frac{x-t}{b-a} \right) \right) f''(t)dt \right) dx \right) d\lambda \right). \qquad (15.93)$$

Using (15.83) we obtain

Theorem 15.37. *All as in Background 15.29. Let $w \in C([a,b])$ such that $\int_a^b w(t)g(t)dt = 1$.*
Then

$$\Gamma_f(\mu_1, \mu_2) = \int_a^b f(t)w(t)dg(t) + \sum_{k=0}^{n-2} \left(\frac{\int_a^b f^{(k+1)}(s_1)dg(s_1)}{g(b) - g(a)} \right)$$

$$\cdot \left\{ \int_X q \left[\int_a^b \cdots \int_a^b w(t)[P(g(p/q), g(s_1)) - P(g(t), g(s_1))] \right. \right.$$

$$\cdot \left. \left. \prod_{i=1}^k [P(g(s_i), g(s_{i+1}))ds_1 \cdots ds_{k+1}dg(t)]d\lambda \right\} \right.$$

$$+ \left\{ \int_X q \left[\int_a^b \cdots \int_a^b w(t)[P(g(p/q), g(s_1)) - P(g(t), g(s_1))] \right. \right.$$

$$\cdot \left. \left. \prod_{i=1}^{n-1} P(g(s_i), g(s_{i+1}))f^{(n)}(s_n)ds_1 \cdots ds_n dg(t) \right] d\lambda \right\}. \qquad (15.94)$$

Note 15.38. All integrals in Theorems 15.31 - 15.37 are well-defined.

Chapter 16

About General Moment Theory

In this chapter we describe the main moment problems and their solution methods from theoretical to applied. In particular we present the standard moment problem, the convex moment problem, and the infinite many conditions moment problem. Moment theory has a lot of important and interesting applications in many sciences and subjects, for a detailed list please see Section 16.4. This treatment follows [28].

16.1   The Standard Moment Problem

Let $g_1, \ldots, g_n$ and h be given real-valued Borel measurable functions on a fixed measurable space $X := (X, A)$. We would like to find the best upper and lower bound on

$$\mu(h) := \int_X h(t)\mu(dt),$$

given that μ is a probability measure on X with prescribed moments

$$\int g_i(t)\mu(dt) = y_i, \quad i = 1, \ldots, n.$$

Here we assume μ such that

$$\int_X |g_i|\mu(dt) < +\infty, \quad i = 1, \ldots, n$$

and

$$\int_X |h|\mu(dt) < +\infty.$$

For each $y := (y_1, \ldots, y_n) \in R^n$, consider the optimal quantities

$$L(y) := L(y \mid h) := \inf_{\mu} \mu(h),$$

$$U(y) := U(y \mid h) := \sup_{\mu} \mu(h),$$

where μ is a probability measure as above with

$$\mu(g_i) = y_i, \quad i = 1, \ldots, n.$$

215

If there is not such probability measure μ we set $L(y) := +\infty$, $U(y) := -\infty$.

If $h := \mathcal{X}_S$ the characteristic function of a given measurable set S of X, then we agree to write

$$L(y \mid \mathcal{X}_S) := L_S(y), \ \ U(y \mid \mathcal{X}_S) := U_S(y).$$

Hence, $L_S(y) \leq \mu(S) \leq U_S(y)$. Consider $g : X \to \mathbb{R}^n$ such that $g(t) := (g_1(t), \ldots, g_n(t))$. Set also $g_0(t) := 1$, all $t \in X$. Here we basically describe Kemperman's (1968)([229]) geometric methods for solving the above main moment problems which were related to and motivated by Markov (1884)([258]), Riesz (1911)([308]) and Krein (1951)([238]). The advantage of the geometric method is that many times is simple and immediate giving us the optimal quantities L, U in a closed-numerical form, on the top of this is very elegant. Here the σ-field A contains all subsets of X.

The next result comes from Richter (1957)([307]), Rogosinsky (1958)([25]) and Mulholland and Rogers (1958)([269]).

Theorem 16.1. *Let $f_1, \ldots, f_N$ be given real-valued Borel measurable functions on a measurable space Ω (such as $g_1, \ldots, g_n$ and h on X). Let μ be a probability measure on Ω such that each f_i is integrable with respect to μ. Then there exists a probability measure μ' of finite support on Ω (i.e., having non-zero mass only at a finite number of points) satisfying*

$$\int_\Omega f_i(t)\mu(dt) = \int_\Omega f_i(t)\mu'(dt),$$

all $i = 1, \ldots, N$.

One can even achieve that the support of μ' has at most $N + 1$ points. So from now on we can consider only about finite supported probability measures.

Set

$$V := conv \ g(X),$$

(conv stands for convex hull) where $g(X) := \{z \in R^n : z = g(t) \ for \ some \ t \in X\}$ is a curve in R^n (if $X = [a, b] \subset R$ or if $X = [a, b] \times [c, d] \subset R^2$).

Let $S \subset X$, and let $M^+(S)$ denote the set of all probability measures on X whose support is finite and contained in S.

The next results come from Kemperman (1968)([229]).

Lemma 16.2. *Given $y \in R^n$, then $y \in V$ iff $\exists \mu \in M^+(X)$ such that*

$$\mu(g) = y \ \ (i.e. \mu(g_i) := \int_X g_i(t)\mu(dt) = y_i, \ i = 1, \ldots, n).$$

Hence $L(y \mid h) < +\infty$ iff $y \in V$ (see that by Theorem 16.1

$$L(y \mid h) = \inf\{\mu(h) : \mu \in M^+(X), \mu(g) = y\}$$

and

$$U(y \mid h) = \sup\{\mu(h) : \mu \in M^+(X), \mu(g) = y\}).$$

Easily one can see that

$$L(y) := L(y \mid h)$$

is a convex function on V, i.e.

$$L(\lambda y' + (1 - \lambda)y'') \leq \lambda L(y') + (1 - \lambda)L(y''),$$

whenever $0 \leq \lambda \leq 1$ and $y', y'' \in V$. Also $U(y) := U(y \mid h) = -L(y \mid -h)$ is a concave function on V.

One can also prove that the following three properties are equivalent:

(i) $\mathrm{int}(V) :=$ interior of $V \neq \phi$;
(ii) $g(X)$ is not the subset of any hyperplane in R^n;
(iii) $1, g_1, g_2, \ldots, g_n$ are linearly independent on X.

From now on we suppose that $1, g_1, \ldots, g_n$ are linearly independent, i.e. $\mathrm{int}(V) \neq \phi$.

Let D^* denote the set of all $(n+1)$-tuples of real numbers $d^* := (d_0, d_1, \ldots, d_n)$ satisfying

$$h(t) \geq d_0 + \sum_{i=1}^{n} d_i g_i(t), \quad \text{all} \quad t \in X. \tag{16.1}$$

Theorem 16.3. *For each $y \in \mathrm{int}(V)$ we have that*

$$L(y \mid h) = \sup \left\{ d_0 + \sum_{i=1}^{n} d_i y_i : d^* = (d_0, \ldots, d_n) \in D^* \right\}. \tag{16.2}$$

Given that $L(y \mid h) > -\infty$, the supremum in (16.2) is even assumed by some $d^* \in D^*$. If $L(y \mid h)$ is finite in $\mathrm{int}(V)$ then for almost all $y \in \mathrm{int}(V)$ the supremum in (16.2) is assumed by a unique $d^* \in D^*$. Thus $L(y \mid h) < +\infty$ in $\mathrm{int}(V)$ iff $D^* \neq \phi$. Note that $y := (y_1, \ldots, y_n) \in \mathrm{int}(V) \subset R^n$ iff $d_0 + \sum_{i=1}^{n} d_i y_i > 0$ for each choice of the real constants d_i not all zero such that $d_0 + \sum_{i=1}^{n} d_i g_i(t) \geq 0$, all $t \in X$. (The last statement comes from Karlin and Shapley (1953), p. 5([222]) and Kemperman (1965), p. 573 ,([228]).)

If h is bounded then $D^* \neq \phi$, trivially.

Theorem 16.4. *Let $d^* \in D^*$ be fixed and put*

$$B(d^*) := \left\{ z = g(t) : d_0 + \sum_{i=1}^{n} d_i g_i(t) = h(t), t \in X \right\} \tag{16.3}$$

Then for each point

$$y \in conv\, B(d^*) \tag{16.4}$$

the quantity $L(y \mid h)$ is found as follows. Set

$$y = \sum_{j=1}^{m} p_j g(t_j)$$

with

$$g(t_j) \in B(d^*),$$

and

$$p_j \geq 0, \quad \sum_{j=1}^{m} p_j = 1. \tag{16.5}$$

Then

$$L(y \mid h) = \sum_{j=1}^{m} p_j h(t_j) = d_0 + \sum_{i=1}^{n} d_i y_i. \tag{16.6}$$

Theorem 16.5. *Let $y \in \mathrm{int}(V)$ be fixed. Then the following are equivalent :*

(i) $\exists \mu \in M^+(X)$ such that $\mu(g) = y$ and $\mu(h) = L(y \mid h)$, i.e. infimum is attained.
(ii) $\exists d^* \in D^*$ satisfying (16.4).

Furthermore for almost all $y \in \mathrm{int}(V)$ there exists at most one $d^* \in D^*$ satisfying (16.4).

In many situations the above infimum is not attained so that Theorem 16.4 is not applicable. The next theorem has more applications. For that put

$$\eta(z) := \lim_{\delta \to 0} \inf_{t} \{ h(t) : t \in X, |g(t) - z| < \delta \}. \tag{16.7}$$

If $\varepsilon \geq 0$ and $d^* \in D^*$, define

$$C_\varepsilon(d^*) := \left\{ z \in g\overline{(T)} : 0 \leq \eta(z) - \sum_{i=0}^{n} d_i z_i \leq \varepsilon \right\}, \quad (z_0 = 1) \tag{16.8}$$

and

$$G(d^*) := \bigcap_{N=1}^{\infty} \overline{\mathrm{conv}}\, C_{1/N}(d^*). \tag{16.9}$$

It is easily proved that $C_\varepsilon(d^*)$ and $G(d^*)$ are closed, furthermore $B(d^*) \subset C_0(d^*) \subset C_\varepsilon(d^*)$, where $B(d^*)$ is defined by (16.3).

Theorem 16.6. *Let $y \in \mathrm{int}(V)$ be fixed.*

(i) *Let $d^* \in D^*$ be such that $y \in G(d^*)$. Then*

$$L(y \mid h) = d_0 + d_1 y_1 + \cdots + d_n y_n. \tag{16.10}$$

(ii) *Suppose that g is bounded. Then there exists $d^* \in D^*$ satisfying*

$$y \in \mathrm{conv}\, C_0(d^*) \subset G(d^*)$$

and

$$L(y \mid h) = d_0 + d_1 y_1 + \cdots + d_n y_n. \tag{16.11}$$

(iii) *We further obtain, whether or not g is bounded, that for almost all $y \in \text{int}(V)$ there exists at most one $d^* \in D^*$ satisfying $y \in G(d^*)$.*

The above results suggest the following practical simple geometric methods for finding $L(y \mid h)$ and $U(y \mid h)$, see Kemperman (1968),([229]).

I. The Method of Optimal Distance

Put

$$M := \text{conv}_{t \in X}(g_1(t), \ldots, g_n(t), h(t)).$$

Then $L(y \mid h)$ is equal to the *smallest* distance between $(y_1, \ldots, y_n, 0)$ and $(y_1, \ldots, y_n, z) \in \overline{M}$. Also $U(y \mid h)$ is equal to the *largest* distance between $(y_1, \ldots, y_n, 0)$ and $(y_1, \ldots, y_n, z) \in \overline{M}$. Here $\overline{M}$ stands for the closure of M. In particular we see that $L(y \mid h) = \inf\{y_{n+1} : (y_1, \ldots, y_n, y_{n+1}) \in M\}$ and

$$U(y \mid h) = \sup\{y_{n+1} : (y_1, \ldots, y_n, y_{n+1}) \in M\}. \tag{16.12}$$

Example 16.7. Let μ denote probability measures on $[0, a]$, $a > 0$. Fix $0 < d < a$. Find

$$L := \inf_{\mu} \int_{[0,a]} t^2 \mu(dt) \quad \text{and} \quad U := \sup_{\mu} \int_{[0,a]} t^2 \mu(dt)$$

subject to

$$\int_{[0,a]} t\mu(dt) = d.$$

So consider the graph $G := \{(t, t^2) : 0 \le t \le a\}$. Call $M := \overline{\text{conv } G} = \text{conv } G$.

A direct application of the optimal distance method here gives us $L = d^2$ (an optimal measure μ is supported at d with mass 1), and $U = da$ (an optimal measure μ here is supported at 0 and a with masses $(1 - \frac{d}{a})$ and $\frac{d}{a}$, respectively).

II. The Method of Optimal Ratio

We would like to find

$$L_S(y) := \inf \mu(S)$$

and

$$U_S(y) := \sup \mu(S),$$

over all probability measures μ such that

$$\mu(g_i) = y_i, \quad i = 1, \ldots, n.$$

Set $S' := X - S$. Call $W_S := \overline{\text{convg}(S)}$, $W_{S'} := \overline{\text{convg}(S')}$ and $W := \overline{\text{convg}(X)}$, where $g := (g_1, \ldots, g_n)$.

Finding $\underline{L_S(y)}$

1) Pick a boundary point z of W and "*draw*" through z a hyperplane H of support to W.

2) Determine the hyperplane H' parallel to H which supports $W_{S'}$ as well as possible, and on the same side as H supports W.

3) Denote

$$A_d := W \cap H = W_S \cap H \quad \text{and}$$
$$B_d := W_{S'} \cap H'.$$

Given that $H' \neq H$, set $G_d := \overline{\text{conv}}(A_d \cup B_d)$. Then we have that

$$L_S(y) = \frac{\Delta(y)}{\Delta}, \tag{16.13}$$

for each $y \in \text{int}(V)$ such that $y \in G_d$. Here, $\Delta(y)$ is the distance from y to H' and Δ is the distance between the distinct parallel hyperplanes H, H'.

<u>Finding $U_S(y)$</u> (Note that $U_S(y) = 1 - L_{S'}(y)$.)

1) Pick a boundary point z of W_S and "*draw*" through z a hyperplane H of support to W_S. Set $A_d := W_S \cap H$.

2) Determine the hyperplane H' parallel to H which supports $g(X)$ and hence W as well as possible, and on the same side as H supports W_S. We are interested only in $H' \neq H$ in which case H is between H' and W_S.

3) Set $B_d := W \cap H' = W_{S'} \cap H'$. Let G_d as above. Then

$$U_S(y) = \frac{\Delta(y)}{\Delta}, \quad \text{for each } y \in \text{int}(V), \text{ where } y \in G_d, \tag{16.14}$$

assuming that H and H' are distinct. Here, $\Delta(y)$ and Δ are defined as above.

Examples here of calculating $L_S(y)$ and $U_S(y)$ tend to be more involved and complicated, however the applications are many.

16.2   The Convex Moment Problem

Definition 16.8. Let $s \geq 1$ be a fixed natural number and let $x_0 \in R$ be fixed. By $m_s(x_0)$ we denote the set of probability measures μ on R such that the associated cumulative distribution function F possesses an $(s-1)$th derivative $F^{(s-1)}(x)$ over $(x_0, +\infty)$ and furthermore $(-1)^s F^{(s-1)}(x)$ is convex in $(x_0, +\infty)$.

Description of the Problem. Let g_i, $i = 1, \ldots, n$; h are Borel measurable functions from R into itself. These are supposed to be locally integrable on $[x_0, +\infty)$ relative to Lebesque measure. Consider $\mu \in m_s(x_0)$, $s \geq 1$ such that

$$\mu(|g_i|) := \int_R |g_i(t)| \mu(dt) < +\infty, \quad i = 1, \ldots, n \tag{16.15}$$

and

$$\mu(|h|) := \int_R |h(t)|\mu(dt) < +\infty. \tag{16.16}$$

Let $c := (c_1, \ldots, c_n) \in R^n$ be such that

$$\mu(g_i) = c_i, \ i = 1, \ldots, n, \mu \in m_s(x_0). \tag{16.17}$$

We would like to find $L(c) := \inf_\mu \mu(h)$ and

$$U(c) := \sup_\mu \mu(h), \tag{16.18}$$

where μ is as above described.

Here, the method will be to transform the above convex moment problem into an ordinary one handled by Section 16.1, see Kemperman (1971)([230]).

Definition 16.9. Consider here another copy of (R, B); B is the Borel σ-field, and further a given function $P(y, A)$ on $R \times B$.

Suppose that for each fixed $y \in R$, $P(y, \cdot)$ is a probability measure on R and for each fixed $A \in B$, $P(\cdot, A)$ is a Borel-measurable real-valued function on R. We call P a *Markov Kernel*. For each probability measure ν on R, let $\mu := T\nu$ denote the probability measure on R given by

$$\mu(A) := (T\nu)(A) := \int_R P(y, A)\nu(dy).$$

T is called a Markov transformation.

In particular: Define the kernel

$$K_s(u, x) := \begin{cases} \frac{s(u-x)^{s-1}}{(u-x_0)^s}, & \text{if } x_0 < x < u, \\ 0, & \text{elsewhere .} \end{cases} \tag{16.19}$$

Notice $K_s(u, x) \geq 0$ and $\int_R K_s(u, x)dx = 1$, all $u > x_0$. Let δ_u be the unit (Dirac) measure at u. Define

$$P_s(u, A) := \begin{cases} \delta_u(A), & \text{if } u \leq x_0; \\ \int_A K_s(u, x)dx, & \text{if } u > x_0. \end{cases} \tag{16.20}$$

Then

$$(T\nu)(A) := \int_R P_s(u, A)\nu(du) \tag{16.21}$$

is a Markov transformation.

Theorem 16.10. *Let $x_0 \in R$ and natural number $s \geq 1$ be fixed. Then the Markov transformation (16.21) $\mu = T\nu$ defines an (1–1) correspondence between the set m^* of all probability measures ν on R and the set $m_s(x_0)$ of all probability measures μ on R as in Definition 16.8. In fact T is a homeomorphism given that m^* and $m_s(x_0)$ are endowed with the weak*-topology.*

Let $\phi : R \to R$ *be a bounded and continuous function. Introducing*

$$\phi^*(u) := (T\phi)(u) := \int_R \phi(x) \cdot P_s(u, dx), \tag{16.22}$$

then

$$\int \phi d\mu = \int \phi^* d\nu. \qquad (16.23)$$

Here ϕ^* is a bounded and continuous function from R into itself.

We find that

$$\phi^*(u) = \begin{cases} \phi(u), & if \ u \leq x_0; \\ \int_0^1 \phi((1-t)u + tx_0)st^{s-1}dt, & if \ u > x_0. \end{cases} \qquad (16.24)$$

In particular

$$\frac{1}{s!}(u - x_0)^s \phi^*(u) = \frac{1}{(s-1)!} \int_{x_0}^u (u-x)^{s-1} \phi(x)dx. \qquad (16.25)$$

Especially, if $r > -1$ we get for $\phi(u) := (u - x_0)^r$ that $\phi^*(u) = \binom{r+s}{s}^{-1}(u - x_0)^r$, for all $u > x_0$. Here $r! := 1 \cdot 2 \cdots r$ and $\binom{r+s}{s} := \frac{(r+1)\cdots(r+s)}{s!}$.

Solving the Convex Moment Problem. Let T be the Markov transformation (16.21) as described above. For each $\mu \in m_s(x_0)$ corresponds exactly one $\nu \in m^*$ such that $\mu = T\nu$. Call $g_i^* := Tg_i$, $i = 1, \ldots, n$ and $h^* := Th$. We have

$$\int_R g_i^* d\mu = \int_R g_i d\mu$$

and

$$\int_R h^* d\nu = \int_R h d\mu.$$

Notice that we obtain

$$\nu(g_i^*) := \int_R g_i^* d\nu = c_i; \ i = 1, \ldots, n. \qquad (16.26)$$

From (16.15), (16.16) we get find

$$\int_R T|g_i|d\nu < +\infty, \ i = 1, \ldots, n$$

and

$$\int_R T|h|d\nu < +\infty. \qquad (16.27)$$

Since T is a positive linear operator we obtain $|Tg_i| \leq T|g_i|$, $i = 1, \ldots, n$ and $|Th| \leq T|h|$, i.e.

$$\int_R |g_i^*|d\nu < +\infty, \ i = 1, \ldots, n$$

and

$$\int_R |h^*|d\nu < +\infty.$$

That is g_i^*, h^* are ν-integrable.

Finally

$$L(c) = \inf_{\nu} \nu(h^*) \tag{16.28}$$

and

$$U(c) = \sup_{\nu} \nu(h^*), \tag{16.29}$$

where $\nu \in m^*$ (probability measure on R) such that (16.26) and (16.27) are true.

Thus the convex moment problem is solved as a standard moment problem, Section 16.1.

Remark 16.11. Here we restrict our probability measures on $[0, +\infty)$ and we consider the case $x_0 = 0$. That is $\mu \in m_s(0)$, $s \geq 1$, i.e. $(-1)^s F^{(s-1)}(x)$ is convex for all $x > 0$ but $\mu(\{0\}) = \nu(\{0\})$ can be positive, $\nu \in m^*$. We have

$$\phi^*(u) = su^{-s} \cdot \int_0^u (u - x)^{s-1} \cdot \phi(x) \cdot dx, \ u > 0. \tag{16.30}$$

Further $\phi^*(0) = \phi(0)$, $(\phi^* = T\phi)$. Especially if

$$\phi(x) = x^r \text{ then } \phi^*(u) = \binom{r+s}{s}^{-1} \cdot u^r, \ (r \geq 0). \tag{16.31}$$

Hence the moment

$$\alpha_r := \int_0^{+\infty} x^r \mu(dx) \tag{16.32}$$

is also expressed as

$$\alpha_r = \binom{r+s}{s}^{-1} \cdot \beta_r, \tag{16.33}$$

where

$$\beta_r := \int_0^{+\infty} u^r \nu(du). \tag{16.34}$$

Recall that $T\nu = \mu$, where ν can be any probability measure on $[0, +\infty)$.

Remark 16.12. Here we restrict our probability measures on $[0, b]$, $b > 0$ and again we consider the case $x_0 = 0$. Let $\mu \in m_s(0)$ and

$$\int_{[0,b]} x^r \mu(dx) := \alpha_r, \tag{16.35}$$

where $s \geq 1$, $r > 0$ are fixed.

Also let ν be a probability measure on $[0, b]$ unrestricted, i.e. $\nu \in m^*$. Then $\beta_r = \binom{r+s}{s}\alpha_r$, where

$$\beta_r := \int_{[0,b]} u^r \nu(du). \tag{16.36}$$

Let $h : [0, b] \to R_+$ be an integrable function with respect to Lebesque measure. Consider $\mu \in m_s(0)$ such that

$$\int_{[0,b]} h d\mu < +\infty. \tag{16.37}$$

i.e.

$$\int_{[0,b]} h^* d\nu < +\infty, \ \nu \in m^*. \tag{16.38}$$

Here $h^* = Th$, $\mu = T\nu$ and

$$\int_{[0,b]} h d\mu = \int_{[0,b]} h^* d\nu.$$

Letting α_r be free, we have that the set of all possible $(\alpha_r, \mu(h)) = (\mu(x^r), \mu(h))$ coincides with the set of all

$$\left(\binom{r+s}{s}^{-1} \cdot \beta_r, \nu(h^*) \right) = \left(\binom{r+s}{s}^{-1} \cdot \nu(u^r), \nu(h^*) \right),$$

where μ as in (16.37) and ν as in (16.38), both probability measures on $[0, b]$. Hence, the set of all possible pairs $(\beta_r, \mu(h)) = (\beta_r, \nu(h^*))$ is precisely the convex hull of the curve

$$\Gamma := \{(u^r, h^*(u)) : 0 \le u \le b\}. \tag{16.39}$$

In order one to determine $L(\alpha_r)$ the infimum of all $\mu(h)$, where μ is as in (16.35) and (16.37), one must determine the lowest point in this convex hull which is on the vertical through $(\beta_r, 0)$. For $U(\alpha_r)$ the supremum of all $\mu(h)$, μ as above, one must determine the highest point of above convex hull which is on the vertical through $(\beta_r, 0)$.

For more on the above see again, Section 16.1.

16.3　Infinite Many Conditions Moment Problem

For that please see also Kemperman (1983),([233]).

Definition 16.13. A finite non-negative measure μ on a compact and Hausdorff space S is said to be *inner regular* when

$$\mu(B) = \sup\{\mu(K) : K \subseteq B; K \text{ compact}\} \tag{16.40}$$

holds for each Borel subset B of S.

Theorem 16.14 (Kemperman, 1983,([233])). *Let S be a compactHausdorff topological space and $a_i : S \to R$ $(i \in I)$ continuous functions (I is an index set of arbitrary cardinality), also let α_i $(i \in I)$ be an associated set of real constants. Call*

$M_0(S)$ *the set of finite non-negative inner regular measures* μ *on* S *which satisfy the moment conditions*

$$\mu(a_i) = \int_S a_i(s)\mu(ds) \leq \alpha_i \text{ all } i \in I. \tag{16.41}$$

Also consider the function $b : S \to R$ *which is continuous and assume that there exist numbers* $d_i \geq 0$ $(i \in I)$, *all but finitely many equal to zero, and further a number* $q \geq 0$ *such that*

$$1 \leq \sum_{i \in I} d_i a_i(s) - qb(s) \text{ all } s \in S. \tag{16.42}$$

Finally assume that $M_0(S) \neq \emptyset$ *and put*

$$U_0(b) = \sup\{\mu(b) : \mu \in M_0(S)\}. \tag{16.43}$$

$(\mu(b) := \int_S b(s)\mu(ds))$. *Then*

$$U_0(b) = \inf \left\{ \sum_{i \in I} c_i \alpha_i \mid c_i \geq 0; \right.$$

$$\left. b(s) \leq \sum_{i \in I} c_i a_i(s) \text{ all } s \in S \right\}, \tag{16.44}$$

here all but finitely many c_i; $i \in I$ *are equal to zero. Moreover,* $U_0(b)$ *is finite and the above supremum is assumed.*

Note 16.15. In general we have: let S be a fixed measurable space such that each 1-point set $\{s\}$ is measurable. Further let $M_0(S)$ denote a fixed non-empty set of finite non-negative measures on S.

For $f : S \to R$ a measurable function we denote

$$L_0(f) := L(f, M_0(S)) :=$$

$$\inf \left\{ \int_S f(s)\mu(ds) : \mu \in M_0(S) \right\}. \tag{16.45}$$

Then we have

$$L_0(f) = -U_0(-f). \tag{16.46}$$

Now one can apply Theorem 16.14 in its setting to find $L_0(f)$.

16.4   Applications – Discussion

The above described moment theory methods have a lot of applications in many sciences. To mention a few of them: Physics, Chemistry, Statistics, Stochastic Processes and Probability, Functional Analysis in Mathematics, Medicine, Material Science, etc. Moment theory could be also considered the theoretical part of linear finite or semi-infinite programming (here we consider discretized finite non-negative measures).

The above described methods have in particular important applications: in the marginal moment problems and the related transportation problems, also in the quadratic moment problem, see Kemperman (1986),([234]).

Other important applications are in Tomography, Crystallography, Queueing Theory, Rounding Problem in Political Science, and Martingale Inequalities in Probability. At last, but not least, moment theory has important applications in estimating the speeds: of the convergence of a sequence of positive linear operators to the unit operator, and of the weak convergence of non-negative finite measures to the unit-Dirac measure at a real number, for that and the solutions of many other important moment problems please see the monograph of Anastassiou (1993),([20]).

Chapter 17

Extreme Bounds on the Average of a Rounded off Observation under a Moment Condition

The moment problem of finding the maximum and minimum expected values of the averages of nonnegative random variables over various sets of observations subject to one simple moment condition is encountered. The solution is presented by means of geometric moment theory (see [229]). This treatment relies on [23].

17.1 Preliminaries

Here $[\cdot]$ and $\lceil \cdot \rceil$ stand for the integral part (floor) and ceiling of the number, respectively.

We consider probability measures μ on $A := [0, a]$, $a > 0$ or $[0, +\infty)$. We would like to find

$$U_\alpha := \sup_\mu \int_A \frac{([t] + \lceil t \rceil)}{2} \mu(dt) \tag{17.1}$$

and

$$L_\alpha := \inf_\mu \int_A \frac{([t] + \lceil t \rceil)}{2} \mu(dt) \tag{17.2}$$

subject to

$$\int_A t^r \mu(dt) = d^r, \quad r > 0, \tag{17.3}$$

where $d > 0$ (here the subscript α in U_α, L_α stands for average).

In this chapter, the underlined probability space is assumed to be nonatomic and thus the space of laws of nonnegative random variables coincides with the space of all Borel probability measures on R_+ (see [302], pp. 17-19).

The solution of the above moment problems are of importance because they permit comparisons to the optimal expected values of the averages of nonnegative random variables, over various sets of observations, subject to one simple and flexible moment condition.

Similar problems have been solved earlier in [20], [83]. Here again the solutions come by the use of tools from the Kemperman geometric moment theory, see [229].

227

Here of interest will be to find

$$U := \sup_{\mu} \int_A ([t] + \lceil t \rceil) \mu(dt) \qquad (17.4)$$

and

$$L := \inf_{\mu} \int_A ([t] + \lceil t \rceil) \mu(dt), \qquad (17.5)$$

subject to the one moment condition (17.3). Then, obviously,

$$U_\alpha = \frac{U}{2}, \quad L_\alpha = \frac{L}{2}. \qquad (17.6)$$

So we use the method of optimal distance in order to find U, L as follows:
Call

$$M := == \operatorname{conv}_{t \in A}(t^r, ([t] + \lceil t \rceil)). \qquad (17.7)$$

When $A = [0, a]$ by Lemma 2 of [229] we obtain that $0 < d \le a$, in order to find finite U and L. Then $U = \sup\{z : (d^r, z) \in M\}$ and $L = \inf\{z : (d^r, z) \in M\}$. Thus U is the largest distance between $(d^r, 0)$ and $(d^r, z) \in \bar{M}$. Also L is the smallest distance between $(d^r, 0)$ and $(d^r, z) \in \bar{M}$. So we are dealing with two-dimensional moment problems over all different variations of $A := [0, a]$, $a > 0$, or $A := [0, +\infty)$ and $r \ge 1$ or $0 < r < 1$.

Description of the graph $(x^r, ([x] + \lceil x \rceil))$, $r > 0$, $a \le +\infty$. This is a stairway made out of the following steps and points, the next step is two units higher than the previous one with a point included in between, namely we have: the lowest part of the graph is $P_0 = (0, 0)$, then open interval $\{P_1 = (0, 1), Q_0 = (1, 1)\}$, the included point $(1, 2)$, open interval $\{P_2 = (1, 3), Q_1 = (2^r, 3)\}$, included point $(2^r, 4)$, open interval $\{P_3 = (2^r, 5), Q_2 = (3^r, 5)\}$, included point $(3^r, 6), \ldots$, $\{$open intervals $\{P_{k+1} := (k^r, 2k + 1), Q_k := ((k + 1)^r, 2k + 1)\}$, included points $(k^r, 2k)$; $k = 0, 1, 2, 3, \ldots\}$, $\ldots$, keep going like this if $a = +\infty$, or if $0 < a \notin \mathbb{N}$ then the last top part, segment of the graph, will be the open closed interval $\{([a]^r, 2[a] + 1), (a^r, 2[a] + 1)\}$, or if $a \in \mathbb{N}$ then the last top part, included point, will be $(a^r, 2a)$. Here $P_{k+2} = ((k + 1)^r, 2k + 3)$. Then

$$\operatorname{slope}(\overline{P_{k+1} P_{k+2}}) = \frac{2}{(k + 1)^r - k^r},$$

which is decreasing in k if $r \ge 1$, and increasing in k if $0 < r < 1$. The closed convex hull of the above graph $\bar{M}$ changes according to values of r and a.

Here X stands for a random variable and E for the expected value.

17.2   Results

The first result follows

Theorem 17.1. *Let $d > 0$, $r \ge 1$ and $0 < a < +\infty$. Consider*

$$U := \sup\{E([X] + \lceil X \rceil) : 0 \le X \le a \text{ a.s.}, \ EX^r = d^r\}. \qquad (17.8)$$

1) *If $k \le d \le k+1$, $k \in \{0, 1, 2, \ldots, [a]-1\}$, $a \notin \mathbb{N}$, then*

$$U = \frac{2d^r + (2k+1)(k+1)^r - (2k+3)k^r}{(k+1)^r - k^r}.\tag{17.9}$$

2) *Let $a \notin \mathbb{N}$ and $[a] \le d \le a$, then*

$$U = 2[a] + 1.\tag{17.10}$$

3) *Let $A = [0, +\infty)$, (i.e., $a = +\infty$) and $k \le d \le k+1$, $k = 0, 1, 2, 3, \ldots$, then U is given by* (17.9).

4) *Let $a \in \mathbb{N} - \{1, 2\}$ and $k \le d \le k+1$, where $k \in \{0, 1, 2, \ldots, a-2\}$, then U is given by* (17.9).

5) *Let $a \in \mathbb{N}$ and $a - 1 \le d \le a$, then*

$$U = \frac{d^r + a^r(2a-1) - (a-1)^r 2a}{a^r - (a-1)^r}.\tag{17.11}$$

Proof. We are working on the (x^r, y) plane. Here the line through the points P_{k+1}, P_{k+2} is given by

$$y = \frac{2x^r + (2k+1)(k+1)^r - (2k+3)k^r}{(k+1)^r - k^r}.\tag{17.12}$$

In the case of $a \in \mathbb{N}$ let $P_a := ((a-1)^r, 2a-1)$. Then the line through P_a and $(a^r, 2a)$ is given by

$$y = \frac{x^r + a^r(2a-1) - (a-1)^r 2a}{a^r - (a-1)^r}.\tag{17.13}$$

The above lines describe the closed convex hull $\bar{M}$ in the various cases mentioned in the statement of the theorem. Then we apply the method of optimal distance described earlier in Section 17.1.

$\square$

Its counterpart follows.

Theorem 17.2. *Let $d > 0$, $0 < r < 1$ and $0 < a < +\infty$. Consider* (17.8),

$$U := \sup\{E([X] + \lceil X \rceil) \colon 0 \le X \le a \text{ a.s.}, EX^r = d^r\}.$$

1) *Let $0 < a \notin \mathbb{N}$ and $[a] \le d \le a$, then*

$$U = 2[a] + 1.\tag{17.14}$$

2) *Let $0 < a \notin \mathbb{N}$ and $0 < d \le [a]$, then*

$$U = \frac{2[a]}{[a]^r} d^r + 1.\tag{17.15}$$

2*) *Let $a = 1$ and $0 < d \le 1$, then $U = d^r + 1$.*

3) *Let $a \in \mathbb{N} - \{1\}$, $0 < d \le a$ and $0 < r \le \frac{1}{2}$, then*

$$U = \left(\frac{2a-1}{a^r}\right) d^r + 1.\tag{17.16}$$

4) *Let $a \in \mathbb{N} - \{1\}$, $0 < d \leq a - 1$, $\frac{1}{2} < r < 1$ and $a \geq \dfrac{(2r)^{\frac{1}{1-r}}}{(2r)^{\frac{1}{1-r}} - 1}$, then*

$$U = 2(a-1)^{1-r} d^r + 1. \tag{17.17}$$

5) *Let $a \in \mathbb{N} - \{1\}$, $a - 1 \leq d \leq a$, $\frac{1}{2} < r < 1$ and $a \geq \dfrac{(2r)^{\frac{1}{1-r}}}{(2r)^{\frac{1}{1-r}} - 1}$, then*

$$U = \frac{d^r + 2a(a^r - (a-1)^r) - a^r}{a^r - (a-1)^r} \tag{17.18}$$

6) *Let $A = [0, +\infty)$, (i.e., $a = +\infty$), then*

$$U = +\infty. \tag{17.19}$$

Proof. We apply again the method of optimal distance. From the graph of $(x^r, ([x] + \lceil x \rceil))$ we have that the slope $(\overline{P_{k+1} P_{k+2}}) = \frac{2}{(k+1)^r - k^r}$ is increasing in k (see Section 17. 1).

Let $0 < a \notin \mathbb{N}$, then the upper envelope of the closed convex hull $\bar{M}$ is formed by the line segments $\{P_1 = (0,1), P_{[a]+1} := ([a]^r, 2[a] + 1)\}$ and $\{P_{[a]+1}, (a^r, 2[a] + 1)\}$. Then case (17.1) is clear.

The line $(P_1 P_{[a]+1})$ in the (x^r, y) plane is given by

$$y = \frac{2[a]}{[a]^r} x^r + 1. \tag{17.20}$$

Therefore case (17.2) follows immediately.

Case (17.6) is obvious by the nature of $\bar{M}$ when $0 < r < 1$.

Next let $a \in \mathbb{N} - \{1\}$ and $0 < r \leq \frac{1}{2}$. That is.,

$$2r \leq 1 < \left(\frac{\xi}{a-1}\right)^{1-r}, \quad \text{for any } \xi \in (a-1, a).$$

That is $2(a-1)^{1-r} < \frac{1}{r\xi^{r-1}}$, any $\xi \in (a-1, a)$. Thus by the mean value theorem we find

$$2(a-1)^{1-r} < \frac{1}{a^r - (a-1)^r}. \tag{17.21}$$

Therefore slope(m_0) < slope(m_1), where m_0 is the line through the points $P_1 = (0,1)$, $P_a = ((a-1)^r, 2a-1)$, and m_1 is the line through the points P_a, $(a^r, 2a)$. In that case the upper envelope of $\bar{M}$ is the line segment with endpoints $P_1 = (0,1)$ and $(a^r, 2a)$. This has equation in the (x^r, y) plane

$$y = \left(\frac{2a-1}{a^r}\right) x^r + 1. \tag{17.22}$$

The last implies case (17.3).

Finally let $a \in \mathbb{N} - \{1\}$ and $\frac{1}{2} < r < 1$, along with

$$a \geq \frac{(2r)^{\frac{1}{1-r}}}{(2r)^{\frac{1}{1-r}} - 1} > 0.$$

Therefore

$$a\left((2r)^{\frac{1}{1-r}} - 1\right) \geq (2r)^{\frac{1}{1-r}} \quad \text{and} \quad a(2r)^{\frac{1}{1-r}} - (2r)^{\frac{1}{1-r}} \geq a.$$

That is $(2r)^{\frac{1}{1-r}}(a-1) \geq a$ and

$$\left(\frac{a}{a-1}\right) \leq (2r)^{\frac{1}{1-r}}.$$

That is, $\left(\frac{a}{a-1}\right)^{1-r} \leq 2r$ and $\xi^{1-r} < a^{1-r} \leq (a-1)^{1-r}2r$, any $\xi \in (a-1, a)$. Thus $\xi^{1-r} \leq 2r(a-1)^{1-r}$, and $\frac{1}{r\xi^{r-1}} \leq 2(a-1)^{1-r}$, any $\xi \in (a-1, a)$. And by the mean value theorem we find that

$$\text{slope}(m_1) = \frac{1}{a^r - (a-1)^r} \leq \text{slope}(m_0) = 2(a-1)^{1-r},$$

where again m_1 is the line through $P_a = ((a-1)^r, 2a-1)$ and $(a^r, 2a)$, while m_0 is the line through $P_1 = (0, 1)$ and P_a. Therefore the upper envelope of $\bar{M}$ is made out of the line segments $\overline{P_1 P_a}$ and $\overline{P_a(a^r, 2a)}$. The line m_0 in the (x^r, y) plane has equation

$$y = 2(a-1)^{1-r}x^r + 1. \tag{17.23}$$

Now case (17.4) is clear.

The line m_1 in the (x^r, y) plane has equation

$$y = \frac{x^r + 2a(a^r - (a-1)^r) - a^r}{a^r - (a-1)^r}. \tag{17.24}$$

At last case (17.5) is obvious. $\qquad\square$

Next we determine L in different cases. We start with the less complicated case where $0 < r < 1$.

Theorem 17.3. *Let $d > 0$, $0 < r < 1$ and $0 < a < +\infty$. Consider*

$$L := \inf\{E(\lfloor X\rfloor + \lceil X\rceil) : 0 \leq X \leq a \text{ a.s.}, \ EX^r = d^r\}. \tag{17.25}$$

1) *Let $1 < a \notin \mathbb{N}$ and $0 < d \leq 1$, then*

$$L = d^r. \tag{17.26}$$

1*) *Let $a < 1$, $0 < d \leq a$, then $L = a^{-r}d^r$.*

2) *Let $2 < a \notin \mathbb{N}$ and $k+1 \leq d \leq k+2$ for $k \in \{0, 1, \ldots, [a]-2\}$, then*

$$L = \frac{2d^r + (2k+1)(k+2)^r - (2k+3)(k+1)^r}{(k+2)^r - (k+1)^r}. \tag{17.27}$$

3) *Let $1 < a \notin \mathbb{N}$ and $[a] \leq d \leq a$, then*

$$L = \frac{2d^r + a^r(2[a]-1) - [a]^r(2[a]+1)}{a^r - [a]^r}. \tag{17.28}$$

4) *Let $a \in \mathbb{N}$ and $0 < d \leq 1$, then*

$$L = d^r. \tag{17.29}$$

5) *Let $a \in \mathbb{N} - \{1\}$ and $k+1 \leq d \leq k+2$, where $k \in \{0, 1, \ldots, a-2\}$, then*

$$L \text{ is given by (17.27).}$$

6) *Let $A := [0, +\infty)$ (i.e., $a = +\infty$) and $0 < d \leq 1$, then*

$$L = d^r. \tag{17.30}$$

7) *Let $A := [0, +\infty)$ and $k+1 \leq d \leq k+2$, where $k \in \{0, 1, 2, \ldots\}$, then L is given by (17.27).*

Proof. The lower-envelope of the closed convex hull $\bar{M}$ is formed by the points $P_0 := (0, 0)$, $\{Q_k := ((k+1)^r, 2k+1), \ k = 0, 1, 2, 3, \ldots\}$, $Q_{[a]-1} := ([a]^r, 2[a] - 1)$, $F := (a^r, 2[a] + 1)$. Here

$$\text{slope}(\overline{Q_k Q_{k+1}}) = \frac{2}{(k+2)^r - (k+1)^r},$$

which is increasing in k since $0 < r < 1$, and $\text{slope}(\overline{P_0 Q_0}) = 1$. Notice that $\text{slope}(\overline{Q_0 Q_1}) = \frac{2}{2^r - 1} > 1$. When $0 < a \notin \mathbb{N}$ we see that

$$\text{slope}(\overline{Q_{[a]-1} F}) > \text{slope}(\overline{Q_{[a]-2} Q_{[a]-1}}).$$

Therefore the lower envelope of $\bar{M}$ is made out of the line segments $\overline{P_0 Q_0}$, $\overline{Q_0 Q_1}$, $\overline{Q_1 Q_2}, \ldots, \overline{Q_k Q_{k+1}}, \ldots, \overline{Q_{[a]-2} Q_{[a]-1}}$ and $\overline{Q_{[a]-1} F}$. In the cases of $a \in \mathbb{N}$ or $a = +\infty$ we have a very similar situation. Cases (17.1), (17.4) and (17.6) are clear.

The line $(Q_k Q_{k+1})$ has equation in the (x^r, y) plane

$$y = \frac{2x^r + (2k+1)(k+2)^r - (2k+3)(k+1)^r}{(k+2)^r - (k+1)^r}. \tag{17.31}$$

Now the cases (17.2), (17.5) and (17.7) follow easily by the principle of optimal distance.

Finally the line $(Q_{[a]-1} F)$ in the (x^r, y) plane has equation

$$y = \frac{2x^r + a^r(2[a] - 1) - [a]^r(2[a] + 1)}{a^r - [a]^r}. \tag{17.32}$$

Thus, case (17.3) now is obvious by applying again the principle of optimal (minimum) distance. $\square$

The more complicated case of L for $r \geq 1$ follows in several steps. Here the lower envelope of the closed convex hull $\bar{M}$ is formed again by the points $P_0 = (0, 0)$, $Q_0 = (1, 1), \ldots, Q_k = ((k+1)^r, 2k+1), \ldots, Q_{[a]-1} = ([a]^r, 2[a]-1)$, $F = (a^r, 2[a]+1)$ or $F^* = (a^r, 2a)$ if $a \in \mathbb{N}$. In the case of $A = [0, +\infty)$, the lower envelope of $\bar{M}$ is formed just by P_0, Q_k; all $k \in Z_+$. Observe that the

$$\text{slope}(\overline{Q_k Q_{k+1}}) = \frac{2}{(k+2)^r - (k+1)^r}$$

is decreasing in k since $r \geq 1$.

Theorem 17.4. *Let $A = [0, +\infty)$, (i.e., $a = +\infty$) $r \geq 1$ and $d > 0$. Consider*

$$L := \inf\{E(\lfloor X \rfloor + \lceil X \rceil) : X \geq 0 \text{ a.s., } EX^r = d^r\}. \tag{17.33}$$

We meet the cases:

    1) *If $r > 1$ we get that $L = 0$.*
    2) *If $r = 1$ and $0 < d \leq 1$, then $L = d$.*
    3) *If $r = 1$ and $1 \leq d < +\infty$, then $L = 2d - 1$.*

Proof. We apply again the principle of optimal (minimum) distance.

When $r > 1$ the lower envelope of $\bar{M}$ is the x^r-axis. Hence case (17.1) is clear.

When $r = 1$, then $\text{slope}(\overline{Q_k Q_{k+1}}) = 2$ any $k \in \mathbb{Z}_+$ and $\text{slope}(\overline{P_0 Q_0}) = 1$. Therefore the lower envelope of $\bar{M}$ here is made by the line segment $(\overline{P_0 Q_0})$ with equation $y = x$, and the line $(Q_0 Q_\infty)$ with equation $y = 2x - 1$. Thus cases (17.2) and (17.3) are established. $\qquad\square$

Theorem 17.5. *Let $a \in \mathbb{N} - \{1\}$, $d > 0$ and $r \geq 1$. Consider L as in (17.25).*
    1) *If $\frac{2(a-1)}{a^r - 1} \geq 1$ and $0 < d \leq 1$, then*

$$L = d^r. \tag{17.34}$$

    2) *If $\frac{2(a-1)}{a^r - 1} \geq 1$ and $1 \leq d \leq a$, then*

$$L = \frac{2(a-1)d^r + (a^r - 2a + 1)}{a^r - 1}. \tag{17.35}$$

    3) *If $\frac{2(a-1)}{a^r - 1} \leq 1$ and $0 < d \leq a$, then*

$$L = \left(\frac{2a - 1}{a^r} \right) d^r. \tag{17.36}$$

Proof. Here $m_0 := \text{line}(P_0 Q_0)$ has equation $y = x^r$ and $\text{slope}(m_0) = 1$. Also $m_1 := \text{line}(Q_0 Q_{a-1})$, where $Q_{a-1} = (a^r, 2a - 1)$, has equation

$$y = \frac{2(a-1)x^r + (a^r - 2a + 1)}{a^r - 1}, \tag{17.37}$$

with $\text{slope}(m_1) = \frac{2(a-1)}{a^r - 1}$.

Notice also that $\text{slope}(Q_k Q_{k+1})$ is decreasing in k. If $\frac{2(a-1)}{a^r - 1} \geq 1$, then the lower envelope of $\bar{M}$ is made out of the line segments $\overline{P_0 Q_0}$ and $\overline{Q_0 Q_{a-1}}$. Hence the cases (17.1) and (17.2) are now clear by application of the principle of minimum distance.

If $\frac{2(a-1)}{a^r - 1} \leq 1$, then the lower envelope of $\bar{M}$ is the line segment $(\overline{P_0 Q_{a-1}})$ with associated line equation

$$y = \left(\frac{2a - 1}{a^r} \right) x^r. \tag{17.38}$$

Thus case (17.3) is established similarly. $\qquad\square$

Theorem 17.6. *Let $2 < a \notin \mathbb{N}$, $n := [a]$, $d > 0$ and $r \geq 1$. Suppose that*

$$1 \leq \frac{2(n-1)}{n^r - 1} \leq \frac{2}{a^r - n^r}. \tag{17.39}$$

Consider L as in (17.25).

1) *If $0 < d \leq 1$, then*

$$L = d^r. \tag{17.40}$$

2) *If $1 \leq d \leq [a]$, then*

$$L = \frac{2([a] - 1)d^r + ([a]^r - 2[a] + 1)}{[a]^r - 1}. \tag{17.41}$$

3) *If $[a] \leq d \leq a$, then*

$$L = \frac{2d^r + a^r(2[a] - 1) - [a]^r(2[a] + 1)}{a^r - [a]^r}. \tag{17.42}$$

Proof. The lower envelope of $\bar{M}$ is made out of the line segments: $\{P_0 = (0,0),\ Q_0 = (1,1)\}$ with equation $y = x^r$, $\{Q_0 = (1,1),\ Q_{[a]-1} = ([a]^r, 2[a] - 1)\}$ with equation

$$y = \frac{2([a] - 1)x^r + ([a]^r - 2[a] + 1)}{[a]^r - 1},$$

and $\{Q_{[a]-1},\ F = (a^r, 2[a] + 1)\}$ with equation

$$y = \frac{2x^r + a^r(2[a] - 1) - [a]^r(2[a] + 1)}{a^r - [a]^r}.$$

Then cases (17.1), (17.2), (17.3) follow immediately by application of the principle of minimum distance as before.

$\square$

Theorem 17.7. *Let $2 < a \notin \mathbb{N}$, $n := [a]$, $d > 0$ and $r \geq 1$. Suppose that*

$$1 \geq \frac{2(n - 1)}{n^r - 1} \geq \frac{2}{a^r - n^r}. \tag{17.43}$$

Consider L as in (17.25). If $0 < d \leq a$, then

$$L = \left(\frac{2[a] + 1}{a^r}\right) d^r. \tag{17.44}$$

Proof. Here the lower envelope of $\bar{M}$ is made out of the line segment $\{P_0 = (0,0),\ F = (a^r, 2[a] + 1)\}$ with equation

$$y = \left(\frac{2[a] + 1}{a^r}\right) x^r,$$

etc. $\square$

Theorem 17.8. *Let $2 < a \notin \mathbb{N}$, $n := [a]$, $d > 0$ and $r \geq 1$. Suppose that*

$$\frac{2(n - 1)}{n^r - 1} \leq \frac{2}{a^r - n^r} \leq 1 \tag{17.45}$$

and

$$\frac{2n - 1}{n^r} < \frac{2}{a^r - n^r}. \tag{17.46}$$

Consider L as in (17.25).

   1) If $0 < d \le [a]$, then

$$L = \left(\frac{2[a] - 1}{[a]^r}\right) d^r. \tag{17.47}$$

   2) If $[a] \le d \le a$, then

$$L = \frac{2d^r + a^r\,(2[a] - 1) - [a]^r\,(2[a] + 1)}{a^r - [a]^r}. \tag{17.48}$$

Proof. Here the lower envelope of $\bar{M}$ is given by the line segments: $\{P_0 = (0,0),$ $Q_{[a]-1} = ([a]^r, 2[a] - 1)\}$ with equation

$$y = \left(\frac{2[a] - 1}{[a]^r}\right) x^r$$

and $\{Q_{[a]-1}, F = (a^r, 2[a] + 1)\}$ with equation

$$y = \frac{2x^r + a^r\,(2[a] - 1) - [a]^r\,(2[a] + 1)}{a^r - [a]^r}. \tag{17.49}$$

$\square$

Theorem 17.9. *Let $2 < a \notin \mathbb{N}$, $n := [a]$, $d > 0$ and $r \ge 1$. Suppose that*

$$\frac{2(n - 1)}{n^r - 1} \le 1 \le \frac{2}{a^r - n^r}. \tag{17.50}$$

Consider L as in (17.25). Then L is given exactly as in cases (17.1) and (17.2) of Theorem 17.8.

Proof. The same as in Theorem 17.8. $\square$

Theorem 17.10. *Let $2 < a \notin \mathbb{N}$, $n := [a]$, $d > 0$ and $r \ge 1$. Suppose that*

$$\frac{2}{a^r - n^r} \le 1 \le \frac{2(n - 1)}{n^r - 1} \tag{17.51}$$

and

$$\frac{2n}{a^r - 1} > 1. \tag{17.52}$$

Consider L as in (17.25).

   1) If $0 < d \le 1$, then

$$L = d^r. \tag{17.53}$$

   2) If $1 \le d \le a$, then

$$L = \frac{2[a]d^r + (a^r - 2[a] - 1)}{a^r - 1}. \tag{17.54}$$

Proof. Here the lower envelope of $\bar{M}$ is made out of the following line segments: $\{P_0 = (0,0), Q_0 = (1,1)\}$ with equation $y = x^r$, and $\{Q_0, F = (a^r, 2[a] + 1)\}$ with equation

$$y = \frac{2[a]x^r + (a^r - 2[a] - 1)}{a^r - 1}, \tag{17.55}$$

etc. $\qquad\qquad\qquad\qquad\qquad\qquad\qquad\qquad\qquad\qquad\qquad\qquad\qquad$ $\square$

Theorem 17.11. *Let* $2 < a \notin \mathbb{N}$, $n := [a]$, $d > 0$ *and* $r \geq 1$. *Suppose that*

$$1 \leq \frac{2}{a^r - n^r} \leq \frac{2(n-1)}{n^r - 1}. \tag{17.56}$$

Consider L as in (17.25). Then L is given exactly as in cases (17.1) and (17.2) of Theorem 17.10.

Proof. Notice that the lower envelope of $\bar{M}$ is the same as in Theorem 17.10, especially here $\frac{2n}{a^r - 1} > 1$. Etc. $\qquad\qquad\qquad\qquad\qquad$ $\square$

Theorem 17.12. *Let* $2 < a \notin \mathbb{N}$, $n := [a]$, $d > 0$ *and* $r \geq 1$. *Suppose that*

$$\frac{2(n-1)}{n^r - 1} \leq \frac{2}{a^r - n^r} \leq 1 \tag{17.57}$$

and

$$\frac{2n - 1}{n^r} \geq \frac{2}{a^r - n^r}. \tag{17.58}$$

Consider L as in (17.25).
    If $0 < d \leq a$, then

$$L = \left(\frac{2[a] + 1}{a^r} \right) d^r. \tag{17.59}$$

Proof. Here the lower envelope of $\bar{M}$ is the line segment $P_0 F$ with equation

$$y = \left(\frac{2[a] + 1}{a^r} \right) x^r, \tag{17.60}$$

etc. $\qquad\qquad\qquad\qquad\qquad\qquad\qquad\qquad\qquad\qquad\qquad\qquad\qquad$ $\square$

Theorem 17.13. *Let* $2 < a \notin \mathbb{N}$, $n := [a]$, $d > 0$ *and* $r \geq 1$. *Suppose that*

$$\frac{2}{a^r - n^r} \leq 1 \leq \frac{2(n-1)}{n^r - 1} \tag{17.61}$$

and

$$\frac{2n}{a^r - 1} \leq 1. \tag{17.62}$$

Consider L as in (17.25).
    If $0 < d \leq a$, then

$$L = \left(\frac{2[a] + 1}{a^r} \right) d^r. \tag{17.63}$$

Proof. Same as in Theorem 17.12. $\qquad\qquad\qquad\qquad\qquad\qquad$ $\square$

Proposition 17.14. *Let $a = 1$, $r \geq 1$ and $0 < d \leq 1$. Then $L = d^r$.*

Proposition 17.15. *Let $0 < a < 1$, $r \geq 1$ and $0 < d \leq a$. Then $L = a^{-r} d^r$.*

Proposition 17.16. *Let $a = 2$, $r \geq \frac{\ln 3}{\ln 2}$ and $0 < d \leq 2$. Then*

$$L = 3 \cdot 2^{-r} d^r.$$

Proposition 17.17. *Let $a = 2$, $1 \leq r \leq \frac{\ln 3}{\ln 2}$.*
  1) *If $0 < d \leq 1$, then $L = d^r$.*
  2) *If $1 \leq d \leq 2$, then $L = \frac{2d^r}{2^r - 1}$.*

Proposition 17.18. *Let $1 < a < 2$, $r \geq 1$ and $2(a^r - 1)^{-1} > 1$.*
  1) *If $0 < d \leq 1$, then $L = d^r$.*
  2) *If $1 \leq d \leq a$, then $L = \frac{2d^r + a^r - 3}{a^r - 1}$.*

Proposition 17.19. *Let $1 < a < 2$, $r \geq 1$ and $2(a^r - 1)^{-1} \leq 1$.*
  If $0 < d \leq a$, then $L = 3a^{-r} d^r$.

Chapter 18

Moment Theory of Random Rounding Rules Subject to One Moment Condition

In this chapter we present the upper and lower bounds for the expected convex combinations of the Adams and Jefferson roundings of a nonnegative random variable with one moment and range conditions. The combination is interpreted as a random rounding rule. We also compare the bounds with the ones for deterministic rounding rules at the respective fractional levels.This treatment follows [89].

18.1 Preliminaries

For a given $\lambda \in (0,1)$, we define a function $f_\lambda(x) = \lambda \lfloor x \rfloor + (1 - \lambda) \lceil x \rceil$, $x \in \mathbb{R}$, where $\lfloor x \rfloor$ and $\lceil x \rceil$ denote the floor and ceiling of x, respectively. Note that $f_\lambda(x)$ can be interpreted as the expectation of rounding x according to the random rule that rounds down and up with probabilities λ and $1 - \lambda$, respectively.

Denote by X an arbitrary element of the class of Borel measurable random variables such that $0 \leq X \leq a$ almost surely and $\mathrm{E}X^r = m_r$ for some $0 < a \leq +\infty$, and $r > 0$, and $0 \leq m_r \leq a^r$. The objective of this chapter is to determine the sharp bounds

$$U(m_r) = U(\lambda, a, r, m_r) = \sup\ Ef_\lambda(X),$$
$$L(m_r) = L(\lambda, a, r, m_r) = \inf\ Ef_\lambda(X),$$

for all possible choices of parameters, where the extrema are taken over the respective class of random variables.

The special cases of the Adams and Jefferson rules, $\lambda = 0$ and 1, respectively, were studied by Anastassiou and Rachev (1992),[83] (see also Anastassiou (1993, Chap.4)),[20] . The analogous results for $\lambda = 1/2$ can be found in Anastassiou (1996),[23] and Chapter 17 here. The method of solution, that will be also implemented here, is based on the optimal distance approach due to Kemperman (1968),[229]. In our case, it consists in determining the extreme values of the second coordinate of points (m_r, x) belonging to the convex hull $H = H(\lambda, a, r)$ of

239

the closure $G = G(\lambda, a, r)$ of the set $\{(t^r, f_\lambda(t)) : 0 \leq t \leq a\}$. Clearly,

$$G = \bigcup_{k=0}^{n-1} \{(t^r, k+1-\lambda) : k \leq t \leq k+1\} \cup \bigcup_{k=0}^{N} \{(k^r, k)\}$$
$$\cup \{(t^r, n+1-\lambda) : n \leq t \leq a\},$$

where $n = \sup\{k \in \mathbb{N} \cup \{0\} : k < a\}$ and $N = \lfloor a \rfloor$ $(= n$ unless $a \in \mathbb{N})$. It will be convenient to use both the notions n and N in describing the results. The planar set G can be visualized as a combination of horizontal closed line segments $\overline{U_k L_{k+1}}$, $0 \leq k < N$, and possibly $\overline{U_N A}$, and separate points S_k, $0 \leq k \leq N$, say, where $U_k = (k^r, k+1-\lambda)$, $L_k = (k^r, k-\lambda)$, $A = (a^r, n+1-\lambda)$, and $S_k = (k^r, k)$. The purpose is to determine, for fixed λ, a and r, the analytic forms of the upper and lower envelopes of H, $U(m_r)$ and $L(m_r)$, $0 \leq m_r \leq a$, respectively. Since $S_k \in \overline{L_k U_k}$, we can immediately observe that

$$H = \begin{cases} \text{conv } \{S_0, U_0, L_1, \ldots, L_N, U_N, A\}, & \text{if } a \in \mathbb{R}_+ \setminus \mathbb{N}, \\ \text{conv } \{S_0, U_0, L_1, \ldots, L_n, U_n, L_a, S_a\}, & \text{if } a \in \mathbb{N}, \\ \text{conv } \{S_0, U_0, L_1, \ldots, L_k, U_k, \ldots\}, & \text{if } a = +\infty. \end{cases} \tag{18.1}$$

A compete analysis of various cases, carried out in Section 18.2, will allow us to simplify further representations (18.1) E.g., the shape of H is strongly affected by the fact that the length of $\overline{U_k L_{k+1}}$ is constant, decreasing and increasing for $r = 1$, $r < 1$ and $r > 1$, respectively, as k increases. The argument makes reasonable considering the cases separately in consecutive subsections of Section 18.2. In each subsection, we derive both the bounds by examining more specific subcases, described by various domains of the end-point a. In Section 18.3 we compare the bounds for $Ef_\lambda(X)$ with those for respective $Ef_\lambda^*(X)$, $0 < \lambda < 1$, where $f_\lambda^*(x) = \lfloor x + 1 - \lambda \rfloor$ is the deterministic λ-rounding rule, and X belongs to a given class of random variables, determined by the moment and support conditions. Note that both the f_λ and f_λ^* round down and up the same proportions of noninteger numbers. The bounds for the latter were precisely described in Anastassiou and Rachev (1992),[83] and Anastassiou (1993),[20].

18.2   Bounds for Random Rounding Rules

18.2.1. Case $r = 1$. Here $L_j \in \overline{L_i L_k}$ and $U_j \in \overline{U_i U_k}$ for all $i < j < k$. Thus we can significantly diminish the number of points generating H (see (18.1)). If a is finite, then H is a hexagon with the vertices $S_0, U_0, U_N, A, L_N, L_1$ for $a \notin \mathbb{N}$ and $S_0, U_0, U_n, S_a, L_a, L_1$ for $a \in \mathbb{N}$. If $a = +\infty$, then H is the part of an infinite strip and has the borders $\overline{U_0 S_0}$, $\overline{S_0 L_1}$ and the parallel halflines $\overrightarrow{U_0 U_1}$ and $\overrightarrow{L_1 L_2}$. In Theorem 18.1 and 18.2 we precisely describe $U(m_1)$ and $L(m_1)$, respectively.

Theorem 18.1. $\widehat{(\text{i})}$ If $\mathbb{N} \not\ni a < \infty$, then

$$U(m_1) = \begin{cases} m_1 + 1 - \lambda, & \text{for } 0 \leq m_1 \leq N, \\ N + 1 - \lambda, & \text{for } N \leq m_1 \leq a. \end{cases} \tag{18.2}$$

(ii) If $\mathbb{N} \ni a < \infty$, then

$$U(m_1) = \begin{cases} m_1 + 1 - \lambda, & for\ 0 \le m_1 \le n, \\ \lambda(m_1 - a) + a, & for\ n \le m_1 \le a. \end{cases} \qquad (18.3)$$

(iii) If $a = +\infty$, then

$$U(m_1) = m_1 + 1 - \lambda, \qquad for\ m_1 \ge 0. \qquad (18.4)$$

Theorem 18.2 . *(i) If $0 < a \le 1$, then*

$$L(m_1) = (1 - \lambda)a^{-1}\, m_1, \qquad for\ 0 \le m_1 \le a. \qquad (18.5)$$

(ii) If $1 < a < \infty$, then

$$L(m_1) = \begin{cases} (1 - \lambda)m_1, & for\ 0 \le m_1 \le 1, \\ m_1 - \lambda, & for\ 1 \le m_1 \le n, \\ \frac{m_1 - a}{a - n} + n + 1 - \lambda, & for\ n \le m_1 \le a. \end{cases} \qquad (18.6)$$

(iii) If $a = +\infty$, then

$$L(m_1) = \begin{cases} (1 - \lambda)m_1, & for\ 0 \le m_1 \le 1, \\ m_1 - \lambda, & for\ m_1 \ge 1. \end{cases} \qquad (18.7)$$

We indicate the supports of distributions providing bounds (18.2), (18.7) (possibly in the limit). We have $U(m_1) = m_1 + 1 - \lambda$ in (18.2), (18.4) for combinations of $t \searrow k$, the limits being integer such that $0 \le k < a$. The latter formulas in (18.2) and (18.3) are obtained for mixtures of any $t \in (n, a]$, and $t \searrow n$ with a, respectively. Two-point distributions supported on 0 and a provide (18.5). We obtain the first bounds in (18.6),(18.7) for mixtures of 0 and $t \nearrow 1$. The second are achieved for those of arbitrary $t \nearrow k$, $k = 1, \ldots, N$. Finally, the last bound in (18.6) is attained once we take combinations of $t \nearrow N$ and a. The supports can be easily determined by analyzing the extreme points of the convex hull H. We leave it to the reader to establish the distributions, possibly limiting, that attain the bounds for $r \ne 1$.

18.2.2. Case $0 < r < 1$. The broken lines determined by $S_0, L_1, \ldots, L_n, A$ and $U_0, U_1, \ldots, U_n$ are convex. This implies that the lower envelope of H coincides with the former one (see (18.13) below). In the trivial case $a \le 1$, this reduces to $\overline{S_0 A}$ (see (18.12)). Also, $U_1, \ldots, U_{n-1}$ do not belong to the graph of $U(m_r)$, which, in consequence, becomes the broken line joining U_0, U_n and A for a finite noninteger a (cf (18.8)). For $a \in \mathbb{N} \setminus \{1\}$, we should consider two subcases. If λ is sufficiently small, then the slope of $\overline{U_0 S_a}$ is less than that of $\overline{U_0 U_n}$ (precisely, if $(a - 1 + \lambda)a^{-r} < n^{1-r}$) and therefore $U(m_r)$ has two linear pieces determined by the line segments $\overline{U_0 U_n}$ and $\overline{U_n S_a}$ (cf (18.10)). Otherwise we obtain a single piece connecting U_0 with S_a (see (18.9)). The same conclusion evidently holds when $a = 1$. Examining the representations of U and L as $a \to +\infty$, we derive the solution in the limiting case $a = +\infty$ (see (18.11) and (18.14)).

Theorem 18.3. *(i) If a finite $a \notin \mathbb{N}$, then*

$$U(m_r) = \begin{cases} N^{1-r}m_r + 1 - \lambda, & for\ 0 \leq m_r \leq N^r, \\ N + 1 - \lambda, & for\ N^r \leq m_r \leq a^r. \end{cases} \tag{18.8}$$

(ii_1) If $a \in \mathbb{N}$ and $\lambda \geq n^{1-r}a^r + 1 - a$, then

$$U(m_r) = (a - 1 + \lambda)a^{-r}m_r + 1 - \lambda, \quad for\ 0 \leq m_r \leq a^r. \tag{18.9}$$

(ii_2) If $a \in \mathbb{N}$ and $\lambda < n^{1-r}a^r + 1 - a$, then

$$U(m_r) = \begin{cases} n^{1-r}m_r + 1 - \lambda, & for\ 0 \leq m_r \leq n^r, \\ \frac{\lambda(m_r - a^r)}{a^r - n^r} + a, & for\ n^r \leq m_r \leq a^r. \end{cases} \tag{18.10}$$

(iii) If $a = +\infty$, then

$$U(m_r) = +\infty, \quad for\ m_r \geq 0. \tag{18.11}$$

Theorem 18.4. *(i) If $0 < a \leq 1$, then*

$$L(m_r) = (1 - \lambda)a^{-r}m_r, \quad for\ 0 \leq m_r \leq a^r. \tag{18.12}$$

(ii) If $1 < a < +\infty$, then

$$L(m_r) = \begin{cases} (1 - \lambda)m_r, & for\ 0 \leq m_r \leq 1, \\ \frac{m_r - k^r}{(k+1)^r - k^r} + k - \lambda, & for\ k^r \leq m_r \leq (k+1)^r, \\ & and\ k = 1, \ldots, n - 1, \\ \frac{m_r - a^r}{a^r - n^r} + n + 1 - \lambda, & for\ n^r \leq m_r \leq a^r. \end{cases} \tag{18.13}$$

(iii) If $a = +\infty$, then

$$L(m_r) = \begin{cases} (1 - \lambda)m_r, & for\ 0 \leq m_r \leq 1, \\ \frac{m_r - k^r}{(k+1)^r - k^r} + k - \lambda, & for\ k^r \leq m_r \leq (k+1)^r, \\ & and\ k = 1, 2, \ldots. \end{cases} \tag{18.14}$$

18.2.3. Case $r > 1$.

Theorem 18.5. *(i) If a is finite noninteger, then*

$$U(m_r) = \begin{cases} \frac{m_r - k^r}{(k+1)^r - k^r} + k + 1 - \lambda, & for\ k^r \leq m_r \leq (k+1)^r, \\ & and\ k = 0, \ldots, N - 1, \\ N + 1 - \lambda, & for\ N^r \leq m_r \leq a^r. \end{cases} \tag{18.15}$$

(ii) If a is finite and integer, then

$$U(m_r) = \begin{cases} \frac{m_r - k^r}{(k+1)^r - k^r} + k + 1 - \lambda, & for\ k^r \leq m_r \leq (k+1)^r, \\ & and\ k = 0, \ldots, n - 1, \\ \frac{\lambda(m_r - a^r)}{a^r - n^r} + a, & for\ n^r \leq m_r \leq a^r. \end{cases} \tag{18.16}$$

(iii) If a is infinite, then

$$U(m_r) = \begin{cases} \frac{m_r - k^r}{(k+1)^r - k^r} + k + 1 - \lambda, & for\ k^r \leq m_r \leq (k+1)^r, \\ & and\ k = 0, 1, \ldots. \end{cases} \tag{18.17}$$

Bounds (18.15), (18.16) and (18.17) are immediate consequences of the fact that the broken lines with the knots $U_0, \ldots, U_N, A$ for $a \notin \mathbb{N}$, $U_0, \ldots, U_n, S_a$ for $a \in \mathbb{N}$, and $U_0, U_1, \ldots$ for $a = +\infty$ are concave. For the special cases $a < 1$ and $a = 1$, yields $U_0 = U_n$, and so the upper bounds reduce to the latter formulas of (18.15) and (18.16), respectively.

Theorem 18.6. *(i) If $0 < a \leq 1$, then (18.12) holds.*
(ii_1) If $1 < a \leq 2$ and $(2 - \lambda)a^{-r} \leq 1 - \lambda$, then

$$L(m_r) = (2 - \lambda)a^{-r}m_r, \quad for\ 0 \leq m_r \leq a^r. \tag{18.18}$$

(ii_2) If $1 < a \leq 2$ and $(2 - \lambda)a^{-r} > 1 - \lambda$, then

$$L(m_r) = \begin{cases} (1 - \lambda)m_r, & for\ 0 \leq m_r \leq 1, \\ \frac{m_r - 1}{a^r - 1} + 1 - \lambda, & for\ 1 \leq m_r \leq a^r. \end{cases} \tag{18.19}$$

Suppose that $2 < a < +\infty$.
(iii_1) If $(n + 1 - \lambda)a^{-r} \leq \min\{1 - \lambda, (n - \lambda)n^{-r}\}$, then

$$L(m_r) = (n + 1 - \lambda)a^{-r}m_r, \quad for\ 0 \leq m_r \leq a^r. \tag{18.20}$$

(iii_2) If $(n - \lambda)n^{-r} < (n + 1 - \lambda)a^{-r}$ and $(n - \lambda)n^{-r} \leq 1 - \lambda$, then

$$L(m_r) = \begin{cases} (n - \lambda)n^{-r}m_r, & for\ 0 \leq m_r \leq n^r, \\ \frac{m_r - a^r}{a^r - n^r} + n + 1 - \lambda, & for\ n^r \leq m_r \leq a^r. \end{cases} \tag{18.21}$$

(iii_3) If $1 - \lambda < \min\{(n + 1 - \lambda)a^{-r}, (n - \lambda)n^{-r}\}$ and $\frac{n}{a^r - 1} \leq \frac{n-1}{n^r - 1}$, then

$$L(m_r) = \begin{cases} (1 - \lambda)m_r, & for\ 0 \leq m_r \leq 1, \\ \frac{n(m_r - 1)}{a^r - 1} + 1 - \lambda, & for\ 1 \leq m_r \leq a^r. \end{cases} \tag{18.22}$$

(iii_4) If $1 - \lambda < \min\{(n + 1 - \lambda)a^{-r}, (n - \lambda)n^{-r}\}$ and $\frac{n}{a^r - 1} > \frac{n-1}{n^r - 1}$, then

$$L(m_r) = \begin{cases} (1 - \lambda)m_r, & for\ 0 \leq m_r \leq 1, \\ \frac{n-1}{n^r - 1}(m_r - 1) + 1 - \lambda, & for\ 1 \leq m_r \leq n^r, \\ \frac{m_r - a^r}{a^r - n^r} + n + 1 - \lambda, & for\ n^r \leq m_r \leq a^r. \end{cases} \tag{18.23}$$

(iv) If $a = +\infty$, then

$$L(m_r) = 0, \quad for\ m_r \geq 0. \tag{18.24}$$

We start by examining case (iii). Since $L_1, \ldots, L_n$ generate a concave broken line, it suffices to analyze the mutual location of $S_0 = (0, 0)$, $L_1 = (1, 1 - \lambda)$, $L_n = (n^r, n - \lambda)$ and $A = (a^r, n + 1 - \lambda)$. This means that we should consider four shapes of the lower envelope determined by the following sets of knots S_0, A, and S_0, L_n, A, and S_0, L_1, A and S_0, L_1, L_n, A, respectively. We show that the all cases are actually possible. First fix λ and n. If $a \searrow n$, then the slope of $\overline{L_n A}$ can be arbitrarily large. This implies that the slope of $\overline{S_0 A}$, equal to $(n + 1 - \lambda)a^{-r}$, becomes greater than that of $\overline{S_0 L_n}$, which is $(n - \lambda)n^{-r}$. The same conclusion holds with S_0 replaced by L_1, for which we derive $n/(a^r - 1) > (n-1)/(n^r - 1)$. If $a \nearrow n + 1$, then $A \to L_{n+1}$, and, in consequence, $\overline{S_0 A}$ and $\overline{L_1 A}$ run beneath $\overline{S_0 L_n}$ and $\overline{L_1 L_n}$, respectively.

If we fix a, and let λ vary, we do not affect the mutual location of $L_1, \ldots, L_n$ and A. For $\lambda \nearrow 1$, the slope of $\overline{S_0 L_1}$, equal to $1 - \lambda$, becomes smaller than either of $\overline{S_0 L_n}$ and $\overline{S_0 A}$. If $\lambda \searrow 0$, this can be arbitrarily close to 1. Then this becomes greater than either of n^{1-r} and $(n+1)a^{-r}(\leq a^{1-r})$, the limiting slopes of $\overline{S_0 L_n}$ and $\overline{S_0 A}$, respectively, since $r > 1$. Letting λ vary from 0 to 1, we meet the cases for which $1 - \lambda$ lies between $(n + 1 - \lambda)a^{-r}$ and $(n - \lambda)n^{-r}$. This shows that all four cases are possible:

(iii_1) the slope of $\overline{S_0 A}$ is minimal among these of $\overline{S_0 A}$, $\overline{S_0 L_n}$ and $\overline{S_0 L_1}$, and (18.20) is the linear function determined by S_0 and A,

(iii_2) the slope of $\overline{S_0 L_n}$ is less that that of $\overline{S_0 A}$ and not greater than that of $\overline{S_0 L_1}$, and (18.21) has two linear pieces, with the knots S_0, L_n and A,

(iii_3) we have three knots S_0, L_1 and A (see (18.22)) if the slope of $\overline{S_0 L_1}$ is minimal, and, moreover, the slope of $\overline{L_1 A}$ is not greater than that of $\overline{L_1 L_n}$,

(iii_4) bound (18.23) has three linear pieces with the ends at S_0, L_1, L_n and A, if the slope of $\overline{S_0 L_1}$ is less than either of $\overline{S_0 L_n}$ and $\overline{S_0 A}$, and the same holds for the pair $\overline{L_1 L_n}$ and $\overline{L_1 A}$.

If $1 < a \leq 2$, then $L_1 = L_n$, and the problem reduces to two cases: we have (18.18) if the slope of $\overline{S_0 A}$ does not exceed the slope of $\overline{S_0 L_1}$, and (18.19) otherwise. If $a \leq 1$, then $n = 0$, and the lower envelope of H is $\overline{S_0 A}$. When the support of X is unbounded, it suffices to take a sequence of two-point distributions with the masses $1 - m_r a^{-r}$ and $m_r a^{-r}$ at 0 and $\mathbb{N} \not\ni a \to +\infty$. Then $E f_\lambda(X)$ coincides with either the right-hand side of (18.20) or the former formula in (18.21), which vanish as $a \to +\infty$ and so (18.24) holds.

18.3 Comparison of Bounds for Random and Deterministic Rounding Rules

In order not to blur the general view, we concentrate on the problems of practical interest for which the right end-point a is sufficiently large. If the rounded variables take on one or two values only, we derive simple results, but usually completely dissimilar to those for larger a. For the same reason, we rather confine ourselves to indicating qualitative differences between the bounds of the random and deterministic roundings. The reader that will be interested in obtaining the precise comparison of the bounds for a specific choice of parameters, is advised to find the respective formulas in Section 18.2 and in Anastassiou (1993, Section 4.1),[20].

Fixing λ, a and r, set $L_k^* = ((k + \lambda)^r, k)$, $U_k^* = ((k + \lambda)^r, k + 1)$, $k \in \mathbb{N} \cup \{0\}$, $A^* = (a^r, n + 1)$ and $A_* = (a^r, n)$. Note that A lies above A_* and below A^*. We first recall the fact that the upper and lower bounds (denoted by U^* and L^*, respectively) for the expectation of the deterministic λ-rounding rule are determined by the upper and lower envelopes of the convex hull H^* of the set

$$G^* = \begin{cases} \overline{S_0 L_0^*} \cup \bigcup_{k=0}^{n-2} \overline{U_k^* L_{k+1}^*} \cup \overline{U_{n-1}^* A_*}, & \text{for } a \leq n + \lambda, \\ \overline{S_0 L_0^*} \cup \bigcup_{k=0}^{n-1} \overline{U_k^* L_{k+1}^*} \cup \overline{U_n^* A^*}, & \text{for } a > n + \lambda. \end{cases}$$

Below we shall repeatedly make use of the facts that all L_k and L_k^* belong to the graph of the function $t \mapsto t^{1/r} - \lambda$, and the same holds for U_k and U_k^* and the function $t \mapsto t^{1/r} + 1 - \lambda$. The shapes of the functions depend on the moment order parameter $r > 0$. As in the previous Section 18.2, we study there cases.

18.3.1. Case $r = 1$. Taking a large enough, we can see that $L = L^*$ and $U = U^*$ in some intervals of m_1. In particular, for $a = +\infty$, the domain of m_1 for which the roundings are equivalent is an infinite interval. We present now some final results, leaving the details of verification to the reader.

We first compare the upper bounds under the condition $a > 1$. It easily follows that $U(m_1) > U^*(m_1)$, when $0 < m_1 < \lambda$. Furthermore, $U(m_1) = U^*(m_1)$ for $\lambda \leq m_1 \leq n+1-\lambda$ when $a \leq n+\lambda$, and for $\lambda \leq m_1 \leq n$ otherwise. (Clearly, $n = +\infty$ if $a = +\infty$.) In the former case, we have $U(m_1) > U^*(m_1)$ for $n + 1 - \lambda < m_1 \leq a$, and the reversed inequality for $n < m_1 \leq a$ in the latter one.

We now treat L and L^* in the case $a \geq 1 + \lambda$. For $0 \leq m_1 < 1$ we have $L(m_1) > L^*(m_1)$. If $a < N + \lambda$, then $L(m_1) = L^*(m_1)$ for $1 \leq m_1 \leq N - 1 + \lambda$, and otherwise the equality holds for $1 \leq m_1 \leq N$. If m_1 is close to the upper support bound, then L and L^* are incomparable, except of the special cases $a = N$ and $a = N + \lambda$ for which $L < L^*$ in $(N + 1 - \lambda, a)$ and $L > L^*$ in (N, a), respectively. If $\mathbb{N} \not\ni a < N + \lambda$, then $L - L^*$ changes the sign from $-$ to $+$ in $(N + 1 - \lambda, a)$. If $a > N + \lambda$, then $L - L^*$ changes the sign from $+$ to $-$ in (N, a).

Observe that U dominates U^* for all possible moment conditions iff either $a \leq n + \lambda$ or $a = +\infty$, and L dominates L^* iff $a = N + \lambda$ and $a = +\infty$. Generally, the deterministic rule f_λ^* has a smaller expectation for small m_1. For large ones, the relations between the bounds strongly depend on the relation between the fractional part of a and the rounding parameter λ.

18.3.2. Case $0 < r < 1$. If $0 < m_r < \lambda^r$, then $0 = L^*(m_r) < L(m_r)$. Observe that L_k and L_k^* belong to the graph of a strictly convex function. This implies that each L_k and L_k^* lie under $\overline{L_{k-1}^* L_k^*}$ and $\overline{L_k L_{k+1}}$, respectively, and, in consequence, $L(m_r)$ and $L^*(m_r)$ cross each other exactly twice in every interval $(k^r, (k+1)^r)$. The former function is smaller in neighborhoods of integers and greater for $m_r \approx (k + \lambda)^r$, with $k \in \mathbb{N} \cup \{0\}$. At the right end of the domain, if $a \leq n + \lambda$, then $L_n^* \notin H^*$ and L^* ends at A_*, situated beneath A. It follows that ultimately $L(m_r) > L^*(m_r)$, as $m_r \nearrow a^r$. If $a < n + \lambda$, then the last linear piece of L^* is $\overline{L_n^* A^*}$ and so L^* is greater in the vicinity of a^r. Generally, the mutual relation of L and L^* is similar to that for $r = 1$ expect of that in the central part of the domain, where the equality is replaced by nearly $2N$ crossings of the functions.

We now examine the upper bounds. Suppose first that $a \in \mathbb{N}$. Then the upper envelope of H is determined by U_0, A and possibly U_n, and that of H^* by S_0, U_n^*, A and possibly U_0^*. Since U_0, U_0^*, U_n and U_n^* are points of the graph of a convex function, $\overline{U_0 U_n}$ crosses $\overline{U_0^* U_n^*}$. Also, the straight lines determined by U_0, U_n and U_0^*, U_n^* run above the points S_0, U_0^* and U_n, A, respectively. Hence, if the upper envelopes of H and H^* actually contain U_n and U_0^*, respectively, they cross exactly once for some $m_r \in (\lambda^r, n)$. If H does and H^* does not, then the line running through S_0, U_n^* and under U_0^* crosses once the concave broken line joining U_0, U_n and A. The same arguments can be used, when U_0^* is an extreme point of the respective hull, and U_n is not. Also, the line segments $\overline{U_0 A}$ and $\overline{S_0 U_n^*}$ have a crossing point. In conclusion, $U > U^*$ for small arguments and $U < U^*$ for large ones.

The same holds true for noninteger $a > n + \lambda$. Here U_n is actually a point generating H and we can drop two of four subcases in the above considerations. If $\mathbb{N} \not\ni a \leq n + \lambda$, the knots of U^* are S_0, U_{n-1}^*, A_* and possibly U_0^*. Now $\overline{U_0^* U_{n-1}^*}$ is situated beneath $\overline{U_0 U_n}$. Since, moreover, S_0 lies under U_0 and A_* lies under A, we immediately deduce that $U > U^*$ for all $m_r \leq a^r$. Finally, for $a = +\infty$, we have $U = U^* = +\infty$.

18.3.3. Case $r > 1$. We first study the upper bounds. Here U_k lies above $\overline{U_{k-1}^* U_k^*}$ and U_k^* lies above $\overline{U_k U_{k+1}}$ for arbitrary integer k, because the above points are elements of the graph of a strictly concave function. In consequence, the intervals in which $U > U^*$ contain $m_r = k^r$, and are separated by the ones containing $m_r = (k + \lambda)^r$, where $U < U^*$. At the left end, $U(m_r) = 1 - \lambda > U^*(m_r) \searrow 0$, as $m_r \searrow 0$, like in the other cases. At the right one, we consider two subcases. For $a > n + \lambda$ (a — possibly integer), $U^*(m_r) = n + 1 > U(m_r)$, as m_r is close to a^r, and the last crossing point of U and U^* belongs to $(n^r, (n + \lambda)^r)$. If $a \leq n + \lambda$, then $U(m_r) = n + 1 - \lambda > U^*(m_r) = n$ for $n^r \leq m_r \leq a^r$, and the last crossing point lies in $((n - 1)^r, (n - 1 + \lambda)^r)$. This is also an obvious analogy with the cases $r \leq 1$.

It remains to discuss the most difficult problem of the lower bounds for $r > 1$. There are various possibilities for the lower envelope of H. This may have four knots S_0, L_1, L_n and A, merely the outer ones being obligatory. Assuming that $a > n + \lambda$, there are two possible envelopes of H^*: one determined by S_0, L_0^*, L_n^* and A^*, and the other with L_n^* dropped. Since $\overline{L_1 L_n}$ is located above $\overline{L_0^* L_n^*}$, in the former case S_0, L_1 and L_n belong to the epigraph of L^*, and A does not. This implies that L and L^* cross once in $(0, a^r)$, for all possible forms of L. The same conclusion holds for the latter case, where the slope of $\overline{L_0^* A^*}$ is not greater than that of $\overline{L_0^* L_n^*}$, and S_0, L_0, L_n remain still separated from A by $\overline{L_0^* A^*}$. Therefore for $a > n + \lambda$, $L(m_r)$ is less than $L^*(m_r)$ for small m_r, and greater for the large ones. In fact, the crossing point is greater than n^r so that the former relation holds in an overwhelming domain.

If $a \leq n + \lambda$, then the lower envelope of H^* is determined by either $S_0, L_0^*, L_{n-1}^*, A_*$ or S_0, L_0^*, A_*. In the four-point case, we can check that S_0, L_1 and A lie above the envelope and L_n does not. Hence, if L_n is an extreme point of H (see Theorem 18.6 $(iii_2), (iii_4)$), then $L - L^*$ changes the sign from $+$ through $-$ to $+$, and is positive everywhere otherwise. In the three-point case, L_n may also be located in the epigraph of L^* as S_0, L_1 and A are. Therefore an additional condition for the two sign changes of $L - L^*$ should be imposed. This is $(n - \lambda)(a^r - \lambda^r) < n(n^r - \lambda^r)$, which is equivalent with locating L_n under $\overline{L_0^* A_*}$. Otherwise L dominates L^* for all $m_r \leq a^r$. The final obvious conclusion is $L(m_r) = L^*(m_r) = 0$ for $0 \leq m_r < +\infty = a$.

Chapter 19

Moment Theory on Random Rounding Rules Using Two Moment Conditions

In this chapter sharp upper and lower bounds are given for the expectation of a randomly rounded nonnegative random variables satisfying a support constraint and two moment conditions. The rounding rule ascribes either the floor or the ceiling to a number due to a given two-point distribution.This treatment follows [85].

19.1 Preliminaries

Let X be a random variable with a support contained in the interval $[0, a]$ for a given $0 < a \leq +\infty$ and two fixed moments $EX = m_1$ and $EX^r = m_r$ for some $0 < r \neq 1$. Assume that the values of X are rounded so that we get the floor

$$\lfloor X \rfloor = \sup\{k \in \mathbb{Z} : k \leq X\}$$

and the ceiling

$$\lceil X \rceil = \inf\{k \in \mathbb{Z} : k \geq X\}$$

with probabilities $\lambda \in [0, 1]$ and $1 - \lambda$, respectively. Our aim is to derive the accurate upper and lower bounds for the expectation of the rounded variable

$$U(\lambda, a, r, m_1, m_r) = \sup E(\lambda \lfloor X \rfloor + (1 - \lambda) \lceil X \rceil), \tag{19.1}$$

$$L(\lambda, a, r, m_1, m_r) = \inf E(\lambda \lfloor X \rfloor + (1 - \lambda) \lceil X \rceil), \tag{19.2}$$

respectively, for all possible combinations of the arguments of the left-hand sides of (19.1) and (19.2). The special cases of the problem with $\lambda = 0$ and $\lambda = 1$, corresponding to well-known Adams and Jefferson rounding rules, respectively, were solved in Anastassiou and Rachev [83] and Rychlik [313] (see also Anastassiou [20,Chapter 4]). In fact, in the above mentioned publications there were examined more general deterministic rules that round down unless the fractional part of a number exceeds a given level $\lambda \in [0, 1]$, which surprisingly leads to radically different bounds from these derived for the respective random ones. The latter were established in the case of one moment constraint $EX^r = m_r$, $r > 0$, by Anastassiou

249

and Rachev [83] for $\lambda = 0, 1$, and Anastassiou [23] for $\lambda = 1/2$ and Anastassiou and Rychlik [89] for general λ.

A general geometric method of determining extreme expectations of a given transformation of a random variable, say $g(X)$ (here $g_\lambda(x) = \lambda\lfloor x \rfloor + (1 - \lambda)\lceil x \rceil$), with fixed expectations $\mu_1 \ldots, \mu_k$ of some other ones $f_1(X), \ldots, f_k(X)$, say, (here $f_1(x) = x$, $f_2(x) = x^r$) was developed by Kemperman [229]. The crucial fact the method relies on is that any random variable X with given generalized moments $E f_i(X) = \mu_i$, $i = 1, \ldots, k$, can be replaced by one supported on $k + 1$ points at most that has identical expectations of each f_i (see Richter [307], Rogosinsky [310]). Therefore it suffices to confine ourselves to establishing the extremes of the$(k + 1)$-st coordinate of the intersection of the convex hull of $(f_1(x), \ldots, f_k(x), g(x))$ for all x from the domain of X, which is the space of all possible generalized moments, with the line $x \mapsto (\mu_1, \ldots, \mu_k, x)$, representing the moment conditions. If X has a compact domain, and all f_i, $i = 1, \ldots, k$, are continuous and g is semicontinuous, then either of extremes is attained (see Kemperman [Theorem 6, 229]). If g is not so, the geometric method would still work if we consider the extremes of g in infinitesimal neighborhoods of the moment points (see Kemperman [Theorem 6, 229]).

We use the Kemperman moment theory for calculating (19.1) and (19.2) in case $a < +\infty$. For $a = +\infty$, we deduce the solutions from these derived for finite supports with $a \to +\infty$. We first note that the problem is well posed iff $m = (m_1, m_r) \in \mathcal{M}(a, r) = conv\mathcal{F}(a, r)$ with $\mathcal{F}(a, r) = \{(t, t^r) : 0 \le t \le a\}$, i.e. $\mathbf{m}$ lies between the straight line $m_r = a^{r-1}m_1$ and the curve $m_r = m_1^r$ for $0 \le m_1 \le a$. The bounds in question are determined by the convex hull of the graph $\mathcal{G}(\lambda, a, r)$ of three-valued function $[0, a] \ni t \mapsto (t, t^r, \lambda\lfloor t \rfloor + (1 - \lambda)\lceil t \rceil)$ for fixed λ, a and r, which will be further ignored in notation for shortness. We easily see that for n denoting the largest integer smaller than a,

$$\mathcal{G} = \bigcup_{k=0}^{n} \{(t, t^r, k + 1 - \lambda) : k < t < k + 1\} \cup \bigcup_{k=0}^{\alpha} \{(k, k^r, k)\},$$

for $a \in \mathbb{N}$, and

$$\mathcal{G} = \bigcup_{k=0}^{n-1} \{(t, t^r, k + 1 - \lambda) : k < t < k + 1\} \cup \bigcup_{k=0}^{n} \{(k, k^r, k)\}$$
$$\cup \{(t, t^r, n + 1 - \lambda) : n < t \le a\},$$

for $a \notin \mathbb{N}$. Since g_λ is not semicontinuous for $0 < \lambda < 1$, we could use the Kemperman method once we consider the closure $\bar{\mathcal{G}}$ that arises by affixing points $U_k = (k, k^r, k + 1 - \lambda)$, $k = 0, \ldots, n$, and $L_k = (k, k^r, k - \lambda)$, $k = 1, \ldots, \lfloor a \rfloor$. Visually, $\bar{G}$ consists of subsequent closed pieces of the planar curve $t \mapsto (t, t^r)$ with integer ends at k and $k + 1$, located horizontally at the respective levels $k + 1 - \lambda$, denoted further by $\widehat{U_k L_{k+1}}$, and separate points $C_k = (c_k, k) = (k, k^r, k) \in \overline{L_k U_k}$, standing for the vertical line segment that joins L_k with U_k. On the left, we start from the origin C_0 and $\widehat{U_0L_1}$, and at the right end we have either $\widehat{U_nL_a}$ and C_a for $a \in \mathbb{N}$ and the trimmed curve $\widehat{U_nA}$ with $A = (a, n + 1 - \lambda) = (a, a^r, n + 1 - \lambda)$ otherwise. We adhere here to the convention introduced in Rychlik [313] of writing points of two- and three-dimensional real space in bold small and capital letters, respectively. Observe that the notation of $a = c_a$ and $A = L_a$ is duplicated for natural a. In the sequel, we shall use a and A then.

In Section 19.2 we present analytic expressions for (19.1) and (19.2) and describe the distributions that attain the respective extremes. E.g., writing $U(\mathbf{m}) = D(t_1, \ldots, t_k)$, we indicate that $\{t_1, \ldots, t_k\} \subset [0, a]$ is the greatest possible support of X with extreme $Eg_\lambda(X)$. The distribution of such an X is unique iff $k \leq 3$, and then the probabilities of all t_k are easily obtainable. However, for some $\mathbf{m} \in \mathcal{M}$ we obtain even infinite support sets, and we drop the formal description of all respective distributions for brevity. If a bound is unattainable, we furnish some t_i with the plus (or minus) sign which would mean that the bound is approached by sequences of distributions with support points tending to the respective arguments of D from the right (or left, respectively). The formal presentation of the results will be supplemented by a geometric interpretation, i.e. the description of the upper and lower envelopes $\overline{\mathcal{M}} = \{\overline{\mathbf{M}} = (\mathbf{m}, \mathbf{U}(\mathbf{m})) : \mathbf{m} \in \mathcal{M}\}$ and $\underline{\mathcal{M}} = \{\underline{\mathbf{M}} = (\mathbf{m}, \mathbf{L}(\mathbf{m})) : \mathbf{m} \in \mathcal{M}\}$, respectively, of the convex hull generated by $\overline{\mathcal{G}} = \overline{\mathcal{G}}(\lambda, \mathbf{a}, \mathbf{r})$. In fact, our proofs consist in solving the dual geometric problems. These will be contained in Section 19.3, together with some purely geometric lemmas whose statements will be repeatedly applied in our reasoning.

19.2   Results

There is a magnitude of different formulas for (19.1) and (19.2) valid in various domains of parameters. For the sake of clearness, we decided to present the upper and lower bound for $r > 1$ and $0 < r < 1$ in separate theorems. In each case, we further choose specific levels of the support range condition a, and finally, for given r and a, we analyze the moment conditions for different regions in $\mathcal{M}(a, r)$. Since U and L are continuous in $\mathbf{m}$, various formulas provide identical values on the borders of neighboring regions. Below we apply the convention of presenting the formulas on the closed parts of $\mathcal{M}$.

It will be further useful to define $t_k = t_k(\mathbf{m}, r) \neq k$ by the equation

$$(m_r - k^r)(t_k - k) = (m_1 - k)(t_k^r - k^r) \tag{19.3}$$

for some $\mathbf{m} \in \mathcal{M}(a, r)$ and $k \in [0, a]$. Note that $t_k = (t_k, t_k^r)$ is the intersection of the straight line running through (k, k^r) and $\mathbf{m}$, and the curve $t \mapsto (t, t^r)$. It is easy to confirm that t_k is a uniquely determined point in $[0, a]$.

Theorem 19.1. *(Upper bounds in case $r > 1$).*

(a) *Suppose that $\mathbb{N} \not\ni a < +\infty$.*

(i) If for some $k = 0, \ldots, n-1$,

$$k \leq m_1 \leq k+1,$$
$$m_1^r \leq m_r \leq [(k+1)^r - k^r](m_1 - k) + k^r, \tag{19.4}$$

then

$$U(m) = D(t_{k+1}, k+1+) = k+2 - \lambda - \frac{k+1-m_1}{k+1-t_{k+1}}. \tag{19.5}$$

(ii) If

$$n \leq m_1 \leq a,$$
$$m_1^r \leq m_r \leq \frac{(a^r - n^r)(m_1 - n)}{a-n} + n^r, \tag{19.6}$$

then

$$U(\mathbf{m}) = D((n, a]) = n+1 - \lambda. \tag{19.7}$$

(iii) If for some $k = 0, \ldots, n-1$,

$$k \leq m_1 \leq k+1,$$
$$[(k+1)^r - k^r](m_1 - k) + k^r \leq m_r \leq n^{r-1} m_1, \tag{19.8}$$

then

$$U(\mathbf{m}) = D(0+, \ldots, n+) = m_1 + 1 - \lambda. \tag{19.9}$$

(iv) If

$$0 \leq m_1 \leq a,$$
$$\max\{n^{r-1} m_1, \tfrac{(a^r - n^r)(m_1 - n)}{a-n} + n^r\} \leq m_r \leq a^{r-1} m_1, \tag{19.10}$$

then

$$U(m) = D(0+, n+, a)$$
$$= 1 - \lambda + \frac{(a^r - n^r)m_1 - (a-n)m_r}{a^r - n^{r-1}a}. \tag{19.11}$$

(b) *Let $\mathbb{N} \ni a < +\infty$.*

    (i) *If (19.6) holds, then*

$$U(m) = D(t_a, a) = a - \lambda \frac{a - m_1}{a - t_a}. \tag{19.12}$$

    (ii) *If (19.10) holds, then*

$$U(m) = D(0+, n+, a)$$
$$= m_1 + 1 - \lambda + \frac{(1-\lambda)(n^{r-1}m_1 - m_r)}{a^r - n^{r-1}a}. \tag{19.13}$$

Otherwise the statements of (ai), (aiii) remain valid for the integer a as well.

(c) *Let $a = +\infty$.*

    (i) *Condition (19.4) for some $k \in \mathbb{N} \cup \{0\}$ implies (19.5).*

(ii) If for $k \in \mathbb{N} \cup \{0\}$

$$k \leq m_1 \leq k+1, \tag{19.14}$$
$$m_r \geq [(k+1)^r - k^r](m_1 - k) + k^r,$$

then

$$U(m) = D(0+, 1+, \ldots) = m_1 + 1 - \lambda. \tag{19.15}$$

Remark 19.2 (*Special cases*). If $a \leq 1$, then (19.6) is satisfied by all $\mathbf{m} \in \mathcal{M}$ and so we find single formulas (19.7) and (19.12) for $a < 1$ and $a = 1$, respectively. For $1 < a \leq 2$, considering (19.8) is superfluous, because it has no interior points. However, as a increases, the upper constraint for m_r in (19.8) tends to $+\infty$ for all m_1 so that in the limiting case we obtain two types of conditions only.

Remark 19.3 (*Geometric interpretation*). Suppose first that a is finite and non-integer. If $a > 2$, none of conditions can be dropped, and the constraints space $\mathcal{M}$ is divided into $n + 3$ pieces. Condition (19.4) states that $\mathbf{m} \in conv\widehat{c_k c_{k+1}}$ for some $k = 0, \ldots, n-1$, whereas (19.5) implies that the respective part of the upper envelope of $conv\mathcal{G}$ coincides with the curved part of the surface of the cone with the horizontal base $conv\widehat{U_k L_{k+1}}$ and vertex U_{k+1}. The part is spread among U_{k+1} and $\widehat{U_k L_{k+1}}$, and will be written $co(U_{k+1}, \widehat{U_k L_{k+1}})$ for short (case (ai)). If $\mathbf{m} \in conv\widehat{c_n a}$ (see (19.6)), the respective piece of $\overline{\mathcal{M}}$ is $conv\widehat{C_n A}$, i.e. loosely speaking, $conv\widehat{c_n a}$ lifted to $n+1-\lambda$ level (case (aii)). The convex polygon $conv c_0 \ldots c_n$ defined in (19.8) corresponds to a fragment of plane slanted to one spanned by $\{(\mathbf{m}, 0) : \mathbf{m} \in \mathcal{M}\}$ and crossing it along the line $m_1 = \lambda - 1$ at the angle $\pi/4$ (case $(aiii)$). Formula (19.11) asserts that $\overline{M}$ is a point of the plane generated by U_0, U_n, and A (plU_0U_nA, for short). Since (19.10) describes the triangle $\Delta c_0 c_n a$ with vertices c_0, c_n and a, it follows that the respective part of $\overline{\mathcal{M}}$ is identical with $\Delta U_0 U_n A$ (case (aiv)).

The case of natural a differs from the above in replacing A by C_a which is situated at the higher level than A. In consequence, for $conv\widehat{c_n a}$ we obtain $co(C_a, \widehat{U_n A})$ (see (19.12)) instead of the flat surface $conv\widehat{C_n A}$ as in (19.7). Also, for $\mathbf{m} \in \Delta c_0 c_n a$, we get $\Delta U_0 U_n C_a$, which is the consequence of (19.10) and (19.13). Note that the line segment $\overline{U_n C_a}$ is sloping in contrast with the horizontal $\overline{U_n A}$. For the remaining $n + 1$ elements of the partition of $\mathcal{M}$, the shape of the upper envelope is identical for integer and noninteger a. The shape of $\overline{\mathcal{M}}$ for $a = +\infty$ can be easily concluded from the above considerations.

Theorem 19.4 (*Lower bounds in case $r > 1$*).

(a) *If $0 < a \leq 1$, then for all $\mathbf{m} \in \mathcal{M}$*

$$L(m) = D(0, t_0) = (1 - \lambda)m_1/t_0. \tag{19.16}$$

(b) *Suppose that $1 < a < +\infty$.*

(i) Conditions

$$0 \le m_1 \le 1,$$
$$m_1^r \le m_r \le m_1, \tag{19.17}$$

imply (19.16).

(ii) If for some $k = 1, \ldots, n-1$, (19.4) holds then

$$L(\mathbf{m}) = D(k-, t_k) = k - \lambda - \frac{t_k - m_1}{t_k - k}. \tag{19.18}$$

(iii) Under the condition (19.6), we have (19.18) with $k = n$.

(iv) If for $k = 1, \ldots, n-1$,

$$k \le m_1 \le k+1,$$
$$[(k+1)^r - k^r](m_1 - k) + k^r \le m_r \le \frac{(n^r-1)(m_1-1)}{n-1} + 1, \tag{19.19}$$

then

$$L(m) = D(1-, \ldots, n-) = m_1 - \lambda. \tag{19.20}$$

Assume, moreover, that

$$(n + 1 - a - \lambda)(n^r - n) \ge \lambda(n^r - 1)(a^r - a). \tag{19.21}$$

(v') If

$$0 \le m_1 \le n,$$
$$\max\{m_1, \frac{(n^r-1)(m_1-1)}{n-1} + 1\} \le m_r \le n^{r-1}m_1, \tag{19.22}$$

then

$$L(\mathbf{m}) = D(0, 1-, n-)$$
$$= m_1 + \lambda \frac{(n-1)m_r - (n^r-1)m_1}{n^r - n}. \tag{19.23}$$

(vi') If

$$0 \le m_1 \le a,$$
$$\max\{n^{r-1}m_1, \frac{(a^r - n^r)(m_1 - n)}{a - n} + n^r\} \le m_r \le a^{r-1}m_1, \tag{19.24}$$

then

$$L(\mathbf{m}) = D(0, n-, a-) = \left(1 - \frac{\lambda}{n}\right) m_1$$
$$+ \frac{[(n-\lambda)(a-n) - n](n^{r-1}m_1 - m_r)}{a^r n - a n^r}. \tag{19.25}$$

If the inequality in (19.21) is reversed, then the following holds.

(v") For

$$0 \le m_1 \le a,$$
$$\max\{m_1, \frac{(a^r - 1)(m_1 - 1)}{a - 1} + 1\} \le m_r \le a^{r-1}m_1, \tag{19.26}$$

we have

$$L(\mathbf{m}) = D(0, 1-, a-)$$
$$= \frac{(1-\lambda)(a^r m_1 - a m_r) + (n+1-\lambda)(m_r - m_1)}{a^r - a}. \tag{19.27}$$

(vi") For

$$1 \leq m_1 \leq a,$$

$$\max\left\{ \frac{(n^r-1)(m_1-1)}{n-1} + 1, \frac{(a^r-n^r)(m_1-n)}{a-n} + n^r \right\} \tag{19.28}$$
$$\leq m_r \leq \frac{(a^r-1)(m_1-1)}{a-1} + 1,$$

yields

$$L(\mathbf{m}) = D(1-,n-,a-) = 1 - \lambda$$
$$+ \frac{[(a^r-1)(n-1)-(n^r-1)n](m_1-1)+(n-1)(n+1-a)(m_r-1)}{(a^r-1)(n-1)-(n^r-1)(a-1)}. \tag{19.29}$$

(c) *Let $a = +\infty$.*

(i) Condition (19.4) implies (19.16).
(ii) If (19.4) holds for some $k \in \mathbb{N}$, then we get (19.18).
(iii) If

$$0 \leq m_1 \leq 1, \tag{19.30}$$
$$m_r \geq m_1,$$

then

$$L(\mathbf{m}) = D(0, 1-, +\infty-) = (1 - \lambda)m_1. \tag{19.31}$$

(iv) If some $k \in \mathbb{N}$ the condition (19.14) is satisfied, then

$$L(m) = D(1-, 2-, \ldots) = m_1 - \lambda. \tag{19.32}$$

Remark 19.5 (*Special cases*). It suffices to consider the case (*biv*) for $a > 3$. If $1 < a \leq 2$, neither of $\mathbf{m} \in \mathcal{M}$ satisfies the respective condition. If $2 < a \leq 3$, (19.19) determines a line segment that can be absorbed by neighboring elements of partition. For $1 < a \leq 2$, we can further reduce the number of subcases. First note that (*bii*) can be dropped. Also, both the (19.22) and (19.28) describe linear relations of m_1 with m_r and so (*bv'*) and (*bvi"*) can be omitted. Furthermore, both the (19.24) and (19.26) are equivalent, and result in the bound

$$L(\mathbf{m}) = D(0, 1 - a-)$$
$$= \frac{(1-\lambda)(a^r m_1 - am_r)-(2-\lambda)(m_1-m_r)}{a^r-a}, \tag{19.33}$$

identical with (19.25) and (19.27). Hence, for $1 < a \leq 2$, it suffices to treat three possibilities (*bi*), (*biii*), and (19.26) with (19.33). Observe that there is no need to examine the integer and fractional a separately, because in the former case, unlike to the supremum problem, we can always replace $a \in \mathbb{N}$ by $a-$ which does not violate moment conditions and decreases the rounding. In fact, we can take the minus away from a in describing the support of extreme distribution in (19.25), (19.27), (19.29) and (19.33) unless $a \in \mathbb{N}$.

Remark 19.6 (*Geometric interpretation*). If $a \leq 1$, then $\underline{\mathcal{M}}$ consists of the line segments $\overline{C_0 T_0}$, with $t_0(\mathbf{m})$ varying between 0 and 1, which constitute the curved conical surface $co(C_0, \widehat{U_0 A})$ of the cone with the vertex C_0 and base $conv\widehat{U_0 A}$.

Putting $a = 1$, we obtain the case (bi). We have analogous solutions in the cases (bii) and $(biii)$, referring to $m \in conv\widehat{c_k c_{k+1}}$, $k = 1, \ldots, n-1$, and $m \in conv\widehat{c_n a}$, respectively. In these domains, $\underline{\mathcal{M}}$ coincides with $n-1$ surfaces $co(L_k, \widehat{U_k L_{k+1}})$, $k = 1, \ldots, n-1$, and $co(L_n, \widehat{U_n A})$, respectively. The condition (19.19) determines the polygon with n vertices $c_1, \ldots, c_n$, and $L(m)$ is the equation of a plane parallel to one including a part of the upper envelope (cf (19.9) and (19.15)). Note that one lies the unit away from the other and all the moment points are situated between them (cf Remark 19.7 below). This implies that the maximal range of the expected rounding with given moments is 1 and this is attained iff the conditions of (biv) (or (civ)) hold.

For finite a, it remains to discuss m of the tetragon $convc_0 c_1 c_n a$. Inequality (19.21) states that A is located either in or above the plane $plC_0 L_1 L_n$. If it is so, then for $m \in \Delta c_0 c_1 c_n$ (see (19.22)) $\underline{M}$ belongs to $plC_0 L_1 L_n$ (cf (19.23)), and, in consequence, to $\Delta C_0 L_1 L_n$. Likewise, formulas (19.24) and (19.25) assert that $\underline{M} = (m, L(m)) \in \Delta C_0 L_n A$ when $m \in \Delta c_0 c_n a$. If A lies on the other side of the plane, we divide $convc_0 c_1 c_n a$ along the other diagonal $\overline{c_1 a}$. Similar arguments lead us to the conclusion that $m \in \Delta c_0 c_1 a$ and $m \in \Delta c_1 c_n a$ imply that $\underline{M} \in \Delta C_0 L_1 A$ and $\underline{M} \in \Delta L_1 L_n A$, respectively.

We leave it to the reader to deduce from Theorem 19.4 that for $a = +\infty$, $\underline{\mathcal{M}}$ is partitioned into the sequence $co(C_0, \widehat{U_0 L_1})$ (cf (ci)), $co(L_k, \widehat{U_k L_{k+1}})$, $k \in \mathbb{N}$, (cf (cii)) and two planar surfaces (see $(ciii)$ and (civ)).

Remark 19.7. Note that the bounds (19.9) and (19.20) are the worst ones, and follow easily from

$$X - \lambda \le \lceil X \rceil - \lambda = \lambda(\lceil X \rceil - 1) + (1 - \lambda)\lceil X \rceil$$
$$\le \lambda \lfloor X \rfloor + (1 - \lambda)\lceil X \rceil$$
$$\le \lambda \lfloor X \rfloor + (1 - \lambda)(\lfloor X \rfloor + 1) \le X + 1 - \lambda.$$

Conclusions for the case $r < 1$ are similar and will not be written down in detail for the sake of brevity. Instead, we merely modify Theorems 19.1 and 19.4.

Theorem 19.8 *(Upper bounds in case $0 < r < 1$). Reverse the inequalities for m_r in (19.4), (19.6), (19.8), (19.10), and (19.14). Replace* max *by* min *in (19.10), and suplement (19.14) with $m_r \ge 0$. Under the above modifications of conditions, the respective statements of Theorem 19.1 hold.*

Theorem 19.9 *(Lower bounds in case $0 < r < 1$). Reverse the inequalities for m_r in (19.4), (19.6), (19.17), (19.19), (19.22), (19.24), (19.26), (19.28) and (19.14), adding $m_r \ge 1$ to the last one. Substitute* min *for* max *in (19.22), (19.24), (19.26) and (19.28). Then the respective conclusions of Theorem 19.4 hold. If, moreover, we change the roles of m_1 and m_r in (19.30), then*

$$L(m) = D(0, 1, +\infty-) = m_1 - \lambda m_r. \tag{19.34}$$

19.3   Proofs

Lemma 19.10 can have wider applications in the geometric moment theory than for the two moment condition problems. At the sacrifice of meaningfulness of three-dimensional geometry arguments, we therefore decided to formulate the result and employ arguments valid for the spaces of any finite dimension.

Lemma 19.10. *Let $\mathcal{M}' \subset \mathcal{M}$ be open and not overlapping with $\mathcal{F}$, and $\partial\mathcal{M}'$ stand for its border. Then each $\overline{M} = (m, U(\mathbf{m})) \in \overline{\mathcal{M}}$ ($\underline{M} = (m, L(m)) \in \underline{\mathcal{M}}$) for $m \in \mathcal{M}'$ can be represented as a convex combination of $\overline{M}_i = (m_i, U(m_i))$ ($\underline{M}_i = (m_i)), U(m_i)$ with $m_i \in \partial\mathcal{M}'$. Accordingly, the subset of $\overline{\mathcal{G}}$ generating the part of the upper (lower) envelope for $m \in \mathcal{M}'$ coincides with that for $m \in \partial\mathcal{M}'$.*

Proof. We only examine the upper envelope case. The other can be handled in the same manner. Take an arbitrary $M = (m, u)$ such that $m \in \mathcal{M}'$ and has a representation

$$M = \sum_i \alpha_i M_i, \qquad M_i = (m_i, u_i) \in \overline{\mathcal{G}} \tag{19.35}$$

for some positive α_i amounting to 1. The condition $\mathcal{M}' \cap \mathcal{F} = \varnothing$ implies that none of $m_i \in \mathcal{M}'$. We can partition all M_i into three, possibly empty, groups: the $M_i = (m_i, U(m_i))$ with $m_i \in \partial\mathcal{M}'$, the $M_i = (m_i, u_i)$ with $m_i \in \partial\mathcal{M}'$ and $u_i < U(m_i)$, and the M_i with respective $m_i \notin \partial\mathcal{M}'$. Our aim is to modify (19.35) so to preserve the moment conditions, despite of replacing some m_i by border points, and not to decrease the criterion functional. To this end, it is sufficient to replace the elements of the second and third groups. In the former case, we simply substitute $\overline{M}_i = (m_i, U(m_i))$ for each M_i.

Less evident arguments are used for eliminating M_i with $m_i \notin \partial\mathcal{M}'$. To be specific, suppose that so is M_1, and we rewrite (19.35) as

$$M = \alpha_1 M_1 + (1 - \alpha_1) \sum_{i \neq 1} \frac{\alpha_i}{1 - \alpha_1} M_i = \alpha_1 M_1 + (1 - \alpha_1) M_{-1}.$$

Note that m and m_1 lie on the other sides of $\partial\mathcal{M}'$. Let $m_1' = \beta m_1 + (1 - \beta)m$, $0 < \beta < 1$, be a crossing point of $\overline{m_1 m}$ with $\partial\mathcal{M}'$, and $M_1' = \beta M_1 + (1 - \beta)M$. Then $m \in \overline{m_1' m_{-1}}$ as well, and likewise

$$\begin{aligned} M &= \frac{\alpha}{\alpha + \beta(1-\alpha)} M_1' + \frac{\beta(1-\alpha_1)}{\alpha + \beta(1-\alpha_1)} M_{-1} \\ &= \gamma M_1' + (1 - \gamma) M_{-1}. \end{aligned} \tag{19.36}$$

Substituting $\overline{M}_1' = (m_1', U(m_1')) \in conv\overline{\mathcal{G}}$ for M_1' yields

$$M' = \gamma \overline{M}_1' + (1 - \gamma) M_{-1},$$

or equivalently

$$(m, u') = \gamma(m_1', U(m_1')) + (1 - \gamma)(m_{-1}, u_{-1}).$$

This has two desired consequences: we replace M_1 with $m_1 \notin \partial\mathcal{M}'$ by $\overline{M}'_1$ with $m'_1 \in \partial\mathcal{M}'$, and derive $u' \geq u$. Repeating the procedure finitely many times, we eliminate all M_i from the third group, improving the approximation of $\overline{M}$ at every step. $\qquad\square$

Lemmas 19.11, 19.12 and 19.13 follow from Lemma 19.10 and will be applied directly in the proofs of our original problems.

Lemma 19.11. *Assume that* $\overline{M_iM_j} \subset \overline{\mathcal{M}}$ *($\underline{\mathcal{M}}$)*, $i,j = 1,2,3$. *Then* $\triangle M_1M_2M_3 \subset \overline{\mathcal{M}}$ *($\underline{\mathcal{M}}$) as well.*

Proof. Project $\triangle M_1M_2M_3$ onto the space of moment conditions. It is evident that the respective projection $\triangle m_1m_2m_3$ has no interior points common with $\mathcal{F}$. Since each of the edge points of $\triangle M_1M_2M_3$ is a convex combination of the vertices, and belongs to $\overline{\mathcal{M}}$ ($\underline{\mathcal{M}}$), then, by Lemma 19.10, for every $\overline{M}$ ($\underline{M}$) with $m \in \triangle m_1m_2m_3$ we have

$$\overline{M}\,(\underline{M}) = \sum_{i=1}^{3} \alpha_i M_i \in \overline{\mathcal{M}}\,(\underline{\mathcal{M}})$$

for some convex combination coefficients α_i, $i = 1,2,3$. All the combinations compose $\triangle M_1M_2M_3$. $\qquad\square$

Lemma 19.12. *Suppose that all the edges* $\overline{M_iM_{i+1}}$, $i = 1,2,3$, *and* $\overline{M_4M_1}$ *of the tetragon* $\mathrm{conv}\,M_1M_2M_3M_4$ *are contained in* $\underline{\mathcal{M}}$. *Then we have* $\triangle M_1M_2M_3 \cup \triangle M_1M_3M_4 \subset \underline{\mathcal{M}}$ *if* M_4 *does not lie beneath* $p\,l\,M_1M_2M_3$, *and* $\triangle M_1M_2M_4 \cup \triangle M_2M_3M_4 \subset \underline{\mathcal{M}}$ *if* M_4 *does not lie above* $p\,l\,M_1M_2M_3$.

Proof. In the standard notation, m_i stands for M_i deprived of the last coordinate. Analysis similar to that in the proof of Lemma 9.11 leads to the conclusion that $(m, L(m))$ is a convex combination of M_i, $i = 1,2,3,4$, for every $m \in \mathrm{conv}\,m_1m_2m_3m_4$.

By the Richter-Rogosinsky theorem, one of M_i is here redundant for a given m. Let m_0 and M_0 denote the crossing points of the diagonals of the respective tetragons. Consider, e.g., the case $m \in \triangle m_1m_2m_0$. The the respective $\underline{M}$ can be written as a combination of either of the triples M_1, M_2, M_3 and M_1, M_2, M_4. The former occurs if M_4 is located over $pl\,M_1M_2M_3$, i.e. $\triangle M_1M_2M_3$ lies beneath $\triangle M_1M_2M_4$. The latter is true in the opposite case. Noting that the former condition is equivalent with locating M_2 over $pl\,M_1M_3M_4$, we similarly analyze three other triangular domains of moment conditions and conclude our claim. In particular, if M_i, $i = 1,2,3,4$, span a plane, then $\mathrm{conv}\,M_1M_2M_3M_4 \in \underline{\mathcal{M}}$. $\qquad\square$

It is easy to obtain a version of Lemma 19.12 for $\overline{M_iM_j} \subset \overline{\mathcal{M}}$. This was not formulated here, because the result will be used for determining the lower bounds only.

Lemma 19.13.(i) *Consider* $\widehat{\overline{U_kL_{k+1}}} \subset \overline{\mathcal{G}}$ *and* $V_{k+1} = (k+1, (k+1)^r, v)$ *for some* $v > k + 1 - \lambda$. *If* $\overline{U_kV_{k+1}} \subset \overline{\mathcal{M}}$, *then* $\mathrm{co}\,(V_{k+1}, \widehat{U_kL_{k+1}}) \subset \overline{\mathcal{M}}$.

(ii) *Let* $\widehat{U_k S} \subset \widehat{U_k L_{k+1}} \cap \overline{\mathcal{G}}$ *and* $V_k = (k, k^r, v)$ *for some* $v < k + 1 - \lambda$. *Then* $\overline{V_k S} \subset \underline{\mathcal{M}}$ *implies* $\mathrm{co}(V_k, \widehat{U_k S}) \subset \underline{\mathcal{M}}$.

The point V_{k+1} defined in Lemma 19.13(i) lies just above L_{k+1} and will be replaced by either U_{k+1} or C_a in the proof of Theorem 19.1. Also, V_k lies under U_k, and will be further replaced by either C_0 or L_k as well as S stands for either L_{k+1} or A.

Proof of Lemma 19.13 (*i*) Since both the $\overline{U_k V_{k+1}}$ and $\widehat{U_k L_{k+1}}$ belong to $\overline{\mathcal{M}}$, due to Lemma 19.10, every $\overline{M}$ with $m \in \widehat{\mathrm{conv} c_k c_{k+1}}$ can be written as a mixture of at most three points from $\widehat{U_k L_{k+1}} \cup \{V_{k+1}\}$. Combining elements of the curve only, we do not exceed $k + 1 - \lambda$, while introducing V_{k+1}, being possible for any $m \in \widehat{\mathrm{conv} c_k c_{k+1}}$, enables us to rise above the level. Assume therefore that $M = (m, u) \in \Delta M_1 M_2 V_{k+1}$ for $M_1, M_2 \in \widehat{U_k L_{k+1}}$, or equivalently $M \in \overline{M_3 V_{k+1}}$ for some $M_3 = (m_3, k + 1 - \lambda) \in \overline{M_1 M_2}$. Note that the line running through m and m_3 crosses $\mathcal{F}$ at c_{k+1} and $t_{k+1} = (t_{k+1}, t_{k+1}^r) \in \widehat{c_k c_{k+1}}$ (see (19.3)). Consider $\overline{T_{k+1} V_{k+1}}$ for $T_{k+1} = (t_{k+1}, k + 1 - \lambda)$. It is easy to see that $\overline{T_{k+1} V_{k+1}}$ lies above $\overline{M_3 V_{k+1}}$ and, consequently, it crosses the vertical line $t \mapsto (m, t)$ at a level $u' > u$, say. In fact, the construction is unique and so it determines $\overline{M}$. Accordingly, for all $m \in \widehat{\mathrm{conv} c_k c_{k+1}}$ the respective $\overline{M} \in \overline{T_{k+1} V_{k+1}} \subset \overline{\mathcal{M}}$, where T_{k+1} runs along the whole $\widehat{U_k L_{k+1}}$ as m varies. The $\overline{T_{k+1} V_{k+1}}$ compose $\mathrm{co}(V_{k+1}, \widehat{U_k L_{k+1}})$.

(*ii*) Analogous. $\qquad\qquad\qquad\qquad\qquad\qquad\qquad\qquad\qquad\qquad\qquad\qquad\square$

Proof of Theorem 19.1 (*a*) It is obvious that the (possibly limiting) distribution of X that attains the upper bound $U(m_1) = \sup E g_\lambda(X)$ under the single moment condition $EX = m_1$, and satisfies $EX^r = m_r$, provides also the respective bound $U(m_1, m_r)$ for the two moment problem. Therefore we directly apply the solution presented in Anastassiou and Rychlik [89]. It follows that $U(m_1) = U(m_1, m_r) = m_1 + 1 - \lambda$ for $(m_1, m_r) \in \mathrm{conv} c_0 \ldots c_n$ and $U(m_1) = U(m_1, m_r) = n + 1 - \lambda$ for $(m_1, m_r) \in \widehat{\mathrm{conv} c_n a}$ which solve (*aiii*) and (*aii*), respectively. The latter can also be derived by noting that $n + 1 - \lambda$ is the greatest possible value of $g_\lambda(X)$ attained when $n < X \leq a$.

Assume now that $m \in \mathrm{conv} \, \widehat{c_k c_{k+1}}$ for some $k = 0, \ldots, n - 1$. Noticing that $\overline{U_k U_{k+1}} \subset \overline{\mathcal{M}}$ and applying Lemma 19.13(i) with $V_{k+1} = U_{k+1}$, we conclude that the respective part of the upper envelope is $\mathrm{co}(U_{k+1}, \widehat{U_k L_{k+1}})$ (case (*ai*)). It remains to consider $m \in \Delta c_0 c_n a$. Observe that $\overline{U_0 A} \subset \overline{\mathcal{M}}$, because $M \in \overline{U_0 A}$ uniquely determine limiting distributions supported on $\{0+, a\}$, and $\overline{U_0 U_n}, \overline{U_n A} \subset \overline{\mathcal{M}}$ (cf (*aiii*) and (*aii*)). By Lemma 19.11, $\triangle U_0 U_n A \subset \overline{\mathcal{M}}$ (case (*aiv*)).

(*b*) From the solution of the single moment problem we derive some parts of $\overline{\mathcal{M}}$. These are $\mathrm{conv} U_0 \ldots U_n$, $\overline{U_n C_a}$, and clearly all the $\widehat{U_k L_{k+1}}$, and $\overline{U_0 C_a}$. The first one provides the solution for $m \in \mathrm{conv} c_0 \ldots c_n$, identical with that for noninteger a, and allows us to repeat the arguments and conclusions of the proof of (*ai*) for $m \in \widehat{c_k c_{k+1}}$, $k = 0, \ldots, n - 1$. In the similar way we use Lemma 19.13 and $\overline{U_n C_a} \subset \overline{\mathcal{M}}$ to deduce that $\overline{M} \in \mathrm{co}(C_a, \widehat{U_n A})$ if $m \in \mathrm{conv} c_n a$ (case (*bi*)). Finally, optimality of $\overline{U_0 U_n}$, $\overline{U_n C_a}$ and $\overline{U_0 C_a}$ and Lemma 19.11 enable us to complete $\overline{\mathcal{M}}$ by adding

$\Delta U_0 U_n C_a$ (case (bii)).

(c) Let $a \to +\infty$. Then extending the range of X does not affect the solution for $m \in conv\widehat{c_k c_{k+1}}$ for a fixed integer $k \geq 0$ (case (ci)). On the other hand, the polygons $conv c_0 \ldots c_n$ gradually appropriate the whole area above the broken line $\bigcup_{k=0}^{\infty} \overline{c_k c_{k+1}}$ (case (cii)). $\square$

Proof of Theorem 19.4. (a) The assertion follows from the direct application of Lemma 19.13 (ii) with C_0 and A standing for V_k and S, respectively.

(b) A partial solution is inherited from the one moment problem. Its geometric interpretation describes the following parts of the lower envelope: $\overline{C_0 L_1}$, $conv L_1 \ldots L_n$ and $\overline{L_n A}$. Combining $\overline{C_0 L_1} \subset \underline{\mathcal{M}}$ with Lemma 19.13(ii) implies $co(C_0, \widehat{U_0 L_1}) \subset \underline{\mathcal{M}}$ which is equivalent with (bi). The polygon with the vertices $L_1, \ldots, L_n$ describes the solution of case (biv). Also, this allows us to deduce from Lemma 19.13 that $co(L_k, \widehat{U_k L_{k+1}}) \subset \underline{\mathcal{M}}$ for $k = 1, \ldots, n-1$, (see (bii)). Yet another application of Lemma 19.13 with $\overline{V_k S} = \overline{L_n A}$ yields $co(L_n, \widehat{U_n A}) \subset \underline{\mathcal{M}}$ (see $(biii)$).

Still remains the task of determining the lower bounds for $m \in conv c_0 c_1 c_n a$. Knowing that $\partial conv C_0 L_1 L_n A \subset \underline{\mathcal{M}}$, we can employ the assertions of Lemma 19.12. If A is situated either on or above $pl C_0 L_1 L_n$, i.e., (19.21) holds, then $\Delta C_0 L_1 L_n$ and $\Delta C_0 L_n A$ are parts of $\underline{\mathcal{M}}$ which correspond to the solutions in cases (bv') and (bvi'), respectively. Otherwise, we have $\Delta C_0 L_1 A \cup \Delta L_1 L_n A \subset \underline{\mathcal{M}}$ (see (bv'') and (bvi'')).

(c) Including arbitrarily large points to the possible support does not change the solution for $m \in conv\widehat{c_k c_{k+1}}$ and a given $k \in \mathbb{N} \cup \{0\}$, which is identical with that for any finite $a > k+1$. Since $\overline{c_1 c_n}$ tends to the vertical halfline starting upwards from c_1 in the plane of $\mathcal{M}$, the domain of the trivial lower estimate $m_1 - \lambda$ ultimately covers the region above $\bigcup_{k=1}^{\infty} \overline{c_k c_{k+1}}$.

The only point remaining is establishing the shape of $\underline{\mathcal{M}}$ for m situated above $\overline{c_0 c_1}$. Take any finite-valued distribution satisfying the moment conditions and represent it as a mixture of ones supported on $[0, 1]$ and $(1, +\infty)$, respectively. The relevant pairs of the first and rth moments satisfy $m_1 \in conv\widehat{c_0 c_1}$ and $m_2 \in conv\widehat{c_1 c_\infty}$, and $m \in \overline{m_1 m_2}$. The respective $\overline{M_1 M_2}$ runs above $\overline{C_0 L_1}$ and the horizontal halfline $t \mapsto (1, t, 1 - \lambda)$, $t \geq 1$, both belonging to $\mathcal{M}$. It can be replaced by $\overline{M_1' M_2'}$ that joins $\overline{C_0 L_1}$ and the halfline, and is located below $\overline{M_1 M_2}$. The family of all such $\overline{M_1' M_2'}$ determines the inclined band of $\underline{\mathcal{M}} = (m_1, m_r, L(m_1, m_r))$ described by (19.30) and (19.31). $\square$

Remark 19.14. (*Proofs of Theorems 19.8 and 19.9*). The analogy between the cases $r > 1$ and $r < 1$ becomes apparent once we realize that both the cases have the same geometric interpretations (cf Remarks 19.3 and 19.6). The only exception is Proposition 19.4 $(ciii)$ and its counterpart (19.34) in Theorem 19.9. Since we use only geometric arguments in the above proofs, there is no need to prove Theorem 19.8 and 19.9 separately. The reader is rather advised to follow again the proofs of Theorem 19.1 and 19.4 as if $r < 1$ were supposed. Also, one can treat the above mentioned difference slightly modifying the last paragraph in the proof of

Theorem 19.4. For $r < 1$, we consider m lying right to $\overline{c_0 c_1}$, and replace the half line $t \mapsto (1, t, 1 - \lambda)$ by $t \mapsto (t, 1, t - \lambda)$ for $t \geq 1$. The latter, together with $\overline{C_0 L_1}$ span the plane described by the equation (19.34).

Chapter 20

Prokhorov Radius Around Zero Using Three Moment Constraints

In this chapter we find exactly the Prokhorov radius of the family of distributions surrounding the Dirac measure at zero whose first, second and fourth moments are bounded by given numbers. This provides the precise relation between the rates of weak convergence to zero and the rate of vanishing of the respective moments. This treatment relies on [88].

20.1 Introduction and Main Result

We start by mentioning the concept of Prokhorov distance of two probability measures μ, ν, which is generally defined on a Polish space with a metric d. This is given by

$$\pi(\mu, \nu) = \inf\{r > 0 : \mu(A) \le \nu(A^r) + r, \ \nu(A) \le \mu(A^r) + r,$$
$$\text{for every closed subset } A\},$$

where $A^r = \{x : d(x, A) < r\}$. Note that in the case of standard real space with the Euclidean metric, the Prokhorov distance of a probability measure μ to the degenerate one δ_0 concentrated at 0 can be written as

$$\pi(\mu, \delta_0) = \inf\{r > 0 : \mu(I_r) \ge 1 - r\}. \tag{20.1}$$

Here and later on $I_r = [-r, r]$, and I_r^c stands for its complement.

For a given triple of positive reals $\mathcal{E} = (\epsilon_1, \epsilon_2, \epsilon_4)$, we consider the family $\mathcal{M}(\mathcal{E})$ of probability measures on the real line such that

$$\mathcal{M}(\mathcal{E}) = \{\mu : \ |\int t^i \, d\mu| < \epsilon_i, \ i = 1, 2, 4\}. \tag{20.2}$$

Theorem 20.1 gives us the precise evaluation of the Prokhorov radius

$$D(\mathcal{E}) = \sup_{\mu \in \mathcal{M}(\mathcal{E})} \pi(\mu, \delta_0)$$

for family of measures (20.2).

Theorem 20.1. *We have*

$$D(\epsilon_1, \epsilon_2, \epsilon_4) = \min\{\epsilon_2^{1/3}, \epsilon_4^{1/5}\}. \tag{20.3}$$

263

This is a refinement of a result in Anastassiou (1992),[19] where the Prokhorov radius $D(\epsilon_1, \epsilon_2) = \epsilon_2^{1/3}$ of the family with constraints on two first moments was established. The problems of determining the Levy and Kantorovich radii under two moment conditions were considered in Anastassiou (1987),[18], and Anastassiou and Rachev (1992),[83],respectively. Anastassiou and Rychlik (1999),[86], studied the Prokhorov radius of measures supported on the positive halfaxis which satisfy conditions on the first three moments. Since the Prokhorov metric induces the topology of weak convergence, formula (20.3) describes the exact rate of weak convergence of measures from $\mathcal{M}(\mathcal{E})$ satisfying the three moment constraints to the Dirac one at zero.

Though our question is stated in an abstract way, it stems straightforwardly from applied probability problems in which rates of convergence of random variables to deterministic ones are evaluated. If we study how fast the random error of a consistent statistical estimate vanishes, then zero is the most natural limiting point. Convergence in probability is implied by that of the first two moments. Adding the fourth one, which has a meaningful interpretation in statistics, allows us to find refined evaluations. These three moments have natural estimates, and so one can easily control their variability. Moreover, the respective power functions form a Tchebycheff system. Convergence of integrals for elements of such systems implies and provides estimates for integrals of general continuous functions. The latter convergence is described by the weak topology, and the presented solution gives a quantitative estimate of uniform weak convergence (expressed in terms of equivalent Prokhorov metric topology) for a large natural class of measures determined by moment conditions.

Formula (20.31) is determined by means of a geometric moment theoretical method of Kemperman (1968),[229], that will be used in Section 20.2 for calculating

$$L_r(M) = \inf_{\mu \in \mathcal{M}(M)} \mu(I_r) \tag{20.4}$$

with given $M = (m_1, m_2, m_4)$ and

$$\mathcal{M}(M) = \{\mu : \int t^i \, d\mu = m_i, \ i = 1, 2, 4\}$$

for all possible m_1, m_2, m_4, and $r > 0$. In Section 20.3 we present the main result of the chapter; having determined (20.4) for various M, we first evaluate respective infima over the boxes in the moment space

$$L_r(\mathcal{E}) = \inf\{L_r(M) : \ |m_i| \leq \epsilon_i, \ i = 1, 2, 4\} \tag{20.5}$$

for every fixed r, and then, letting r vary, we determine

$$D(\mathcal{E}) = \inf\{r > 0 : \ L_r(\mathcal{E}) \geq 1 - r\}. \tag{20.6}$$

In the short last section we sketch possible directions for a further research.

20.2 Auxiliary Moment Problem

Fixing $r > 0$, we now confine ourselves on solving moment problem (20.4). This is well stated iff

$$M \in \mathcal{W} = \{(m_1, m_2, m_4) : m_1 \in \mathbb{R}, m_2 \geq m_1^2, m_4 \geq m_2^2\}.$$

Note that $\mathcal{W} = \mathrm{conv}\mathcal{T} = \mathrm{conv}\{T = (t, t^2, t^4) : t \in \mathbb{R}\}$, the convex hull of the graph of function $\mathbb{R} \ni t \mapsto (t, t^2, t^4)$. Geometrically, $\mathcal{W}$ is a set unbounded above whose bottom $\underline{\mathcal{W}}$ is a membrane spanned by $\mathcal{T}$. The membrane can be represented as $\underline{\mathcal{W}} = \bigcup_{t \geq 0} \overline{T_- T_+}$, where $T_- = (-t, t^2, t^4)$, $T_+ = (t, t^2, t^4)$, and $\overline{AB}$ denotes the line segment with end-points A and B. The side surface consists of vertical halflines $T^\uparrow$ running upwards from the points $T \in \mathcal{T}$. Consider the following surfaces in $\mathcal{W}$:

$\triangle \mathbf{0} R_+ R_-$ — the triangle with vertices $\mathbf{0} = (0, 0, 0)$, $R_+ = (r, r^2, r^4)$ and
     $R_- = (-r, r^2, r^4)$,
$\mathrm{mem}(R_+, \widehat{\mathbf{0}R_+}) = \bigcup_{0 \leq t \leq r} \overline{TR_+}$, and $\mathrm{mem}(R_-, \widehat{\mathbf{0}R_-}) = \bigcup_{-r \leq t \leq 0} \overline{TR_-}$ the mem-
     branes connecting R_+ and R_- with the points of the curves $\widehat{\mathbf{0}R_+} = \{(t, t^2, t^4) :$
     $0 \leq t \leq r\}$ and $\widehat{\mathbf{0}R_-} = \{(-t, t^2, t^4) : 0 \leq t \leq r\}$, respectively,
$\overline{R_- R_+}^\uparrow$, $\overline{\mathbf{0}R_+}^\uparrow$, and $\overline{\mathbf{0}R_-}^\uparrow$ — the infinite bands above the line segments $\overline{R_- R_+}$,
     $\overline{\mathbf{0}R_+}$ and $\overline{\mathbf{0}R_-}$, respectively.

They partition $\mathcal{W}$ into five closed subsets with nonoverlapping interiors:

$\mathcal{W}_1$ —   the set of points situated on and above $\triangle \mathbf{0} R_+ R_-$,
$\mathcal{W}_2$ —   the moment points on and above $\mathrm{mem}(R_+, \widehat{\mathbf{0}R_+})$,
$\mathcal{W}_3$ —   the points on and above $\mathrm{mem}(R_-, \widehat{\mathbf{0}R_-})$,
$\mathcal{W}_4$ —   the points between $\triangle \mathbf{0} R_+ R_-$, $\mathrm{mem}(R_+, \widehat{\mathbf{0}R_+})$, $\mathrm{mem}(R_-, \widehat{\mathbf{0}R_-})$, and $\underline{\mathcal{W}}^{I_r} =$
     $\bigcup_{0 \leq t \leq r} \overline{T_- T_+}$, the last surface being a part of the bottom of the moment space,
$\mathcal{W}_5$ —   the moment points lying on and above $\underline{\mathcal{W}}^{I_r^c} = \bigcup_{t \geq r} \overline{T_- T_+}$.

The solution to (20.4) is expressed by different formulae for the elements of the above partition.

Theorem 20.2. *The solution to (20.4) is described by*

$$L_r(M) = \begin{cases} 1 - m_2/r^2, & \text{if } M \in \mathcal{W}_1, \\ (r - |m_1|)^2/(r^2 - 2|m_1|r + m_2), & \text{if } M \in \mathcal{W}_2 \cup \mathcal{W}_3, \\ (r^2 - m_2)^2/(r^4 - 2m_2 r^2 + m_4), & \text{if } M \in \mathcal{W}_4, \\ 0, & \text{if } M \in \mathcal{W}_5. \end{cases} \tag{20.7}$$

One can easily confirm that the formulae for neighboring regions coincide on their common borders. In particular, this implies continuity of L_r.

Proof of Theorem 20.2. First notice that $\mathcal{W}_5$ is the closure of the convex hull of $\mathcal{T}(I_r^c) = \{(t, t^2, t^4) : |t| > r\}$. The inner elements of $\mathcal{W}_5$ are the moment points for measures supported on I_r^c and therefore $L_r(M) = 0$ for all $M \in \mathcal{W}_5$.

The other formulae in (20.7) will be determined by means of the optimal ratio method due to Kemperman (1968),[229], that allows us to find sharp lower and upper bounds for probability measures of a given set (here: the lower one for those of I_r) under the conditions that the integrals of some given functions with respect to the measures take on prescribed values (here: $\int t^i \, d\mu = m_i$, $i = 1, 2, 4$). The method can be used under mild assumptions about the structure of probability space and functions appearing in the moment conditions (cf Kemperman (1968, Section 5),[229]). These are satisfied in the case we consider and therefore we merely present a version adapted to our problem instead of the general description. Given a boundary point W of $\mathcal{W}$, $W \notin \mathcal{W}_5$, we use a hyperplane $\mathcal{H}$ supporting $\mathcal{W}$ at W, and another one $\mathcal{H}'$ supporting $\mathcal{W}_5$ that is the closest one parallel to $\mathcal{H}$. Then for every moment point M in the closure of $conv(\mathcal{W} \cap \mathcal{H}) \cup (\mathcal{W}_5 \cap \mathcal{H}')$, we have

$$L_r(M) = \frac{d(M, \mathcal{H}')}{d(\mathcal{H}, \mathcal{H}')}, \tag{20.8}$$

where the numerator and denominator in (20.8) denote the distances from the moment point M and hyperplane $\mathcal{H}$ to $\mathcal{H}'$, respectively.

We shall therefore take into account the hyperplanes $\mathcal{H}$ supporting points $W \in \bigcup_{|t|<r} T^{\uparrow} \cup \underline{\mathcal{W}}^{I_r}$. First consider the vertical plane $\mathcal{H} : m_2 = 0$ that supports $\mathcal{W}$ at all points of $\mathbf{0}^{\uparrow}$. Then $\mathcal{H}' : m_2 - r^2 = 0$ is the closest and parallel to $\mathcal{H}$ plane that supports $\mathcal{W}_5$. Since $\mathcal{H} \cap \mathcal{W} = \mathbf{0}^{\uparrow}$, and $\mathcal{H}' \cap \mathcal{W}_5 = \overline{R_- R_+}^{\uparrow}$, then for every $M \in conv \mathbf{0}^{\uparrow} \cup \overline{R_- R_+}^{\uparrow} = \mathcal{W}_1$, we apply (20.8) to find $L_r(M) = 1 - m_2/r^2$.

Consider now a side hyperplane $\mathcal{H}$ such that $\mathcal{H} \cap \mathcal{W} = T^{\uparrow}$ for some $0 < t < r$, described by the formula $\mathcal{H} : m_2 - 2tm_1 + t^2 = 0$. We can easily observe that $\mathcal{H}'$ is the plane parallel to $\mathcal{H}$ that supports $\mathcal{H}_5$ along $R_+^{\uparrow}$. This can be written as $\mathcal{H}' : m_2 - 2tm_1 + 2tr - r^2 = 0$. Applying the standard formula

$$d(Y, \mathcal{A}) = \left| \sum_{i=1}^{n} a_i y_i + b \right| \Big/ \left(\sum_{i=1}^{n} a_i^2 \right)^{1/2}$$

measuring the Euclidean distance between a point $Y = (y_1, \ldots, y_n)$ and a hyperplane $\mathcal{A} : \sum_{i=1}^{n} a_i x_i + b = 0$ in $\mathbb{R}^n$, we obtain

$$d(M, \mathcal{H}') = |m_2 - 2m_1 t + 2tr - r^2|/(1 + 4t^2)^{1/2},$$
$$d(\mathcal{H}, \mathcal{H}') = d(\mathcal{H}, R_+) = (r - t)^2/(1 + 4t^2)^{1/2},$$

and, in consequence,

$$L_r(M) = \frac{|m_2 - 2m_1 t + 2tr - r^2|}{(r - t)^2} \tag{20.9}$$

for all $M \in conv T_+^{\uparrow} \cup R_+^{\uparrow} = \overline{T_+ R_+}^{\uparrow}$. Representing M as a point of the plane containing $\overline{T_+ R_+}^{\uparrow}$, we get $m_2 = (r + t)(m_1 - r) + r^2$, which enables us to express t in terms of m_1 and m_2 as

$$t = \frac{m_1 r - m_2}{r - m_1}.$$

This substituted into (20.9) yields

$$L_r(M) = \frac{(r - m_1)^2}{r^2 - 2m_1 r + m_2}. \tag{20.10}$$

Note that this is valid for all $M \in \mathcal{W}_2 = \bigcup_{0 \le t \le r} \overline{T_+ R_+}^{\uparrow}$.

The respective formula for $M \in \mathcal{W}_3$ is obtained by replacing m_1 by $-m_1$ in (20.10). This is justified by the fact that the arguments of the optimal ratio method are purely geometric, and both $\mathcal{W}$ and $\mathcal{W}_5$ are symmetric about the plane $m_1 = 0$.

Consider now a plane $\mathcal{H}$ that touches the bottom side of $\mathcal{W}$ along $\overline{T_- T_+}$ for some $0 < t < r$. This is defined by the formula $\mathcal{H} : m_4 - 2t^2 m_2 + t^4 = 0$. Then $\mathcal{H}' : m_4 - 2t^2 m_2 - r^4 + 2t^2 r^2 = 0$ is the closest parallel hyperplane to $\mathcal{H}$ that supports $\mathcal{W}_5$ along $\overline{R_- R_+}$. Arguments similar to those applied in the analysis of the side hyperplanes produce

$$d(M, \mathcal{H}') = |m_4 - 2m_2 t^2 + 2t^2 r^2 - r^4| / (1 + 4t^4)^{1/2}, \tag{20.11}$$

$$d(\mathcal{H}, \mathcal{H}') = (r^2 - t^2)^2 / (1 + 4t^4)^{1/2}, \tag{20.12}$$

where

$$t^2 = \frac{m_2 r^2 - m_4}{r^2 - m_2} \tag{20.13}$$

is determined from the equation $m_4 = (r^2 + t^2)(m_2 - r^2) + r^4$, defining the plane that contains both $\overline{T_- T_+}$ and $\overline{R_- R_+}$. Dividing (20.11) by (20.12) and substituting (20.13) for t^2 gives the penultimate formula in (20.7). Observe that this is valid for the moment points of the trapezoids $conv \overline{T_- T_+} \cup \overline{R_- R_+}$, $0 \le t \le r$, whose union forms $\mathcal{W}_4$. This ends the proof of Theorem 20.2. $\qquad\square$

20.3   Proof of Theorem 20.1.

We first confirm that fixing m_2 and m_4 we minimize $L_r(m_1, m_2, m_4)$ at $m_1 = 0$. Note that $L_r(M)$ for $M \in \mathcal{W}_1 \cup \mathcal{W}_4 \cup \mathcal{W}_5$ does not depend on the value of m_1, and $(m_1, m_2, m_4) \in \mathcal{W}_i$ implies $(0, m_2, m_4) \in \mathcal{W}_i$, $i = 1, 4, 5$. Differentiating the second formula of (20.7) with respect to $|m_1|$, we derive

$$\frac{\partial L_r(M)}{\partial |m_1|} = \frac{2(r - |m_1|)(|m_1| r - m_2)}{(r^2 - 2|m_1| r + m_2)^2},$$

which is nonnegative for $M \in \mathcal{W}_2 \cup \mathcal{W}_3$, because $|m_1| \le r$ and $m_2 \le |m_1| r$ there. Therefore we decrease $L_r(M)$ moving $M \in \mathcal{W}_2 \cup \mathcal{W}_3$ perpendicularly towards the plane $m_1 = 0$ until we reach the border. Then we can move further entering either $\mathcal{W}_1$ or $\mathcal{W}_4$ that would not result in change of $L_r(M)$ until we finally arrive at $(0, m_2, m_4)$.

Evaluating (20.5) we can therefore concentrate on the moment points from the rectangular

$$\mathcal{R}_0 = \{(0, m_2, m_4) : \; m_i \le \epsilon_i, \; i = 2, 4\}. \tag{20.14}$$

The points of $\mathcal{R}_0 \cap \mathcal{W}$ may generally belong to any of $\mathcal{W}_1, \mathcal{W}_4$ and $\mathcal{W}_5$. However, if some $M \in \mathcal{R}_0 \cap \mathcal{W}_5$, which is possible when $\epsilon_2 \geq r^2$ and $\epsilon_4 \geq r^4$, then $L_r(M) = L_r(\mathcal{E}) = 0$, which is useless in determining (20.6). Otherwise the moment points of $\mathcal{R}_0$ belong to either $\mathcal{W}_1$ or $\mathcal{W}_4$. In the former case L_r is evidently decreasing in m_2 and does not depend on m_4 (20.7). In the latter, L_r is decreasing in m_4, and increasing in m_2, because $m_2 \leq r^2$, $m_4 \leq m_2 r^2$, and so

$$\frac{\partial L_r(M)}{\partial m_2} = \frac{2(r^2 - m_2)(m_2 r^2 - m_4)}{(r^4 - 2m_2 r^2 + m_4)^2} \geq 0$$

for $M \in \mathcal{R}_0 \cap \mathcal{W}_4$.

We now claim that L_r is minimized on (14) at $E = (0, \epsilon, \epsilon r^2)$ with $\epsilon = \min\{\epsilon_2, \epsilon_4/r^2\}$ so that

$$L_r(\mathcal{E}) = L_r(E) = 1 - \epsilon^2/r^2. \tag{20.15}$$

The last equation follows from the fact that $E \in \mathcal{W}_1$. We prove the former using the following arguments. First see that if $\epsilon_4 > \epsilon_2 r^2$, we can exclude from considerations all points situated above $\overline{E_0 E}$ for $E_0 = (0, 0, \epsilon r^2)$. Indeed, any point $M = (0, m_2, m_4)$ of this area can be replaced by $M' = (0, m_2, \epsilon r^2) \in \overline{E_0 E}$ so that $L_r(M') = L_r(M)$. Then we exclude all points of $\mathcal{R}_0$ below $\overline{0E}$, which belong to $\mathcal{W}_4$. Keeping m_4 fixed and decreasing m_2 until we reach $\overline{0E}$, we actually decrease L_r. What still remains to analyze is $\triangle 0 E_0 E$. Without increase of L_r we vertically raise all $M \in \triangle 0 E_0 E$ to the level $\overline{E_0 E}$, and finally move them right to E which results in decreasing L_r.

Now we are only left with the task of determining (20.6) which, by (20.15), consists in solving the equation $1 - \epsilon^2/r^2 = 1 - r$, or, equivalently,

$$\min\{r^2 \epsilon_2, \epsilon_4\} = r^5. \tag{20.16}$$

If $\epsilon_4^{1/5} \leq \epsilon_2^{1/3}$ then the graphs of both sides of (20.16) cross each other at level ϵ_4, and the solution is $\epsilon_4^{1/5}$. Otherwise they meet below ϵ_4 for $r = \epsilon_2^{1/3}$. These conclusions establish the assertion of Theorem 20.1. $\qquad\square$

20.4　Concluding Remarks

A natural extension of the above problem consists in analyzing distributions tending to points different from zero. However, by reference to Anastassiou and Rychlik (1999),[86], in this case one can hardly expect obtaining final results in form of nice explicit formulae. Another question of interest is the Prokhorov radius described by other moments. Also, one can replace Tchebycheff systems of specific powers by elements of general families of functions, e.g. convex and symmetric ones. A next step of this study is determining radii of classes described by moment conditions in other metrices which induce the topology of weak convergence (see Anastassiou (1987),[18], Anastassiou and Rachev (1992),[83]). Comparing rates of convergence of radii of given classes of measures in various metrices would shed some new light on mutual relations of the metrices.

Chapter 21

Precise Rates of Prokhorov Convergence Using Three Moment Conditions

In this chapter we consider families of life distributions with the first three moments belonging to small neighborhoods of respective powers of a positive number.For various shapes of the neighborhoods, we determine exact convergence rates of their Prokhorov radii to zero. This gives a refined evaluation of the effect of specific moment convergence on the weak one. This treatment relies on [90].

21.1 Main Result

We start with mentioning the notion of the Prokhorov distance π of two Borel probability measures μ, ν defined on a Polish space with a metric ρ:

$$\pi(\mu, \nu) = \inf\{r > 0 : \mu(\mathcal{A}) \leq \nu(\mathcal{A}^r) + r, \ \nu(\mathcal{A}) \leq \mu(\mathcal{A}^r) + r \quad \text{for all closed sets } \mathcal{A}\}, \tag{21.1}$$

where $\mathcal{A}^r = \{x : \ \rho(x, \mathcal{A}) < \epsilon\}$. The Prokhorov distance generates the weak topology of probability measures. In particular, if we consider probability measures on the real axis with the standard Euclidean metric and $\nu = \delta_a$ is a Dirac measure concentrated at point a, then (21.1) takes on the form

$$\pi(\mu, \delta_a) = \inf\{r > 0 : \ \mu(I_r) \geq 1 - r\}$$

with $I_r = [a - r, a + r]$. For fixed positive a and ϵ_i, $i = 1, 2, 3$, we consider the family $\mathcal{M}(\mathcal{E})$, $\mathcal{E} = (\epsilon_1, \epsilon_2, \epsilon_3)$, of all probability measures supported on $[0, +\infty)$ with finite first, second and third moments $m_i = \int t^i \, \mu(dt)$, $i = 1, 2, 3$, respectively, such that $|m_i - a^i| \leq \epsilon_i$, $i = 1, 2, 3$. The probability distributions concentrated on the positive half line are called the life distributions. The first standard moments provide a meaningful parametric description of statistical properties of distributions: location, dispersion and skewness, respectively. Note that functions t^i, $i = 1, 2, 3$, form a Chebyshev system on $[0, +\infty)$. The Chebyshev systems are of considerable importance in the approximation theory, because they allow one to estimate integrals of arbitrary continuous functions by means of respective expectations of the elements of the system (see Karlin and Studden (1966),[223]). The objective of this chapter is to establish the exact evaluation of the Prokhorov radius

$$\Pi(\varepsilon) = \sup_{\mu \in \mathcal{M}(\mathcal{E})} \pi(\mu, \delta_a) \tag{21.2}$$

of the parametric neighborhood $\mathcal{M}(\mathcal{E})$ of the degenerate measure δ_a. We are motivated by the problem of describing the rate of uniform weak convergence of measures with respect to that of its several moments. We therefore suppose that all ϵ_i are small in comparison with a. However, it is worth pointing out that our results are not asymptotic and we derive precise value of (21.2) for fixed nonzero ϵ_i. Presentation of some asymptotic approximations will follow Theorem 21.1 containing the main result. Note that $\epsilon_i \to 0$, $i = 1, 2$, implies that the respective measure converges weakly to δ_a. Anastassiou (1992),[19], calculated the Prokhorov radius of the class of measures satisfying two moment bounds $|m_i - a^i| \le \epsilon_i$, $i = 1, 2$. The Levy and Kantorovich radii of the class were determined in Anastassiou (1987),[18] and Anastassiou and Rachev (1992),[83], respectively. In the last, the Prokhorov radius problem for the Chebyshev system $\{\sin t, \cos t\}$ on $[0, 2\pi]$ was also solved. Anastassiou and Rychlik (1997),[88], analyzed the Prokhorov radius of classes of measures surrounding zero satisfying first, second and fourth moments conditions. Below we give a refinement of the result by Anastassiou (1992),[19].

Theorem 21.1. *If* $0 < \Pi(\mathcal{E}) < a^* = \min\{a, 1\}$, *then it can be determined by means of one of the following algorithms:*
(i) If

$$2a\epsilon_1 + \epsilon_2 < 1, \tag{21.3}$$

$$\epsilon_3 \ge 3a^2\epsilon_1 + 3a\epsilon_2 - (2a\epsilon_1 + \epsilon_2)^{2/3}\epsilon_1, \tag{21.4}$$

then

$$\Pi(\mathcal{E}) = (2a\epsilon_1 + \epsilon_2)^{1/3}. \tag{21.5}$$

(ii) Assume that there is $r = r(a, \epsilon_1, \epsilon_3) \in (0, a^*)$ *such that*

$$(3a^2 + r^2)\epsilon_1 + \epsilon_3 = ar^2(1 + 2r) - 2(1 - r)[(a^2 + r^2/3)^{3/2} - a^3]. \tag{21.6}$$

If

$$|\epsilon_1 + (1 - r)[(a^2 + r^2/3)^{1/2} - a]| \le r^2, \tag{21.7}$$

$$\epsilon_2 \ge |2a\epsilon_1 + 2a(1 - r)[(a^2 + r^2/3)^{1/2} - a] - (1 + 2r)r^2/3|, \tag{21.8}$$

then $\Pi(\mathcal{E}) = r(a, \epsilon_1, \epsilon_3)$.
(iii) Assume that there is $r = r(a, \epsilon_1, \epsilon_3) \in (0, a^*)$ *such that*

$$(a - ar + r^2 - \epsilon_1)^3 = (1 - r)^2[a^3 - r(a - r)^3 + \epsilon_3]. \tag{21.9}$$

If

$$r^2 - (1 - r)[(a^2 + r^2/3)^{1/2} - a] \le \epsilon_1 \le r^2, \tag{21.10}$$

$$\epsilon_2 \ge |(a - ra + r^2 - \epsilon_1)/(1 - r) + r(a - r)^2 - a^2|, \tag{21.11}$$

then $\Pi(\mathcal{E}) = r(a, \epsilon_1, \epsilon_3)$.

(iv)Assume that there is $r = r(a, \epsilon_2, \epsilon_3) \in (0, a^*)$ *such that*

$$(3a^2 + r^2)\epsilon_2 + 2a\epsilon_3 = 3a^2 r^2 - (1 + 2r)r^4/3 - (1 - r)r^6/(27a^2). \qquad (21.12)$$

If

$$|\epsilon_2 + (2 + r)r^2/3 + (1 - r)r^4/(9a^2)| \leq 2ar^2, \qquad (21.13)$$

$$2a\epsilon_1 \geq \epsilon_2 + r^3 + (1 - r)r^4/(9a^2), \qquad (21.14)$$

then $\Pi(\mathcal{E}) = r(a, \epsilon_2, \epsilon_3)$.

(v) Assume that there is $r = r(a, \epsilon_2, \epsilon_3) \in (0, a^*)$ *such that*

$$[a - r(a - r)^2 - \epsilon_2]^2 = (1 - r)[a^3 - r(a - r)^3 + \epsilon_3]^2. \qquad (21.15)$$

If

$$2ar^2 - (2 + r)r^2/3 - r^4/(9a^2) \leq \epsilon_2 \leq 2ar^2 - r^3, \qquad (21.16)$$

$$\epsilon_1 \geq a - ra + r^2 - (1 - r)^{1/2}[a^2 - r(a - r)^2 - \epsilon_2]^{1/2}, \qquad (21.17)$$

then $\Pi(\mathcal{E}) = r(a, \epsilon_2, \epsilon_3)$.

(vi)Assume that there is $r = r(a, \epsilon_1, \epsilon_2, \epsilon_3) \in (0, a^*)$ *such that*

$$(1 - r)^{1/2}[(3a^2 - r^2)\epsilon_1 + 3a\epsilon_2 - \epsilon_3] = (r^2 - 2a\epsilon_1 - \epsilon_2)(2a\epsilon_1 + \epsilon_2 - r^3)^{1/2}. \qquad (21.18)$$

If

$$2a\epsilon_1 + \epsilon_2 \leq r^2, \qquad (21.19)$$

$$[a^2 - r^2 + 2a(a^2 + r^2/3)^{1/2}]\epsilon_1 + [2a + (a^2 + r^2/3)^{1/2}]\epsilon_2$$
$$-[(a^2 + r^2/3)^{1/2} - a]r^2 \leq \epsilon_3 \leq (3a^2 - r^2)\epsilon_1 + 3a\epsilon_2, \qquad (21.20)$$

then $\Pi(\mathcal{E}) = r(a, \epsilon_1, \epsilon_2, \epsilon_3)$.

(vii)Assume that there is $r = r(a, \epsilon_1, \epsilon_2, \epsilon_3) \in (0, a^*)$ *such that*

$$(1 - r)^{1/2}[(3a^2 - r^2)\epsilon_1 - 3a\epsilon_2 - \epsilon_3] = [r^2 - 2a\epsilon_1 + \epsilon_2](2a\epsilon_1 - \epsilon_2 - r^3)^{1/2}. \qquad (21.21)$$

If

$$2a\epsilon_1 - r^2 \leq \epsilon_2$$
$$\leq [(a^2 + r^2/3)^{1/2} + a - r]\epsilon_1 - [(a^2 + r^2/3)^{1/2} - a]r, \qquad (21.22)$$
$$(3a^2 - r^2/3)\epsilon_1 + [3a + r^2/(3a)]\epsilon_2 - r^4/(3a)$$
$$\leq \epsilon_3 \leq [a^2 - r^2 + 2a(a^2 + r^2/3)^{1/2}]\epsilon_1 - [2a + (a^2 + r^2/3)^{1/2}]\epsilon_2$$
$$-[(a^2 + r^2/3)^{1/2} - a]r^2, \qquad (21.23)$$

then $\Pi(\mathcal{E}) = r(a, \epsilon_1, \epsilon_2, \epsilon_3)$.

We have only considered the Prokhorov radii that satisfy $\Pi(\mathcal{E}) \leq \min\{a, 1\}$ which is of interest in examining the rate of convergence. The conditions for $\Pi(\mathcal{E}) = 1 < a$ can be concluded from the proof of Theorem 21.1. The case $a \leq \Pi(\mathcal{E}) \leq 1$ can be analyzed by means of tools used in our proof. The solution has a simpler form than that of Theorem 21.1, but we do not include it here. Although our formulas for determining the Prokhorov radius look complicated at the first glance, they provide simple numerical algorithms. Case (i) has an explicit solution. In the remaining ones, one should determine solutions of polynomial equations (21.6), (21.9), (21.12), (21.15), (21.18) and (21.21) in r of degree from 6 in case (iii) to 9 in case (v) with fixed a and ϵ_i, $i = 1, 2, 3$, and check if they satisfy respective inequality constraints. If the Prokhorov radius is determined by one of the polynomial equations, there are no other solutions to the equations in $(0, a^*)$.

One can establish that (21.9) and (21.15) can be used in describing arbitrarily small Prokhorov radii which is of interest in evaluating convergence rates iff $a < 1/6$ and $1/3$, respectively. For $1/6 \leq a < 1 - 3\sqrt{3}/8$ and $1/3 \leq a < 37/64$, respectively, the formulae enable us to determine Prokhorov radii belonging to some right neighborhoods of a that are separated from 0. For larger a, (21.9) and (21.15) are useless. The remaining five cases allow us to determine $\Pi(\mathcal{E})$ in the whole possible range $(0, a^*)$ for arbitrary $a > 0$.

In comparison with analogous two moment problems, Theorem 21.1 reveals abundant representations of Prokhorov radius, depending on relations among ϵ_i, $i = 1, 2, 3$. Formula (21.5) coincides with the Prokhorov radius of the neighborhood of δ_a described by conditions on the first and second moments only (cf Anastassiou (1992),[19]). It follows that if the conditions on the third one are not very restrictive (see 21.4), then they are fulfilled by the measures that determine the Prokhorov radius in the two moment case. One can also confirm that equations (21.6), (21.9), and (21.12), (21.15) determine the Prokhorov radii for classes of measures with other choices of two moment conditions. The former two refer to the first and third moments case and the latter to that of the second and third moments, respectively. Again, (21.8), (21.11), and (21.14), (21.17) indicate that admitting sufficiently large deviations of the second and first moments, respectively, we obtain the Prokhorov radii depending only on the remaining pairs. One can deduce from our proof that cases (iii) and (v) refer to smaller values of ϵ_3/ϵ_1 than cases (ii) and (iv), respectively. Using (21.10) and (21.16) we can find some lower estimates $\Pi(\mathcal{E}) \geq \epsilon_1^{1/2}$ and $\Pi(\mathcal{E}) \geq (\epsilon_2/2a)^{1/2}$ for the Prokhorov radius determined by equations (21.9) and (21.16), respectively. In the last two cases the radius depends directly on all ϵ_i, $i = 1, 2, 3$, and the respective inequality constraints represent upper bounds for the range of each ϵ_i. We can find some explicit evaluations of $\Pi(\mathcal{E})$ there making use of the fact that all factors in (21.18) and (21.21) are positive. E.g., from (21.18) and (21.21) we conclude that

$$(2a\epsilon_1 + \epsilon_2)^{1/2} \leq \quad \Pi(\varepsilon) \quad \leq (2a\epsilon_1 + \epsilon_2)^{1/3}, \tag{21.24}$$

$$(2a\epsilon_1 - \epsilon_2)^{1/2} \leq \quad \Pi(\varepsilon) \quad \leq (2a\epsilon_1 - \epsilon_2)^{1/3}, \tag{21.25}$$

(cf (21.5)).

In cases (ii) and (iv) there are no simple estimates of $\Pi(\mathcal{E})$ in terms of ϵ_1 and ϵ_2 that could be compared with (21.5), (21.24) and (21.25), because either ϵ_2 or ϵ_1 can be arbitrarily large then. Nevertheless, assuming that the parameters ϵ_i are small and so is $\Pi(\mathcal{E})$, we can solve the approximates of (21.6) and (21.12) neglecting all powers of r except of the lowest ones. This produces

$$\Pi(\mathcal{E}) \approx 1 - \frac{3a^2\epsilon_1 + \epsilon_3}{4a^3},$$

$$\Pi(\mathcal{E}) \approx \left(\frac{3a^2\epsilon_2 + 2a\epsilon_3}{3a^2 - \epsilon_2}\right)^{1/2},$$

respectively. Analogous approximations of (21.9), (21.15), (21.18), and (21.21) derive

$$\Pi(\mathcal{E}) \approx \frac{a^3 + \epsilon_3 - (a - \epsilon_1)^3}{6a^2\epsilon_1 - 3a\epsilon_1^2 + 2\epsilon_3},$$

$$\Pi(\mathcal{E}) \approx \frac{(a^3 + \epsilon_3)^2 - (a^2 - \epsilon_2)^3}{6a^4\epsilon_2 - 3a^2\epsilon_2^2 + 4a^3\epsilon_3 - \epsilon_3^2},$$

$$\Pi(\mathcal{E}) \approx 1 - \frac{(2a\epsilon_1 + \epsilon_2)^3}{(3a^2\epsilon_1 + 3a\epsilon_2 - \epsilon_3)^2},$$

$$\Pi(\mathcal{E}) \approx 1 - \frac{(2a\epsilon_1 - \epsilon_2)^3}{(3a^2\epsilon_1 - 3a\epsilon_2 - \epsilon_3)^2},$$

respectively.

21.2   Outline of Proof

The proof of Theorem 21.1 is lengthy for full and complete details see,[86] and Chapter 22 next. Below we only sketch the main ideas. A general idea of proof of Theorem 21.1 comes from Anastassiou (1992),[19]. We first determine

$$L_r(M) = \inf\left\{\mu(I_r) : \int_0^\infty t^i\,\mu(dt) = m_i,\ i = 1, 2, 3\right\} \qquad (21.26)$$

for fixed $0 < r < a$ and all

$$M = (m_1, m_2, m_3) \in \mathcal{W} = \left\{M : \int_0^\infty t^i\,\mu(dt) = m_i,\ i = 1, 2, 3\right\}. \qquad (21.27)$$

Thus

$$\Pi(\mathcal{E}) = \inf\{r > 0 : \mathcal{L}_r(\mathcal{E}) \geq 1 - r\}, \qquad (21.28)$$

where

$$\mathcal{L}_r(\mathcal{E}) = \inf\{L_r(M) : M \in \mathcal{B}(\mathcal{E}) \cap \mathcal{W}\}, \qquad (21.29)$$

$$\mathcal{B}(\mathcal{E}) = \{M = (m_1, m_2, m_3) : |m_i - a^i| \leq \epsilon_i,\ i = 1, 2, 3\}. \qquad (21.30)$$

Further rectangular parallelepipeds (21.30) will be called shortly boxes. The advantage of the method is that having solved (21.26), we can further analyze function $L_r(M)$ of four real parameters rather than the original problem (21.2) with an infinitely dimensional domain. We shall see that the function is continuous although it is defined by eight different formulae on various parts of $\mathcal{W}$. In particular, it follows that the infimum in (21.29) is attained at some $M = M_{r\mathcal{E}}$ for every fixed r. Moreover, $M_{r\mathcal{E}}$ changes continuosly as we let r vary, which yields the continuity of

$$\mathcal{L}_r(\mathcal{E}) = L_r(M_{r\mathcal{E}}) \tag{21.31}$$

in r, and accordingly (21.28) is a solution to the equation

$$L_r(M_{r\mathcal{E}}) = \mathcal{L}_r(\mathcal{E}) = 1 - r. \tag{21.32}$$

The proof of Theorem 21.1 consists in determining all $M_{r\mathcal{E}}$ that satisfy both requirements of (21.32) for all $\mathcal{E}$ such that $0 < r = \Pi(\mathcal{E}) < a^*$. It occurs that the problem has seven types of solutions, depending on mutual relations among ϵ_i and r. These provide seven respective statements in Theorem 21.1. In Theorem 21.2 we solve auxiliary moment problem (21.26) for all M of the moment space (21.27) and arbitrary fixed $r < a$. In the proof we use arguments of the geometric moment theory developed in Kemperman (1968),[229], in a general setup and some results of Rychlik (1997),[314], useful in examining three moment problems. The conclusions of Theorem 21.2 are applied in Theorem 21.3 for determining all moment points such that (21.31) holds for some r and $\mathcal{E}$. Then the proof of Theorem 21.1 consists in selecting the points that additionally satisfy (21.32).

It is widely known that (21.27) has the explicit representation

$$\mathcal{W} = \{M = (m_1, m_2, m_3) : \; m_1 \geq 0, \; m_2 \geq m_1^2, \; m_1 m_3 \geq m_2^2\} \tag{21.33}$$

(cf e.g., Karlin and Studden (1966),[223]). Theorem 21.2 below states that (21.26) has eight different representations in different subregions of (21.33). We define the subregions using geometric notions. These are both more concise and illustrative than analytic ones that involve complicated inequality relations in m_i, $i = 1, 2, 3$. We first introduce some notation. We shall use small letters for points on the line and capital ones for those in the three-dimensional space, e.g., $M = (m_1, m_2, m_3) \in \mathbb{R}^3$. Pieces of the graph of function $t \mapsto (t, t^2, t^3)$, $t \geq 0$, will be denoted by $\widehat{SV} = \{T = (t, t^2, t^3) : \; s \leq t \leq v\}$. We distinguish some specific points of the curve: O, A, B, C, A_1, A_2, and D are generated by arguments 0, a, $b = a - r$, $c = a + r$, $a_1 = (a^2 + r^2/3)^{1/2}$, $a_2 = a + r^2/(3a)$ and $d = [2a + (a^2 + 3r^2)^{1/2}]/3$, respectively. We shall write $conv\{\cdot\}$ for the convex hull of a family of points and/or sets, using a simpler notation $\overline{ST} = conv\{S, T\}$ and $\triangle STV = conv\{S, T, V\}$ for line segments and triangles, respectively. The plane spanned by S, T, V will be written as $pl\{STV\}$. We shall also use a notion of membrane spanned by a point and a piece of curve

$$mem\{V, \widehat{SU}\} = \bigcup_{s \leq t \leq u} \overline{TV}.$$

Note that

$$conv\{V, W, \widehat{SU}\} = \bigcup_{s \leq t \leq u} \triangle TVW. \tag{21.34}$$

At last, for a set $\mathcal{A}$ we define the family $\mathcal{A}^{\uparrow}$ of points lying above $\mathcal{A}$ by

$$\mathcal{A}^{\uparrow} = \{(m_1, m_2, m_3 + t) : (m_1, m_2, m_3) \in \mathcal{A}, \; t \geq 0\}.$$

Observe that (21.33) is alternatively described by

$$\mathcal{W} = mem\{O, \widehat{O\infty}\}^{\uparrow} = conv\{\widehat{O\infty}\}. \tag{21.35}$$

We can now formulate the solution to our auxiliary moment problem (21.26). In some cases, when this is provided by a unique measure, we display the respective representations.

Theorem 21.2. *For fixed $0 < r < a$ and $M = (m_1, m_2, m_3) \in \mathcal{W}$, (21.26) can be explicitly written as follows.*

(i) If $M \in \mathcal{V}_r = mem\{O, \widehat{OB}^{\uparrow}\} \cup \Delta OBC^{\uparrow} \cup mem\{O, \widehat{C\infty}\}^{\uparrow}$, then

$$L_r(M) = 0. \tag{21.36}$$

(ii) If $M \in \Delta ABC^{\uparrow}$, then

$$L_r(M) = 1 - (m_2 - 2am_1 + a^2)/r^2. \tag{21.37}$$

(iii) If $M \in mem\{C, \widehat{AC}\}^{\uparrow}$, then

$$L_r(M) = \frac{(c - m_1)^2}{c^2 - 2cm_1 + m_2}. \tag{21.38}$$

(iv) If $M \in mem\{B, \widehat{BA}\}^{\uparrow}$, then

$$L_r(M) = \frac{(m_1 - b)^2}{m_2 - 2bm_1 + b^2}. \tag{21.39}$$

(v) If $M \in conv\{O, B, D, C\}$, then

$$L_r(M) = \frac{-m_3 + (b + c)m_2 - bcm_1}{d(d - b)(c - d)}. \tag{21.40}$$

This is attained by a unique measure supported on $0, b, d$, and c.
(vi) If $M \in conv\{O, C, \widehat{DC}\}$, then

$$L_r(M) = \frac{cm_1 - m_2}{t(c - t)} = \frac{(cm_1 - m_2)^3}{(cm_2 - m_3)(c^2 m_1 - 2cm_2 + m_3)}. \tag{21.41}$$

This is attained by a unique measure supported on $0, c$, and

$$t = \frac{cm_2 - m_3}{cm_1 - m_2} \in [d, c]. \tag{21.42}$$

(vii) If $M \in conv\{O, B, \widehat{BD}\}$, then

$$L_r(M) = \frac{m_2 - bm_1}{t(t - b)} = \frac{(m_2 - bm_1)^3}{(m_3 - bm_2)(m_3 - 2bm_2 + b^2 m_1)}. \tag{21.43}$$

This is attained by a unique measure supported on $0, b$, and

$$t = \frac{m_3 - bm_2}{m_2 - bm_1} \in [b, d]. \tag{21.44}$$

(viii) If $M \in conv\{B, C, \widehat{AD}\}$, then

$$L_r(M) = \frac{-m_2 + (b+c)m_1 - bc}{(t-b)(c-t)}$$

$$= \frac{(-m_2 + (b+c)m_1 - bc)^3}{[-m_3 + (2b+c)m_2 - b(b+2c)m_1 + b^2 c]}$$

$$\times \frac{1}{[m_3 - (b+2c)m_2 + c(2b+c)m_1 - bc^2]}. \tag{21.45}$$

This is attained by a unique measure supported on b, c, and

$$t = \frac{-m_3 + (b+c)m_2 - bcm_1}{-m_2 + (b+c)m_1 - bc} \in [a, d]. \tag{21.46}$$

One can see that for every $0 < r < a^*$ there exists M such that $L_r(M) = 1 - r$ (cf (21.32)) in each region of the partition of the moment space presented in Theorem 21.2, except of the first one. E.g., the equation holds true for $M \in \Delta ABC^{\uparrow}$ iff $M \in \overline{B_r^A C_r^A}^{\uparrow}$ with $B_r^A = rB + (1-r)A$ and $C_r^A = rC + (1-r)A$. In $conv\{B, C, \widehat{AD}\}$, this is fulfilled by $M \in \bigcup_{a \leq t \leq d} \overline{B_r^T C_r^T}$.

In Theorem 21.3 we describe all moment pints $M_{r\mathcal{E}}$ at which the minimal positive value of function (21.26) over boxes $\mathcal{B}(\mathcal{E})$ is attained for arbitrary $0 < r < a^*$ and $\epsilon_i > 0$, $i = 1, 2, 3$. Let $V_1 = (a - \epsilon_1, a^2 + \epsilon_2, a^3 + \epsilon_3)$, $V_2 = (a - \epsilon_1, a^2 - \epsilon_2, a^3 + \epsilon_3)$, $V_3 = (a + \epsilon_1, a^2 - \epsilon_2, a^3 + \epsilon_3)$ denote three of vertices of the top rectangular side $\mathcal{R} = \{(m_1, m_2, a^3 + \epsilon_3) : |m_i - a^i| \leq \epsilon_i, \ i = 1, 2\}$ of $\mathcal{B}(\mathcal{E})$.

Theorem 21.3. *For all $0 < r < a^*$ and each of the following cases:*

$$M_{r\mathcal{E}} = V_1 \in \Delta ABC^{\uparrow}, \tag{21.47}$$

$$\{M_{r\mathcal{E}}\} = \overline{V_1 V_2} \cap \Delta BA_1 C, \tag{21.48}$$

$$\{M_{r\mathcal{E}}\} = \overline{V_1 V_2} \cap mem\{B, \widehat{AA_1}\}, \tag{21.49}$$

$$\{M_{r\mathcal{E}}\} = \overline{V_2 V_3} \cap \Delta BA_2 C, \tag{21.50}$$

$$\{M_{r\mathcal{E}}\} = \overline{V_2 V_3} \cap mem\{B, \widehat{AA_2}\}, \tag{21.51}$$

$$M_{r\mathcal{E}} = V_1 \in conv\{B, C, \widehat{AA_1}\}, \tag{21.52}$$

$$M_{r\mathcal{E}} = V_2 \in conv\{B, C, \widehat{AA_2}\}, \tag{21.53}$$

there exist $\epsilon_i > 0$, $i = 1, 2, 3$ such that

$$\mathcal{L}_r(\mathcal{E}) = L_r(M_{r\mathcal{E}}) > 0. \tag{21.54}$$

The proof of Theorem 21.1 relies on the results of Theorem 21.3. We consider various forms (21.47)- (21.53)of moment points solving (21.31) and indicate ones that satisfy (21.32) for some $r < a^*$. Solutions to (21.32) with respect to r derive values of Prokhorov radius for various combinations of ϵ_i, $i = 1, 2, 3$.

Chapter 22

On Prokhorov Convergence of Probability Measures to the Unit under Three Moments

In this chapter for an arbitrary positive number, we consider the family of probability measures supported on the positive halfaxis with the first three moments belonging to small neighborhoods of respective powers of the number. We present exact relations between the rate of uniform convergence of the moments and that of the Prokhorov radius of the family to the respective Dirac measure, dependent on the shape of the moment neighborhoods. This treatment follows [86].

22.1   Main Result

We start with by mentioning the notion of the Prokhorov distance π of two Borel probability measures μ, ν defined on a Polish space with a metric ρ:

$$\pi(\mu, \nu) =$$

$$\inf\left\{r > 0 : \mu(\mathcal{A}) \leq \nu(\mathcal{A}^r) + r, \ \nu(\mathcal{A}) \leq \mu(\mathcal{A}^r) + r \ \text{for all closed sets } \mathcal{A}\right\}, \quad (22.1)$$

where $\mathcal{A}^r = \{x : \rho(x, \mathcal{A}) < \epsilon\}$. The Prokhorov distance generates the weak topology of probability measures. In particular, if we consider probability measures on the real axis with the standard Euclidean metric and $\nu = \delta_a$ is a Dirac measure concentrated at point a, then (22.1) takes on the form

$$\pi(\mu, \delta_a) = \inf\left\{r > 0 : \mu(I_r) \geq 1 - r\right\},$$

with $I_r = [a - r, a + r]$. For fixed positive a and ϵ_i, $i = 1, 2, 3$, we consider the family $\mathcal{M}(\mathcal{E})$, $\mathcal{E} = (\epsilon_1, \epsilon_2, \epsilon_3)$, of all probability measures supported on $[0, +\infty)$ with finite first, second, and third moments $m_i = \int t^i \mu(dt)$, $i = 1, 2, 3$, respectively, such that $|m_i - a^i| \leq \epsilon_i$, $i = 1, 2, 3$. The probability distributions concentrated on the positive halfline are called the life distributions. The first standard moments provide a meaningful parametric description of statistical properties of distributions: location, dispersion, and skewness, respectively. Note that functions t^i, $i = 1, 2, 3$, form a Chebyshev system on $[0, +\infty)$. The Chebyshev systems are of considerable importance in the approximation theory, because they allow one to estimate integrals of arbitrary continuous functions by means of respective expectations of the elements of the system (see [223]). The objective of this chapter is to give the exact evaluation of the Prokhorov radius

$$\Pi(\mathcal{E}) = \sup_{\mu \in \mathcal{M}(\mathcal{E})} \pi(\mu, \delta_a) \tag{22.2}$$

of the parametric neighborhood $\mathcal{M}(\mathcal{E})$ of the degenerate measure δ_a. We are inspired by the problem of describing the rate of uniform weak convergence of measures with respect to that of its several moments. We therefore, suppose that all ϵ_i are small in comparison with a. However, it is worth pointing out that the results are not asymptotic and we derive precise value of (22.2) for fixed nonzero ϵ_i. Presentation of some asymptotic approximations will follow Theorem 22.1 containing the main result. Note that $\epsilon_i \to 0$, $i = 1, 2$, implies that the respective measure converges weakly to δ_a. Anastassiou [19] calculated the Prokhorov radius of the class of measures satisfying two moment bounds $|m_i - a^i| \leq \epsilon_i$, $i = 1, 2$. The Levy and Kantorovich radii of the class were determined in [18,83], respectively. In the latter, the Prokhorov radius problem for the Chebyshev system $\{\sin t, \cos t\}$ on $[0, 2\pi]$ was also solved. Anastassiou and Rychlik [88] analyzed the Prokhorov radius of classes of measures surrounding zero satisfying first, second, and fourth moments conditions. Below we give a refinement of the result by Anastassiou [19].

Theorem 22.1. *If $0 < \Pi(\mathcal{E}) < a^* = \min\{a, 1\}$, then it can be determined by means of one of the following algorithms.*

(i) If

$$2a\epsilon_1 + \epsilon_2 < 1, \tag{22.3}$$

$$\epsilon_3 \geq 3a^2\epsilon_1 + 3a\epsilon_2 - (2a\epsilon_1 + \epsilon_2)^{2/3}\epsilon_1, \tag{22.4}$$

then

$$\Pi(\mathcal{E}) = (2a\epsilon_1 + \epsilon_2)^{1/3}. \tag{22.5}$$

(ii) Assume that there is $r = r(a, \epsilon_1, \epsilon_3) \in (0, a^*)$ such that

$$\left(3a^2 + r^2\right)\epsilon_1 + \epsilon_3 = ar^2(1 + 2r) - 2(1 - r)\left[\left(a^2 + \frac{r^2}{3}\right)^{3/2} - a^3\right]. \tag{22.6}$$

If

$$\left|\epsilon_1 + (1 - r)\left[\left(a^2 + \frac{r^2}{3}\right)^{1/2} - a\right]\right| \leq r^2, \tag{22.7}$$

$$\epsilon_2 \geq \left|2a\epsilon_1 + 2a(1 - r)\left[\left(a^2 + \frac{r^2}{3}\right)^{1/2} - a\right] - (1 + 2r)\frac{r^2}{3}\right|, \tag{22.8}$$

then $\Pi(\mathcal{E}) = r(a, \epsilon_1, \epsilon_3)$.

(iii) Assume that there is $r = r(a, \epsilon_1, \epsilon_3) \in (0, a^*)$ such that

$$\left(a - ar + r^2 - \epsilon_1\right)^3 = (1 - r)^2\left[a^3 - r(a - r)^3 + \epsilon_3\right]. \tag{22.9}$$

If

$$r^2 - (1 - r) \left[\left(a^2 + \frac{r^2}{3} \right)^{1/2} - a \right] \leq \epsilon_1 \leq r^2, \tag{22.10}$$

$$\epsilon_2 \geq \left| \frac{(a - ra + r^2 - \epsilon_1)}{(1 - r)} + r(a - r)^2 - a^2 \right|, \tag{22.11}$$

then $\Pi(\mathcal{E}) = r(a, \epsilon_1, \epsilon_3)$.

(iv) Assume that there is $r = r(a, \epsilon_2, \epsilon_3) \in (0, a^*)$ *such that*

$$\left(3a^2 + r^2 \right) \epsilon_2 + 2a\epsilon_3 = 3a^2 r^2 - (1 + 2r)\frac{r^4}{3} - (1 - r)\frac{r^6}{(27a^2)}. \tag{22.12}$$

If

$$\left| \epsilon_2 + (2 + r)\frac{r^2}{3} + (1 - r)\frac{r^4}{(9a^2)} \right| \leq 2ar^2, \tag{22.13}$$

$$2a\epsilon_1 \geq \epsilon_2 + r^3 + (1 - r)\frac{r^4}{(9a^2)}, \tag{22.14}$$

then $\Pi(\mathcal{E}) = r(a, \epsilon_2, \epsilon_3)$.

(v) Assume that there is $r = r(a, \epsilon_2, \epsilon_3) \in (0, a^*)$ *such that*

$$\left[a - r(a - r)^2 - \epsilon_2 \right]^2 = (1 - r) \left[a^3 - r(a - r)^3 + \epsilon_3 \right]^2. \tag{22.15}$$

If

$$2ar^2 - (2 + r)\frac{r^2}{3} - \frac{r^4}{(9a^2)} \leq \epsilon_2 \leq 2ar^2 - r^3, \tag{22.16}$$

$$\epsilon_1 \geq a - ra + r^2 - (1 - r)^{1/2} \left[a^2 - r(a - r)^2 - \epsilon_2 \right]^{1/2}, \tag{22.17}$$

then $\Pi(\mathcal{E}) = r(a, \epsilon_2, \epsilon_3)$.

(vi) Assume that there is $r = r(a, \epsilon_1, \epsilon_2, \epsilon_3) \in (0, a^*)$ *such that*

$$(1 - r)^{1/2} \left[\left(3a^2 - r^2 \right) \epsilon_1 + 3a\epsilon_2 - \epsilon_3 \right] = \left(r^2 - 2a\epsilon_1 - \epsilon_2 \right) \left(2a\epsilon_1 + \epsilon_2 - r^3 \right)^{1/2}. \tag{22.18}$$

If

$$2a\epsilon_1 + \epsilon_2 \leq r^2, \tag{22.19}$$

$$\left[a^2 - r^2 + 2a \left(a^2 + \frac{r^2}{3} \right)^{1/2} \right] \epsilon_1 + \left[2a + \left(a^2 + \frac{r^2}{3} \right)^{1/2} \right] \epsilon_2$$

$$- \left[\left(a^2 + \frac{r^2}{3} \right)^{1/2} - a \right] r^2 \leq \epsilon_3 \leq \left(3a^2 - r^2 \right) \epsilon_1 + 3a\epsilon_2, \tag{22.20}$$

then $\Pi(\mathcal{E}) = r(a, \epsilon_1, \epsilon_2, \epsilon_3)$.

(vii) Assume that there is $r = r(a, \epsilon_1, \epsilon_2, \epsilon_3) \in (0, a^*)$ such that

$$(1 - r)^{1/2} \left[(3a^2 - r^2)\epsilon_1 - 3a\epsilon_2 - \epsilon_3 \right] = (r^2 - 2a\epsilon_1 + \epsilon_2)\left(2a\epsilon_1 - \epsilon_2 - r^3\right)^{1/2}. \tag{22.21}$$

If

$$2a\epsilon_1 - r^2 \leq \epsilon_2$$

$$\leq \left[\left(a^2 + \frac{r^2}{3}\right)^{1/2} + a - r \right] \epsilon_1 - \left[\left(a^2 + \frac{r^2}{3}\right)^{1/2} - a \right] r, \tag{22.22}$$

$$\left(3a^2 - \frac{r^2}{3}\right)\epsilon_1 + \left[3a + \frac{r^2}{(3a)}\right]\epsilon_2 - \frac{r^4}{(3a)}$$

$$\leq \epsilon_3 \leq \left[a^2 - r^2 + 2a\left(a^2 + \frac{r^2}{3}\right)^{1/2}\right]\epsilon_1 - \left[2a + \left(a^2 + \frac{r^2}{3}\right)^{1/2}\right]\epsilon_2$$

$$- \left[\left(a^2 + \frac{r^2}{3}\right)^{1/2} - a\right]r^2, \tag{22.23}$$

then $\Pi(\mathcal{E}) = r(a, \epsilon_1, \epsilon_2, \epsilon_3)$.

We have only considered the Prokhorov radii that satisfy $\Pi(\mathcal{E}) \leq \min\{a, 1\}$ which is of interest in examining the rate of convergence. The conditions for $\Pi(\mathcal{E}) = 1 < a$ can be concluded from the proof of Theorem 22.1. The case $a \leq \Pi(\mathcal{E}) \leq 1$ can be analyzed by means of tools used in our proof. The solution has a simpler form than that of Theorem 22.1, but we do not include it here. Although the formulas for determining the Prokhorov radius look complicated at the first glance, they provide simple numerical algorithms. Case (i) has an explicit solution. In the remaining ones, one should determine solutions of polynomial equations (22.6), (22.9), (22.12), (22.15), (22.18), and (22.21) in r of degree from 6 in Case (iii) to 9 in Case (v) with fixed a and ϵ_i, $i = 1, 2, 3$, and check if they satisfy respective inequality constraints. If the Prokhorov radius is determined by one of the polynomial equations, there are no other solutions to the equations in $(0, a^*)$.

In comparison with analogous two moment problems, Theorem 22.1 reveals abundant representations of Prokhorov radius, depending on relations among ϵ_i, $i = 1, 2, 3$. Formula (22.5) coincides with the Prokhorov radius of the neighborhood of δ_a described by conditions on the first and second moments only (cf. [19]). It follows that if the conditions on the third one are not very restrictive (see (22.4)), then they are fulfilled by the measures that determine the Prokhorov radius in the two moment case. One can also confirm that equations (22.6), (22.9) and (22.12), (22.15) determine the Prokhorov radii for classes of measures with other choices of two moment conditions. The former two refer to the first and third moments case and the latter to that of the second and third moments, respectively. Again, (22.8),

(22.11) and (22.14), (22.17) indicate that admitting sufficiently large deviations of the second and first moments, respectively, we obtain the Prokhorov radii depending only on the remaining pairs. One can deduce from our proof that Cases (iii) and (v) refer to smaller values of ϵ_3/ϵ_1 than Cases (ii) and (iv), respectively. Using (22.10) and (22.16) we can find some lower estimates $\Pi(\mathcal{E}) \geq \epsilon_1^{1/2}$ and $\Pi(\mathcal{E}) \geq (\epsilon_2/2a)^{1/2}$ for the Prokhorov radius determined by equations (22.9) and (22.16), respectively. In the last two cases, the radius depends directly on all $\epsilon_i, i = 1, 2, 3$, and the respective inequality constraints represent upper bounds for the range of each ϵ_i. We can get some explicit evaluations of $\Pi(\mathcal{E})$ there making use of the fact that all factors in (22.18) and (22.21) are positive. For example, from (22.18) and (22.21) we conclude that

$$(2a\epsilon_1 + \epsilon_2)^{1/2} \leq \Pi(\mathcal{E}) \leq (2a\epsilon_1 + \epsilon_2)^{1/3}, \qquad (22.24)$$

$$(2a\epsilon_1 - \epsilon_2)^{1/2} \leq \Pi(\mathcal{E}) \leq (2a\epsilon_1 - \epsilon_2)^{1/3} \qquad (22.25)$$

(cf. (22.5)).

In Cases (ii) and (iv) there are no simple estimates of $\Pi(\mathcal{E})$ in terms of ϵ_1 and ϵ_2 that could be compared with (22.5), (22.24) and (22.25), because either ϵ_2 or ϵ_1 can be arbitrarily large then. Nevertheless, assuming that the parameters ϵ_i are small and so is $\Pi(\mathcal{E})$, we can solve the approximates of (22.6) and (22.12) neglecting all powers of r except of the lowest ones. This yields

$$\Pi(\mathcal{E}) \approx 1 - \frac{3a^2\epsilon_1 + \epsilon_3}{4a^3},$$

$$\Pi(\mathcal{E}) \approx \left(\frac{3a^2\epsilon_2 + 2a\epsilon_3}{3a^2 - \epsilon_2} \right)^{1/2},$$

respectively. Analogous approximations of (22.9), (22.15), (22.18), and (22.21) give

$$\Pi(\mathcal{E}) \approx \frac{a^3 + \epsilon_3 - (a - \epsilon_1)^3}{6a^2\epsilon_1 - 3a\epsilon_1^2 + 2\epsilon_3},$$

$$\Pi(\mathcal{E}) \approx \frac{\left(a^3 + \epsilon_3\right)^2 - \left(a^2 - \epsilon_2\right)^3}{6a^4\epsilon_2 - 3a^2\epsilon_2^2 + 4a^3\epsilon_3 - \epsilon_3^2},$$

$$\Pi(\mathcal{E}) \approx 1 - \frac{(2a\epsilon_1 + \epsilon_2)^3}{(3a^2\epsilon_1 + 3a\epsilon_2 - \epsilon_3)^2},$$

$$\Pi(\mathcal{E}) \approx 1 - \frac{(2a\epsilon_1 - \epsilon_2)^3}{(3a^2\epsilon_1 - 3a\epsilon_2 - \epsilon_3)^2},$$

respectively.

A general idea of proof of Theorem 22.1 comes from [19]. We first find

$$L_r(M) = \inf \left\{ \mu(I_r) : \int_0^\infty t^i \, \mu\,(dt) = m_i, \ i = 1, 2, 3 \right\}, \qquad (22.26)$$

for fixed $0 < r < a$ and all

$$M = (m_1, m_2, m_3) \in \mathcal{W} = \left\{ M : \int_0^\infty t^i \, \mu \, (dt) = m_i, \ i = 1, 2, 3 \right\}. \qquad (22.27)$$

Then

$$\Pi(\mathcal{E}) = \inf \left\{ r > 0 : \mathcal{L}_r(\mathcal{E}) \geq 1 - r \right\}, \qquad (22.28)$$

where

$$\mathcal{L}_r(\mathcal{E}) = \inf \left\{ L_r(M) : M \in \mathcal{B}(\mathcal{E}) \cap \mathcal{W} \right\}, \qquad (22.29)$$

$$\mathcal{B}(\mathcal{E}) = \left\{ M = (m_1, m_2, m_3) : \left| m_i - a^i \right| \leq \epsilon_i, \ i = 1, 2, 3 \right\}. \qquad (22.30)$$

Further rectangular parallelepipeds (22.30) will be called shortly boxes. The advantage of the method is that having solved (22.26), we can further analyze function $L_r(M)$ of four real parameters rather than the original problem (22.2) with an infinitely-dimensional domain. We shall see that the function is continuous although it is defined by eight different formulae on various parts of $\mathcal{W}$. In particular, it follows that the infimum in (22.29) is attained at some $M = M_{r\mathcal{E}}$ for every fixed r. Moreover, $M_{r\mathcal{E}}$ changes continuously as we let r vary, which yields the continuity of

$$\mathcal{L}_r(\mathcal{E}) = L_r \left(M_{r\mathcal{E}} \right) \qquad (22.31)$$

in r, and accordingly (22.28) is a solution to the equation

$$L_r \left(M_{r\mathcal{E}} \right) = \mathcal{L}_r(\mathcal{E}) = 1 - r. \qquad (22.32)$$

The proof of Theorem 22.1 consists in determining all $M_{r\mathcal{E}}$ that satisfy both requirements of (22.32) for all $\mathcal{E}$ such that $0 < r = \Pi(\mathcal{E}) < a^*$. It occurs that the problem has seven types of solutions, depending on mutual relations among ϵ_i and r. These provide seven respective statements in Theorem 22.1.

In Section 22.2, we solve auxiliary moment problem (22.26) for all M of the moment space (22.27) and arbitrary fixed $r < a$. In Section 22.3, we analyze variability of the solution over different subregions of the moment space. The results are applied in Section 22.4 for determining all moment points such that (22.31) holds for some r and $\mathcal{E}$. In Section 22.5, we choose those that additionally satisfy (22.32).

22.2   An Auxiliary Moment Problem

It is widely known that (22.27) has the explicit representation

$$\mathcal{W} = \left\{ M = (m_1, m_2, m_3) : m_1 \geq 0, \ m_2 \geq m_1^2, \ m_1 m_3 \geq m_2^2 \right\} \qquad (22.33)$$

(cf., e.g., [314]). Theorem 22.2 below states that (22.26) has eight different representations in different subregions of (22.33). We define the subregions using geometric notions. These are both more concise and illustrative than analytic ones that involve complicated inequality relations in m_i, $i = 1, 2, 3$. We first introduce some notation. We shall use small letters for points on the line and capital ones for those in the three-dimensional space, e.g., $M = (m_1, m_2, m_3) \in \mathbb{R}^3$. Pieces of the graph of function $t \mapsto (t, t^2, t^3)$, $t \geq 0$, will be denoted by $\widehat{SV} = \{T = (t, t^2, t^3) : s \leq t \leq v\}$. We distinguish some specific points of the curve: O, A, B, C, A_1, A_2, and D are generated by arguments 0, a, $b = a - r$, $c = a + r$, $a_1 = (a^2 + r^2/3)^{1/2}$, $a_2 = a + r^2/(3a)$ and $d = [2a + (a^2 + 3r^2)^{1/2}]/3$, respectively. We shall write $\operatorname{conv}\{\cdot\}$ for the convex hull of a family of points and/or sets, using a simpler notation $\overline{ST} = \operatorname{conv}\{S, T\}$ and $\triangle STV = \operatorname{conv}\{S, T, V\}$ for line segments and triangles, respectively. The plane spanned by S, T, V will be written as $\operatorname{pl}\{STV\}$. We shall also apply a notion of membrane spanned by a point and a piece of curve

$$mem\left\{V, \widehat{SU}\right\} = \bigcup_{s \leq t \leq u} \overline{TV}.$$

Notice that

$$conv\left\{V, W, \widehat{SU}\right\} = \bigcup_{s \leq t \leq u} \triangle TVW. \tag{22.34}$$

Finally, for a set $\mathcal{A}$ we define the family $\mathcal{A}^\uparrow$ of points lying above $\mathcal{A}$ by

$$\mathcal{A}^\uparrow = \{(m_1, m_2, m_3 + t) : (m_1, m_2, m_3) \in \mathcal{A}, \, t \geq 0\}.$$

Observe that (22.33) is alternatively described by

$$\mathcal{W} = mem\{O, \widehat{O\infty}\}^\uparrow = conv\left\{\widehat{O\infty}\right\}. \tag{22.35}$$

We can now formulate the solution to our auxiliary moment problem (22.26). In some cases, when this is provided by a unique measure, we display the respective representations, because they will be further used in our study.

Theorem 22.2 *For fixed $0 < r < a$ and $M = (m_1, m_2, m_3) \in \mathcal{W}$, (22.26) can be explicitly written as follows.*

  (i) If $M \in \mathcal{V}_r = mem\{O, \widehat{OB}\}^\uparrow \cup \triangle OBC^\uparrow \cup mem\{O, \widehat{C\infty}\}^\uparrow$, then

$$L_r(M) = 0. \tag{22.36}$$

(ii) If $M \in \triangle ABC^\uparrow$, then

$$L_r(M) = 1 - \frac{\left(m_2 - 2am_1 + a^2\right)}{r^2}. \tag{22.37}$$

(iii) If $M \in mem\{C, \widehat{AC}\}^\uparrow$, then

$$L_r(M) = \frac{\left(c - m_1\right)^2}{c^2 - 2cm_1 + m_2}. \tag{22.38}$$

(iv) If $M \in mem\{B, \widehat{BA}\}^\uparrow$, then

$$L_r(M) = \frac{\left(m_1 - b\right)^2}{m_2 - 2bm_1 + b^2}. \tag{22.39}$$

(v) If $M \in conv\{O, B, D, C\}$, then

$$L_r(M) = \frac{-m_3 + (b+c)m_2 - bcm_1}{d(d-b)(c-d)}. \tag{22.40}$$

This is attained by a unique measure supported on 0, b, d, and c.

(vi) If $M \in conv\{O, C, \widehat{DC}\}$, then

$$L_r(M) = \frac{cm_1 - m_2}{t(c-t)} = \frac{(cm_1 - m_2)^3}{(cm_2 - m_3)\left(c^2m_1 - 2cm_2 + m_3\right)}. \tag{22.41}$$

This is attained by a unique measure supported on $0, c$, and

$$t = \frac{cm_2 - m_3}{cm_1 - m_2} \in [d, c]. \tag{22.42}$$

(vii) If $M \in conv\{O, B, \widehat{BD}\}$, then

$$L_r(M) = \frac{m_2 - bm_1}{t(t-b)} = \frac{(m_2 - bm_1)^3}{(m_3 - bm_2)\left(m_3 - 2bm_2 + b^2m_1\right)}. \tag{22.43}$$

This is attained by a unique measure supported on $0, b$, and

$$t = \frac{m_3 - bm_2}{m_2 - bm_1} \in [b, d]. \tag{22.44}$$

(viii) If $M \in conv\{B, C, \widehat{AD}\}$, then

$$L_r(M) = \frac{-m_2 + (b+c)m_1 - bc}{(t-b)(c-t)}$$

$$= \frac{(-m_2 + (b+c)\,m_1 - bc)^3}{\left[-m_3 + (2b+c)m_2 - b(b+2c)m_1 + b^2c\right]}$$

$$\times \frac{1}{\left[m_3 - (b+2c)m_2 + c(2b+c)m_1 - bc^2\right]}. \tag{22.45}$$

This is attained by a unique measure supported on b, c, and

$$t = \frac{-m_3 + (b+c)m_2 - bcm_1}{-m_2 + (b+c)m_1 - bc} \in [a, d]. \tag{22.46}$$

In order to simplify further analysis of (22.37)–(22.46), we represented them so that all factors of products and quotients are nonnegative there. Also, we defined the subregions of the moment space as its closed subsets. One can confirm that the formulae for $L_r(M)$ for neighboring subregions coincide on their common borders. Because each formula is continuous in M in the respective subregion, so is $L_r(M)$ in the whole domain. Observe also that letting r vary, we change continuously the shapes of subregions. Combining this fact with the continuity of each above formulae with respect to r, we conclude that $L_r(M)$ is continuous in r as well.

The proof is based on the geometric moment theory developed in [229] in a general setup. The theory provides tools for determining extreme integrals of a criterion function over the class of probability measures such that the respective integrals of some other functions satisfy some equality constraints. Kemperman [229] used the fact that each measure with given expectations of some functions has a discrete counterpart with the same expectations such that its support contains at most one more point than the number of conditions, and replaced the original problem by analyzing the images of integrals which coincide with convex hulls of the respective integrands. E.g., in view of the theory the latter representation in (22.35) is evident. For the special case of indicator criterion function that we examine here, Kemperman [229] described an optimal ratio method for solving respective moment problems. A version of the method directly applicable in the problem is presented in Lemma 22.3. It is obvious that $L_r(M) = 0$ iff $M \in \mathcal{V}_r$ which denotes here the closure of the family of moment points $conv\{\widehat{OB}, \widehat{C\infty}\}$ for the measures supported on the complement of I_r. The optimal ratio method allows to determine $L_r(M)$ for other elements of the moment space (22.27).

Lemma 22.3. *Let $\mathcal{H}$ be a hyperplane supporting the closure of the moment space and let $\mathcal{H}'$ be the closest parallel hyperplane to $\mathcal{H}$ supporting $\mathcal{V}_r$. Then for all $M \in conv\{\mathcal{W} \cap \mathcal{H}, \mathcal{V}_r \cap \mathcal{H}'\}$, yields*

$$L_r(M) = \frac{\rho(\mathcal{H}', M)}{\rho(\mathcal{H}', \mathcal{H})}, \tag{22.47}$$

where the numerator and denominator stand for the distances of hyperplane $\mathcal{H}'$ from moment point M and hyperplane $\mathcal{H}$, respectively.

Since explicit determining the supporting hyperplanes poses serious problems especially in multidimensional spaces, we also apply some other tools useful for studying three moment problems. Lemmas 22.4 and 22.5 give adaptations of auxiliary results of Rychlik [314] to our problem.

Lemma 22.4. *Let $0 \leq t_1 < t_2 < t_3 < t_4$. Then $T_i = (t_i, t_i^2, t_i^3)$ lies below (above) the plane spanned by T_j, $j \neq i$, iff $i = 1$ or 3 (2 or 4, respectively).*

Lemma 22.5. *A measure attaining $L_r(M)$ for a moment point M of an open set $\mathcal{W}' \subset \mathcal{W}$, $\mathcal{W}' \cap \widehat{O\infty} = \emptyset$, is a mixture of measures attaining $L_r(M_i)$ for some moment points M_i of the border of $\mathcal{W}'$. Accordingly, the subset of support points of the measures attaining $L_r(M)$ for border moment points contains that for the inner ones.*

In the rest, we shall repeatedly use analytic representations of moment points as combinations of image points $T_i \in \widehat{O\infty}$. A four-point representation $M = \sum_{i=1}^{4} \alpha_i T_i$ has the coefficients

$$\alpha_i = \frac{m_3 - m_2 \sum_{j \neq i} t_j + m_1 \sum_{j \neq i \neq k} t_j t_k - \prod_{j \neq i} t_j}{\prod_{j \neq i} (t_i - t_j)}, \qquad 1 \leq i, j, k \leq 4. \tag{22.48}$$

Also, $M = \sum\limits_{i=1}^{3} \alpha_i T_i \in \mathrm{pl}\{T_1, T_2, T_3\}$ with coefficients

$$\alpha_i = \frac{m_2 - m_1 \sum\limits_{j \neq i} t_j + \prod\limits_{j \neq i} t_j}{\prod\limits_{j \neq i} (t_i - t_j)}, \qquad 1 \leq i, j \leq 3, \tag{22.49}$$

iff it satisfies the equation

$$m_3 - m_2 (t_1 + t_2 + t_3) + m_1 (t_1 t_2 + t_1 t_3 + t_2 t_3) - t_1 t_2 t_3 = 0. \tag{22.50}$$

If the left-hand side of (22.50) is less or greater than 0, then M lies below or above the plane. Generally, formulae (22.48) and (22.49) represent coefficients of linear combinations. If all $\alpha_i \geq 0$, then the combinations are convex.

Proof of Theorem 22.2. We first determine $\mathcal{V}_r = \mathrm{conv}\{\widehat{OB}, \widehat{C\infty}\}$, where L_r vanishes. We claim that

$$\mathcal{V}_r = mem\{O, \widehat{OB}\}^\uparrow \cup \triangle OBC^\uparrow \cup mem\{O, \widehat{C\infty}\}^\uparrow$$

One can easily check that all $M \in \mathcal{V}_r$ satisfy $m_1 \geq 0$ with $m_2 \geq \max\{m_1^2, (b+c)m_1 - bc\}$, and we determine the range of the third coordinate. We start with showing that this is unbounded from above. If either $m_1 \leq b$ or $m_1 > c$ and $m_2 > m_1^2$, then (m_1, m_2) has a convex representation

$$m_1 = \alpha t + (1 - \alpha)s, \tag{22.51}$$

$$m_2 = \alpha t^2 + (1 - \alpha)s^2, \tag{22.52}$$

for $t \to \infty$ and $s \to m_1$ from the left. Substituting $s = (m_1 - \alpha t)/(1 - \alpha)$ from (22.51) into (22.52), we obtain

$$\alpha t^2 = m_2 - m_1^2 + 2 (tm_1 - m_2) \alpha > m_2 - m_1^2 > 0$$

and

$$m_3 > \alpha t^3 > \left(m_2 - m_1^2\right) t \to \infty, \qquad \text{as } t \to \infty.$$

If $b < m_1 \leq c$ and $m_2 > (b + c)m_1 - bc$, then (m_1, m_2) is a convex combination of (b, b^2), (c, c^2), and (t, t^2) for sufficiently large t. By (22.49), the coefficient of t is

$$\alpha = \frac{m_2 - (b + c)m_1 + bc}{(t - b)(t - c)}$$

and so $m_3 > \alpha t^3 \to \infty$, as $t \to \infty$. Since $\mathcal{V}_r$ is a convex body, we need to show that $mem\{O, \widehat{OB}\}$, $\triangle OBC$ and $mem\{O, \widehat{C\infty}\}$ form its bottom. The surfaces consist of moment points generated by $0, t \in (0, b]$, and $0, b, c$, and $0, t \in [c, \infty)$, respectively, which belong to $[0, b] \cup [c, \infty)$. Moreover, the membranes are parts of the lower envelope of $\mathcal{W}$. Consequently, it suffices to prove that no moment point of a measure supported on $[0, b] \cup [c, \infty)$ lies beneath $\triangle OBC$. We can confine ourselves on three-point distributions. Due to Lemma 22.5, if $0 \leq t_i \leq b$ or $t_i \geq c$, $i = 1, 2, 3$, then the respective T_i lie above $\triangle OBC$ and so do all $M \in \triangle T_1 T_2 T_3$. This completes the proof of Part (i).

The next three parts are proved by means of the optimal ratio method. Consider the vertical hyperplane $\mathcal{H}_a : m_2 - 2am_1 + a^2 = 0$ supporting $\mathcal{W}$ along halfline $A^\uparrow$. Then $\mathcal{H}'_a : m_2 - (b+c)m_1 + bc = 0$ obeys the requirements of Lemma 22.3, and $\mathcal{H}'_a \cap \mathcal{V}_r = \overline{BC}^\uparrow$. Therefore, for $M \in \text{conv}\{A^\uparrow, \overline{BC}^\uparrow\} = \triangle ABC^\uparrow$, we have

$$L_r(M) = \frac{\rho\left(\mathcal{H}'_a, M\right)}{\rho\left(\mathcal{H}'_a, A\right)} = \frac{|m_2 - (b+c)m_1 + bc|}{r^2} = 1 - \frac{m_2 - 2am_1 + a^2}{r^2},$$

as stated in (22.37). For proving (22.38), we consider $\mathcal{H}_t : m_2 - 2tm_1 + t^2 = 0$ for some $t \in (a,c)$. Then $\mathcal{H}_t \cap \mathcal{W} = T^\uparrow$, and $\mathcal{H}'_t : m_2 - 2tm_1 + 2tc - c^2 = 0$ is the closest parallel plane to $\mathcal{H}_t$ that supports $\mathcal{V}_r$ along $C^\uparrow$. Then for $M \in \text{conv}\{T^\uparrow, C^\uparrow\} = \overline{TC}^\uparrow$, we obtain

$$L_r(M) = \frac{\rho(\mathcal{H}'_t, M)}{\rho(\mathcal{H}'_t, T)} = \frac{|m_2 - 2tm_1 + 2tc - c^2|}{(c-t)^2}. \tag{22.53}$$

Condition $M \in \overline{TC}^\uparrow$ implies $m_2 = (t+c)m_1 - tc$. This allows us to replace t with $(cm_1 - m_2)/(c - m_1)$ in (22.53) and leads to (22.38). Note that the formula holds true for all $M \in \bigcup_{a \leq t \leq c} \overline{TC}^\uparrow = mem\{C, \widehat{AC}\}^\uparrow$. In the same manner we can prove part (iv) by studying $\mathcal{H}_t$ for $t \in (b,a)$ with $\mathcal{H}'_t : m_2 - 2tm_1 + 2tb - b^2 = 0$ that touch $\mathcal{V}_r$ along $B^\uparrow$.

We are left with the task of determining $L_r(M)$ for $\text{conv}\{O, B, C, \widehat{BC}\}$ that is bounded below by $mem\{O, \widehat{BC}\}$, and above by $\triangle OBC$, $\triangle ABC$, $mem\{C, \widehat{AC}\}$ and $mem\{B, \widehat{BA}\}$. Observe that the optimal convex representations for M from the border surfaces are attained for the support points 0 with $t \in (b,c)$, and 0, b, c, and b, a, c, and c with $t \in (a,c)$, and b with $t \in (b,a)$, respectively. Due to Lemma 22.5, examining $L_r(M)$ for the inner points, we concentrate on the measures supported on 0, b, c and some $t_i \in (b,c)$.

We claim that the optimal support contains a single $t \in (b,c)$. To show this, suppose that $M \in \overline{PQ}$ for some $P \in \triangle OBC$ and $Q \in \text{conv}\{T_i : b < t_i < c\}$. All T_i lie below $\text{pl}\{OBC\}$ (see Lemma 22.4), and so do Q and M. The ray passing from P through M towards Q runs down through $\text{conv}\{\widehat{BC}\} \ni Q$. We decrease the contribution of Q which is of interest once we take the most distant point of $\text{conv}\{\widehat{BC}\}$. This is an element of the bottom envelope $mem\{B, \widehat{BC}\}$ of $\text{conv}\{\widehat{BC}\}$ (cf. [314]). Accordingly, an arbitrary combination of $T_i \in \widehat{BC}$ can be replaced by a pair B and $T \in \widehat{BC}$, which is the desired conclusion.

For given $M \in \text{conv}\{O, B, C, \widehat{BC}\}$, we now aim at minimizing

$$\alpha(t) = \frac{-m_3 + (b+c)m_2 - bcm_1}{t(t-b)(c-t)} \tag{22.54}$$

(cf. (22.48)) which is the coefficient of $T \in \widehat{BC}$ in the linear representation of M in terms of O, B, C, and T. If $b < t < c$, both the numerator and denominator of (22.54) are positive, the former being constant. By differentiating, we conclude that the denominator has two local extremes: a negative minimum at $d' < b$ and a positive maximum at $d = [2a + (a^2 + 3r^2)^{1/2}]/3 \in (a, c)$, which is the desired solution. We complete the proof of part (v) by noticing that $L_r(M) = \alpha(d)$ iff the representation is actually convex, i.e., when $M \in \text{conv}\{O, B, C, D\}$.

Moment points of $\text{conv}\{O, C, \widehat{DC}\}$ with borders $mem\{O, \widehat{DC}\}$, $\triangle ODC$ and $mem\{C, \widehat{DC}\}$ are treated in what follows. The solutions of the moment problem on the respective parts of the border are provided by combinations of 0 with some $t \in [d, c]$, and 0, d, c, and c with some $t \in [d, c]$, respectively. Referring again to Lemma 22.5, we study combinations of 0, d, c and some $t \in (d, c)$ when we evaluate (22.26) for the inner points. We showed above that combinations with a single t reduce $\mu(I_r)$. Analyzing (22.54) for $t > d$ we see that decreasing t provides a further reduction of the measure. However, we can proceed so as long as $M \in \text{conv}\{O, D, T, C\}$ and stop when $M \in \triangle OTC$. By (22.50), the condition is equivalent to $m_3 - (t + c)m_2 + tcm_1 = 0$, which allows us to determine (22.42). Combining (22.42) with (22.49), we conclude (22.44) for $M \in \bigcup_{d \leq t \leq c} \triangle OTC = \text{conv}\{O, C, \widehat{DC}\}$ (cf. (22.34)).

What remains to treat is the region situated above $mem\{O, \widehat{BD}\}$, $\triangle BDC$, and below $mem\{C, \widehat{AD}\}$, $\triangle ABC$, $mem\{B, \widehat{BA}\}$, and $\triangle OBD$. Analysis similar to that in the proof of the previous two cases leads to minimization of (22.54) with respect to $t \in (b, d)$. For a fixed M, we can reduce (22.54) by increasing t until M belongs to the border of the tetrahedron $\text{conv}\{O, B, T, C\}$. There are two possibilities here: either $M \in \text{conv}\{O, B, \widehat{BD}\}$, and then $M \in \triangle OBT$ ultimately, or $M \in \text{conv}\{B, C, \widehat{AD}\}$, and then M becomes a point of $\triangle BCT$. Analytic representations of $M \in \text{pl}\{OBT\}$ and $M \in \text{pl}\{BCT\}$ (cf. (22.50)) makes it possible to write (22.44) and (22.46), respectively. Final formulae (22.43) and (22.45) are obtained by plugging respective support points into (22.49). Verification of details is left to the reader.   □

One observes that for every $0 < r < a^*$ there exists M such that $L_r(M) = 1 - r$ (cf. (22.32)) in each region of the partition of the moment space presented in Theorem 22.2, except of the first one. E.g., the equation holds true for $M \in \triangle ABC^{\uparrow}$ iff $M \in \overline{B_r^A C_r^A}^{\uparrow}$ with $B_r^A = rB + (1 - r)A$ and $C_r^A = rC + (1 - r)A$. In $\text{conv}\{B, C, \widehat{AD}\}$, this is satisfied by $M \in \bigcup_{a \leq t \leq d} \overline{B_r^T C_r^T}$. A more difficult problem that will be studied in next two sections is determining M that minimize L_r over boxes (22.30). The moment points that obey both the conditions are needed for the final determination of the Prokhorov radius.

22.3   Further Auxiliary Results

Lemma 22.6. $L_r(M)$ *is nonincreasing in* m_3 *as* m_1, m_2 *are fixed and* $M \in \mathcal{W}$.

Proof. We prove the property for every element of the partition of the moment space presented in Theorem 22.2 and refer to continuity of the function for justifying the claim. The result is trivial for moment points in the first five subregions.

Consider $M \in \text{conv}\{O, C, \widehat{DC}\}$. Note that (22.42) is decreasing in m_3, and examine the former representation in (22.41). This is the ratio of two positive values: the numerator is constant, and the denominator decreases for $t \in [c/2, c] \supset [d, c]$. Hence, by increasing m_3 we decrease t, and increase the denominator, and decrease the ratio, as required.

In $\text{conv}\{O, B, \widehat{BD}\}$, increase of m_3 implies that of t (see (22.44)). As $t > b$ increases, so does the denominator of the first representation in (22.43). Since the numerator is a positive constant, the result follows. Likewise, from (22.14) we deduce that $t(m_3)$ is decreasing in $\text{conv}\{B, C, \widehat{AD}\}$, and (22.13) is increasing in t as $t \in [a, c] \supset [a, d]$. $\qquad\square$

Lemma 22.7. *$L_r(M)$ is nondecreasing in m_1 as m_2, m_3 are fixed, and nonincreasing in m_2 as m_1, m_3 are fixed, when $M \in \triangle ABC^\uparrow \cup mem\{C, \widehat{AC}\}^\uparrow \cup mem\{B, \widehat{BA}\}^\uparrow$.*

Proof. Glancing at (22.37)–(22.39), we immediately verify the latter statement. The former is also evident for $M \in \triangle ABC^\uparrow$. Differentiating (22.6), we get

$$\frac{\partial L_r(M)}{\partial m_1} = \frac{2\,(c - m_1)\,(cm_1 - m_2)}{\left(c^2 - 2cm_1 + m_2\right)^2} \geq 0$$

in $mem\{C, \widehat{AC}\}^\uparrow$, because both factors of the numerator are nonnegative there. In the same manner we conclude the statement for $M \in mem\{B, \widehat{BA}\}^\uparrow$. $\qquad\square$

Lemma 22.8. *$L_r(M)$ is nonincreasing in m_1 as m_2, m_3 are fixed, and nondecreasing in m_2 as m_1, m_3 are fixed, when*

$$M \in \text{conv}\{O, B, D, C\} \cup \text{conv}\{O, C, \widehat{DC}\} \cup \text{conv}\{O, B, \widehat{BD}\}.$$

Proof. By (22.40), the assertion is obvious for $\text{conv}\{O, B, D, C\}$. Assume now that $M \in \text{conv}\{O, B, \widehat{BD}\}$, and let first m_1 vary and keep $m_1 m_2$ fixed. Increase of m_1 implies that of t and $t(t - b)$ (cf. (22.44)). Therefore, the central term of (22.43) is the ratio of two nonnegative factors: the former is decreasing in m_1, and the latter is increasing. Also,

$$\frac{\partial t}{\partial m_2} = \frac{b^2 m_1 - m_3}{(m_2 - bm_1)^2}. \tag{22.55}$$

Observe that $m_3 - b^2 m_1 = 0$ is the plane supporting $\text{conv}\{O, B, \widehat{BD}\}$ along $\overline{OB}$, and the rest of the region lies above the plane. Therefore, (22.55) is nonnegative, and both t and $t(t - b)$ are nonincreasing in m_2. Using (22.43) again, we confirm the opposite for $L_r(M)$.

The last case requires a more subtle analysis. Expressing (22.41) and (22.42) in terms of $\beta = \beta(M) = cm_2 - m_3 \geq 0$ and $\gamma = \gamma(M) = cm_1 - m_2 \geq 0$, we obtain $L_r(M) = \gamma^3 / [\beta(c\gamma - \beta)]$ and $t = \beta/\gamma$, respectively. Differentiating the former formally, we find

$$L_r'(M) = \frac{\gamma^2\,\left(2c\beta\gamma\gamma' + 2\beta\beta'\gamma - 3\beta^2\gamma' - c\beta'\gamma^2\right)}{\beta^2(c\gamma - \beta)^2}. \tag{22.56}$$

Since $\frac{\partial \beta}{\partial m_1} = 0$ and $\frac{\partial \gamma}{\partial m_1} = c$,

$$\frac{\partial L_r(M)}{\partial m_1} = \frac{c\gamma^2(2c\gamma - 3\beta)}{\beta(c\gamma - \beta)^2} = \frac{3c\gamma^3(2c/3 - t)}{\beta(c\gamma - \beta)^2}. \tag{22.57}$$

The sign of (22.57) is the same as that of $(2/3)c - t$. We shall see that $(2/3)c < t$ for all $t \in (d, c)$. It suffices to check that $(2/3)c < d = (1/3)[2a + (a^2 + 3r^2)^{1/2}]$, and a simple algebra shows that this is equivalent to $r < a$. Therefore, (22.57) is negative, and L_r actually decreases in m_1. Combining (22.56) with $\frac{\partial \beta}{\partial m_2} = c$ and $\frac{\partial \gamma}{\partial m_2} = -1$, we have

$$\frac{\partial L_r(M)}{\partial m_2} = \frac{\gamma^2\left(3\beta^2 - c^2\gamma^2\right)}{\beta^2(c\gamma - \beta)^2} = \frac{3\gamma^4\left(t^2 - c^2/3\right)}{\beta^2(c\gamma - \beta)^2} > 0,$$

because $t > (2/3)c > c/\sqrt{3}$, as we checked above. This ends the proof. $\qquad\square$

Analyzing variability of $L_r(M)$ in $\mathrm{conv}\{B, C, \widehat{AD}\}$, we split the region into three parts.

Lemma 22.9. *Set $a < a_1 = (a^2 + r^2/3)^{1/2} < a_2 = a + r^2/(3a) < d$.*

(i) $L_r(M)$ is nondecreasing in m_1 as m_2, m_3 are fixed, and nonincreasing in m_2 as m_1, m_3 are fixed, when $M \in \mathrm{conv}\{B, C, \widehat{AA_1}\}$.

(ii) $L_r(M)$ is nondecreasing in m_1 as m_2, m_3 are fixed, and in m_2 as m_1, m_3 are fixed, when $M \in \mathrm{conv}\{B, C, \widehat{A_1A_2}\}$.

(iii) $L_r(M)$ is nonincreasing in m_1 as m_2, m_3 are fixed, and nondecreasing in m_2 as m_1, m_3 are fixed, when $M \in \mathrm{conv}\{B, C, \widehat{A_2D}\}$.

Proof. Rewrite (22.45) and (22.46) as $t = \beta/\gamma \in [a, d]$ and $L_r(M) = \gamma^3/[(\beta - b\gamma)(c\gamma - \beta)]$ for

$$\beta = \beta(M) = -m_3 + (b + c)m_2 - bcm_1 \geq 0,$$

$$\gamma = \gamma(M) = -m_2 + (b + c)m_1 - bc \geq 0.$$

Therefore

$$L_r'(M) = \frac{\gamma^2\left[2(b + c)\beta\gamma\gamma' + 2\beta\beta'\gamma - (b + c)\beta'\gamma^2 - bc\gamma^2\gamma' - 3\beta^2\gamma'\right]}{(\beta - b\gamma)^2(c\gamma - \beta)^2}.$$

Because $\frac{\partial \beta}{\partial m_1} = -bc$ and $\frac{\partial \gamma}{\partial m_1} = b + c$,

$$\frac{\partial L_r(M)}{\partial m_1} = \frac{\beta\gamma^2\left[2\left(b^2 + bc + c^2\right)\gamma - 3(b + c)\beta\right]}{(\beta - b\gamma)^2(c\gamma - \beta)^2}$$

$$= \frac{\beta\gamma^3\left(a_2 - t\right)}{3(b + c)(\beta - b\gamma)^2(c\gamma - \beta)^2}. \tag{22.58}$$

Similarly, due to $\frac{\partial \beta}{\partial m_2} = b + c$ and $\frac{\partial \gamma}{\partial m_2} = -1$, we derive

$$\frac{\partial L_r(M)}{\partial m_2} = \frac{\gamma^2\left[3\beta^2 - \left(b^2 + bc + c^2\right)\gamma^2\right]}{(\beta - b\gamma)^2(c\gamma - \beta)^2} = \frac{3\gamma^4\left(t^2 - a_1^2\right)}{(\beta - b\gamma)^2(c\gamma - \beta)^2}. \tag{22.59}$$

Accordingly, if $M \in \triangle BCT$ for some $t \in [a, d]$, then (22.58) and (22.59) hold. Analyzing the signs of the formulae, we arrive to the final conclusion. $\qquad\square$

22.4   Minimizing Solutions in Boxes

In Theorem 22.3, we describe all moment pints $M_{r\mathcal{E}}$ at which the minimal positive value of function (22.26) over boxes $\mathcal{B}(\mathcal{E})$ is attained for arbitrary $0 < r < a^*$ and $\epsilon_i > 0$, $i = 1, 2, 3$. In contrast with Lemmas 22.6–22.9 proved by means analytic techniques, we here use geometric arguments mainly. Let $V_1 = (a - \epsilon_1, a^2 + \epsilon_2, a^3 + \epsilon_3)$, $V_2 = (a - \epsilon_1, a^2 - \epsilon_2, a^3 + \epsilon_3)$, $V_3 = (a + \epsilon_1, a^2 - \epsilon_2, a^3 + \epsilon_3)$ denote three of vertices of the top rectangular side $\mathcal{R}(\mathcal{E}) = \{(m_1, m_2, a^3 + \epsilon_3) : |m_i - a^i| \leq \epsilon_i, i = 1, 2\}$ of $\mathcal{B}(\mathcal{E})$.

Theorem 22.10. For all $0 < r < a^*$ and each of the following cases:

$$M_{r\mathcal{E}} \;=\; V_1 \in \triangle ABC^\uparrow, \tag{22.60}$$

$$\{M_{r\mathcal{E}}\} \;=\; \overline{V_1 V_2} \cap \triangle BA_1C, \tag{22.61}$$

$$\{M_{r\mathcal{E}}\} \;=\; \overline{V_1 V_2} \cap mem\{B, \widehat{AA_1}\}, \tag{22.62}$$

$$\{M_{r\mathcal{E}}\} \;=\; \overline{V_2 V_3} \cap \triangle BA_2C, \tag{22.63}$$

$$\{M_{r\mathcal{E}}\} \;=\; \overline{V_2 V_3} \cap mem\{B, \widehat{AA_2}\}, \tag{22.64}$$

$$M_{r\mathcal{E}} \;=\; V_1 \in \mathrm{conv}\left\{B, C, \widehat{AA_1}\right\}, \tag{22.65}$$

$$M_{r\mathcal{E}} \;=\; V_2 \in \mathrm{conv}\left\{B, C, \widehat{AA_2}\right\}, \tag{22.66}$$

there exist $\epsilon_i > 0$, $i = 1, 2, 3$ such that

$$\mathcal{L}_r(\mathcal{E}) = L_r(M_{r\mathcal{E}}) > 0. \tag{22.67}$$

Proof. By Lemma 22.6, $\mathcal{L}_r(\mathcal{E})$ is attained on the upper rectangular side $\mathcal{R}(\mathcal{E})$ of $\mathcal{B}(\mathcal{E})$. We consider all possible horizontal sections $\underline{\mathcal{W}} = \underline{\mathcal{W}}(\epsilon_3) = \{M \in \mathcal{W} : m_3 = a^3 + \epsilon_3\}$, $\epsilon_3 > 0$, of the moment space and use Lemmas 22.7 - 22.9 for describing points minimizing $L_r(M)$ over all $M \in \mathcal{R}(\mathcal{E}) \cap \underline{\mathcal{W}}$. We shall follow the convention of denoting by $\underline{\mathcal{A}}$ the horizontal section of a set $\mathcal{A}$ at the level $m_3 = a^3 + \epsilon_3$. Also, we set $\underline{A} = (a, a^2, a^3 + \epsilon_3)$, and define $\underline{O}$, $\underline{B}$, $\underline{C}$ analogously. By (22.33), every section $\underline{\mathcal{W}}$ is a convex surface bounded below and above by parabolas $m_2 = m_1^2$ and $m_2 = (m_3 m_1)^{1/2}$, $0 \leq m_1 \leq m_3^{1/3}$, respectively. The last belongs to a vertical side surface of $\mathcal{W}$, and the latter is a part of its bottom. On the right, they meet at $P = (m_3^{1/3}, m_3^{2/3}, m_3) \in \widehat{O\infty}$.

We start with the simplest problem of analyzing high level sections, and then decrease ϵ_3 to include more complicated cases. We provide detailed descriptions of the sections for three ranges of levels $m_3 \geq c^3$, $d^3 \leq m_3 < c^3$, and $a^3 < m_3 < d^3$ that contain different number of subregions defined in Theorem 22.2. For each range, we further study shape variations of some subregion slices caused by decreasing the level in order to conclude various statements of Theorem 22.10. We shall see that the number of cases increases as $\epsilon_3 \to 0$, and these which occur on higher levels are still valid for the lower ones. Since arguments justifying the particular cases for different levels are in principle similar, we present them in detail once respective solution appear for the first time and omit later.

Suppose first that $m_3 = a^3 + \epsilon_3 \geq c^3$. Only point of first four elements of partition of Theorem 22.2 attain the level. Moreover, no bottom points of $\triangle ABC^{\uparrow}$, $mem\{C, \widehat{AC}\}^{\uparrow}$, and $mem\{B, \widehat{BA}\}^{\uparrow}$ belong to $\underline{\mathcal{W}}$. It follows that the shapes of the sections $\triangle \underline{A}\,\underline{B}\,\underline{C}$, $\mathrm{conv}\{\widehat{\underline{A}\underline{C}}\}$, $\mathrm{conv}\{\widehat{\underline{B}\underline{A}}\}$ coincide with the projections of respective sets onto (m_1, m_2)-plane. By Lemma 22.7, L_r decreases as we move left and up in each subregion on the level (here we refer to directions on the horizontal plane, but no confusion should arise if we use the same notions for describing movements in the space). The minimum over $\mathcal{R}(\mathcal{E})$ is so attained at its right upper corner V_1. Excluding cases $V_1 \in \underline{\mathcal{V}_r}$ and $V_1 \notin \underline{\mathcal{W}}$ that imply $\mathcal{L}_r(\mathcal{E}) = 0$, we see that (22.60) is the only possibility.

If $d^3 \leq m_3 < c^3$, all regions except of $\mathrm{conv}\{O, B, \widehat{BD}\}$ should be examined. It is only $\underline{mem\{B, \widehat{BA}\}}$ whose shape remains unchanged with respect to the previous case. The section of $\triangle ABC^{\uparrow}$, formerly triangular, is now deprived of its upper vertex $\underline{C}$, being cut off by a line segment $\overline{QS}$, say, with $Q \in \overline{BC}$ and $S \in \overline{AC}$. The upper part of the linear border of $\underline{mem\{C, \widehat{AC}\}^{\uparrow}}$ is replaced by a curve $\widehat{SP} \subset mem\{C, \widehat{AC}\}$. There is a point U, say, such that $\{U\} = \widehat{SP} \cap \overline{DC}$. Lines $\widehat{SP}$, and $\overline{QS} \in \triangle ABC$, and $\overline{QU} \subset \triangle BDC$ are the borders of $\underline{\mathrm{conv}\{B, C, \widehat{AD}\}}$. The last one, together with $\overline{XQ}$ and $\overline{XU}$ for some $X \in \overline{OC}$ form the borders of the respective triangular section of the tetrahedron $\mathrm{conv}\{O, B, D, C\}$. Note that X is a bottom point of $\mathcal{W}$ and belongs to the upper parabolic bound of $\underline{\mathcal{W}}$. Also, $\overline{XU} \subset \triangle ODC$ and $\widehat{UP} \subset mem\{C, \widehat{DC}\}$ separate the points of $\underline{\mathrm{conv}\{O, C, \widehat{DC}\}}$ from the rest of the section. Finally, $\underline{\mathcal{V}_r}$ is located below $\underline{\mathrm{conv}\{O, B, D, C\}}$ and right to $\underline{\triangle ABC^{\uparrow}}$, among $\widehat{OQ}$, $\widehat{OB}$, and $\overline{BQ} \in \overline{BC}^{\uparrow}$ and $\overline{XQ} \in \triangle OBC$. It can be verified that the slopes of all curved and straight lines mentioned above are positive. Also, writing $\overline{T_1 T_2}$ and $\widehat{T_1 T_2}$, we adopted the convention that T_1 has the first two coordinates less than T_2.

A crucial fact for our analysis is the mutual location of $\underline{A} \in A^{\uparrow}$ and $Q \in \overline{BC}$. We consider a more general problem replacing $Q \in \overline{BC}$ by $M \in \overline{BT}$ for some $t > a$. Two-point representation of M gives

$$M = M(m_3) = \left(\frac{m_3 + bt(b+t)}{b^2 + bt + t^2}, \frac{(b+t)m_3 + b^2 t^2}{b^2 + bt + t^2}, m_3 \right). \tag{22.68}$$

As m_3 varies between b^3 and t^3, then all m_i increase, and M slides up from B to T along $\overline{BT}$. It follows that M is above and right to $\underline{A}$ for sufficiently large m_3 and below and left for small ones. For $m_3 = a^3 + \epsilon_3$, we can check that $m_1 < a$ and $m_2 \leq a^2$ are equivalent to

$$\epsilon_3 < r(t - a)(a + b + t), \tag{22.69}$$

$$\epsilon_3 \leq r(t - a) \frac{ab + at + bt}{b + t}, \tag{22.70}$$

respectively. The right-hand side of (22.70) is smaller than that of (22.69). It follows that as ϵ_3 decreases from $t^3 - a^3$ to 0, M is first located above and right to $\underline{A}$, then moves to the left and ultimately below $\underline{A}$. These changes of mutual location occur on higher levels for larger t.

In particular, we show that $Q \in \overline{BC}$ is left and above $\underline{A}$ on the section $m_3 = d^3$. To this end we set $\epsilon_3 = d^3 - a^3$ and $t = c$ and check that (22.69) is true and (22.70) should be reversed. Indeed, the former can now be written as $a^3 + 3ar^2 > d^3$, and further transformed into $(t-1)(t^2 - 20t - 8) < 0$ under the change of variables $t = (1 + 3r^2/a^2)^{1/2} \in (1,2)$. The cubic inequality holds true for $t \in (-\infty, 10 - 6\sqrt{3}) \cup (1, 10 + 6\sqrt{3}) \supset (1,2)$. Negation of (22.70) is equivalent to each of the following:

$$2\left[2 + (1 + 3t)^{1/2}\right]^3 > 54 + 81t - 27t^2,$$

$$(26 + 6t)(1 + 3t)^{1/2} > 26 + 45t - 27t^2,$$

$$27t^2\left(13 + 94t - 27t^2\right) > 0,$$

for $t = r^2/a^2 \in (0,1)$, and the last one is obviously true.

It follows that for $m_3 = a^3 + \epsilon_3$ varying from c^3 to d^3 there are two possible locations of Q with respect to $\underline{A}$. For $\epsilon_3 \geq 3ar^2$, Q is right and above, and (22.67) enforces (22.60) like in the previous case. Otherwise this is only one of three possibilities. For the other we should assume that $\overline{V_1V_2}$ lies right to Q. Let S_i, $i = 1, 2$, stand for the intersections of $\widehat{SU}$ and $\overline{A_iC}$, respectively. Each element of the sequence A, S, S_1, S_2 and U lies above and right to the preceding. Consider $\overline{\text{conv}\{B, C, \widehat{AA_1}\}}$ located among $\overline{QS}$, $\overline{QS_1}$ and $\widehat{SS_1}$. If V_1 belongs there, then

$$\mathcal{R}(\mathcal{E}) \cap \underline{\mathcal{W}} \subset \triangle ABC^\uparrow \cup mem\{C, \widehat{AC}\}^\uparrow$$

$$\cup mem\{B, \widehat{BA}\}^\uparrow \cup \text{conv}\left\{B, C, \widehat{AA_1}\right\}. \tag{22.71}$$

By Lemmas 22.7 and 22.9(i), L_r is increasing in m_1 and decreasing in m_2 in this part of the horizontal slice, and (22.65) can be concluded.

Assume now that V_1 is right to Q and above $\overline{QS_1}$. Then $\overline{V_1V_2}$ and $\overline{QS_1} \subset \triangle BA_1C$ cross each other, because V_2 lies below $\underline{A}$. We establish that (22.61) holds. If $\mathcal{R}(\mathcal{E})$ contains any points of the sets $\text{conv}\{O, C, \widehat{DC}\}$, $\text{conv}\{O, B, D, C\}$, and $\text{conv}\{B, C, \widehat{A_2D}\}$, i.e., ones lying above $\overline{OS_2}$ and $\widehat{S_2P}$, then by Lemmas 22.8 and 22.9(iii) they can be replaced by ones of $\mathcal{R}(\mathcal{E}) \cap (\overline{OS_2} \cup \widehat{S_2P})$, and the minimum of L_r will not be affected. Similarly, by Lemmas 22.7 and 22.9(i), all points of $\mathcal{R}(\mathcal{E})$ located below $\overline{QS_1}$ and $\widehat{S_1P}$ can be replaced by some of $\mathcal{R}(\mathcal{E}) \cap (\overline{QS_1} \cup \widehat{S_1P})$. We could exclude points of $\widehat{S_2P}$ from consideration once we show that L_r increases along $\widehat{SP}$. Indeed, the points of $\widehat{SP}$ have a parametric representation $M = \alpha T + (1 - \alpha)C$ with $m_3 = a^3 + \epsilon_3 < c^3$ fixed and $t \in (a, m_3^{1/3})$, and so all

$$L_r(M) = \alpha = \frac{c^3 - m_3}{c^3 - t^3}, \tag{22.72}$$

$$m_1 = c - \frac{c^3 - m_3}{c^2 + ct + t^2}, \tag{22.73}$$

$$m_2 = c^2 - \frac{\left(c^3 - m_3\right)(c + t)}{c^2 + ct + t^2}, \tag{22.74}$$

are decreasing. Accordingly, it suffices to restrict ourselves to analyzing $L_r(M)$ in $\mathcal{R}(\mathcal{E}) \cap \mathrm{conv}\{B, C, \widehat{A_1 A_2}\}$, which by Lemma 22.9(ii) is minimized at the lowest left point of the surface, i.e., the one defined by (22.61).

Suppose now that $m_3 = a^3 + \epsilon_3 < d^3$. The most apparent difference from the previous case is presence of $\mathrm{conv}\{O, B, \widehat{BD}\}$. This contains bottom moment points including P and is located along the upper border of $\mathcal{W}$. Precisely, this is separated from $\mathrm{conv}\{O, B, D, C\}$ and $\mathrm{conv}\{B, C, \widehat{AD}\}$ by a line segment $\overline{YZ}$, say, with $Y \in \overline{OD}$ and $Z \in \overline{BD}$, and a curve $\widehat{ZP}$, respectively. Writing P, we refer to the notion introduced above, although now the point is defined for a different level m_3. It will cause no confusion if we use it and some other intersection points of horizontal sections with given segments in the new context. Furthermore, appearance of $\mathrm{conv}\{B, C, \widehat{AD}\}$ changes radically. Now its border has four parts: two linear $\overline{QS}$ and $\overline{QZ}$ and two curved ones $\widehat{SP}$ and $\widehat{ZP}$. These are borders with $\triangle ABC^\uparrow$, $\mathrm{conv}\{O, B, D, C\}$, $mem\{C, \widehat{AC}\}^\uparrow$, and $\mathrm{conv}\{O, B, \widehat{BD}\}$, respectively. On this level, the region is spread until the upper right vertex P of $\mathcal{W}$. We can also observe that the slice of $\mathrm{conv}\{O, B, D, C\}$ has four linear edges with vertices Q, X, Y, and Z. On the left to the tetrahedron, behind $\overline{XY}$, we can find $\mathrm{conv}\{O, C, \widehat{DC}\}$, which is the convex hull $\mathrm{conv}\{\widehat{XY}\}$ of the piece of parabola $m_2 = (m_3 m_1)^{1/2}$ between X and Y. In contrast with the last case, the region is separated from P here. The remaining regions look in much the same way as above. There are $\mathcal{V}_r = \mathrm{conv}\{Q, \widehat{QX}, \widehat{QB}\}$ below $\mathrm{conv}\{O, B, D, C\}$, and $\triangle ABC^\uparrow = \mathrm{conv}\{\underline{B}, \underline{A}, Q, S\}$ below $\mathrm{conv}\{B, C, \widehat{AD}\}$, with the common border $\underline{BQ} \in \overline{BC}^\uparrow$. Beneath $\underline{AB}$, there is $mem\{B, \widehat{BA}\}^\uparrow$ with the lower border $\underline{\widehat{BA}}$. The right part of parabola $m_2 = m_1^2$, $a \le m_1 \le m_3^{1/3}$, is contained in $mem\{C, \widehat{AC}\}^\uparrow$ that borders upon $\triangle ABC^\uparrow$ and $\mathrm{conv}\{B, C, \widehat{AD}\}$ through $\overline{AS}$ and $\widehat{SP}$, respectively.

Finally we examine slices of three parts of $\mathrm{conv}\{B, C, \widehat{AD}\}$ described in Lemma 22.9. $\mathrm{conv}\{B, C, \widehat{AA_1}\}$ and $\mathrm{conv}\{B, C, \widehat{A_1 A_2}\}$ are separated by $\overline{QS_1}$ if A_2 is below the slice. The latter borders upon $\mathrm{conv}\{B, C, \widehat{A_2 D}\}$ along $\overline{QS_2}$. If we go down through and below A_2, and A_1, then S_i, $i = 2, 1$, are consecutively identical with P and then are replaced by some $Z_i \in \overline{BA_i} \cap \widehat{ZP}$, respectively, that move left and down so that for sufficiently small ϵ_3 they become situated below and right to $\underline{A}$ (see (22.69) and (22.70)). Note that for $M = \alpha T + (1 - \alpha)B$, formulae (22.72)–(22.74) hold with c replaced by b. These decrease in t as well, and therefore, $L_r(M)$ becomes smaller as we slide M along $\widehat{ZP}$ down and left.

The above observations are essential for detecting solutions to (22.31) for the case $V_1 \notin \triangle ABC^\uparrow \cup \mathrm{conv}\{B, C, \widehat{AA_1}\}$. Otherwise we could use previous arguments for obtaining (22.71) and concluding either (22.60) or (22.65). Under the assumption, we study possible locations of $M_{r\mathcal{E}}$ for three ranges of levels of horizontal slices.

If $a_2^3 \leq m_3 < d^3$, then there are some moment points in $\mathcal{R}(\mathcal{E})$ located above $\overline{QS_1} \cup \widehat{S_1 P}$. Ones that are below can be eliminated by Lemmas 22.7 and 22.9(i), and this will not increase minimal value of L_r. If there are any $M \in \mathcal{R}(\mathcal{E})$ located above and right to $\overline{QS_2}$ and $\widehat{S_2 P}$, we reduce them using Lemmas 22.8 and 22.9(iii). It follows that $M_{r\mathcal{E}} \in \mathrm{conv}\{B, C, \widehat{A_1 A_2}\} \cup \widehat{S_2 P}$. Note that $\widehat{S_2 P}$ lies above and right to $\underline{A}$, has a positive direction, and L_r increases as M moves along $\widehat{S_2 P}$ to the right. We can therefore, eliminate $\widehat{S_2 P}$ from the further analysis. Referring to Lemma 22.9(ii), we assert that $M_{r\mathcal{E}}$ is the point of $\mathrm{conv}\{B, C, \widehat{A_1 A_2}\} \cap \mathcal{R}(\mathcal{E})$ with the smallest possible m_1 and m_2. For $a_1^3 \leq m_3 < a_2^3$, we use arguments of Lemmas 22.7–22.9 again for eliminating all points situated below $\overline{QS_1} \cup \widehat{S_1 P}$ and above $\overline{QZ_2} \cup \widehat{Z_2 P}$. What remains is $\mathrm{conv}\{B, C, \widehat{A_1 A_2}\}$, and we eventually arrive to the conclusion of the previous case. Similar arguments applied in the case $a^3 < m_3 < a_1^3$ enable us to assert that $M_{r\mathcal{E}} \in \mathrm{conv}\{B, C, \widehat{A_1 A_2}\} \cup \widehat{Z_1 P}$. Generally, the curve cannot be eliminated here, because Z_1 may lie left to $\underline{A}$, and there exist rectangles containing a piece of $\widehat{Z_1 P}$ and no points of $\mathrm{conv}\{B, C, \widehat{A_1 A_2}\}$. See that in either part of the union function L_r is increasing in m_1 and m_2. Consequently, the minimum is attained at the left lower corner of $(\mathrm{conv}\{B, C, \widehat{A_1 A_2}\} \cup \widehat{Z_1 P}) \cap \mathcal{R}(\mathcal{E})$. Note that the borders of $\mathrm{conv}\{B, C, \widehat{A_1 A_2}\}$ consist of lines with positive slopes and so is that of $\widehat{Z_1 P}$. Therefore, in all three cases it makes sense indicating the lowest and farthest left point of respective intersections with rectangulars whose sides are parallel to m_1- and m_2-axes.

Now we are in a position to formulate final conclusions, analyzing possible locations of V_2. Assume that V_2 lies either below or on $\overline{QS_1}$ in the case $m_3 \geq a_1^3$. Then $\overline{V_1 V_2}$ and $\overline{QS_1}$ cross each other at the lowest and farthest to the left point of $\mathrm{conv}\{B, C, \widehat{A_1 A_2}\} \cap \mathcal{R}(\mathcal{E})$, and (22.61) follows. The same conclusion holds if V_2 lies either below or on $\overline{QZ_1}$ in the case $m_3 < a_1^3$. If V_2 is below $\widehat{Z_1 P}$ that is only possible when Z_1 lies left to $\underline{A}$ for some $m_3 < a_1^3$, then (22.62) holds, because $\mathcal{R}(\mathcal{E})$ does not contain any points of $\mathrm{conv}\{B, C, \widehat{A_1 A_2}\}$, and one that is the farthest left element of $\widehat{Z_1 P}$ is that of intersection with $\overline{V_1 V_2}$.

Assuming that $V_2 \in \mathrm{conv}\{B, C, \widehat{A_1 A_2}\}$, which is possible for Q lying below $\underline{A}$, we obtain (22.67). Indeed, V_2 is the left lower vertex of $\mathcal{R}(\mathcal{E})$, and so this is also the left farthest and lowest point of the respective intersection with the domain of arguments minimizing L_r for every level between a^3 and d^3. For $m_3 \geq a_2^3$, it suffices to consider V_2 lying right to $\overline{QS_2}$. Like in the other cases considered below, this is only possible when Q is located below $\underline{A}$. It is clear to see that the lowest and farthest left point of $\mathrm{conv}\{B, C, \widehat{A_1 A_2}\} \cap \mathcal{R}(\mathcal{E})$ is that at which $\overline{V_2 V_3}$ and $\overline{QS_2}$ intersect each other. If $m_3 < a_2^3$ and V_2 lies right to $\overline{QZ_2}$, the same conclusion holds with S_2 replaced by Z_2. Since $\overline{QS_2}, \overline{QZ_2} \subset \triangle BA_2 C$, we can summarize both the cases by writing (22.63).

The last possible solution (22.64) to (22.31) occurs when $m_3 < a_2^3$ and V_2 lies right to $\widehat{Z_2 P}$. In fact, the points $M \in \widehat{Z_2 P}$ with $m_2 < a^2$ are of interest. If $a_1^3 \le m_3 \le a_2^3$, then $\widehat{Z_2 P}$ is a part of the left border of $\mathrm{conv}\{B, C, \widehat{A_1 A_2}\}$ and the other $\overline{QZ_2}$ is located below the level of $\overline{V_2 V_3}$. Therefore, L_r is minimized at the intersection of $\widehat{Z_2 P}$ and $\overline{V_2 V_3}$. If $m_3 < a_1^3$ and Z_1 lies below $\underline{A}$, then it is also possible that $M_{r\mathcal{E}} \in \widehat{Z_1 P} \cap \mathcal{R}(\mathcal{E})$. $\qquad\qquad\square$

22.5 Conclusions

Proof of Theorem 22.1. This relies on the results of Theorem 22.10. We consider various forms (22.60)–(22.67) of moment points solving (22.31) and indicate ones that satisfy (22.32) for some $r < a^*$. Solutions to (22.32) with respect to r provide values of Prokhorov radius for various combinations of ϵ_i, $i = 1, 2, 3$. However, only (22.60) enables us to determine an explicit formula for the radius. In this case, evaluating (22.37) at $V_1 = (a - \epsilon_1, a^2 + \epsilon_2, a^3 + \epsilon_3)$, we rewrite (22.32) as $1 - (2a\epsilon_1 + \epsilon_2)/r^2 = 1 - r$, and easily determine (22.5). Relation (22.3) expresses analytically the fact that V_1 lies on the same side of the vertical plane $m_2 - (b + c)m_1 + bc = 0$ containing $\overline{BC}^\uparrow$ as $\widehat{BC}$ does. Condition (22.4) is equivalent to locating V_1 above $\mathrm{pl}\{ABC\}$ (cf. (22.50)). Both, combined with positivity of all ϵ_i, ensure that $V_1 \in \triangle ABC^\uparrow$.

If $M_{r\mathcal{E}} \in \overline{V_1 V_2}$, then $M_{r\mathcal{E}} = (a - \epsilon_1, m_2, a^3 + \epsilon_3)$ for some $|m_2 - a^2| \le \epsilon_2$. Conditions (22.32) and (22.61) enforce the representation

$$M_{r\mathcal{E}} = (1 - r)A_1 + \alpha r(C - B) + rB,$$

for some $0 \le \alpha \le 1$, that gives

$$(1 - r)\left(a^2 + \frac{r^2}{3}\right)^{1/2} + 2\alpha r^2 + r(a - r) = a - \epsilon_1, \qquad (22.75)$$

$$(1 - r)\left(a^2 + \frac{r^2}{3}\right)^{3/2} + 2\alpha r^2 \left(3a^2 + r^2\right) + r(a - r)^3 = a^3 + \epsilon_3, \qquad (22.76)$$

in particular. By (22.75),

$$2\alpha r^2 = a - r(a - r) - (1 - r)\left(a^2 + \frac{r^2}{3}\right)^{1/2} - \epsilon_1 \in \left[0, 2r^2\right] \qquad (22.77)$$

enables us to describe the range of ϵ_1 by (22.7). Plugging (22.77) into (22.76), we obtain (22.6). Furthermore, we apply

$$(1 - r)\left(a^2 + \frac{r^2}{3}\right) + 4a\alpha r^2 + r(a - r)^2 = m_2,$$

combine it with (22.77) and $|m_2 - a^2| \le \epsilon_2$, and eventually obtain (22.8).

Under (22.62), $M_{r\mathcal{E}} = (a - \epsilon_1, m_2, a^3 + \epsilon_3)$ fulfills

$$(1 - r)t + r(a - r) = a - \epsilon_1, \tag{22.78}$$

$$(1 - r)t^2 + r(a - r)^2 = m_2, \tag{22.79}$$

$$(1 - r)t^3 + r(a - r)^3 = a^3 + \epsilon_3, \tag{22.80}$$

for some

$$a \leq t \leq \left(a^2 + \frac{r^2}{3}\right)^{1/2}, \tag{22.81}$$

$$\left|m_2 - a^2\right| \leq \epsilon_2. \tag{22.82}$$

We determine t from (22.78) and substitute it into (22.79)- (22.81). Then the first one combined with (22.82) yields (22.11). Condition (22.80) provides (22.9) defining a relation among ϵ_1, ϵ_3, and the Prokhorov radius. The last one coincides with (22.10).

Cases (iv) and (v) can be handled in much the same way as (ii) and (iii), respectively. For the former, we use the relations

$$(1 - r)\left[a + \frac{r^2}{(3a)}\right]^i + \alpha r\left(c^i - b^i\right) + rb^i = m_i, \qquad i = 1, 2, 3, \tag{22.83}$$

with $0 \leq \alpha \leq 1$, $m_1 \geq a - \epsilon_1$, $m_2 = a^2 - \epsilon_2$, and $m_3 = a^3 + \epsilon_3$. In the latter, we put the same m_i, $i = 1, 2, 3$, in the equations

$$(1 - r)t^i + rb^i = m_i, \qquad i = 1, 2, 3, \tag{22.84}$$

with $a \leq t \leq a + r^2/(3a)$. Here we use the second equations of (22.83) and (22.84) for determining parameters α and t, respectively. Geometric investigations carried out in the proof of Theorem 22.10 make it possible to deduce that $m_2 < a^2$ implies $m_1 < a$ for both (22.83) and (22.84). Therefore, $M_{r\mathcal{E}}$ is actually located on the left half of $\overline{V_2 V_3}$.

For the proof of last two statements we need a formula defining r in (22.31) for $M_{r\mathcal{E}} \in \operatorname{conv}\{B, C, \widehat{AA_2}\}$. Direct application of (22.45) leads to more complicated one than that derived from equations

$$2\alpha r^2 + (1 - r)t + r(a - r) = m_1, \tag{22.85}$$

$$4\alpha r^2 a + (1 - r)t^2 + r(a - r)^2 = m_2, \tag{22.86}$$

$$2\alpha r^2\left(3a^2 + r^2\right) + (1 - r)t^3 + r(a - r)^3 = m_3 \tag{22.87}$$

for $0 \leq \alpha \leq 1$ and $t \in [a, a + r^2/(3a)]$. Subtracting (22.85) multiplied by $2a$ from (22.86) and adding $(1 - r)a^2$, we find

$$(1 - r)(t - a)^2 = m_2 - 2am_1 + a^2 - r^3.$$

Since $t \geq a$, we can write it as

$$t = a + \left(\frac{m_2 - 2am_1 + a^2 - r^3}{1 - r}\right)^{1/2} = a + \tau, \tag{22.88}$$

say. Using the formula and a representation of $2\alpha r^2$ derived from (22.85), we can transform (22.87) as follows:

$$m_3 = \left(3a^2 + r^2\right)\left[m_1 - a + r^2 - (1 - r)\tau\right] + (1 - r)(a + \tau)^3$$

$$= \left(3a^2 + r^2\right)m_1 - \left(3a^2 + r^2\right)\left(a - r^2\right) + (1 - r)a^3 - (1 - r)r^2\tau$$

$$+(3a + \tau)\left(m_2 - 2am_1 + a^2 - r^3\right)$$

$$= 3am_2 - \left(3a^2 - r^2\right)m_1 + a^3 - ar^2 + \left(m_2 - 2am_1 + a^2 - r^2\right)\tau.$$

Referring again to (22.88), we ultimately derive

$$\frac{-m_3 + 3am_2 - \left(3a^2 - r^2\right)m_1 + a^3 - ar^2}{-m_2 + 2am_1 - a^2 + r^2} = \left(\frac{m_2 - 2am_1 + a^2 - r^3}{1 - r}\right)^{1/2}. \tag{22.89}$$

Observe that for all $M \in \text{conv}\{B, C, \widehat{AD}\}$, the denominator of the left-hand side is positive and so is the numerator. If (22.65) holds, we can further simplify (22.89) and conclude (22.18). For asserting that $V_1 \in \text{conv}\{B, C, \widehat{AA_1}\}$, it suffices to assume that it is located in the halfspace $m_2 - 2am_1 + a^2 - r^2 \leq 0$, above pl$\{BA_1C\}$ and below pl$\{ABC\}$. The first requirement coincides with (22.19), and the others, by (22.50), represent conditions (22.20) on the range of ϵ_3. If $M_{r\mathcal{E}} = V_2$, then (22.89) implies (22.21). Condition $V_2 \in \text{conv}\{B, C, \widehat{A_1A_2}\}$ can be equivalently expressed by saying that V_2 lies right to $B^\uparrow$ between the planes containing $\overline{BC}^\uparrow$, $\overline{BA}^\uparrow$, $\triangle BA_2C$ and $\triangle BA_1C$. Analytic descriptions of the conditions are presented in (22.22) and (22.23). This completes the proof of Theorem 22.1. $\qquad\square$

It is of interest if all representations of Prokhorov radius given in Theorem 22.1 are possible for arbitrary $a > 0$ assuming various shapes of moment boxes $\mathcal{B}(\mathcal{E})$. Applying statements of Theorem 22.10, we assert that for all $0 < r < a^*$ there are ϵ_i, $i = 1, 2, 3$, such that $\Pi(\mathcal{E}) = r$ in each of Cases (i), (ii), (iv), (vi), and (vii). It follows from the fact that there are moment points M that belong to any of $\triangle ABC^\uparrow$, $\triangle BA_iC$, $\text{conv}\{B, C, \widehat{AA_i}\}$, $i = 1, 2$, which are convex combinations of points situated above and right to A. Therefore, for every $0 < r < a^*$ it is possible to construct a box such that $M = M_{r\mathcal{E}}$ satisfies conditions of (22.60), (22.61), (22.63), (22.65), or (22.66).

The remaining two cases need a more thorough treatment. Formulae (22.9) and (22.15) can be used in describing the Prokhorov radius of moment neighborhoods of some δ_a iff $(1 - r)A_i + rB$, $i = 1, 2$, respectively, lie above the level $m_3 = a^3$ for some $0 < r < a^*$. The conditions coincide with positivity of

$$\beta_i(r) = (1 - r)\left[1 + \frac{r^2}{(3a)}\right]^{3i/2} + r\left(1 - \frac{r}{a}\right)^3 - 1, \qquad 0 < r < a^*, \quad i = 1, 2. \tag{22.90}$$

We can check that $\beta_1 \leq \beta_2$, $\beta_i(0) = \beta_i'(0) = 0$ for all $a > 0$, and $\beta_i''(0) > 0$ iff $a < i/6$, $i = 1, 2$. Accordingly, (22.9) and (22.15) are applicable in evaluating convergence rates for $a < 1/6$ and $1/3$, respectively. A numerical analysis shows that these are sufficient conditions for positivity of (22.90) on the whole domains. On the other hand, β_i, $i = 1, 2$, are nonpositive everywhere if $a \geq 1 - (3/4)^{3i/2} \approx$ either 0.3505 or 0.5781, respectively. For intermediate a, the functions are negative about 0 and positive about $a^* = a$, which means that (22.9), (22.15) are of use for sufficiently large boxes. For example, for $a = 0.25$, β_1 changes the sign at $r \approx 0.1196$ ($\beta_2 > 0$), and so does β_2 at $r \approx 0.1907$ ($\beta_1 < 0$) for $a = 0.45$. If $a = 0.34$, then $r_1 \approx 0.3162$ and $r_2 \approx 0.01004$ are the roots of β_i, $i = 1, 2$.

Chapter 23

Geometric Moment Methods Applied to Optimal Portfolio Management

In this chapter we start with the brief description of the basics of Geometric Moment theory method for Optimization of integrals due to Kemperman [229], [232], [234]. Then we give solutions to several new Moment problems with applications to Stock Market and Financial Mathematics. That is we give methods for optimal allocation of funds over stocks and bonds at maximum return. More precisely we describe here the Optimal Portfolio Management under Optimal selection of securities so to maximize profit. The above are done within the models of Optimal frontier and Optimizing concavity. This treatment follows [44].

23.1 Introduction

The main problem we deal with and solve here is: the optimal allocation of funds over stocks and bonds and at the same time given certain level of expectation, best choice of securities on the purpose to maximize return. The results are very general so that they stand by themselves as "formulas" to treat other similar stochastic situations and structures far away from the stock market and financial mathematics. The answers to the above described problem are given under two models of investing, the Optimal frontier and Optimizing concavity, as being the most natural.

There are presented many examples all motivated from financial mathematics and of course fitting and working well there. The method of proof derives from the Geometric Moment theory of Kemperman, see [229], [232], [234], and several new Moment results of very general nature are given here. We start the chapter with basic Geometric Moment review and we show the proving tool we use next repeatedly.

The continuation of this material will be one to derive algorithms out of this theory and create computer software of implementation and work with actual numerical data of the Stock Market.

301

23.2   Preliminaries

Geometric Moment Theory (see [229], [232], [234], of J. Kemperman). Let $g_1, \ldots, g_n$ and h be given real-valued $(\mathcal{B} \cap \mathcal{A})$ (Borel and $\mathcal{A}$)-measurable functions on a fixed measurable space $T = (T, \mathcal{A})$. We also suppose that all 1-point sets $\{t\}$, $t \in T$ are $(\mathcal{B} \cap \mathcal{A})$-measurable. We would like to find the best upper and lower bounds on the integral

$$\mu(h) = \int_T h(t)\mu(dt),$$

given that μ is a (with respect to $\mathcal{A}$) probability measure on T with given moments

$$\mu(g_j) = y_j, \qquad j = 1, \ldots, n.$$

We denote by $m^+ = m^+(T)$ the collection of all $\mathcal{A}$-probability measures on T such that $\mu(|g_j|) < \infty$ $(j = 1, \ldots, n)$ and $\mu(|h|) < \infty$. For each $y = (y_1, \ldots, y_n) \in \mathbb{R}^n$, consider the bounds $L(y) = L(y \mid h) = \inf \mu(h)$, $U(y) = U(y \mid h) = \sup \mu(h)$, such that

$$\mu \in m^+(T); \quad \mu(g_j) = y_j, \qquad j = 1, \ldots, n.$$

If there is not such measure μ we set $L(y) = \infty$, $U(y) = -\infty$. Let $M^+(T)$ be the set of all probability measures on T that are finitely supported. By the next Theorem 23.1 we find that

$$L(y \mid h) = \inf\{\mu(h) \colon \mu \in M^+(T), \mu(g) = y\}, \tag{23.1}$$

and

$$U(y \mid h) = \sup\{\mu(h) \colon \mu \in M^+(T), \mu(g) = y\}. \tag{23.2}$$

Here $\mu(g) = y$ means $\mu(g_j) = y_j$, $j = 1, \ldots, n$.

Theorem 23.1. (Richter [307], Rogosinsky [310], Mulholland and Rogers [269]). *Let $f_1, \ldots, f_N$ be given real-valued Borel measurable functions on a measurable space $\Omega = (\Omega, \mathcal{F})$. Let μ be a probability measure on Ω such that each f_i is integrable with respect to μ. Then there exists a probability measure μ' of finite support on Ω satisfying*

$$\mu'(f_j) = \mu(f_j) \quad \text{for all} \ \ j = 1, \ldots, N.$$

One can even attain that the support of μ' has at most $N + 1$ points.

Hence from now on we deal only with finitely supported probability measures on T. Consequently our initial problem is restated as follows.

Let $T \neq \emptyset$ set and $g \colon T \to \mathbb{R}^n$, $h \colon T \to \mathbb{R}$ be given $(\mathcal{B} \cap \mathcal{A})$-measurable functions on T, where $g(t) = (g_1(t), \ldots, g_n(t))$. We want to find $L(y \mid h)$ and $U(y \mid h)$ defined by (23.1) and (23.2).

Here a very important set is

$$V = \operatorname{conv} g(T) \subseteq \mathbb{R}^n,$$

where "conv" means convex hull, and the range

$$g(T) = \{z \in \mathbb{R}^n \colon z = g(t) \text{ for some } t \in T\}.$$

Clearly g is a curve in n-space (if T is an one-dimensional interval) or a two-dimensional surface in n-space (if T is a square).

Lemma 23.2. *We have* $y \in V$ *iff there exists* $\mu \in M^+(T)$ *such that* $\mu(g) = y$. *Hence, by (23.1),*

$$L(y \mid h) < \infty \quad \text{iff} \quad y \in V.$$

Similarly, from (23.2),

$$U(y \mid h) > -\infty \quad \text{iff} \quad y \in V.$$

We get easily that the interior of V, $int(V) \neq \emptyset$ iff $1, g_1, \ldots, g_n$ are linearly independent on T.

So without loss of generality we suppose that $\{1, g_1, \ldots, g_n\}$ is a linearly independent set of functions on T, hence $int(V) \neq \emptyset$. Also clearly we have $U(y \mid h) = -L(y \mid -h)$, and $L(y \mid h)$ is a convex function for $y \in V$, while $U(y \mid h)$ is a concave function for $y \in V$. We consider $V^* = \text{conv } g^*(T) \subseteq \mathbb{R}^{n+1}$, where $g^*(t) = (g_1(t), \ldots, g_n(t), h(t))$, $t \in T$. By Lemma 23.2 we have that $(y_1, \ldots, y_n, y_{n+1}) \in V^*$ iff there exists $\mu \in M^+(T)$ such that $\mu(g_j) = y_j, j = 1, \ldots, n$ and $\mu(h) = y_{n+1}$. Clearly $L(y \mid h)$ equals to $y_{n+1} \in \mathbb{R} \cup \{-\infty\}$ which corresponds to the lowest point (in the infimum sense) of V^* along the vertical line through $(y_1, y_2, \ldots, y_n, 0)$.

Similarly $U(y \mid h)$ equals to $\overline{y}_{n+1} \in \mathbb{R} \cup \{\infty\}$ which corresponds to the highest point (in the supremum sense) of V^* along the vertical line through $(y_1, \ldots, y_n, 0)$. Clearly the above optimal points of V^* could be boundary points of V^* since V^* is not always closed.

Example. Let T be a subset of $\mathbb{R}^n$ and $g(t) = t$ for all $t \in T$. Then the graphs of the functions $L(y \mid h)$ and $U(y \mid h)$ on $V = \text{conv}(T)$ correspond to the bottom part and top part, respectively, of the convex hull of the graph $h \colon T \to \mathbb{R}$.

The analytic method (based on the geometry of moment space) to evaluate $L(y \mid h)$ and $U(y \mid h)$ is highly more difficult and complicated and not mentioned here, see [229]. But it is necessary to use it when it is much more difficult to describe convex hulls in $\mathbb{R}^n$, a quite common fact for $n \geq 4$. Still in $\mathbb{R}^3$ it is already difficult but possible to describe precisely the convex hulls. Typically the convex hull V^* decomposes into many different regions, each with its own analytic formulae for $L(y \mid h)$ and $U(y \mid h)$.

So the application of the above described geometric moment theory method is preferable for solving moment problems when only one or at the most two moment conditions are prescribed. Here we only use the geometric moment theory method. The derived results are elegant, very precise and simple, and the proving method relies on the basic geometric properties of related figures.

Claim: We can use in the same way and equally the expectations of random variables (r.v.) defined on a non-atomic probability space or the associated integrals against probability measures over a real domain for the standard moment problem defined below.

Let $(\varrho, \mathcal{F}, P)$ be a nonatomic probability space. Let X be r.v.'s on ϱ taking values (a.s.) in a finite or infinite interval of $\mathbb{R}$ which we call it a "real domain" denoted by $\mathcal{J}$. Here E denotes the expectation. We want to find $\inf_X E(h(X))$, and $\sup_X E(h(X))$, such that

$$Eg_i(X) = d_i, \qquad i = 1,\ldots,n, \quad d_i \in \mathbb{R}, \tag{23.3}$$

where h, g_i are Borel measurable functions from $\mathcal{J}$ into $\mathbb{R}$ with $E(|h(X)|)$, $E(|g_i(X)|) < \infty$. Clearly here $h(X)$, $g_i(X)$ are $\mathcal{F}$-measurable real valued functions, that is r.v.'s on ϱ. Naturally each X has a probability distribution function F which corresponds uniquely to a Borel probability measure μ, i.e., $Law(X) = \mu$. Hence

$$Eg_i(X) = \int_{\mathcal{J}} g_i(x)\, dF(x) = \int_{\mathcal{J}} g_i(x)\, d\mu(x) = d_i, \qquad i = 1,\ldots,n$$

and

$$Eh(X) = \int_{J} h(x)\, dF(x) = \int_{J} h(x)\, d\mu(x). \tag{23.4}$$

Next consider the related moment problem in the integral form: calculate

$$\inf_{\mu} \int_{J} h(t)\, d\mu(t), \quad \sup_{\mu} \int_{J} h(t)\, d\mu(t),$$

where μ are all the Borel probability measures such that

$$\int_{J} g_i(t)\, d\mu(t) = d_i, \qquad i = 1,\ldots,n, \tag{23.5}$$

where h, g_i are Borel measurable functions such that

$$\int_{J} |h(t)|\, d\mu(t), \quad \int_{J} |g_i(t)|\, d\mu(t) < \infty.$$

We state

Theorem 23.3. *It holds*

$$\inf_{\substack{(\sup) \\ \text{all } X \text{ as in (23.3)}}} Eh(X) = \inf_{\substack{(\sup) \text{ all } \mu \text{ as in (23.5)}, \\ \textit{Borel probability measures}}} \int_{J} h(t)\, d\mu(t). \tag{23.6}$$

Proof. Easy. $\qquad\qquad\qquad\qquad\qquad\qquad\qquad\qquad\qquad\qquad\qquad\square$

Note. One can easily and similarly see that the equivalence of moment problems is also valid either these are written in expectations form involving functions of random vectors or these are written in integral form involving multidimensional functions and products of probability measures over real domains of $\mathbb{R}^k$, $k > 1$. Of course again the underlied probability space should be non-atomic.

23.3 Main Results

Part I.

We give the following moment result to be used a lot in our work.

Theorem 23.4. *Let $(\varrho, \mathcal{F}, P)$ be a non-atomic probability space. Let X, Y be random variables on ϱ taking values (a.s.) in $[-a, a]$ and $[-\beta, \beta]$, respectively, $a, \beta > 0$. Denote $\gamma := E(XY)$. We would like to find $\sup\limits_{X,Y} \gamma$, $\inf\limits_{X,Y} \gamma$ such that the moments*

$$EX = \mu_1, \quad EY = \nu_1, \quad \mu_1 \in [-a, a], \quad \nu_1 \in [-\beta, \beta]$$

are given.

Consider the triangles

$$T_1 := \langle\{(-a, -\beta), (a, -\beta), (a, \beta)\}\rangle,$$
$$T_2 := \langle\{(-a, -\beta), (-a, \beta), (a, \beta)\}\rangle,$$
$$T_3 := \langle\{(a, -\beta), (a, \beta), (-a, \beta)\}\rangle,$$
$$T_4 := \langle\{(a, -\beta), (-a, \beta), (-a, -\beta)\}\rangle.$$

Then it holds (Here $\lambda, \rho, \varphi \geq 0: \lambda + \rho + \varphi = 1$.)

I) If $(\mu_1, \nu_1) \in T_1 \cap T_4$, i.e. $\mu_1 = a(-\lambda + \rho)$, $\nu_1 = \beta(\varphi - 1)$, then

$$\sup_{X,Y} \gamma = -\beta\mu_1 + a\nu_1 + a\beta,$$
$$\inf_{X,Y} \gamma = -\beta\mu_1 - a\nu_1 - a\beta. \tag{23.7}$$

II) If $(\mu_1, \nu_1) \in T_1 \cap T_3$, i.e. $\mu_1 = a(1 - \varphi)$, $\nu_1 = \beta(-\lambda + \rho)$, then

$$\sup_{X,Y} \gamma = -\beta\mu_1 + a\nu_1 + a\beta,$$
$$\inf_{X,Y} \gamma = \beta\mu_1 + a\nu_1 - a\beta. \tag{23.8}$$

III) If $(\mu_1, \nu_1) \in T_2 \cap T_3$, i.e. $\mu_1 = a(\lambda - \rho)$, $\nu_1 = \beta(1 - \varphi)$, then

$$\sup_{X,Y} \gamma = \beta\mu_1 - a\nu_1 + a\beta,$$
$$\inf_{X,Y} \gamma = \beta\mu_1 + a\nu_1 - a\beta. \tag{23.9}$$

IV) If $(\mu_1, \nu_1) \in T_2 \cap T_4$, i.e. $\mu_1 = a(\varphi - 1)$, $\nu_1 = \beta(\lambda - \rho)$, then

$$\sup_{X,Y} \gamma = \beta\mu_1 - a\nu_1 + a\beta,$$
$$\inf_{X,Y} \gamma = -\beta\mu_1 - a\nu_1 - a\beta. \tag{23.10}$$

Proof. Here $\gamma := EXY$, such that

$$-a \leq X \leq a, \quad -\beta \leq Y \leq \beta, \quad a, \beta > 0,$$

and the underlied probability space is non-atomic. That is,

$$\gamma = \int_{[-a,a]\times[-\beta,\beta]} xy \, d\mu, \tag{23.11}$$

where μ is a probability measure on $[-a,a] \times [-\beta,\beta]$. We suppose here that

$$\int_{[-a,a]\times[-\beta,\beta]} x \, d\mu = \mu_1, \qquad \int_{[-a,a]\times[-\beta,\beta]} y \, d\mu = \nu_1, \tag{23.12}$$

where $\mu_1 \in [-a,a]$, $\nu_1 \in [-\beta,\beta]$. Clearly we must have always $-a \le \mu_1 \le a$, $-\beta \le \nu_1 \le \beta$.

Here we consider and study the surface $S\colon z = xy$, $|x| \le a$, $|y| \le \beta$. We consider also the tetrahedron V with vertices $A := (a, \beta, a\beta)$, $B := (-a, -\beta, a\beta)$, $C := (a, -\beta, -a\beta)$, $D = (-a, \beta, -a\beta)$. Notice $(0,0,0) \in V$ and we establish that V is the convex hull of the surface S.

Clearly we have

1) Line segment (ℓ_1) through points B and C:

$$\left. \begin{array}{l} x = -a + 2at \\ y = -\beta \\ z = a\beta - 2a\beta t \end{array} \right\}, \qquad 0 \le t \le 1. \tag{23.13}$$

Notice that $xy = z$ so that (ℓ_1) is on surface S.

2) Line segment (ℓ_2) through points A, C:

$$\left. \begin{array}{l} x = a \\ y = -\beta + 2\beta t \\ z = -a\beta + 2a\beta t \end{array} \right\}, \qquad 0 \le t \le 1. \tag{23.14}$$

Notice that $xy = z$ so that (ℓ_2) is on surface S.

3) Line segment (ℓ_3) through points A, D:

$$\left. \begin{array}{l} x = -a + 2at \\ y = \beta \\ z = -a\beta + 2a\beta t \end{array} \right\}, \qquad 0 \le t \le 1. \tag{23.15}$$

Notice that $xy = z$ so that (ℓ_3) is on surface S.

4) Line segment (ℓ_4) through points B, D:

$$\left. \begin{array}{l} x = -a \\ y = -\beta + 2\beta t \\ z = a\beta - 2a\beta t \end{array} \right\}, \qquad 0 \le t \le 1. \tag{23.16}$$

Notice that $xy = z$ so that (ℓ_4) is on surface S.

We consider also the following line segments.

5) Line segment (ℓ_5) through the points C, D:

$$\begin{array}{l} x = a - 2at \\ y = -\beta + 2\beta t \, , \qquad 0 \le t \le 1. \\ z = -a\beta \end{array} \tag{23.17}$$

6) Line segment (ℓ_6) through the points A, B:

$$\left.\begin{array}{l} x = a - 2at \\ y = \beta - 2\beta t \\ z = a\beta \end{array}\right\}, \quad 0 \le t \le 1. \tag{23.18}$$

Therefore the surface of V consists on the top by the triangles: $(A \stackrel{\triangle}{B} D)$ and $(A \stackrel{\triangle}{B} C)$, and in the bottom by the triangles $(B \stackrel{\triangle}{C} D)$ and $(A \stackrel{\triangle}{C} D)$.

In one case we have $(\mu_1, \nu_1) \in$ triangle $T_1 := \langle\{(-a, -\beta), (a, -\beta), (a, \beta)\}\rangle$, and in another case $(\mu_1, \nu_1) \in$ triangle $T_2 := \langle\{(-a, -\beta), (-a, \beta), (a, \beta)\}\rangle$. Thus $(\mu_1, \nu_1) \in T_1$ iff $\exists \lambda, \rho, \varphi \ge 0$: $\lambda + \rho + \varphi = 1$ with $\mu_1 = a(1 - 2\lambda)$ and $\nu_1 = \beta(2\varphi - 1)$. Also $(\mu_1, \nu_1) \in T_2$ iff $\exists \lambda^*, \rho^*, \varphi^* \ge 0$: $\lambda^* + \rho^* + \varphi^* = 1$ with $\mu_1 = a(2\varphi^* - 1)$ and $\nu_1 = \beta(1 - 2\lambda^*)$.

Description of the faces of V:

1) Plane P_1 through A, B, C has equation

$$z = -\beta x + ay + a\beta, \tag{23.19}$$

and it corresponds to triangle T_1.

2) Plane P_2 through A, B, D has equation

$$z = \beta x - ay + a\beta, \tag{23.20}$$

and it corresponds to triangle T_2.

3) Plane P_3 through A, C, D has equation

$$z = \beta x + ay - a\beta \tag{23.21}$$

and it corresponds to triangle T_3. Here

$$T_3 := \langle\{(a, -\beta), (a, \beta), (-a, \beta)\}\rangle.$$

In fact $(\mu_1, \nu_1) \in T_3$ iff $\exists \lambda, \rho, \varphi \ge 0$: $\lambda + \rho + \varphi = 1$ with

$$\mu_1 = a(1 - 2\varphi), \quad \nu_1 = \beta(1 - 2\lambda).$$

4) Plane P_4 through B, C, D has equation

$$z = -\beta x - ay - a\beta \tag{23.22}$$

and it corresponds to triangle T_4. Here

$$T_4 := \langle\{(a, -\beta), (-a, \beta), (-a, -\beta)\}\rangle.$$

In fact $(\mu_1, \nu_1) \in T_4$ iff $\exists \lambda^*, \rho^*, \varphi^* \ge 0$: $\lambda^* + \rho^* + \varphi^* = 1$ with

$$\mu_1 = a(2\lambda^* - 1) \quad \text{and} \quad \nu_1 = \beta(2\rho^* - 1).$$

The justification of V is clear.

Conclusion. We have seen that the tetrahedron V is the convex hull of surface S: $z = xy$ over $[-a, a] \times [-\beta, \beta]$. Call $O := (o, o)$, $A_1 := (a, \beta)$, $A_2 := (-a, \beta)$,

$A_3 := (-a, -\beta)$, $A_4 := (a, -\beta)$. Clearly $(A_1 A_2 A_3 A_4)$ is an orthogonal parallelogram with sides of lengths $2a$ and 2β.

We notice the following for the triangles

$$T_1 = \left(A_3 \stackrel{\triangle}{A_4} O\right) \cup \left(A_4 \stackrel{\triangle}{O} A_1\right),$$

$$T_2 = \left(A_3 \stackrel{\triangle}{O} A_2\right) \cup \left(A_2 \stackrel{\triangle}{O} A_1\right),$$

$$T_3 = \left(A_4 \stackrel{\triangle}{O} A_1\right) \cup \left(A_1 \stackrel{\triangle}{O} A_2\right),$$

$$T_4 = \left(A_4 \stackrel{\triangle}{O} A_3\right) \cup \left(O \stackrel{\triangle}{A_3} A_2\right). \tag{23.23}$$

That is

$$T_1 \cap T_4 = \left(A_3 \stackrel{\triangle}{A_4} O\right), \quad T_1 \cap T_3 = \left(A_4 \stackrel{\triangle}{O} A_1\right),$$

$$T_2 \cap T_3 = \left(A_1 \stackrel{\triangle}{O} A_2\right), \quad T_2 \cap T_4 = \left(A_3 \stackrel{\triangle}{O} A_2\right). \tag{23.24}$$

Also we see the following:

1) Let $(x, y) \in \left(A_3 \stackrel{\triangle}{A_4} O\right)$ iff $(x, y) \in T_1 \cap T_4$ iff $x = a(-\lambda + \rho)$, $y = \beta(\varphi - 1)$, where

$$\lambda, \rho, \varphi \geq 0: \lambda + \rho + \varphi = 1.$$

2) Let $(x, y) \in \left(A_4 \stackrel{\triangle}{O} A_1\right)$ iff $(x, y) \in T_1 \cap T_3$ iff $x = a(1 - \varphi)$, $y = \beta(-\lambda + \rho)$, where

$$\lambda, \rho, \varphi \geq 0: \lambda + \rho + \varphi = 1.$$

3) Let $(x, y) \in \left(A_1 \stackrel{\triangle}{A_2} O\right)$ iff $(x, y) \in T_2 \cap T_3$ iff $x = a(\lambda - \rho)$, $y = \beta(1 - \varphi)$, where

$$\lambda, \rho, \varphi \geq 0: \lambda + \rho + \varphi = 1.$$

4) Let $(x, y) \in \left(A_2 \stackrel{\triangle}{A_3} O\right)$ iff $(x, y) \in T_2 \cap T_4$ iff $x = a(\varphi - 1)$, $y = \beta(\lambda - \rho)$, where

$$\lambda, \rho, \varphi \geq 0: \lambda + \rho + \varphi = 1.$$

Finally applying the geometric moment theory method presented in Section 23.2 we derive

I) Let $(\mu_1, \nu_1) \in T_1 \cap T_4$, then

$$\sup_{\mu} \int_{[-a,a] \times [-\beta,\beta]} xy \, d\mu = -\beta \mu_1 + a\nu_1 + a\beta,$$

$$\inf_{\mu} \int_{[-a,a] \times [-\beta,\beta]} xy \, d\mu = -\beta \mu_1 - a\nu_1 - a\beta. \tag{23.25}$$

II) Let $(\mu_1, \nu_1) \in T_1 \cap T_3$, then

$$\sup_{\mu} \int_{[-a,a] \times [-\beta,\beta]} xy \, d\mu = -\beta \mu_1 + a\nu_1 + a\beta$$

$$\inf_{\mu} \int_{[-a,a] \times [-\beta,\beta]} xy \, d\mu = \beta \mu_1 + a\nu_1 - a\beta. \tag{23.26}$$

III) Let $(\mu_1, \nu_1) \in T_2 \cap T_3$, then

$$\sup_{\mu} \int_{[-a,a] \times [-\beta,\beta]} xy \, d\mu = \beta\mu_1 - a\nu_1 + a\beta,$$

$$\inf_{\mu} \int_{[-a,a] \times [-\beta,\beta]} xy \, d\mu = \beta\mu_1 + a\nu_1 - a\beta. \tag{23.27}$$

IV) Let $(\mu_1, \nu_1) \in T_2 \cap T_4$, then

$$\sup_{\mu} \int_{[-a,a] \times [-\beta,\beta]} xy \, d\mu = \beta\mu_1 - a\nu_1 + a\beta$$

$$\inf_{\mu} \int_{[-a,a] \times [-\beta,\beta]} xy \, d\mu = -\beta\mu_1 - a\nu_1 - a\beta. \tag{23.28}$$

$\square$

23.3.1. The Financial Portfolio Problem

Let the r.v.'s X giving the annual return of one risky asset (say a stock) and Y giving the annual return of another risky asset (say a corporate bond) such that $EX = \mu_1$, $EX^2 = \mu_2$, $EY = \nu_1$, $EY^2 = \nu_2$ are known, and $EXY = \gamma$ may not be known. Here X, Y are not necessarily independent.

Denote by R the *return of a portfolio* of the stock and bond $R = wX + (1-w)Y$, where w is the fraction of funds allocated to the stock. Here we would like to find the optimal w^* such that the *utility function*

$$u(R) = ER - \lambda \operatorname{Var} R \tag{23.29}$$

is maximized, where $\lambda > 0$ is the *risk-avertion parameter* which is fixed.

The above is the *optimal frontier problem*. Notice that X, Y are not necessarily normal random variables, typically they have variances much larger than means. So the *not short selling* case is

$$\max u(R) \colon 0 \le w \le 1. \tag{23.30}$$

In general may be $\ell \le w \le u$, $-\infty < \ell < u < \infty$. If $\ell < 0$, it means one can *sell* the stock and *put* it all in the bond.

In Part II of the chapter we deal in the same manner with another utility function

$$u(R) = Eg(R), \tag{23.31}$$

where $g(x)$ is a strictly concave function. This is the Optimizing concavity problem. However typically we deal with many random variables X_i, Y_i, $i = 1, \ldots, n$ that are the annual returns of n stocks and bonds, respectively. Thus in general in portfolio we have the return $R_p = \sum_{i=1}^{N} w_i X_i$, where X_i is the annual return from stock i or bond i.

Therefore we need to calculate

$$\max_{w_1, \ldots, w_N} u(R_p), \quad \text{such that} \quad u_i \le w_i \le \ell_i, \quad i = 1, \ldots, N, \tag{23.32}$$

and u_i, ℓ_i are fixed known limits. To maximize $u(R_p)$ is not yet known even if $i = 2$ and $0 \le w \le 1$. We present these for the first time here.

Typically here the distributions of X_i are not known. We work first the special important case of

$$R = wX + (1 - w)Y, \quad 0 \le w \le 1 \ \text{ or } \ \ell \le w \le u, \tag{23.33}$$

and

$$u(R) = ER - \lambda \, \text{Var} \, R, \quad \lambda > 0. \tag{23.34}$$

One can easily observe that

$$u(R) = wEX + (1 - w)EY - \lambda(E(R^2) - (ER)^2)$$
$$= w\mu_1 + (1 - w)\nu_1 - \lambda\{w^2 EX^2 + (1 - w)^2 EY^2 + 2w(1 - w)EXY$$
$$- w^2\mu_1^2 - (1 - w)^2\nu_1^2 - 2w(1 - w)\mu_1\nu_1\}.$$

So after calculations we find that

$$f(w) := u(R) = Tw^2 + Mw + K, \tag{23.35}$$

where $\lambda > 0$,

$$T := -\lambda\mu_2 - \lambda\nu_2 + 2\lambda\gamma + \lambda\mu_1^2 + \lambda\nu_1^2 - 2\lambda\mu_1\nu_1, \tag{23.36}$$
$$M := \mu_1 - \nu_1 + 2\lambda\nu_2 - 2\lambda\gamma - 2\lambda\nu_1^2 + 2\lambda\mu_1\nu_1, \tag{23.37}$$

and

$$K := \nu_1 - \lambda\nu_2 + \lambda\nu_1^2. \tag{23.38}$$

To have maximum for $f(w)$ we need $T < 0$.

To find $\max\limits_{w,\gamma} u(R)$ is very difficult and not as important as the following modifications of the problem. More precisely we would like to calculate

$$\max_{w}\left(\max_{\gamma} u(R)\right), \tag{23.39}$$

the *best allocation with the best possible dependence between assets*, and

$$\max_{w}\left(\min_{\gamma} u(R)\right), \tag{23.40}$$

the *best allocation under the worst possible dependence between the assets*. According to Theorem 23.4 we obtain

$$\gamma_{\inf} \le \gamma \le \gamma_{\sup}, \quad \text{i.e. } \ -\gamma_{\sup} \le -\gamma \le -\gamma_{\inf}. \tag{23.41}$$

So in the last modified problem γ could be known and fixed, or unknown and variable between these bounds. Hence for fixed γ we find $\max\limits_{w} u(R)$, and for variable γ, using Theorem 23.4, we find tight upper and lower bounds for it, both cases are being treated in a similar manner.

Denote

$$A(w) := (-\lambda\mu_2 - \lambda\nu_2 + \lambda\mu_1^2 + \lambda\nu_1^2 - 2\lambda\mu_1\nu_1)w^2$$
$$+ (\mu_1 - \nu_1 + 2\lambda\nu_2 - 2\lambda\nu_1^2 + 2\lambda\mu_1\nu_1)w + (\nu_1 - \lambda\nu_2 + \lambda\nu_1^2). \quad (23.42)$$

Therefore from (23.35) we obtain

$$u(R) = A(w) + 2\lambda w(w-1)\gamma. \quad (23.43)$$

In case of $0 \le w \le 1$ we have

$$\min_{\substack{\max \\ \gamma}} u(R) = A(w) + 2\lambda w(w-1)\left(\max_{\min}\gamma\right), \quad (23.44)$$

i.e.

$$\min_{\substack{\max \\ \gamma}} u(R) = A(w) + 2\lambda w(1-w)\left(-\left(\sup_{\inf}\gamma\right)\right). \quad (23.45)$$

Using Theorem 23.4 in the last case (23.45) we find tight upper and lower bounds for $\max_w u(R)$.

Proposition 23.5. *Let γ be fixed and $w \in [0,1]$ or $w \in [\ell, u]$, $-\infty < \ell < u < \infty$. Suppose $T < 0$ and $-\frac{M}{2T} \in [0,1]$ or $-\frac{M}{2T} \in [\ell, u]$. Then the maximum in (23.35) is*

$$f\left(-\frac{M}{2T}\right) = \max_w u(R) = \frac{-M^2 + 4TK}{4T}. \quad (23.46)$$

E.g. $\lambda = 1$, $\gamma = 6$, $\mu_2 = 5$, $\nu_2 = 10$, $\mu_1 = 2$, $\nu_1 = 3$. Then $T = -2 < 0$, $M = 1$, $K = 2$, $-\frac{M}{2T} = 0.25 \in [0,1]$, and $\max_w u(R) = \frac{17}{8}$.

In another example we can take

$$\lambda = 1, \ \mu_1 = -10, \ \nu_1 = -20, \ \mu_2 = 4, \ \nu_2 = 6, \ \gamma = 2,$$

then $T = 94 > 0$.

Proposition 23.6. *Let γ be fixed and $w \in [0,1]$ or $w \in [\ell, u]$, $-\infty < \ell < u < \infty$. Here $T > 0$ or $T < 0$ and $-\frac{M}{2T} \notin [0,1]$ or $T < 0$ and $-\frac{M}{2T} \notin [\ell, u]$. Then the maximum in (23.35) is*

$$\max_{w\in[0,1]} u(R) = \max\{f(0), f(1)\} = \max\{K, \lambda(\mu_1^2 - \mu_2) + \mu_1\}, \quad (23.47)$$

and

$$\max_{w\in[u,\ell]} u(R) = \max\{f(u), f(\ell)\}. \quad (23.48)$$

Application of γ-moment problem (Theorem 23.4) and Propositions 23.5 and 23.6. Let $0 \le w \le 1$, from (23.43) we find

$$u(R) = A(w) + 2\lambda w(1-w)(-\gamma), \quad (23.49)$$

where $\gamma = EXY$ and $A(w)$ as in (23.42). Then

$$u(R) \le A(w) + 2\lambda w(1-w)(-\gamma_{\inf}) =: \theta(w), \quad (23.50)$$

and

$$u(R) \geq A(w) + 2\lambda w(1 - w)(-\gamma_{\sup}) =: L(w). \tag{23.51}$$

Hence one can maximize $\theta(w)$ and $L(w)$ in w using Propositions 23.5, 23.6. That is finding tight estimates for the crucial quantities (23.39) and (23.40). That is, we obtain

$$\max_{w \in [0,1]} \left(\max_{\gamma} u(R) \right) \leq \max_{w \in [0,1]} \theta(w), \tag{23.52}$$

and

$$\max_{w \in [0,1]} \left(\min_{\gamma} u(R) \right) \geq \max_{w \in [0,1]} L(w). \tag{23.53}$$

Notice that in expressing $u(R)$ we use all moments μ_1, ν_1, μ_2, ν_2.

23.3.2. The Portfolio Problem Involving Three Securities (similarly one can treat the problem of more than three securities).

Let X_1, X_2, X_3 be r.v.'s giving the annual return of three risky assets (stocks or bonds).

Let $R := w_1 X_1 + w_2 X_2 + w_3 X_3$, $\ell_i \leq w_i \leq u_i$, $i = 1, 2, 3$ the return of a portfolio, and the *utility function* $u(R) = ER - \lambda \operatorname{Var} R$, $\lambda > 0$ *the risk-avertion parameter* being fixed. Again we would like to find the *optimal* (w_1, w_2, w_3) such that $u(R)$ *is maximized*. Clearly

$$u(R) = w_1 EX_1 + w_2 EX_2 + w_3 EX_3 - \lambda\big(E(R^2) - (ER)^2\big).$$

Set $\mu_{1i} := EX_i$, $\mu_{2i} := EX_i^2$, $i = 1, 2, 3$. We have $\mu_{1i} \leq \mu_{2i}^{1/2}$, $i = 1, 2, 3$. Hence we obtain

$$
\begin{aligned}
u(R) = {}& w_1 \mu_{11} + w_2 \mu_{12} + w_3 \mu_{13} - \lambda\big[w_1^2 EX_1^2 + w_2^2 EX_2^2 + w_3^2 EX_3^2 \\
& + 2w_1 w_2 EX_1 X_2 + 2w_2 w_3 EX_2 X_3 + 2w_3 w_1 EX_3 X_1 - w_1^2 \mu_{11}^2 - w_2^2 \mu_{12}^2 - w_3^2 \mu_{13}^2 \\
& - 2w_1 w_2 \mu_{11} \mu_{12} - 2w_2 w_3 \mu_{12} \mu_{13} - 2w_1 w_3 \mu_{11} \mu_{13}\big].
\end{aligned}
$$

Call $\gamma_1 := EX_1 X_2$, $\gamma_2 := EX_2 X_3$, $\gamma_3 = EX_3 X_1$, then

$$\gamma_1 \leq \mu_{21}^{1/2} \mu_{22}^{1/2}, \quad \gamma_2 \leq \mu_{22}^{1/2} \mu_{23}^{1/2}, \quad \gamma_3 \leq \mu_{23}^{1/2} \mu_{21}^{1/2}.$$

Therefore it holds

$$
\begin{aligned}
u(R) = {}& w_1 \mu_{11} + w_2 \mu_{12} + w_3 \mu_{13} - w_1^2 \lambda \mu_{21} - w_2^2 \lambda \mu_{22} - w_3^2 \lambda \mu_{23} \\
& - 2w_1 w_2 \lambda \gamma_1 - 2w_2 w_3 \lambda \gamma_2 - 2w_3 w_1 \lambda \gamma_3 + w_1^2 \lambda \mu_1^2 + w_2^2 \lambda \mu_{12}^2 + w_3^2 \lambda \mu_{13}^2 \\
& + 2w_1 w_2 \lambda \mu_{11} \mu_{12} + 2w_2 w_3 \lambda \mu_{12} \mu_{13} + 2w_1 w_3 \lambda \mu_{11} \mu_{13}.
\end{aligned}
$$

Consequently we have

$$
\begin{aligned}
g(w_1, w_2, w_3) := u(R) = {}& (-\lambda \mu_{21} + \lambda \mu_{11}^2) w_1^2 + (-\lambda \mu_{22} + \lambda \mu_{12}^2) w_2^2 \\
& + (-\lambda \mu_{23} + \lambda \mu_{13}^2) w_3^2 + (\mu_{11} w_1 + \mu_{12} w_2 + \mu_{13} w_3) \\
& + (-2\lambda \gamma_1 + 2\lambda \mu_{11} \mu_{12}) w_1 w_2 + (-2\lambda \gamma_2 + 2\lambda \mu_{12} \mu_{13}) w_2 w_3 \\
& + (-2\lambda \gamma_3 + 2\lambda \mu_{11} \mu_{13}) w_1 w_3. \tag{23.54}
\end{aligned}
$$

Here we would like first to study about the absolute maximum of g over $\mathbb{R}^3$. For that we set the first partials of g equal to zero, etc. Namely we have

$$D_1 g(w_1, w_2, w_3) = (-2\lambda\mu_{21} + 2\lambda\mu_{11}^2)w_1 + \mu_{11}$$
$$+ (-2\lambda\gamma_1 + 2\lambda\mu_{11}\mu_{12})w_2 + (-2\lambda\gamma_3 + 2\lambda\mu_{11}\mu_{13})w_3 = 0,$$
$$(23.55)$$

$$D_2 g(w_1, w_2, w_3) = (-2\lambda\mu_{22} + 2\lambda\mu_{12}^2)w_2 + \mu_{12}$$
$$+ (-2\lambda\gamma_1 + 2\lambda\mu_{11}\mu_{12})w_1 + (-2\lambda\gamma_2 + 2\lambda\mu_{12}\mu_{13})w_3 = 0,$$
$$(23.56)$$

and

$$D_3 g(w_1, w_2, w_3) = (-2\lambda\mu_{23} + 2\lambda\mu_{13}^2)w_3 + \mu_{13}$$
$$+ (-2\lambda\gamma_2 + 2\lambda\mu_{12}\mu_{13})w_2 + (-2\lambda\gamma_3 + 2\lambda\mu_{11}\mu_{13})w_1 = 0.$$
$$(23.57)$$

Furthermore we obtain

$$D_{11} g(w_1, w_2, w_3) = (-2\lambda\mu_{21} + 2\lambda\mu_{11}^2), \qquad (23.58)$$
$$D_{22} g(w_1, w_2, w_3) = (-2\lambda\mu_{22} + 2\lambda\mu_{12}^2), \qquad (23.59)$$

and

$$D_{33} g(w_1, w_2, w_3) = (-2\lambda\mu_{23} + 2\lambda\mu_{13}^2). \qquad (23.60)$$

Also we have

$$D_{12} g(w_1, w_2, w_3) = D_{21} g(w_1, w_2, w_3) = (-2\lambda\gamma_1 + 2\lambda\mu_{11}\mu_{12}), \qquad (23.61)$$
$$D_{31} g(w_1, w_2, w_3) = D_{13} g(w_1, w_2, w_3) = (-2\lambda\gamma_3 + 2\lambda\mu_{11}\mu_{13}), \qquad (23.62)$$

and

$$D_{23} g(w_1, w_2, w_3) = D_{32} g(w_1, w_2, w_3) = (-2\lambda\gamma_2 + 2\lambda\mu_{12}\mu_{13}). \qquad (23.63)$$

Next we form the determinant of second partials of g

$$\Delta_3(w) := \begin{vmatrix} -2\lambda\mu_{21} + 2\lambda\mu_{11}^2, & -2\lambda\gamma_1 + 2\lambda\mu_{11}\mu_{12}, & -2\lambda\gamma_3 + 2\lambda\mu_{11}\mu_{13} \\ -2\lambda\gamma_1 + 2\lambda\mu_{11}\mu_{12}, & -2\lambda\mu_{22} + 2\lambda\mu_{12}^2, & -2\lambda\gamma_2 + 2\lambda\mu_{12}\mu_{13} \\ -2\lambda\gamma_3 + 2\lambda\mu_{11}\mu_{13}, & -2\lambda\gamma_2 + 2\lambda\mu_{12}\mu_{13}, & -2\lambda\mu_{23} + 2\lambda\mu_{13}^2 \end{vmatrix}, \qquad (23.64)$$

where $w := (w_1, w_2, w_3) \in \mathbb{R}^3$.

One can have true $\Delta_3(w) \neq 0$. So we can solve the linear system over $\mathbb{R}^3$

$$\begin{cases} (-2\lambda\mu_{21} + 2\lambda\mu_{11}^2)w_1 + (-2\lambda\gamma_1 + 2\lambda\mu_{11}\mu_{12})w_2 \\ \quad + (-2\lambda\gamma_3 + 2\lambda\mu_{11}\mu_{13})w_3 = -\mu_{11}, \\ (-2\lambda\gamma_1 + 2\lambda\mu_{11}\mu_{12})w_1 + (-2\lambda\mu_{22} + 2\lambda\mu_{12}^2)w_2 \\ \quad + (-2\lambda\gamma_2 + 2\lambda\mu_{12}\mu_{13})w_3 = -\mu_{12}, \\ (-2\lambda\gamma_3 + 2\lambda\mu_{11}\mu_{13})w_1 + (-2\lambda\gamma_2 + 2\lambda\mu_{12}\mu_{13})w_2 \\ \quad + (-2\lambda\mu_{23} + 2\lambda\mu_{13}^2)w_3 = -\mu_{13}. \end{cases} \qquad (23.65)$$

Here $\Delta_3(w)$ is the basic coefficient matrix of system (23.65).

Given that $\Delta_3(w) \neq 0$ the system (23.65) has a unique solution $w^* := (w_1^*, w_2^*, w_3^*) \in \mathbb{R}^3$. That is, w^* is a critical point for g.

Assuming that g is a concave function (it is also differentiable) then it *has an absolute maximum at w^* over $\mathbb{R}^3$* (see [196], p. 116, Theorem 3.7 and comments of "Functions of several variables" by Wendell Fleming). *If $w^* \in \prod_{i=1}^{3} [\ell_i, u_i]$ then the global maximum of g there will be $g(w^*)$.*

Note. In general let $Q(x, h) := \sum_{i,j=1}^{n} g_{ij}(x) h_i h_j$, where $g \in C^{(2)}(K)$, K-open convex $\subset \mathbb{R}^n$, $h := (h_1, \ldots, h_n)$, $x = (x_1, \ldots, x_n)$.

We need

Theorem 23.7 (p. 114, [196], W. Fleming).

(a) *g is convex on K iff $Q(x, \cdot) \geq 0$ (positive semidefinite), $\forall x \in K$,*
(b) *g is concave on K iff $Q(x, \cdot) \leq 0$ (negative semidefinite), $\forall x \in K$,*
(c) *if $Q(x, \cdot) > 0$ (positive definite), $\forall x \in K$, then g is strictly convex on K,*
(d) *if $Q(x, \cdot) < 0$ (negative definite), $\forall x \in K$, then g is strictly concave on K.*

Let $\Delta(x) := \det[g_{ij}(x)]$ and $\Delta_{n-k}(x)$ denote the determinant with $n - k$ rows, obtained by deleting the last k rows and columns of $\Delta(x)$. Put $\Delta_0(x) = 1$. For our particular function (23.54) we do have

$$g_{ij}(x) = g_{ji}(x), \qquad i, j = 1, 2, 3.$$

That is $Q(x, \cdot)$ is a symmetric form. From the theory of quadratic forms we have equivalently that $Q(x, \cdot)$ is *positive definite* iff the $(n + 1)$ numbers $\Delta_0(x)$, $\Delta_1(x), \ldots, \Delta_n(x) > 0$. Also $Q(x, \cdot)$ is *negative definite* iff the above $(n+1)$ numbers alternate in signs from positive to negative.

In our problem we have

$$\Delta_0(w) = 1, \tag{23.66}$$

$$\Delta_1(w) = -2\lambda\mu_{21} + 2\lambda\mu_{11}^2, \tag{23.67}$$

$$\Delta_2(w) = \begin{vmatrix} -2\lambda\mu_{21} + 2\lambda\mu_{11}^2, & -2\lambda\gamma_1 + 2\lambda\mu_{11}\mu_{12} \\ -2\lambda\gamma_1 + 2\lambda\mu_{11}\mu_{12}, & -2\lambda\mu_{22} + 2\lambda\mu_{12}^2 \end{vmatrix}, \tag{23.68}$$

and $\Delta_3(w)$ as before, *all independent* of any $w \in \mathbb{R}^3$. We can have examples of $\Delta_1(w) < 0$, $\Delta_2(w) > 0$, $\Delta_3(w) < 0$, $\forall w \in \mathbb{R}^3$, see next. Therefore in that case $Q(w, \cdot)$ is negative definite, i.e. $Q(w, \cdot) < 0$, $\forall w \in \mathbb{R}^3$. Thus (by Theorem 3.6, [196]) g is strictly concave in $\mathbb{R}^3$.

Note. Let f be a (strictly) concave function from $\mathbb{R}$ into $\mathbb{R}$. The easily one can prove that $g(w_1, w_2, w_3) := f(w_1) + f(w_2) + f(w_3)$ is (strictly) concave function from $\mathbb{R}^3$ into $\mathbb{R}$.

Example. Clearly $f(x) := -0.75x^2 + 0.5x$ is strictly concave. Hence

$$g(w_1, w_2, w_3) := (-0.75)(w_1^2 + w_2^2 + w_3^2) + (0.5)(w_1 + w_2 + w_3) \qquad (23.69)$$

is strictly concave too. This function g is related to the following setting.

Take $\lambda = 1$, $\mu_{11} = \mu_{12} = \mu_{13} = 1/2$, $\mu_{21} = \mu_{22} = \mu_{23} = 1$, $\gamma_1 = \gamma_2 = \gamma_3 = 1/4$. Then $\Delta_0(w) = 1$, $\Delta_1(w) = -1.5 < 0$, $\Delta_2(w) = 2.25 > 0$, $\Delta_3(w) = -3.375 < 0$. The linear system (23.65) reduces to $-1.5w_i = -0.5$, $i = 1, 2, 3$. Then $w_i^* = w_i = 0.33$, $i = 1, 2, 3$. That is, $(0.33, 0.33, 0.33)$ is the critical number of g as above (23.69). Hence

$$\max\big[(-0.75)(w_1^2 + w_2^2 + w_3^2) + (0.5)(w_1 + w_2 + w_3)\big] = 0.249975,$$

and $w_i^* \in [-1, 1]$ or in $[0, 1]$, $i = 1, 2, 3$, etc.

Comment. By Maximum and Minimum value theorem since $\chi := \prod_{i=1}^{3} [\ell_i, u_i]$, $\ell_i, u_i \in \mathbb{R}$ is a compact subset of $\mathbb{R}^3$ and since $g(w) = u(R)$ is a continuous real valued function over χ, then there are points w^* and w_* in χ such that

$$g(w^*) = \sup\{g(w) \colon w \in \chi\},$$
$$g(w_*) = \inf\{g(w) \colon w \in \chi\}. \qquad (23.70)$$

Remark 23.8. Let here $w_i \in [0, 1]$, $i = 1, 2, 3$. Then one can write

$$u(R) = B(w) + (2\lambda w_1 w_2)(-\gamma_1) + (2\lambda w_2 w_3)(-\gamma_2) + (2\lambda w_1 w_3)(-\gamma_3), \qquad (23.71)$$

where

$$\begin{aligned}
B(w) := & (-\lambda\mu_{21} + \lambda\mu_{11}^2)w_1^2 + (-\lambda\mu_{22} + \lambda\mu_{12}^2)w_2^2 \\
& + (-\lambda\mu_{23} + \lambda\mu_{13}^2)w_3^2 + \mu_{11}w_1 + \mu_{12}w_2 + \mu_{13}w_3 \\
& + 2\lambda\mu_{11}\mu_{12}w_1 w_2 + 2\lambda\mu_{12}\mu_{13}w_2 w_3 + 2\lambda\mu_{11}\mu_{13}w_1 w_3. \qquad (23.72)
\end{aligned}$$

By Theorem 23.4 again we find that

$$-\gamma_{i\,\sup} \le -\gamma_i \le -\gamma_{i\,\inf}, \qquad i = 1, 2, 3. \qquad (23.73)$$

Consequently we get

$$\begin{aligned}
u(R) \le & B(w) + (2\lambda w_1 w_2)(-\gamma_{1\,\inf}) + (2\lambda w_2 w_3)(-\gamma_{2\,\inf}) \\
& + (2\lambda w_1 w_3)(-\gamma_{3\,\inf}) =: \theta(w), \qquad (23.74)
\end{aligned}$$

also we obtain

$$\begin{aligned}
u(R) \ge & B(w) + (2\lambda w_1 w_2)(-\gamma_{1\,\sup}) + (2\lambda w_2 w_3)(-\gamma_{2\,\sup}) \\
& + (2\lambda w_1 w_3)(-\gamma_{3\,\sup}) =: L(w). \qquad (23.75)
\end{aligned}$$

Now we can maximize $\theta(w)$, $L(w)$ over $w \in [0, 1]^3$ as described earlier in this Section 23.3.2. That is finding tight estimates for the crucial quantities (23.39) and (23.40). That is, we find

$$\max_{w \in [0,1]^3} \left(\max_{(\gamma_1, \gamma_2, \gamma_3)} u(R) \right) \le \max_{w \in [0,1]^3} \theta(w), \qquad (23.76)$$

and

$$\max_{w \in [0,1]^3} \left(\min_{(\gamma_1, \gamma_2, \gamma_3)} u(R) \right) \geq \max_{w \in [0,1]^3} L(w), \tag{23.77}$$

Notice that in expressing $u(R)$ we use all moments μ_{1i}, μ_{2i}, $i = 1, 2, 3$.

23.3.3. Applications of γ-Moment Problem to Variance

Let $-a \leq X \leq a$, $-\beta \leq Y \leq \beta$ be r.v.'s, $a, \beta > 0$. Set $\gamma := EXY$, $\mu_1 := EX$, $\nu_1 := EY$, $\gamma, \mu_1, \nu_1 \in \mathbb{R}$. The covariance is

$$\text{cov}(X, Y) = \gamma - \mu_1 \nu_1 \tag{23.78}$$

and in our case it exists as it is easily observed by

$$\text{Var}(X) \leq 2a^2, \quad \text{Var}(Y) \leq 2\beta^2.$$

Then by Theorem 23.4 we have

$$\sup_{X,Y} \text{cov}(X, Y) = \sup_{X,Y} \gamma - \mu_1 \nu_1, \tag{23.79}$$

and

$$\inf_{X,Y} \text{cov}(X, Y) = \inf_{X,Y} \gamma - \mu_1 \nu_1. \tag{23.80}$$

We need to mention

Corollary 23.9 (see [235], Kingman and Taylor, p. 366). *It holds*

$$\text{Var}(X + Y) = \text{Var}(X) + \text{Var}(Y) + 2 \, \text{cov}(X, Y). \tag{23.81}$$

Thus we obtain

$$\sup_{X,Y} \text{Var}(X + Y) \leq \sup_{X,Y} \text{Var}(X) + \sup_{X,Y} \text{Var}(Y) + 2 \sup_{X,Y} \text{cov}(X, Y)$$

$$\leq \sup_{X} \text{Var}(X) + \sup_{Y} \text{Var}(Y) + 2 \sup_{X,Y} \text{cov}(X, Y). \tag{23.82}$$

And also we have

$$\inf_{X,Y} \text{Var}(X + Y) \geq \inf_{X,Y} \text{Var}(X) + \inf_{X,Y} \text{Var}(Y) + 2 \inf_{X,Y} \text{cov}(X, Y)$$

$$\geq \inf_{X} \text{Var}(X) + \inf_{Y} \text{Var}(Y) + 2 \inf_{X,Y} \text{cov}(X, Y). \tag{23.83}$$

Above we have $EX = \mu_1$, $EY = \nu_1$ as prescribed moments. Notice that $\text{Var}(X) = EX^2 - \mu_1^2$ and $\text{Var}(Y) = EY^2 - \nu_1^2$. So we would like to find $\inf_{\mu}^{\sup} \int_{-a}^{a} x^2 \, d\mu$ such that

$$\int_{-a}^{a} x \, d\mu = \mu_1, \qquad a > 0,$$

where μ is a probability measure on $[-a, a]$, etc. Applying the basic geometric moment theory one derives that

$$\sup_{X} \text{Var}(X) = a^2 - \mu_1^2,$$

$$\sup_{Y} \text{Var}(Y) = \beta^2 - \nu_1^2, \tag{23.84}$$

and

$$\inf_{X} \operatorname{Var}(X) = 0, \qquad \inf_{Y} \operatorname{Var}(Y) = 0. \tag{23.85}$$

We have proved

Proposition 23.10. *Let the r.v.'s X, Y be such that $-a \le X \le a$, $-\beta \le Y \le \beta$, $a, \beta > 0$ and $EX = \mu_1$, $EY = \nu_1$ are prescribed, where $\mu_1, \nu_1 \in \mathbb{R}$. Set $\gamma := EXY$. Then*

$$\sup_{X,Y} \operatorname{Var}(X + Y) \le a^2 + \beta^2 - (\mu_1 + \nu_1)^2 + 2\gamma_{\sup}, \tag{23.86}$$

and

$$\inf_{X,Y} \operatorname{Var}(X + Y) \ge 2\big(\gamma_{\inf} - \mu_1\nu_1\big). \tag{23.87}$$

Here $\gamma_{\sup}$ and $\gamma_{\inf}$ are given by Theorem 23.4.

(A) Next we generalize (23.86) and (23.87). Namely let $X_1, X_2, \ldots, X_n$ be r.v.'s such that $-a_i \le X_i \le a_i$, $a_i > 0$, $i = 1, \ldots, n$. Thus $E|X_i| \le a_i$ and $EX_i^2 \le a_i^2$. We write $v_{ij} := \operatorname{cov}(X_i, X_j)$ which do exist. Let $0 \le \ell_i \le w_i \le u_i$, $i = 1, \ldots, n$, usually $w_i \in [0, 1]$ be such that $\sum_{i=1}^{n} w_i = 1$. Here w_i's are fixed, they could be the percentages of allocations of funds to bonds and stocks.

Set

$$Z = \sum_{i=1}^{n} w_i X_i. \tag{23.88}$$

Then by Theorem 14.4, p. 366, [235], we obtain

$$\operatorname{Var}(Z) = \sum_{i=1}^{n} \sum_{j=1}^{n} w_i w_j v_{ij}. \tag{23.89}$$

That is, we have

$$\operatorname{Var}(Z) = \sum_{i=1}^{n} w_i^2 v_{ii} + \sum_{\substack{i=1 \\ i \ne j}}^{n} \sum_{j=1}^{n} w_i w_j \operatorname{cov}(X_i, X_j)$$

$$= \sum_{i=1}^{n} w_i^2 \operatorname{Var}(X_i) + \sum_{\substack{i=1 \\ i \ne j}}^{n} \sum_{j=1}^{n} w_i w_j \operatorname{cov}(X_i, X_j). \tag{23.90}$$

Set

$$\mu_{1i} := EX_i, \quad \gamma_{ij} := EX_i X_j, \qquad i, j = 1, \ldots, n. \tag{23.91}$$

Proposition 23.11. *We obtain*

$$\operatorname{Var}(Z) \le \sum_{i=1}^{n} w_i^2 (a_i^2 - \mu_{1i}^2) + \sum_{\substack{i=1 \\ i \ne j}}^{n} \sum_{j=1}^{n} w_i w_j \left(\sup_{\mu_{1i}, \mu_{1j}} \gamma_{ij} - \mu_{1i}\mu_{1j} \right), \tag{23.92}$$

and

$$\text{Var}(Z) \geq \sum_{\substack{i=1\\i\neq j}}^{n} \sum_{j=1}^{n} w_i w_j \left(\inf_{\mu_{1i},\mu_{1j}} \gamma_{ij} - \mu_{1i}\mu_{1j} \right). \tag{23.93}$$

(B) Let again $X_1,\ldots,X_n$ be r.v.'s such that $-a_i \leq X_i \leq a_i$, $a_i > 0$, $i = 1,\ldots,n$. Set $v_{ij} := \text{cov}(X_i, X_j)$. Let fixed w_i be such that $\ell_i \leq w_i \leq u_i$, $\ell_i, u_i \in \mathbb{R}$, $i = 1,\ldots,n$.

Set

$$Z = \sum_{i=1}^{n} w_i X_i. \tag{23.94}$$

Then by Theorem 14.4, p. 366 of [235] we get

$$\text{Var}(Z) = \sum_{i=1}^{n} \sum_{j=1}^{n} w_i w_j v_{ij}$$

$$= \sum_{i=1}^{n} w_i^2 \, \text{Var}(X_i) + \sum_{\substack{i=1\\i\neq j}}^{n} \sum_{j=1}^{n} w_i w_j \text{cov}(X_i, X_j). \tag{23.95}$$

Furthermore we obtain

$$\text{Var}(Z) = \sum_{i=1}^{n} w_i^2 \, \text{Var}(X_i) + \sum \sum_{\substack{i,j \in \{1,\ldots,n\}\\ i \neq j\\ w_i w_j \geq 0}} w_i w_j \, \text{cov}(X_i, X_j)$$

$$+ \sum \sum_{\substack{i,j \in \{1,\ldots,n\}\\ i \neq j\\ w_i w_j < 0}} w_i w_j \, \text{cov}(X_i, X_j). \tag{23.96}$$

Here we are given $EX_i = \mu_{1i}$, $EX_j = \mu_{1j}$, we set also $\gamma_{ij} := EX_iX_j$. By Theorem 23.4 we have

$$\gamma_{ij\,\text{inf}} \leq \gamma_{ij} \leq \gamma_{ij\,\text{sup}}. \tag{23.97}$$

Obviously we derive

$$\inf_{\mu_{1i},\mu_{1j}} \text{cov}(X_i, X_j) \leq \text{cov}(X_i, X_j) \leq \sup_{\mu_{1i},\mu_{1j}} \text{cov}(X_i, X_j). \tag{23.98}$$

We get

Proposition 23.12. *It holds*

$$\text{Var}(Z) \leq \sum_{i=1}^{n} w_i^2 (a_i^2 - \mu_{1i}^2) + \sum \sum_{\substack{i,j \in \{1,\ldots,n\}\\ i \neq j\\ w_i w_j \geq 0}} w_i w_j \left(\sup_{\mu_{1i},\mu_{1j}} \gamma_{ij} - \mu_{1i}\mu_{1j} \right)$$

$$+ \sum \sum_{\substack{i,j \in \{1,\ldots,n\}\\ w_i w_j < 0}} w_i w_j \left(\inf_{\mu_{1i},\mu_{1j}} \gamma_{ij} - \mu_{1i}\mu_{1j} \right), \tag{23.99}$$

and

$$\operatorname{Var}(Z) \geq \sum \sum_{\substack{i,j \in \{1,\ldots,n\} \\ i \neq j \\ w_i w_j \geq 0}} w_i w_j \left(\inf_{\mu_{1i},\mu_{1j}} \gamma_{ij} - \mu_{1i}\mu_{1j} \right)$$

$$+ \sum \sum_{\substack{i,j \in \{1,\ldots,n\} \\ i \neq j \\ w_i w_j < 0}} w_i w_j \left(\sup_{\mu_{1i},\mu_{1j}} \gamma_{ij} - \mu_{1i}\mu_{1j} \right). \tag{23.100}$$

Part II

Here we treat the *utility function*

$$u(R) = Eg(R), \tag{23.101}$$

where $g \colon \mathcal{K} \subseteq \mathbb{R} \to \mathbb{R}^+$, $\mathcal{K}$ is an interval, g is a concave function, e.g. $g(x) = \ell n(2 + x)$, $x \geq -1$, and $g(x) = (1 + x)^\alpha$, $0 < \alpha < 1$, $x \geq -1$. Also again

$$R := wX + (1 - w)Y, \qquad 0 \leq w \leq 1, \tag{23.102}$$

and X, Y, r.v.'s as in Section 23.3.1.

That is, it is

$$u(R) = Eg(wX + (1 - w)Y). \tag{23.103}$$

Here the objectives are as in Sections 23.3.1, 23.3.2, etc.

Lemma 23.13. *Let $g \colon \mathcal{K} \subseteq \mathbb{R} \to \mathbb{R}$, $\mathcal{K}$ interval, be a concave function. Then*

$$G(x,y) := g(wx + (1 - w)y), \qquad 0 \leq w \leq 1 \text{ fixed}, \tag{23.104}$$

is a concave function in (x, y).

Proof. Let $0 \leq \lambda \leq 1$, then we observe

$$\begin{aligned}
G(\lambda(x_1, y_1) + (1 - \lambda)(x_2, y_2)) &= G(\lambda x_1 + (1 - \lambda)x_2, \lambda y_1 + (1 - \lambda)y_2) \\
&= g(w(\lambda x_1 + (1 - \lambda)x_2) + (1 - w)(\lambda y_1 + (1 - \lambda)y_2)) \\
&= g(\lambda(wx_1 + (1 - w)y_1) + (1 - \lambda)(wx_2 + (1 - w)y_2)) \\
&\geq \lambda g(wx_1 + (1 - w)y_1) + (1 - \lambda)g(wx_2 + (1 - w)y_2)) \\
&= \lambda G(x_1, y_1) + (1 - \lambda)G(x_2, y_2).
\end{aligned} \qquad \square$$

So in our case $G(x, y)$ is concave and nonnegative. We would like to calculate $\inf_{\sup} u(R)$, i.e. to find

$$\inf_{\substack{\sup \\ \mu}} \int_{\mathcal{K}^2} g(wx + (1 - w)y) \, d\mu(x, y) \tag{23.105}$$

such that the moments

$$\int_{\mathcal{K}^2} x \, d\mu(x,y) = \mu_1, \qquad \int_{\mathcal{K}^2} y \, d\mu(x,y) = \nu_1 \tag{23.106}$$

are prescribed, $\mu_1, \nu_1 \in \mathbb{R}$. Here μ is a probability measure on $\mathcal{K}^2$.

We easily obtain via geometric moment theory that

$$\sup_{\mu} u(R) = g(w\mu_1 + (1-w)\nu_1), \quad g \geq 0, \tag{23.107}$$

and

$$\max_{w \in [0,1]} \left(\sup_{(X,Y)} u(R) \right) = \max_{w \in [0,1]} g(w\mu_1 + (1-w)\nu_1). \tag{23.108}$$

Examples

(i) Let $g(x) = \ell n(2+x)$, $x \geq -1$, then

$$\sup_{\mu} u(R) = \ell n(2 + w\mu_1 + (1-w)\nu_1). \tag{23.109}$$

Suppose here $\mu_1 \geq \nu_1 \geq -1$. Then $w\mu_1 + (1-w)\nu_1 \geq -1$, $0 \leq w \leq 1$. We would like to calculate

$$\max_{w \in [0,1]} \left(\sup_{\mu} u(R) \right) = \max_{w \in [0,1]} (\ell n(2 + w\mu_1 + (1-w)\nu_1).$$

We notice that

$$\max_{w \in [0,1]} (w\mu_1 + (1-w)\nu_1) = \mu_1 \quad \text{at } w = 1.$$

Consequently, since ℓn is a strictly increasing function, we obtain

$$\max_{w \in [0,1]} \left(\sup_{(X,Y)} u(R) \right) = \ell n(2 + \mu_1). \tag{23.110}$$

(ii) Let $g(x) = (1+x)^\alpha$, $0 < \alpha < 1$, $x \geq -1$. Then

$$\sup_{\mu} u(R) = (1 + w\mu_1 + (1-w)\nu_1)^\alpha. \tag{23.111}$$

Suppose again $\mu_1 \geq \nu_1 \geq -1$, then $w\mu_1 + (1-w)\nu_1 \geq -1$, $0 \leq w \leq 1$. The function $(1+x)^\alpha$ is strictly increasing. Therefore, similarly as in (i), we get

$$\max_{w \in [0,1]} \left(\sup_{(X,Y)} u(R) \right) = (1 + \mu_1)^\alpha. \tag{23.112}$$

Next we give another important moment result.

Theorem 23.14. *Let $a, \beta > 1$ and $T := \text{conv}\langle\{(0,\beta), (a,-1), (-a,-1)\}\rangle$, and the random pairs (X,Y) taking values on T, i.e.*

$$X = a(\rho - \varphi), \quad Y = \lambda(\beta + 1) - 1, \quad \lambda, \rho, \varphi \geq 0: \lambda + \rho + \varphi = 1. \tag{23.113}$$

Let $g\colon \mathbb{R} \to \mathbb{R}^+$ be a concave function and $0 \le w \le 1$. We would like to find

$$\inf_{(X,Y)} u(R) = \inf_{(X,Y)} Eg(wX + (1-w)Y) \tag{23.114}$$

such that

$$Ex = \mu_1, \qquad EY = \nu_1, \tag{23.115}$$

where $(\mu_1, \nu_1) \in [-1,1]^2$ is given. Also to determine

$$\max_{w \in [0,1]} \left(\inf_{(X,Y)} u(R) \right). \tag{23.116}$$

We prove that

$$\inf_{(X,Y)} u(R) = \left\{ g(w(1-a) - 1) \right.$$
$$+ \frac{(g(w(a+1) - 1) - g(w(1-a) - 1))}{2a}(\mu_1 + a) \tag{23.117}$$
$$\left. + \frac{(2g((1-w)\beta) - g(w(a+1) - 1) - g(w(1-a) - 1))}{2(\beta+1)}(\nu_1 + 1) \right\}.$$

One can apply $\max_{w \in [0,1]}$ to both sides of (23.117) to get (23.116).

Proof. Let $a, \beta > 1$ and the triangle T with vertices $\{(0,\beta), (a,-1), (-a,-1)\}$. Thus $(x,y) \in T$ iff $(x,y) = \lambda(0,\beta) + \rho(a,-1) + \varphi(-a,-1)$, where $\lambda, \rho, \varphi \ge 0$: $\lambda + \rho + \varphi = 1$, iff $x = \rho a - \varphi a$, $y = \lambda\beta - \rho - \varphi$. That is, $(x,y) \in T$ iff $x = a(\rho - \varphi)$ and $y = \lambda(\beta+1) - 1$. For example, if $a = 2$ then $\beta = 3$ by similar triangles. Clearly, $[-1,1]^2 \subset T$.

Let $g\colon \mathbb{R} \to \mathbb{R}^+$ be a concave function, then $G(x,y) := g(wx + (1-w)y)$, $0 \le w \le 1$ is concave in (x,y) and nonnegative, the same is true for G over T. We would like to find

$$\inf_{\mu} u(R) = \inf_{\mu} \int_T g(wx + (1-w)y) \, d\mu(x,y) \tag{23.118}$$

under

$$\int_T x \, d\mu(x,y) = \mu_1, \qquad \int_T y \, d\mu(x,y) = \nu_1, \tag{23.119}$$

where $(\mu_1, \nu_1) \in [-1,1]^2$ is given, and μ is a probability measure on T. Call V the convex hull of $G\big|_T$. The lower part of V will be the triangle

$$T^* = \text{conv}\langle \{(-a,-1,G(-a,-1)), (a,-1,G(a,-1)), (0,\beta,G(0,\beta))\} \rangle$$
$$= \text{conv}\langle \{A, B, C\} \rangle, \tag{23.120}$$

where

$$A := (-a,-1,g(w(1-a) - 1)),$$
$$B := (a,-1,g(w(a+1) - 1)),$$
$$C := (0,\beta,g((1-w)\beta)).$$

We can find that the plane (ABC) has an equation

$$z = g(w(1-a)-1) + \left(\frac{g(w(a+1)-1) - g(w(1-a)-1)}{2a}\right) \cdot (x+a)$$

$$+ \left(\frac{2g((1-w)\beta) - g(w(a+1)-1) - g(w(1-a)-1)}{2(\beta+1)}\right) \cdot (y+1).$$

$$(23.121)$$

Applying the basic geometric moment theory we get

$$\inf_{\mu} u(R) = \left\{ g(w(1-a)-1) + \frac{(g(w(a+1)-1) - g(w(1-a)-1))}{2a}(\mu_1 + a) \right.$$

$$\left. + \frac{(2g((1-w)\beta) - g(w(a+1)-1) - g(w(1-a)-1))}{2(\beta+1)}(\nu_1 + 1) \right\}.$$

$$(23.122)$$

Then one can apply $\max\limits_{w\in[0,1]}$ to both sides of (23.122). $\qquad\square$

Application 23.16 to (23.117). Here $0 \le w \le 1$. Take $a = 2$ then we get $\beta = 3$. We consider $g(x) = ln(3+x) \ge 0$ for $x \ge -2$. Notice $g(-w-1) = ln(2-w) \ge 0$, and g is a strictly concave function.

Applying (23.117) we obtain

$$\max_{w\in[0,1]}\left(\inf_{(X,Y)} u(R)\right) = \max_{w\in[0,1]}\left\{ ln(2-w) + \frac{(ln(2+3w) - ln(2-w))}{4}(\mu_1 + 2) \right.$$

$$\left. + \frac{(2ln3 + ln(2-w) - ln(2+3w))}{8}(\nu_1 + 1) \right\}. \quad (23.123)$$

Let $\mu_1 = \nu_1 = 0$, then

$$\max_{w\in[0,1]}\left(\inf_{(X,Y)} u(R)\right) = \max_{w\in[0,1]}\left\{ \frac{ln3}{4} + ln(2-w) + ln\left(\frac{2+3w}{2-w}\right)^{1/2} \right.$$

$$\left. + ln\left(\frac{2-w}{2+3w}\right)^{1/8} \right\} = \max_{w\in[0,1]}\left\{ \frac{ln3}{4} + ln(\tau(w)) \right\}, \quad (23.124)$$

where

$$\tau(w) := (2-w)^{5/8}(2+3w)^{3/8} > 0, \qquad 0 \le w \le 1. \quad (23.125)$$

We set

$$\tau'(w) = (2-w)^{-3/8}(2+3w)^{-5/8}(1-3w) = 0.$$

Here $w = 1/3$ is the only critical number for $\tau(w)$.

If $w < \frac{1}{3}$ then $\tau'(w) > 0$.

If $w > \frac{1}{3}$ then $\tau'(w) < 0$.

Thus $\tau(w)$, $0 \leq w \leq 1$ has a local maximum at $w = 1/3$. Notice $\tau(0) = 2$, $\tau(1) = (125)^{1/8} = 1.8285791$ and

$$\tau\left(\frac{1}{3}\right) = (347.22222)^{1/8} = 2.077667.$$

Thus the global maximum over $[0, 1]$ of the continuous function $\tau(w)$ is 2.077667. That is, $\tau(w) \leq 2.077667$, for all $0 \leq w \leq 1$. Since ℓn is a strictly increasing function we obtain that

$$\ell n \tau(w) \leq \ell n(2.077667) = 0.7312456,$$

for all $0 \leq w \leq 1$, with equality at $w = 1/3$. Consequently we have

$$\max_{w \in [0,1]} \left(\inf_{(X,Y)} u(R) \right) = 1.0058987, \quad \text{where } \mu_1 = \nu_1 = 0. \tag{23.126}$$

Application 23.16 to (23.117). Here again $0 \leq w \leq 1$, $a = 2$, $\beta = 3$, $\mu_1 = \nu_1 = 0$. We consider $g(x) := (2 + x)^{\alpha}$, $0 < \alpha < 1$, $x \geq -2$. Notice g is strictly concave and nonnegative. Also $g(-w - 1) = (1 - w)^{\alpha} \geq 0$. Thus

$$\max_{w \in [0,1]} \left(\inf_{(X,Y)} u(R) \right) = \max_{w \in [0,1]} \left\{ (1 - w)^{\alpha} + \left(\frac{(1 + 3w)^{\alpha} - (1 - w)^{\alpha}}{2} \right) \right.$$
$$\left. + \left(\frac{2(5 - 3w)^{\alpha} - (1 + 3w)^{\alpha} - (1 - w)^{\alpha}}{8} \right) \right\}$$
$$= \frac{1}{8} \max_{w \in [0,1]} \chi(w), \tag{23.127}$$

where

$$\chi(w) := 2(5 - 3w)^{\alpha} + 3(1 + 3w)^{\alpha} + 3(1 - w)^{\alpha}, \quad 0 \leq w \leq 1, \ 0 < \alpha < 1, \tag{23.128}$$

is strictly positive and continuous.

We have

$$\chi'(w) = -6\alpha(5 - 3w)^{\alpha - 1} + 9\alpha(1 + 3w)^{\alpha - 1} - 3\alpha(1 - w)^{\alpha - 1}, \tag{23.129}$$

which does not exist at $w = 1$, that is a critical point for $\chi(w)$. Also we obtain

$$\chi''(w) = 18\alpha(\alpha - 1)(5 - 3w)^{\alpha - 2} + 27\alpha(\alpha - 1)(1 + 3w)^{\alpha - 2}$$
$$+ 3\alpha(\alpha - 1)(1 - w)^{\alpha - 2} < 0, \quad 0 \leq w < 1. \tag{23.130}$$

That is $\chi'(w)$ is strictly decreasing on $[0, 1)$, i.e. $\chi'(1-) < \chi'(0)$. Also $\chi(w)$ is strictly concave on $[0, 1]$. We rewrite and observe that

$$\chi'(w) = a \left[6 \left[\frac{(5 - 3w)^{1-\alpha} - (1 + 3w)^{1-\alpha}}{(1 + 3w)^{1-\alpha}(5 - 3w)^{1-\alpha}} \right] + 3 \left[\frac{(1 - w)^{1-\alpha} - (1 + 3w)^{1-\alpha}}{(1 - w)^{1-\alpha}(1 + 3w)^{1-\alpha}} \right] \right] < 0, \tag{23.131}$$

if $\frac{2}{3} \leq w < 1$. That is, $\chi(w)$ is strictly decreasing over $\left[\frac{2}{3}, 1\right)$. Notice that

$$\chi'(0) = 6\alpha(1 - 5^{\alpha - 1}) > 0. \tag{23.132}$$

That is, χ at least close to zero is strictly increasing. Clearly there exists unique $w_0 \in (0,1)$ such that $\chi'(w_0) = 0$, the only critical number of χ there.

We find $\chi(0) = 2(5^\alpha + 3)$,

$$\chi(1) = 2 \cdot 2^\alpha + 3 \cdot 4^\alpha. \tag{23.133}$$

Consequently we found

$$\max_{w \in [0,1]} \left(\inf_{(X,Y)} u(R) \right) = \frac{1}{8} \max_{w \in [0,1]} \left\{ 2(5^\alpha + 3), 2^\alpha(2 + 3 \cdot 2^\alpha), \chi(w_0) \right\}. \tag{23.134}$$

Next we present the most natural moment result here, and we give again applications.

Theorem 23.17. *Let $\beta > 0$, and the triangles*

$$T_1 := \mathrm{conv}\langle \{(\beta,\beta), (\beta,-\beta), (-\beta,-\beta)\} \rangle,$$
$$T_2 := \mathrm{conv}\langle \{(\beta,\beta), (-\beta,-\beta), (-\beta,\beta)\} \rangle.$$

Let the random variables X, Y taking values in $[-\beta,\beta]$. Let $g\colon [-\beta,\beta] \to \mathbb{R}^+$ concave function and $0 \le w \le 1$. We suppose that

$$g(\beta(1 - 2w)) + g(\beta(2w - 1)) \ge (g(-\beta) + g(\beta)), \tag{23.135}$$

for all $0 \le w \le 1$. We would like to find

$$\inf_{(X,Y)} u(R) = \inf_{(X,Y)} Eg(wX + (1 - w)Y) \tag{23.136}$$

such that

$$EX = \mu_1, \qquad EY = \nu_1 \tag{23.137}$$

are prescribed moments, $\mu_1, \nu_1 \in [-\beta,\beta]$. Also to determine

$$\max_{w \in [0,1]} \left(\inf_{(X,Y)} u(R) \right). \tag{23.138}$$

We establish that:

If $(\mu_1, \nu_1) \in T_1$ (iff $\mu_1 = \beta(\lambda + \rho - \varphi)$, $\nu_1 = \beta(\lambda - \rho - \varphi)$, $\lambda, \rho, \varphi \ge 0\colon \lambda + \rho + \varphi = 1$) then it holds

$$\inf_{(X,Y)} u(R) = \left\{ g(\beta) + \frac{(g(\beta(2w - 1)) - g(-\beta)))}{2\beta}(\mu_1 - \beta) \right.$$
$$\left. + \frac{(g(\beta) - g(\beta(2w - 1)))}{2\beta}(\nu_1 - \beta) \right\}. \tag{23.139}$$

If $(\mu_1, \nu_1) \in T_2$ (iff $\mu_1 = \beta(\lambda - \rho - \varphi)$, $\nu_1 = \beta(\lambda - \rho + \varphi)$, $\lambda, \rho, \varphi \ge 0\colon \lambda + \rho + \varphi = 1$) then it holds

$$\inf_{(X,Y)} u(R) = \left\{ g(\beta) + \frac{(g(\beta) - g(\beta(1 - 2w)))}{2\beta}(\mu_1 - \beta) \right.$$
$$\left. + \frac{(g(\beta(1 - 2w)) - g(-\beta))}{2\beta}(\nu_1 - \beta) \right\}. \tag{23.140}$$

One can apply $\max\limits_{w\in[0,1]}$ to both sides of (23.139) and (23.140) to get (23.138).

Proof. Let $g\colon [-\beta,\beta] \to \mathbb{R}^+$ be a concave function, $\beta > 0$. Then $G(x,y) := g(wx + (1-w)y) \geq 0$, $x,y \in [-\beta,\beta]$, $0 \leq w \leq 1$, is concave in (x,y) over $[-\beta,\beta]^2$. We call $\mathcal{R} := [-\beta,\beta]^2$ the square with vertices (β,β), $(-\beta,\beta)$, $(-\beta,-\beta)$, $(\beta,-\beta)$. Clearly $(0,0) \in \mathcal{R}$. On the surface $z = G(x,y)$ consider the set of points $\theta := \{A,B,C,D\}$, where

$$A := (\beta,\beta,g(\beta)), \quad B := (\beta,-\beta,g(\beta(2w-1))),$$
$$C := (-\beta,-\beta,g(-\beta)), \quad D := (-\beta,\beta,g(\beta(1-2w))). \tag{23.141}$$

These are the corner-extreme points of this surface over $\mathcal{R}$.

Put $V := \operatorname{conv}\theta$. The lower part of V is the *same* as the lower part of V^*—the convex hull of $G(\mathcal{R})$. We call this lower part W, which is a convex surface made out of two triangles. We will describe them.

We need first describe the line segments $\ell_1 := \overline{AC}$, $\ell_2 := \overline{BD}$. We have for $\ell_1 := \overline{AC}$ that

$$
\begin{aligned}
x &= \beta - 2\beta t \\
y &= \beta - 2\beta t \\
z &= g(\beta) + (g(-\beta) - g(\beta))t, \qquad 0 \leq t \leq 1.
\end{aligned}
\tag{23.142}
$$

And for $\ell_2 := \overline{BD}$ we have

$$
\begin{aligned}
x &= -\beta + 2\beta t \\
y &= \beta - 2\beta t \\
z &= g(\beta(1-2w)) + (g(\beta(2w-1)) - g(\beta(1-2w)))t, \qquad 0 \leq t \leq 1.
\end{aligned}
\tag{23.143}
$$

When $t = 1/2$, then $x = y = 0$ for both ℓ_1, ℓ_2 line segments. Furthermore their z coordinates are

$$z(\ell_1) = \frac{g(\beta) + g(-\beta)}{2}, \quad \text{and}$$
$$z(\ell_2) = \frac{g(\beta(2w-1)) + g(\beta(1-2w))}{2}. \tag{23.144}$$

Since G is concave, $0 \leq w \leq 1$, it is natural to suppose that

$$z(\ell_2) \geq z(\ell_1). \tag{23.145}$$

Thus the line segment $\overline{BD}$ is above the line segment $\overline{AC}$.

Therefore the lower part W of V^* consists of the union of the triangles

$$P := (A \overset{\triangle}{B} C) \quad \text{and} \quad Q := (A \overset{\triangle}{C} D).$$

The equation of the plane through $P = (A \overset{\triangle}{B} C)$ is

$$z = g(\beta) + \frac{g(\beta(2w-1)) - g(-\beta)}{2\beta}(x-\beta) + \frac{(g(\beta) - g(\beta(2w-1)))}{2\beta}(y-\beta), \tag{23.146}$$

and the equation of the plane through $Q = (A \overset{\triangle}{C} D)$ is

$$z = g(\beta) + \frac{(g(\beta) - g(\beta(1 - 2w))))}{2\beta}(x - \beta) + \frac{(g(\beta(1 - 2w)) - g(-\beta))}{2\beta}(y - \beta).$$

$$(23.147)$$

We also call

$$T_1 := \mathrm{proj}_{xy}(A \overset{\triangle}{B} C) = \mathrm{conv}\langle\{(\beta, \beta), (\beta, -\beta), (-\beta, -\beta)\}\rangle,$$

$$T_2 := \mathrm{proj}_{xy}(A \overset{\triangle}{C} D) = \mathrm{conv}\langle\{(\beta, \beta), (-\beta, -\beta), (-\beta, \beta)\}\rangle. \qquad (23.148)$$

We see that

$$(x, y) \in T_1 \quad \text{iff} \quad \exists \lambda, \rho, \varphi \geq 0 \colon \lambda + \rho + \varphi = 1$$

with $(x, y) = \lambda(\beta, \beta) + \rho(\beta, -\beta) + \varphi(-\beta, -\beta)$ iff

$$x = \beta(1 - 2\varphi), \quad y = \beta(2\lambda - 1). \qquad (23.149)$$

And we have

$$(x, y) \in T_2 \quad \text{iff} \quad \exists \lambda, \rho, \varphi \geq 0 \colon \lambda + \rho + \varphi = 1$$

with $(x, y) = \lambda(\beta, \beta) + \rho(-\beta, -\beta) + \varphi(-\beta, \beta)$, iff

$$x = \beta(2\lambda - 1), \quad y = \beta(1 - 2\rho). \qquad (23.150)$$

We would like to calculate

$$L := \inf_{\mu} \int_{[-\beta,\beta]^2} g(wx + (1 - w)y)\, d\mu(x, y) \qquad (23.151)$$

such that

$$\int_{[-\beta,\beta]^2} x\, d\mu(x, y) = \mu_1, \quad \int_{[-\beta,\beta]^2} y\, d\mu(x, y) = \nu_1, \qquad (23.152)$$

where $(\mu_1, \nu_1) \in T_1$ iff $\mu_1 = \beta(\lambda + \rho - \varphi)$,

$$\nu_1 = \beta(\lambda - \rho - \varphi), \ \lambda, \rho, \varphi \geq 0 \colon \lambda + \rho + \varphi = 1, \qquad (23.153)$$

or $(\mu_1, \nu_1) \in T_2$ iff $\mu_1 = \beta(\lambda - \rho - \varphi)$, $\nu_1 = \beta(\lambda - \rho + \varphi)$, are prescribed first moments. Here μ is a probability measure on $[-\beta, \beta]^2$, $\beta > 0$.

The basic assumption needed here is

$$g(\beta(1 - 2w)) + g(\beta(2w - 1)) \geq (g(-\beta) + g(\beta)), \quad \text{for all } 0 \leq w \leq 1. \qquad (23.154)$$

Using basic geometric moment theory we obtain: If $(\mu_1, \nu_1) \in T_1$ we obtation

$$L = g(\beta) + \frac{(g(\beta(2w - 1)) - g(-\beta)))}{2\beta}(\mu_1 - \beta) + \frac{(g(\beta) - g(\beta(2w - 1)))}{2\beta}(\nu_1 - \beta).$$

$$(23.155)$$

When $(\mu_1, \nu_1) \in T_2$ we get

$$L = g(\beta) + \frac{(g(\beta) - g(\beta(1 - 2w)))}{2\beta}(\mu_1 - \beta) + \frac{(g(\beta(1 - 2w)) - g(-\beta))}{2\beta}(\nu_1 - \beta).$$

$$(23.156)$$

At the end one can apply $\max_{w\in[0,1]}$ to both sides of (23.155) and (23.156) to find (23.138). $\qquad\square$

Application 23.18 to (23.139) and (23.140). Let

$$g(x) := (1+x)^\alpha, \quad 0 < \alpha < 1, \quad x \geq -1, \quad 0 \leq w \leq 1,$$

which is a nonnegative concave function. Take $\beta = 1$ and see $g(1) = 2^\alpha$, $g(-1) = 0$. If $(\mu_1, \nu_1) \in T_1$ then from (23.139) we obtain

$$\inf_{(X,Y)} u(R) = 2^\alpha + 2^{\alpha-1} w^\alpha (\mu_1 - 1) + 2^{\alpha-1}(1 - w^\alpha)(\nu_1 - 1). \tag{23.157}$$

If $(\mu_1, \nu_1) \in T_2$ we get from (23.140) that

$$\inf_{(X,Y)} u(R) = 2^\alpha + 2^{\alpha-1}(1 - (1-w)^\alpha)(\mu_1 - 1) + 2^{\alpha-1}(1-w)^\alpha(\nu_1 - 1). \tag{23.158}$$

Notice that assumption (23.135) is fulfilled:

It is

$$(1 + (1 - 2w))^\alpha + (1 + (2w - 1))^\alpha = 2^\alpha((1-w)^\alpha + w^\alpha) \geq 2^\alpha, \quad 0 \leq w \leq 1, \tag{23.159}$$

true by

$$(1 - w)^\alpha + w^\alpha \geq ((1 - w) + w)^\alpha = 1.$$

Next take $\mu_1 = \nu_1 = 0$, i.e. $(0,0) \in T_1 \cap T_2$. Then

$$\inf_{(X,Y)} u(R) = 2^\alpha - 2^{\alpha-1}, \tag{23.160}$$

and

$$\max_{w\in[0,1]} \left(\inf_{(X,Y)} u(R) \right) = 2^\alpha - 2^{\alpha-1}. \tag{23.161}$$

Application 23.19 to (23.139) and (23.140). Let

$$g(x) := \ell n(2 + x), \quad x > -2, \quad 0 \leq w \leq 1$$

which is an increasing concave function. Take $\beta = \frac{1}{2}$. Clearly $g(x) > 0$ for $-\frac{1}{2} \leq x \leq \frac{1}{2}$. If $(\mu_1, \nu_1) \in T_1$ we obtain via (23.139) that

$$\inf_{(X,Y)} (u(R)) = \ell n(2.5) + (\ell n(w + 1.5) - \ell n(1.5))\left(\mu_1 - \frac{1}{2}\right)$$
$$+ (\ell n(2.5) - \ell n(w + 1.5))\left(\nu_1 - \frac{1}{2}\right). \tag{23.162}$$

If $(\mu_1, \nu_1) \in T_2$ we obtain via (23.140) that

$$\inf_{(X,Y)} (u(R)) = \ell n(2.5) + (\ell n(2.5) - \ell n(2.5 - w))\left(\mu_1 - \frac{1}{2}\right)$$
$$+ (\ell n(2.5 - w) - \ell n(1.5))\left(\nu_1 - \frac{1}{2}\right). \tag{23.163}$$

We need to establish that g fulfills (23.135).

That is, that it holds

$$\ell n(2.5 - w) + \ell n(1.5 + w) \geq \ell n(1.5) + \ell n(2.5), \tag{23.164}$$

as true.

Equivalently we need to have true that

$$\ell n(3.75 + w(1 - w)) \geq \ell n(3.75), \qquad 0 \leq w \leq 1. \tag{23.165}$$

The last inequality (23.165) is valid since ℓn is a strictly increasing function. So assumption (23.135) is fulfilled.

Next take $\mu_1 = \nu_1 = 0$, i.e. $(0,0) \in T_1 \cap T_2$. Then we find

$$\inf_{(X,Y)} u(R) = 0.6608779, \tag{23.166}$$

and

$$\max_{w \in [0,1]} \left(\inf_{(X,Y)} u(R) \right) = 0.6608779. \tag{23.167}$$

23.3.4. The Multivariate Case

Let X_i be r.v.'s, $w_i \geq 0$ be such that $\sum_{i=1}^{n} w_i = 1$ and $R := \sum_{i=1}^{n} w_i X_i$. Let g be a concave function from an interval $\mathcal{K} \subseteq \mathbb{R}$ into $\mathbb{R}^+$, and

$$u(R) := Eg(R).$$

We need

Lemma 23.20. *Let $g \colon \mathcal{K} \to \mathbb{R}$ be a concave function where $\mathcal{K}$ is an interval of $\mathbb{R}$, and let*

$$G(x_1, \ldots, x_n) = g\left(\sum_{i=1}^{n} w_i x_i \right).$$

Then $G(x_1, \ldots, x_n)$ is concave over $\mathcal{K}^n$.

Proof. Indeed we obtain $(0 \leq \lambda \leq 1)$

$$
\begin{aligned}
G(\lambda(x_1', &\ldots, x_n') + (1 - \lambda)(x_1'', \ldots, x_n'')) \\
&= G(\lambda x_1' + (1 - \lambda)x_1'', \ldots, \lambda x_n' + (1 - \lambda)x_n'') \\
&= g\left(\sum_{i=1}^{n} w_i(\lambda x_i' + (1 - \lambda)x_i'') \right) \\
&= g\left(\lambda \left(\sum_{i=1}^{n} w_i x_i' \right) + (1 - \lambda) \left(\sum_{i=1}^{n} w_i x_i'' \right) \right) \\
&\geq \lambda g\left(\sum_{i=1}^{n} w_i x_i' \right) + (1 - \lambda)g\left(\sum_{i=1}^{n} w_i x_i'' \right) \\
&= \lambda G(x_1', \ldots, x_n') + (1 - \lambda)G(x_1'', \ldots, x_n'').
\end{aligned}
$$

$\square$

We would like to determine

$$\sup_{(X_1,\ldots,X_n)} u(R) = \sup_{(X_1,\ldots,X_n)} Eg\left(\sum_{i=1}^{n} w_i X_i\right) = \sup_{\mu} \int_{\mathcal{K}^n} g\left(\sum_{i=1}^{n} w_i x_i\right) d\mu(x_1,\ldots,x_n),$$

$$(23.168)$$

such that

$$EX_i = \int_{\mathcal{K}^n} x_i\, d\mu(x_1,\ldots,x_n) = \mu_i \in \mathcal{K}, \quad i = 1,\ldots,n,$$

$$(23.169)$$

are given first moments.

Here μ is a probability measure on $\mathcal{K}^n$. Clearly the basic geometric moment theory says that

$$\sup_{(X_1,\ldots,X_n)} u(R) = g\left(\sum_{i=1}^{n} w_i \mu_i\right),$$

$$(23.170)$$

where $g \geq 0$ is concave.

We further would like to find

$$\max_{\vec{w}:=(w_1,\ldots,w_n)}\left(\sup_{(X_1,\ldots,X_n)} u(R)\right) = \max_{\vec{w}} g\left(\sum_{i=1}^{n} w_i \mu_i\right),$$

$$(23.171)$$

where $w_i \geq 0$ and $\sum_{i=1}^{n} w_i = 1$.

Examples 23.21.

i) Let $g(x) := \ell n(2 + x)$, $x \geq -1$, then

$$\sup_{(X_1,\ldots,X_n)} u(R) = \ell n\left(2 + \sum_{i=1}^{n} w_i \mu_i\right),$$

$$(23.172)$$

where $\mu_1 \geq \mu_2 \geq \cdots \geq \mu_n \geq -1$, that is giving $\sum_{i=1}^{n} w_i \mu_i \geq -1$. Notice that $\sum_{i=1}^{n} w_i \mu_i \leq \mu_1$. Hence

$$\max_{\vec{w}}\left(\sum_{i=1}^{n} w_i \mu_i\right) = \mu_1, \quad \text{when } w_1 = 1,\ w_i = 0,\ i = 2,\ldots,n.$$

$$(23.173)$$

Since ℓn is strictly increasing we obtain

$$\max_{(w_1,\ldots,w_n)}\left(\sup_{(X_1,\ldots,X_n)} u(R)\right) = \ell n(2 + \mu_1).$$

$$(23.174)$$

ii) Let $g(x) := (1 + x)^{\alpha}$, $x \geq -1$, $0 < \alpha < 1$. Then

$$\sup_{(X_1,\ldots,X_n)} u(R) = \left(1 + \sum_{i=1}^{n} w_i \mu_i\right)^{\alpha}, \quad \text{where } \mu_1 \geq \mu_2 \geq \cdots \geq \mu_n \geq -1, \quad (23.175)$$

giving $\sum_{i=1}^{n} w_i \mu_i \geq -1$, where $w_i \geq 0$ such that $\sum_{i=1}^{n} w_i = 1$.

Thus

$$\max_{\vec{w}} \left(\sup_{(X_1,\ldots,X_n)} u(R) \right) = \max_{\vec{w}} \left(1 + \sum_{i=1}^{n} w_i \mu_i \right)^{\alpha}. \qquad (23.176)$$

Since $(1 + x)^{\alpha}$ is strictly increasing we get that

$$\max_{(w_1,\ldots,w_n)} \left(\sup_{(X_1,\ldots,X_n)} u(R) \right) = (1 + \mu_1)^{\alpha}, \qquad (23.177)$$

at $w_1 = 1$, $w_i = 0$, $i = 2, \ldots, n$.

Chapter 24

Discrepancies Between General Integral Means

In this chapter are presented sharp estimates for the difference of general integral means with respect to even different finite measures. This is accomplished by the use of Ostrowski and Fink inequalities and the Geometric Moment Theory Method. The produced inequalities are with respect to supnorm of a derivative of the involved function. This treatment follows [48].

24.1 Introduction

This chapter is motivated by the works of J. Duoandikoetxea [186] and P. Cerone [123]. We use Ostrowski's ([277]) and Fink's ([195]) inequalities along with the Geometric Moment Theory Method, see [229], [20], [28], to prove our results.

We compare general averages of functions with respect to various finite measures over different subintervals of a domain, even disjoint. The estimates are sharp and the inequalities are attained. They are with respect to the supnorm of a derivative of the involved function f.

24.2 Results

Part A

As motivation we mention

Proposition 24.1. *Let* μ_1, μ_2 *finite Borel measures on* $[a,b] \subseteq \mathbb{R}$, $[c,d]$, $[\tilde{e},g] \subseteq [a,b]$, $f \in C^1([a,b])$. *Denote* $\mu_1([c,d]) = m_1 > 0$, $\mu_2([\tilde{e},g]) = m_2 > 0$. *Then*

$$\left| \frac{1}{m_1} \int_c^d f(x)d\mu_1 - \frac{1}{m_2} \int_{\tilde{e}}^g f(x)d\mu_2 \right| \leq \|f'\|_\infty (b-a). \tag{24.1}$$

Proof. From mean value theorem we have

$$|f(x) - f(y)| \leq \|f'\|_\infty (b-a) =: \gamma, \quad \forall x, y \in [a,b].$$

That is,

$$-\gamma \le f(x) - f(y) \le \gamma, \quad \forall x, y \in [a, b],$$

and by fixing y we obtain

$$-\gamma \le \frac{1}{m_1} \int_c^d f(x) d\mu_1 - f(y) \le \gamma.$$

The last is true $\forall y \in [\tilde{e}, g]$. Thus

$$-\gamma \le \frac{1}{m_1} \int_c^d f(x) d\mu_1 - \frac{1}{m_2} \int_{\tilde{e}}^g f(x) d\mu_2 \le \gamma,$$

proving the claim. $\square$

As a related result we have

Corollary 24.2. *Let $f \in C^1([a, b])$, $[c, d]$, $[\tilde{e}, g] \subseteq [a, b] \subseteq \mathbb{R}$. Then we have*

$$\left| \frac{1}{d - c} \int_c^d f(x) dx - \frac{1}{g - \tilde{e}} \int_{\tilde{e}}^g f(x) dx \right| \le \|f'\|_\infty \cdot (b - a). \tag{24.2}$$

We use the following famous Ostrowski inequality, see [277], [21].

Theorem 24.3. *Let $f \in C^1([a, b])$, $x \in [a, b]$. Then*

$$\left| f(x) - \frac{1}{b - a} \int_a^b f(x) dx \right| \le \frac{\|f'\|_\infty}{2(b - a)} \left((x - a)^2 + (x - b)^2 \right), \tag{24.3}$$

and inequality (24.3) is sharp, see [21].

We also have

Corollary 24.4. *Let $f \in C^1([a, b])$, $x \in [c, d] \subseteq [a, b] \subseteq \mathbb{R}$. Then*

$$\left| f(x) - \frac{1}{b - a} \int_a^b f(x) dx \right| \le \frac{\|f'\|_\infty}{2(b - a)} \text{Max}\{((c - a)^2 + (c - b)^2), ((d - a)^2 + (d - b)^2)\}. \tag{24.4}$$

Proof. Obvious. $\square$

We denote by $\mathcal{P}([a, b])$ the power set of $[a, b]$. We give the following

Theorem 24.5. *Let $f \in C^1([a, b])$, μ be a finite measure on $([c, d], \mathcal{P}([c, d]))$, where $[c, d] \subseteq [a, b] \subseteq \mathbb{R}$ and $m := \mu([c, d]) > 0$. Then*
1)

$$\left| \frac{1}{m} \int_{[c, d]} f(x) d\mu - \frac{1}{b - a} \int_a^b f(x) dx \right|$$

$$\le \frac{\|f'\|_\infty}{2(b - a)} \text{Max}\{((c - a)^2 + (c - b)^2), ((d - a)^2 + (d - b)^2)\}. \tag{24.5}$$

2) *Inequality (24.5) is attained when $d = b$.*

Proof. 1) By (24.4) integrating against μ/m.

2) Here (24.5) reduces to

$$\left| \frac{1}{m} \int_{[c,b]} f(x)d\mu - \frac{1}{b-a} \int_a^b f(x)dx \right| \leq \frac{\|f'\|_\infty}{2}(b-a). \tag{24.6}$$

We prove that (24.6) is attained. Take

$$f^*(x) = \frac{2x - (a+b)}{b-a}, \qquad a \leq x \leq b.$$

Then $f^{*\prime}(x) = \frac{2}{b-a}$ and $\|f^{*\prime}\|_\infty = \frac{2}{b-a}$, along with

$$\int_a^b f^*(x)dx = 0.$$

Therefore (24.6) becomes

$$\left| \frac{1}{m} \int_{[c,b]} f^*(x)d\mu \right| \leq 1. \tag{24.7}$$

Finally pick $\frac{\mu}{m} = \delta_{\{b\}}$ the Dirac measure supported at $\{b\}$, then (24.7) turns to equality. $\qquad\square$

We further have

Corollary 24.6. *Let* $f \in C^1([a,b])$ *and* $[c,d] \subseteq [a,b] \subseteq \mathbb{R}$. *Let* $M(c,d) := \{\mu\colon \mu$ *a measure on* $([c,d], \mathcal{P}([c,d]))$ *of finite positive mass}, denoted* $m := \mu([c,d])$. *Then*

1) *It holds*

$$\sup_{\mu \in M(c,d)} \left| \frac{1}{m} \int_{[c,d]} f(x)d\mu - \frac{1}{b-a} \int_a^b f(x)dx \right|$$

$$\leq \frac{\|f'\|_\infty}{2(b-a)} \operatorname{Max}\{((c-a)^2 + (c-b)^2), ((d-a)^2 + (d-b)^2)\} \tag{24.8}$$

$$= \frac{\|f'\|_\infty}{2(b-a)} \times \begin{cases} (d-a)^2 + (d-b)^2, & \text{if } d+c \geq a+b \\ (c-a)^2 + (c-b)^2, & \text{if } d+c \leq a+b \end{cases}$$

$$\leq \frac{\|f'\|_\infty}{2}(b-a). \tag{24.9}$$

Inequality (24.9) becomes equality if $d = b$ *or* $c = a$ *or both.*

2) *It holds*

$$\sup_{\substack{\text{all} c,d \\ a \leq c < d \leq b}} \left(\sup_{\mu \in M(c,d)} \left| \frac{1}{m} \int_{[c,d]} f(x)d\mu - \frac{1}{b-a} \int_a^b f(x)dx \right| \right) \leq \frac{\|f'\|_\infty}{2}(b-a). \tag{24.10}$$

Next we restrict ourselves to a subclass of $M(c,d)$ of finite measures μ with given first moment and by the use of the Geometric Moment Theory Method, see [229], [20], [28], we produce a sharper than (24.8) inequality. For that we need

Lemma 24.7. *Let* ν *be a probability measure on* $([a,b], \mathcal{P}([a,b]))$ *such that*

$$\int_{[a,b]} x\,d\nu = d_1 \in [a,b] \tag{24.11}$$

is given. Then

i)

$$U_1 := \sup_{\nu \ as \ in \ (24.11)} \int_{[a,b]} (x-a)^2 d\nu = (b-a)(d_1-a), \qquad (24.12)$$

and

ii)

$$U_2 := \sup_{\nu \ as \ in \ (24.11)} \int_{[a,b]} (x-b)^2 d\nu = (b-a)(b-d_1). \qquad (24.13)$$

Proof. i) We observe the graph

$$G_1 = \{(x, (x-a)^2): a \le x \le b\},$$

which is a convex arc above the x-axis. We form the closed convex hull of G_1 and we name it $\widehat{G}_1$ which has as an upper concave envelope the line segment ℓ_1 from $(a, 0)$ to $(b, (b-a)^2)$. We consider the vertical line $x = d_1$ which cuts ℓ_1 at the point Q_1. Then U_1 is the distance from $(d_1, 0)$ to Q_1. By using the equal ratios property of related here similar triangles we find

$$\frac{d_1 - a}{b - a} = \frac{U_1}{(b-a)^2},$$

which proves the claim.

ii) We observe the graph

$$G_2 = \{(x, (x-b)^2): a \le x \le b\},$$

which is a convex arc above the x-axis. We form the closed convex hull of G_2 and we name it $\widehat{G}_2$ which has as an upper concave envelope the line segment ℓ_2 from $(b, 0)$ to $(a, (b-a)^2)$. We consider the vertical line $x = d_1$ which intersects ℓ_2 at the point Q_2.

Then U_2 is the distance from $(d_1, 0)$ to Q_2. By using the equal ratios property of the related similar triangles we derive

$$\frac{U_2}{(b-a)^2} = \frac{b - d_1}{b - a},$$

which proves the claim. $\qquad \square$

Furthermore we need

Lemma 24.8. *Let $[c, d] \subseteq [a, b] \subseteq \mathbb{R}$ and let ν be a probability measure on $([c, d], \mathcal{P}([c, d]))$ such that*

$$\int_{[c,d]} x \, d\nu = d_1 \in [c, d] \qquad (24.14)$$

is given. Then

(i)

$$U_1 := \sup_{\nu \ as \ in \ (24.14)} \int_{[c,d]} (x-a)^2 d\nu = d_1(c+d-2a) - cd + a^2, \qquad (24.15)$$

and

(ii)

$$U_2 := \sup_{\nu \text{ as in } (24.14)} \int_{[c,d]} (x-b)^2 d\nu = d_1(c+d-2b) - cd + b^2. \tag{24.16}$$

(iii) *It holds*

$$\sup_{\nu \text{ as in } (24.14)} \int_{[c,d]} \left[(x-a)^2 + (x-b)^2 \right] d\nu = U_1 + U_2. \tag{24.17}$$

Proof. (i) We notice that

$$\int_c^d (x-a)^2 d\nu = (c-a)^2 + 2(c-a)(d_1-c) + \int_c^d (x-c)^2 d\nu.$$

Using (24.12) which is applied on $[c,d]$, we find

$$\sup_{\nu \text{ as in } (24.14)} \int_c^d (x-a)^2 d\nu = (c-a)^2 + 2(c-a)(d_1-c)$$

$$+ \sup_{\nu \text{ as in } (24.14)} \int_c^d (x-c)^2 d\nu$$

$$= (c-a)^2 + 2(c-a)(d_1-c) + (d-c)(d_1-c)$$

$$= d_1(c+d-2a) - cd + a^2,$$

proving the claim.

(ii) We notice that

$$\int_c^d (x-b)^2 d\nu = (b-d)^2 + 2(b-d)(d-d_1) + \int_c^d (x-d)^2 d\nu.$$

Using (24.13) which is applied on $[c,d]$, we get

$$\sup_{\nu \text{ as in } (24.14)} \int_c^d (x-b)^2 d\nu = (b-d)^2 + 2(b-d)(d-d_1)$$

$$+ \sup_{\nu \text{ as in } (24.14)} \int_c^d (x-d)^2 d\nu$$

$$= (b-d)^2 + 2(b-d)(d-d_1) + (d-c)(d-d_1)$$

$$= d_1(c+d-2b) - cd + b^2,$$

proving the claim.

(iii) Similar to Lemma 24.7 and above and clear, notice that $(x-a)^2 + (x-b)^2$ is convex, etc. $\square$

Now we are ready to present

Theorem 24.9. *Let $[c,d] \subseteq [a,b] \subseteq \mathbb{R}$, $f \in C^1([a,b])$, μ a finite measure on $([c,d], \mathcal{P}([c,d]))$ of mass $m := \mu([c,d]) > 0$. Suppose that*

$$\frac{1}{m} \int_c^d x \, d\mu = d_1, \quad c \le d_1 \le d, \tag{24.18}$$

is given.

   Then

$$\sup_{\mu \ as \ above} \left| \frac{1}{m} \int_c^d f(x)d\mu - \frac{1}{b-a} \int_a^b f(x)dx \right|$$

$$\leq \frac{\|f'\|_\infty}{(b-a)} \left[d_1\big((c+d)-(a+b)\big) - cd + \frac{a^2+b^2}{2} \right]. \qquad (24.19)$$

Proof. Denote

$$\beta(x) := \frac{\|f'\|_\infty}{2(b-a)} \big((x-a)^2 + (x-b)^2\big),$$

then by Theorem 24.3 we have

$$-\beta(x) \leq f(x) - \frac{1}{b-a} \int_a^b f(x)dx \leq \beta(x), \quad \forall x \in [c,d].$$

Thus

$$-\frac{1}{m} \int_c^d \beta(x)d\mu \leq \frac{1}{m} \int_c^d f(x)d\mu - \frac{1}{b-a} \int_a^b f(x)dx \leq \frac{1}{m} \int_c^d \beta(x)d\mu,$$

and

$$\left| \frac{1}{m} \int_c^d f(x)d\mu - \frac{1}{b-a} \int_a^b f(x)dx \right| \leq \frac{1}{m} \int_c^d \beta(x)d\mu =: \theta.$$

   Here $\nu := \frac{\mu}{m}$ is a probability measure subject to (24.18) on $([c,d], \mathcal{P}([c,d]))$ and

$$\theta = \frac{\|f'\|_\infty}{2(b-a)} \left(\int_c^d (x-a)^2 \frac{d\mu}{m} + \int_c^d (x-b)^2 \frac{d\mu}{m} \right)$$

$$= \frac{\|f'\|_\infty}{2(b-a)} \left(\int_c^d (x-a)^2 d\nu + \int_c^d (x-b)^2 d\nu \right).$$

Using (24.14), (24.15), (24.16) and (24.17) we derive

$$\theta \leq \frac{\|f'\|_\infty}{2(b-a)} \left\{ \big(d_1(c+d-2a) - cd + a^2\big) + \big(d_1(c+d-2b) - cd + b^2\big) \right\}$$

$$= \frac{\|f'\|_\infty}{(b-a)} \left[d_1\big((c+d)-(a+b)\big) - cd + \frac{a^2+b^2}{2} \right],$$

proving the claim. $\qquad\qquad\qquad\qquad\qquad\qquad\qquad\qquad\qquad\qquad\qquad\qquad\square$

   We make

Remark 24.10. On Theorem 24.9.

   1) Case of $c + d \geq a + b$, using $d_1 \leq d$ we get

$$d_1\big((c+d)-(a+b)\big) - cd + \frac{a^2+b^2}{2} \leq \frac{(d-a)^2+(d-b)^2}{2}. \qquad (24.20)$$

2) Case of $c + d \leq a + b$, using $d_1 \geq c$ we find that

$$d_1\big((c+d) - (a+b)\big) - cd + \frac{a^2 + b^2}{2} \leq \frac{(c-a)^2 + (c-b)^2}{2}. \tag{24.21}$$

Hence under (24.18) inequality (24.19) is sharper than (24.8).

We also give

Corollary 24.11. *All assumptions as in Theorem 24.9. Then*

$$\left| \frac{1}{m} \int_c^d f(x)d\mu - \frac{1}{b-a} \int_a^b f(x)dx \right|$$

$$\leq \frac{\|f'\|_\infty}{(b-a)} \left[d_1\big((c+d) - (a+b)\big) - cd + \frac{a^2 + b^2}{2} \right]. \tag{24.22}$$

By Remark 24.10 inequality (24.22) is sharper than (24.5).

Part B

Here we follow Fink's work [195]. We need

Theorem 24.12 ([195]). *Let* $f \colon [a, b] \to \mathbb{R}$, $f^{(n-1)}$ *is absolutely continuous on* $[a, b]$, $n \geq 1$. *Then*

$$f(x) = \frac{n}{b-a} \int_a^b f(t)dt + \sum_{k=1}^{n-1} \left(\frac{n-k}{k!} \right) \left(\frac{f^{(k-1)}(b)(x-b)^k - f^{(k-1)}(a)(x-a)^k}{b-a} \right)$$

$$+ \frac{1}{(n-1)!(b-a)} \int_a^b (x-t)^{n-1} k(t, x) f^{(n)}(t)dt, \tag{24.23}$$

where

$$k(t, x) := \begin{cases} t - a, & a \leq t \leq x \leq b, \\ t - b, & a \leq x < t \leq b. \end{cases} \tag{24.24}$$

For $n = 1$ *the sum in (24.23) is taken as zero.*

We also need Fink's inequality

Theorem 24.13 ([195]). *Let* $f^{(n-1)}$ *be absolutely continuous on* $[a, b]$ *and* $f^{(n)} \in L_\infty(a, b)$, $n \geq 1$. *Then*

$$\left| \frac{1}{n} \left(f(x) + \sum_{k=1}^{n-1} F_k(x) \right) - \frac{1}{b-a} \int_a^b f(x)dx \right|$$

$$\leq \frac{\|f^{(n)}\|_\infty}{n(n+1)!(b-a)} \left[(b-x)^{n+1} + (x-a)^{n+1} \right], \quad \forall x \in [a, b], \tag{24.25}$$

where

$$F_k(x) := \left(\frac{n-k}{k!} \right) \left(\frac{f^{(k-1)}(a)(x-a)^k - f^{(k-1)}(b)(x-b)^k}{b-a} \right). \tag{24.26}$$

Inequality (24.25) is sharp, in the sense that is attained by an optimal f *for any* $x \in [a, b]$.

We get

Corollary 24.14. *Let $f^{(n-1)}$ be absolutely continuous on $[a,b]$ and $f^{(n)} \in L_\infty(a,b)$, $n \geq 1$. Then $\forall x \in [c,d] \subseteq [a,b]$ we have*

$$\left| \frac{1}{n} \left(f(x) + \sum_{k=1}^{n-1} F_k(x) \right) - \frac{1}{b-a} \int_a^b f(x) dx \right|$$

$$\leq \frac{\|f^{(n)}\|_\infty}{n(n+1)!(b-a)} \left[(b-x)^{n+1} + (x-a)^{n+1} \right]$$

$$\leq \frac{\|f^{(n)}\|_\infty}{n(n+1)!} (b-a)^n. \tag{24.27}$$

Also we have

Proposition 24.15. *Let $f^{(n-1)}$ be absolutely continuous on $[a,b]$ and $f^{(n)} \in L_\infty(a,b)$, $n \geq 1$. Let μ be a finite measure of mass $m > 0$ on*

$$([c,d], \mathcal{P}([c,d])), \quad [c,d] \subseteq [a,b] \subseteq \mathbb{R}.$$

Then

$$K := \left| \frac{1}{n} \left(\frac{1}{m} \int_{[c,d]} f(x) d\mu + \sum_{k=1}^{n-1} \frac{1}{m} \int_{[c,d]} F_k(x) d\mu \right) - \frac{1}{b-a} \int_a^b f(x) dx \right|$$

$$\leq \frac{\|f^{(n)}\|_\infty}{n(n+1)!(b-a)} \left[\frac{1}{m} \int_{[c,d]} (b-x)^{n+1} d\mu + \frac{1}{m} \int_{[c,d]} (x-a)^{n+1} d\mu \right]$$

$$\leq \frac{\|f^{(n)}\|_\infty}{n(n+1)!} (b-a)^n. \tag{24.28}$$

Proof. By (24.27). $\qquad\qquad\qquad\qquad\qquad\qquad\qquad\qquad\qquad\qquad\qquad\square$

Similarly, based on Theorem A of [195] we also conclude

Proposition 24.16. *Let $f^{(n-1)}$ be absolutely continuous on $[a,b]$ and $f^{(n)} \in L_p(a,b)$, where $1 < p < \infty$, $n \geq 1$. Let μ be a finite measure of mass $m > 0$ on $([c,d], \mathcal{P}([c,d]))$, $[c,d] \subseteq [a,b] \subseteq \mathbb{R}$.*
Here $p' > 1$ such that $\frac{1}{p} + \frac{1}{p'} = 1$. Then

$$\left| \frac{1}{n} \left(\frac{1}{m} \int_{[c,d]} f(x) d\mu + \sum_{k=1}^{n-1} \frac{1}{m} \int_{[c,d]} F_k(x) d\mu \right) - \frac{1}{b-a} \int_a^b f(x) dx \right|$$

$$\leq \left(\frac{B((n-1)p'+1, p'+1)^{1/p'} \|f^{(n)}\|_p}{n!(b-a)} \right)$$

$$\cdot \left(\frac{1}{m} \int_{[c,d]} ((x-a)^{np'+1} + (b-x)^{np'+1})^{1/p'} d\mu \right)$$

$$\leq \left(\frac{B((n-1)p'+1, p'+1)^{1/p'} (b-a)^{n-1+\frac{1}{p'}}}{n!} \right) \|f^{(n)}\|_p. \tag{24.29}$$

We make the following

Remark 24.17. Clearly we have the following for

$$g(x) := (b - x)^{n+1} + (x - a)^{n+1} \le (b - a)^{n+1}, \quad a \le x \le b, \tag{24.30}$$

where $n \ge 1$. Here $x = \frac{a+b}{2}$ is the only critical number of g and

$$g''\left(\frac{a+b}{2}\right) = n(n+1)\frac{(b-a)^{n-1}}{2^{n-2}} > 0,$$

giving that $g\left(\frac{a+b}{2}\right) = \frac{(b-a)^{n+1}}{2^n} > 0$ is the global minimum of g over $[a,b]$. Also g is convex over $[a,b]$. Therefore for $[c,d] \subseteq [a,b]$ we have

$$M := \max_{c \le x \le d}\left\{(x - a)^{n+1} + (b - x)^{n+1}\right\}$$

$$= \max\left\{(c - a)^{n+1} + (b - c)^{n+1}, (d - a)^{n+1} + (b - d)^{n+1}\right\}. \tag{24.31}$$

We find further that

$$M = \begin{cases} (d - a)^{n+1} + (b - d)^{n+1}, & \text{if } c + d \ge a + b \\ (c - a)^{n+1} + (b - c)^{n+1}, & \text{if } c + d \le a + b. \end{cases} \tag{24.32}$$

If $d = b$ or $c = a$ or both then

$$M = (b - a)^{n+1}. \tag{24.33}$$

Based on Remark 24.17 we present

Theorem 24.18. *All assumptions, terms and notations as in Proposition 24.15. Then*
1)

$$K \le \frac{\|f^{(n)}\|_\infty}{n(n+1)!(b-a)} \max\{(c - a)^{n+1} + (b - c)^{n+1},$$

$$(d - a)^{n+1} + (b - d)^{n+1}\} \tag{24.34}$$

$$= \frac{\|f^{(n)}\|_\infty}{n(n+1)!(b-a)} \times \left\{\begin{array}{ll} (d - a)^{n+1} + (b - d)^{n+1}, & \text{if } c + d \ge a + b, \\ (c - a)^{n+1} + (b - c)^{n+1}, & \text{if } c + d \le a + b \end{array}\right\}$$

$$\le \frac{\|f^{(n)}\|_\infty}{n(n+1)!}(b - a)^n, \tag{24.35}$$

where K is as in (24.28). If $d = b$ or $c = a$ or both, then (24.35) becomes equality. When $d = b$, $\frac{\mu}{m} = \delta_{\{b\}}$ and $f(x) = \frac{(x-a)^n}{n!}$, $a \le x \le b$, then inequality (24.34) is attained, i.e. it becomes equality, proving that (24.34) is a sharp inequality.

2) We also have

$$\sup_{\mu \in M(c,d)} K \le \text{R.H.S.}(24.34), \tag{24.36}$$

and

$$\sup_{\substack{\text{all } c,d \\ a \le c \le d \le b}} \left(\sup_{\mu \in M(c,d)} K\right) \le \text{R.H.S.}(24.35). \tag{24.37}$$

Proof. It remains to prove only the sharpness, via attainability of (24.34) when $d = b$. In that case (24.34) reduces to

$$\left| \frac{1}{n}\left(\frac{1}{m}\int_{[c,d]} f(x)d\mu + \sum_{k=1}^{n-1} \frac{1}{m}\int_{[c,b]} F_k(x)d\mu \right) - \frac{1}{b-a}\int_a^b f(x)dx \right|$$

$$\leq \frac{\|f^{(n)}\|_\infty}{n(n+1)!}(b-a)^n. \tag{24.38}$$

The optimal measure here will be $\frac{\mu}{m} = \delta_{\{b\}}$ and then (24.38) becomes

$$\left| \frac{1}{n}\left(f(b) + \sum_{k=1}^{n-1} F_k(b) \right) - \frac{1}{b-a}\int_a^b f(x)dx \right| \leq \frac{\|f^{(n)}\|_\infty}{n(n+1)!}(b-a)^n. \tag{24.39}$$

The optimal function here will be

$$f^*(x) = \frac{(x-a)^n}{n!}, \quad a \leq x \leq b.$$

Then we observe that

$$f^{*(k-1)}(x) = \frac{(x-a)^{n-k+1}}{(n-k+1)!}, \quad k-1 = 0, 1, \ldots, n-2,$$

and $f^{*(k-1)}(a) = 0$ for $k-1 = 0, 1, \ldots, n-2$. Clearly here $F_k(b) = 0$, $k = 1, \ldots, n-1$. Also we have

$$\frac{1}{b-a}\int_a^b f^*(x)dx = \frac{(b-a)^n}{(n+1)!} \quad \text{and} \quad \|f^{*(n)}\|_\infty = 1.$$

Putting all these elements in (24.39) we have

$$\left| \frac{(b-a)^n}{nn!} - \frac{(b-a)^n}{(n+1)!} \right| = \frac{(b-a)^n}{n(n+1)!},$$

proving the claim. $\qquad\qquad\qquad\qquad\qquad\qquad\qquad\qquad\qquad\qquad\qquad\qquad\quad \square$

Again next we restrict ourselves to the subclass of $M(c,d)$ of finite measures μ with given first moment and by the use of Geometric Moment Theory Method, see [229], [20], [28], we produce a sharper than (24.36) inequality. For that we need

Lemma 24.19. *Let* $[c,d] \subseteq [a,b] \subseteq \mathbb{R}$ *and* ν *be a probability measure on* $([c,d],$ $\mathcal{P}([c,d]))$ *such that*

$$\int_{[c,d]} x\, d\nu = d_1 \in [c,d] \tag{24.40}$$

is given, $n \geq 1$. *Then*

$$W_1 := \sup_{\nu \text{ as in (24.40)}} \int_{[c,d]} (x-a)^{n+1} d\nu$$

$$= \left(\sum_{k=0}^n (d-a)^{n-k}(c-a)^k \right)(d_1 - d) + (d-a)^{n+1}. \tag{24.41}$$

Proof. We observe the graph

$$G_1 = \{(x, (x-a)^{n+1}) \colon c \leq x \leq d\},$$

which is a convex arc above the x-axis. We form the closed convex hull of G_1 and we name it $\widehat{G}_1$, which has as an upper concave envelope the line segment $\overline{\ell}_1$ from $(c, (c-a)^{n+1})$ to $(d, (d-a)^{n+1})$. Call ℓ_1 the line through $\overline{\ell}_1$. The line ℓ_1 intersects x-axis at $(t, 0)$, where $a \leq t \leq c$. We need to determine t: The slope of ℓ_1 is

$$\tilde{m} = \frac{(d-a)^{n+1} - (c-a)^{n+1}}{d-c} = \sum_{k=0}^{n}(d-a)^{n-k}(c-a)^k.$$

The equation of line ℓ_1 is

$$y = \tilde{m} \cdot x + (d-a)^{n+1} - \tilde{m}d.$$

Hence $\tilde{m}t + (d-a)^{n+1} - \tilde{m}d = 0$ and

$$t = d - \frac{(d-a)^{n+1}}{\tilde{m}}.$$

Next we consider the moment right triangle with vertices $(t, 0)$, $(d, 0)$ and $(d, (d-a)^{n+1})$. Clearly $(d_1, 0)$ is between $(t, 0)$ and $(d, 0)$. Consider the vertical line $x = d_1$, it intersects $\overline{\ell}_1$ at Q. Clearly then $W_1 = \text{length}(\overline{(d_1, 0), Q})$, the line segment of which length we find by the formed two similar right triangles with vertices $\{(t, 0), (d_1, 0), Q\}$ and $\{(t, 0), (d, 0), (d, (d-a)^{n+1})\}$. We have the equal ratios

$$\frac{d_1 - t}{d - t} = \frac{W_1}{(d-a)^{n+1}},$$

i.e.

$$W_1 = (d-a)^{n+1}\left(\frac{d_1 - t}{d - t}\right). \qquad \square$$

We also need

Lemma 24.20. *Let $[c, d] \subseteq [a, b] \subseteq \mathbb{R}$ and ν be a probability measure on $([c, d],$ $\mathcal{P}([c, d]))$ such that*

$$\int_{[c,d]} x\, d\nu = d_1 \in [c, d] \tag{24.42}$$

is given, $n \geq 1$. Then
   1)

$$W_2 := \sup_{\nu \text{ as in } (24.42)} \int_{[c,d]} (b-x)^{n+1}\, d\nu$$

$$= \left(\sum_{k=0}^{n}(b-c)^{n-k}(b-d)^k\right)(c-d_1) + (b-c)^{n+1}. \tag{24.43}$$

   2) *It holds*

$$\sup_{\nu \text{ as in } (24.42)} \int_{[c,d]} \left[(x-a)^{n+1} + (b-x)^{n+1}\right] d\nu = W_1 + W_2, \tag{24.44}$$

where W_1 as in (24.41).

Proof. 1) We observe the graph

$$G_2 = \left\{ (x, (b-x)^{n+1}) \colon c \le x \le d \right\},$$

which is a convex arc above the x-axis. We form the closed convex hull of G_2 and we name it $\widehat{G}_2$, which has as an upper concave envelope the line segment $\overline{\ell}_2$ from $(c, (b-c)^{n+1})$ to $(d, (b-d)^{n+1})$. Call ℓ_2 the line through $\overline{\ell}_2$. The line ℓ_2 intersects x-axis at $(t^*, 0)$, where $d \le t^* \le b$. We need to determine t^*: The slope of ℓ_2 is

$$\tilde{m}^* = \frac{(b-c)^{n+1} - (b-d)^{n+1}}{c-d} = -\left(\sum_{k=0}^{n} (b-c)^{n-k}(b-d)^k \right).$$

The equation of line ℓ_2 is

$$y = \tilde{m}^* x + (b-c)^{n+1} - \tilde{m}^* c.$$

Hence

$$\tilde{m}^* t^* + (b-c)^{n+1} - \tilde{m}^* c = 0$$

and

$$t^* = c - \frac{(b-c)^{n+1}}{\tilde{m}^*}.$$

Next we consider the moment right triangle with vertices $(c, (b-c)^{n+1})$, $(c, 0)$, $(t^*, 0)$. Clearly $(d_1, 0)$ is between $(c, 0)$ and $(t^*, 0)$. Consider the vertical line $x = d_1$, it intersects $\overline{\ell}_2$ at Q^*. Clearly then

$$W_2 = \text{length}\overline{((d_1, 0), Q^*)},$$

the line segment of which length we find by the formed two similar right triangles with vertices $\{Q^*, (d_1, 0), (t^*, 0)\}$ and $\{(c, (b-c)^{n+1}), (c, 0), (t^*, 0)\}$. We have the equal ratios

$$\frac{t^* - d_1}{t^* - c} = \frac{W_2}{(b-c)^{n+1}},$$

i.e.

$$W_2 = (b-c)^{n+1} \left(\frac{t^* - d_1}{t^* - c} \right).$$

2) Similar to above and clear. $\square$

We make the useful

Remark 24.21. By Lemmas 24.19, 24.20 we get

$$\lambda := W_1 + W_2 = \left(\sum_{k=0}^{n} (d-a)^{n-k}(c-a)^k \right) (d_1 - d)$$

$$+ \left(\sum_{k=0}^{n} (b-c)^{n-k}(b-d)^k \right) (c - d_1) + (d-a)^{n+1} + (b-c)^{n+1} > 0, \quad (24.45)$$

$n \geq 1$.

We present the important

Theorem 24.22. *Let $f^{(n-1)}$ be absolutely continuous on $[a,b]$ and $f^{(n)} \in L_\infty(a,b)$, $n \geq 1$. Let μ be a finite measure of mass $m > 0$ on $([c,d], \mathcal{P}([c,d]))$, $[c,d] \subseteq [a,b] \subseteq \mathbb{R}$. Furthermore we suppose that*

$$\frac{1}{m} \int_{[c,d]} x \, d\mu = d_1 \in [c,d] \tag{24.46}$$

is given. Then

$$\sup_{\substack{\mu \text{ as above}}} K \leq \frac{\|f^{(n)}\|_\infty}{n(n+1)!(b-a)} \lambda, \tag{24.47}$$

and

$$K \leq \text{R.H.S.}(24.47), \tag{24.48}$$

where K is as in (24.28) and λ as in (24.45).

Proof. By Proposition 24.15 and Lemmas 24.19 and 24.20. $\qquad\square$

We make

Remark 24.23. We compare M as in (24.31) and (24.32) and λ as in (24.45). We easily obtain that

$$\lambda \leq M. \tag{24.49}$$

As a result we have that (24.48) is sharper than (24.34) and (24.47) is sharper than (24.36). That is reasonable since we restricted ourselves to a subclass of $M(c,d)$ of measures μ by assuming the moment condition (24.46).

We finish chapter with

Remark 24.24. I) When $c = a$ and $d = b$ then d_1 plays no role in the best upper bounds we found with the Geometric Moment Theory Method. That is, the restriction on measures μ via the first moment d_1 has no effect in producing sharper estimates as it happens when $a < c < d < b$. More precisely we notice that:

1)

$$\text{R.H.S.}(24.19) = \frac{\|f'\|_\infty}{2}(b-a) = \text{R.H.S.}(24.9), \tag{24.50}$$

2) by (24.45) here $\lambda = (b-a)^{n+1}$ and

$$\text{R.H.S.}(24.47) = \frac{\|f^{(n)}\|_\infty}{n(n+1)!}(b-a)^n = \text{R.H.S.}(24.35). \tag{24.51}$$

II) Further differences of general means over any $[c_1, d_1]$ and $[c_2, d_2]$ subsets of $[a,b]$ (even disjoint) with respect to μ_1 and μ_2, respectively, can be found by the above results and triangle inequality as a straightforward application, etc.

Chapter 25

Grüss Type Inequalities Using the Stieltjes Integral

In this chapter we establish sharp inequalities and close to sharp, for the Chebychev functional with respect to the Stieltjes integral. We give applications to Probability. The estimates here are with respect to moduli of continuity of the involved functions. This treatment relies on [55].

25.1 Motivation

The main inspiration here comes from

Theorem 25.1.(G. Grüss, 1935, [205]) *Let f, g integrable functions from $[a, b] \to \mathbb{R}$, such that $m \leq f(x) \leq M$, $\rho \leq g(x) \leq \sigma$, $\forall\ x \in [a, b]$, where $m, M, \rho, \sigma \in \mathbb{R}$. Then*

$$\left| \frac{1}{b-a} \int_a^b f(x)g(x)dx - \frac{1}{(b-a)^2} \left(\int_a^b f(x)dx \right) \left(\int_a^b g(x)dx \right) \right| \leq \frac{1}{4}(M-m)(\sigma-\rho).$$

(25.1)

The constant $\dfrac{1}{4}$ is the best, so inequality (25.1) is sharp.

Other very important motivation comes from [91], p.245 and S. Dragomir articles [155], [154], [162]. The main feature here is that Grüss type inequalities are presented with respect to Stieljes integral and the modulus of continuity.

25.2 Main Results

We mention

Definition 25.2. (see also [317], p.55) Let $f : [a, b] \to \mathbb{R}$ be a continuous function.
  We denote by

$$\omega_1(f, \delta) := \sup_{\substack{x,y \in [a,b] \\ |x-y| \leq \delta}} |f(x) - f(y)|, \quad 0 < \delta \leq b - a,$$

(25.2)

the first modulus of continuity of f.

Terms and Assumptions. From now on in this chapter we consider continuous functions $f, g : [a, b] \to \mathbb{R}$, $a \neq b$, α is a function of bounded variation from $[a, b]$ into $\mathbb{R}$, such that $\alpha(a) \neq \alpha(b)$. We denote by α^* the total variation function of α. Clearly here $\alpha^*(b) > 0$.

We denote by

$$D(f, g) := \frac{1}{\alpha(b) - \alpha(a)} \int_a^b f(x) g(x) d\alpha(x)$$

$$- \frac{1}{(\alpha(b) - \alpha(a))^2} \left(\int_a^b f(x) d\alpha(x) \right) \left(\int_a^b g(x) d\alpha(x) \right), \tag{25.3}$$

this is Chebychev functional for Stieltjes integral.

We derive

Theorem 25.3. *It holds*

   i)

$$|D(f, g)| \leq \frac{1}{2(\alpha(b) - \alpha(a))^2} \int_a^b \left(\int_a^b |f(y) - f(x)||g(y) - g(x)| d\alpha^*(y) \right) d\alpha^*(x),$$
$$\tag{25.4}$$

and

$$|D(f, g)| \leq \frac{1}{2(\alpha(b) - \alpha(a))^2} \int_a^b \left(\int_a^b \omega_1(f, |x - y|) \omega_1(g, |x - y|) d\alpha^*(y) \right) d\alpha^*(x).$$
$$\tag{25.5}$$

Inequalities (25.4) and (25.5) are sharp, in fact attained, when $f(x) = g(x) = id(x) := x$, $\forall \ x \in [a, b]$ and α is increasing (i.e. $d\alpha^* = d\alpha$).

   ii) Let $m \leq f(x) \leq M$, $\rho \leq g(x) \leq \sigma$, $\forall \ x \in [a, b]$. Then

$$|D(f, g)| \leq \frac{1}{2} \left(\frac{\alpha^*(b)}{\alpha(b) - \alpha(a)} \right)^2 (M - m)(\sigma - \rho). \tag{25.6}$$

When α is increasing it holds

$$|D(f, g)| \leq \frac{1}{2}(M - m)(\sigma - \rho). \tag{25.7}$$

We continue with

Theorem 25.4. *Assume f, g be of Lipschitz type of the form $\omega_1(f, \delta) \leq L_1 \delta^\beta$, $\omega_1(g, \delta) \leq L_2 \delta^\beta$, $0 < \beta \leq 1$, $L_1, L_2 > 0$, $\forall \ \delta > 0$.*

   Then

   i)

$$|D(f, g)| \leq \frac{L_1 L_2}{2(\alpha(b) - \alpha(a))^2} \left(\int_a^b \left(\int_a^b (x - y)^{2\beta} d\alpha^*(y) \right) d\alpha^*(x) \right). \tag{25.8}$$

If α is increasing then again in (25.8) we have $d\alpha^* = d\alpha$.
Additionally suppose that α is Lipschitz type, i.e. $\exists\ \lambda > 0$ such that

$$|\alpha(x) - \alpha(y)| \leq \lambda|x - y|, \ \forall\ x, y \in [a, b].$$

Then
ii)

$$|D(f, g)| \leq \frac{L_1 L_2 \lambda^2 (b - a)^{2(1+\beta)}}{2(1 + \beta)(1 + 2\beta)(\alpha(b) - \alpha(a))^2}. \tag{25.9}$$

Next we have

Theorem 25.5. *For $0 < \delta \leq b - a$ it holds*

$$|D(f, g)| \leq \frac{1}{2(\alpha(b) - \alpha(a))^2} \left[\int_a^b \left(\int_a^b \left\lceil \frac{|x - y|}{\delta} \right\rceil^2 d\alpha^*(y) \right) d\alpha^*(x) \right] \omega_1(f, \delta)\omega_1(g, \delta). \tag{25.10}$$

When α is increasing then again $d\alpha^* = d\alpha$.

We derive

Theorem 25.6. *It holds*
i)

$$|D(f, g)| \leq 2 \left(\frac{\alpha^*(b)}{\alpha(b) - \alpha(a)} \right)^2 \omega_1 \left(f, \frac{1}{\alpha^*(b)} \left(\int_a^b \left(\int_a^b (x - y)^2 d\alpha^*(y) \right) d\alpha^*(x) \right)^{1/2} \right)$$

$$\times \omega_1 \left(g, \frac{1}{\alpha^*(b)} \left(\int_a^b \left(\int_a^b (x - y)^2 d\alpha^*(y) \right) d\alpha^*(x) \right)^{1/2} \right). \tag{25.11}$$

ii)

$$|D(f, f)| = \left| \frac{1}{\alpha(b) - \alpha(a)} \int_a^b f^2(x) d\alpha(x) - \frac{1}{(\alpha(b) - \alpha(a))^2} \left(\int_a^b f(x) d\alpha(x) \right)^2 \right|$$

$$\leq 2 \left(\frac{\alpha^*(b)}{\alpha(b) - \alpha(a)} \right)^2 \omega_1^2 \left(f, \frac{1}{\alpha^*(b)} \left(\int_a^b \left(\int_a^b (x - y)^2 d\alpha^*(y) \right) d\alpha^*(x) \right)^{1/2} \right). \tag{25.12}$$

We also have

Theorem 25.7. *Here α is increasing. Then*
i)

$$|D(f, g)| \leq 2\omega_1 \left(f, \frac{1}{\alpha(b) - \alpha(a)} \left(\int_a^b \left(\int_a^b (x - y)^2 d\alpha(y) \right) d\alpha(x) \right)^{1/2} \right)$$

$$\times \omega_1 \left(g, \frac{1}{\alpha(b) - \alpha(a)} \left(\int_a^b \left(\int_a^b (x-y)^2 d\alpha(y) \right) d\alpha(x) \right)^{1/2} \right), \qquad (25.13)$$

and

ii)

$$|D(f,f)| \leq 2\omega_1^2 \left(f, \frac{1}{\alpha(b) - \alpha(a)} \left(\int_a^b \left(\int_a^b (x-y)^2 d\alpha(y) \right) d\alpha(x) \right)^{1/2} \right). \quad (25.14)$$

Proof of Theorems 25.3 - 25.7. Since f, g are continuous functions and α is of bounded variation from $[a, b] \to \mathbb{R}$, by [91], p.245 we have

$$K := K(f,g) := (\alpha(b) - \alpha(a)) \int_a^b f(x)g(x)d\alpha(x) - \left(\int_a^b f(x)d\alpha(x) \right) \left(\int_a^b g(x)d\alpha(x) \right)$$

$$= \frac{1}{2} \int_a^b \left(\int_a^b (f(y) - f(x))(g(y) - g(x))d\alpha(y) \right) d\alpha(x). \qquad (25.15)$$

Hence by (25.15) we derive that

$$|K| \leq \frac{1}{2} \int_a^b \left| \int_a^b (f(y) - f(x))(g(y) - g(x))d\alpha(y) \right| d\alpha^*(x)$$

$$\leq \frac{1}{2} \int_a^b \left(\int_a^b |f(y) - f(x)||g(y) - g(x)|d\alpha^*(y) \right) d\alpha^*(x). \qquad (25.16)$$

So we find

$$|K(f,g)| \leq \frac{1}{2} \int_a^b \left(\int_a^b |f(y) - f(x)||g(y) - g(x)|d\alpha^*(y) \right) d\alpha^*(x) =: \rho_1, \quad (25.17)$$

proving (25.4).

By $m \leq f(x) \leq M$, $\rho \leq g(x) \leq \sigma$, all $x \in [a, b]$, we obtain

$$|f(x) - f(y)| \leq M - m,$$

$$|g(x) - g(y)| \leq \sigma - \rho, \ \forall \ x, y \in [a, b].$$

Consequently, we see that

$$\rho_1 \leq \frac{1}{2}(M - m)(\sigma - \rho)(\alpha^*(b))^2, \qquad (25.18)$$

i.e.

$$|K(f,g)| \leq \frac{1}{2}(\alpha^*(b))^2(M - m)(\sigma - \rho), \qquad (25.19)$$

proving (25.6).

That is, for α increasing it holds

$$|K(f,g)| \le \frac{(\alpha(b) - \alpha(a))^2}{2}(M - m)(\sigma - \rho), \qquad (25.20)$$

proving (25.7).

Let now $f = g = id$, $id(x) := x$, $\forall\, x \in [a,b]$ and α increasing, then by (25.17) we have

$$|K(id, id)| = K(id, id) = \frac{1}{2}\int_a^b \left(\int_a^b (y - x)^2 d\alpha(y) \right) d\alpha(x)$$

$$= \frac{1}{2}\int_a^b \left(\int_a^b |y - x||y - x| d\alpha^*(y) \right) d\alpha^*(x), \qquad (25.21)$$

by $\alpha^*(x) = \alpha(x) - \alpha(a)$.

That is proving inequality (25.17) is attained, that is a sharp inequality.

We further derive

$$\rho_1 \le \frac{1}{2}\int_a^b \left(\int_a^b \omega_1(f, |x - y|)\omega_1(g, |x - y|) d\alpha^*(y) \right) d\alpha^*(x). \qquad (25.22)$$

That is,

$$|K(f,g)| \le \frac{1}{2}\int_a^b \left(\int_a^b \omega_1(f, |x - y|)\omega_1(g, |x - y|) d\alpha^*(y) \right) d\alpha^*(x) = \rho_2, \qquad (25.23)$$

proving (25.5).

Since $\omega_1(id, \delta) = \delta$, $\forall\, \delta: 0 < \delta \le b - a$, the last inequality (25.23) is attained and sharp as (25.17) above.

Let now f, g be Lipschitz type of the form

$$\omega_1(f, \delta) \le L_1\delta^\beta, \quad 0 < \beta \le 1,$$

$$\omega_1(g, \delta) \le L_2\delta^\beta, \quad L_1, L_2 > 0.$$

Then

$$\rho_2 \le \frac{L_1 L_2}{2}\left(\int_a^b \left(\int_a^b |x - y|^\beta |x - y|^\beta d\alpha^*(y) \right) d\alpha^*(x) \right). \qquad (25.24)$$

That is,

$$|K(f,g)| \le \frac{L_1 L_2}{2}\left(\int_a^b \left(\int_a^b (x - y)^{2\beta} d\alpha^*(y) \right) d\alpha^*(x) \right) =: \rho_3, \qquad (25.25)$$

proving (25.8).

In the case of $\alpha : [a,b] \to \mathbb{R}$ being Lipschitz, i.e, there exists $\lambda > 0$ such that

$$|\alpha(x) - \alpha(y)| \le \lambda|x - y|, \ \forall\, x, y \in [a,b].$$

We have for say $x \geq y$ that

$$\alpha^*(x) - \alpha^*(y) = V_\alpha[y, x] \leq \lambda(x - y),$$

i.e.

$$|\alpha^*(x) - \alpha^*(y)| \leq \lambda|x - y|, \ \forall \ x, y \in [a, b],$$

hence α^* is same type of Lipschitz function.

Thus, see (25.25), we derive

$$\rho_3 \leq \frac{L_1 L_2 \lambda^2}{2} \left(\int_a^b \left(\int_a^b (x - y)^{2\beta} dy \right) dx \right) \tag{25.26}$$

$$= \frac{L_1 L_2 \lambda^2 (b - a)^{2(\beta+1)}}{2(2\beta + 1)(\beta + 1)}. \tag{25.27}$$

Hence it holds

$$|K(f, g)| \leq \frac{L_1 L_2 \lambda^2}{2(\beta + 1)(2\beta + 1)} (b - a)^{2(\beta+1)}, \tag{25.28}$$

proving (25.9).

Next noticing, $(0 < \delta \leq b - a)$,

$$\rho_2 = \frac{1}{2} \int_a^b \left(\int_a^b \omega_1 \left(f, \frac{\delta|x - y|}{\delta} \right) \omega_1 \left(g, \frac{\delta|x - y|}{\delta} \right) d\alpha^*(y) \right) d\alpha^*(x)$$

(by [317], p.55) (here $\lceil \cdot \rceil$ is the ceiling of number)

$$\leq \frac{1}{2} \left[\int_a^b \left(\int_a^b \left\lceil \frac{|x - y|}{\delta} \right\rceil^2 d\alpha^*(y) \right) d\alpha^*(x) \right] \tag{25.29}$$

$$\omega_1(f, \delta)\omega_1(g, \delta) \leq \frac{1}{2} \left[\int_a^b \left(\int_a^b \left(1 + \frac{|x - y|}{\delta} \right)^2 d\alpha^*(y) \right) d\alpha^*(x) \right] \tag{25.30}$$

$$\omega_1(f, \delta)\omega_1(g, \delta) =$$

$$\frac{1}{2} \left[\int_a^b \left(\int_a^b \left(1 + \frac{(x - y)^2}{\delta^2} + \frac{2|x - y|}{\delta} \right) d\alpha^*(y) \right) d\alpha^*(x) \right] \omega_1(f, \delta)\omega_1(g, \delta)$$

$$= \frac{1}{2} \left[(\alpha^*(b))^2 + \frac{1}{\delta^2} \left(\int_a^b \left(\int_a^b (x - y)^2 d\alpha^*(y) \right) d\alpha^*(x) \right) \right.$$

$$\left. + \frac{2}{\delta} \left(\int_a^b \left(\int_a^b |x - y| d\alpha^*(y) \right) d\alpha^*(x) \right) \right] \omega_1(f, \delta)\omega_1(g, \delta) \tag{25.31}$$

$$\leq \frac{1}{2} \left[(\alpha^*(b))^2 + \frac{1}{\delta^2} \left(\int_a^b \left(\int_a^b (x - y)^2 d\alpha^*(y) \right) d\alpha^*(x) \right) \right]$$

$$+\frac{2}{\delta}\left(\int_a^b\left(\int_a^b(x-y)^2 d\alpha^*(y)\right)d\alpha^*(x)\right)^{1/2}\alpha^*(b)\right]\omega_1(f,\delta)\omega_1(g,\delta) \qquad (25.32)$$

(letting

$$\delta=\delta^*:=\frac{1}{\alpha^*(b)}\left(\int_a^b\left(\int_a^b(x-y)^2 d\alpha^*(y)\right)d\alpha^*(x)\right)^{1/2}>0, \qquad (25.33)$$

here $\alpha^*(b)>0$, $0<\delta^*\le b-a$)

$$=\frac{1}{2}[(\alpha^*(b))^2+(\alpha^*(b))^2+2(\alpha^*(b))^2]$$

$$\omega_1(f,\delta^*)\omega_1(g,\delta^*)=2(\alpha^*(b))^2\omega_1(f,\delta^*)\omega_1(g,\delta^*). \qquad (25.34)$$

That is, we have proved (25.10) and

$$|K(f,g)|\le 2(\alpha^*(b))^2\omega_1\left(f,\frac{1}{\alpha^*(b)}\left(\int_a^b\left(\int_a^b(x-y)^2 d\alpha^*(y)\right)d\alpha^*(x)\right)^{1/2}\right)$$

$$\times\omega_1\left(g,\frac{1}{\alpha^*(b)}\left(\int_a^b\left(\int_a^b(x-y)^2 d\alpha^*(y)\right)d\alpha^*(x)\right)^{1/2}\right), \qquad (25.35)$$

establishing (25.11).

In particular, we get

$$|K(f,f)|=\left|(\alpha(b)-\alpha(a))\int_a^b f^2(x)d\alpha(x)-\left(\int_a^b f(x)d\alpha(x)\right)^2\right|$$

$$\le 2(\alpha^*(b))^2\omega_1^2\left(f,\frac{1}{\alpha^*(b)}\left(\int_a^b\left(\int_a^b(x-y)^2 d\alpha^*(y)\right)d\alpha^*(x)\right)^{1/2}\right), \qquad (25.36)$$

proving (25.12).

In case of α being increasing we obtain, respectively,

$$|K(f,g)|\le 2(\alpha(b)-\alpha(a))^2$$

$$\omega_1\left(f,\frac{1}{\alpha(b)-\alpha(a)}\left(\int_a^b\left(\int_a^b(x-y)^2 d\alpha(y)\right)d\alpha(x)\right)^{1/2}\right)$$

$$\times\omega_1\left(g,\frac{1}{\alpha(b)-\alpha(a)}\left(\int_a^b\left(\int_a^b(x-y)^2 d\alpha(y)\right)d\alpha(x)\right)^{1/2}\right), \qquad (25.37)$$

proving (25.13), and

$$|K(f,f)|\le 2(\alpha(b)-\alpha(a))^2\omega_1^2\left(f,\frac{1}{\alpha(b)-\alpha(a)}\left(\int_a^b\left(\int_a^b(x-y)^2 d\alpha(y)\right)d\alpha(x)\right)^{1/2}\right),$$

$$(25.38)$$

proving (25.14).

We have established all claims. $\qquad\square$

25.3   Applications

Let $(\Omega, \mathcal{A}, P)$ be a probability space, $X : \Omega \to [a, b]$, a.e., be a random variable with distribution function F.

Then here

$$D_X(f, g) := D(f, g) = \int_a^b f(x)g(x)dF(x) - \left(\int_a^b f(x)dF(x) \right) \left(\int_a^b g(x)dF(x) \right),$$
(25.39)

i.e.

$$D_X(f, g) = E(f(X)g(X)) - E(f(X))E(g(X)).$$
(25.40)

By [91], pp. 245-246,

$$D_X(f, g) \geq 0,$$

if f, g both are increasing, or both are decreasing over $[a, b]$, while

$$D_X(f, g) \leq 0,$$

if f is increasing and g is decreasing, or f is decreasing and g is increasing over $[a, b]$.

We give

Corollary 25.8. *It holds*
i)

$$|D_X(f, g)| \leq \frac{1}{2} \int_a^b \left(\int_a^b |f(y) - f(x)||g(y) - g(x)|dF(y) \right) dF(x),$$
(25.41)

and

$$|D_X(f, g)| \leq \frac{1}{2} \int_a^b \left(\int_a^b \omega_1(f, |x - y|)\omega_1(g, |x - y|)dF(y) \right) dF(x).$$
(25.42)

Inequalities (25.41), (25.42) are sharp.
ii) Let $m \leq f(x) \leq M$, $\rho \leq g(x) \leq \sigma$, $\forall\, x \in [a, b]$. Then

$$|D_X(f, g)| \leq \frac{1}{2}(M - m)(\sigma - \rho).$$
(25.43)

Proof. By Theorem 25.3. $\qquad\square$

Corollary 25.9. *Assume $f, g : \omega_1(f, \delta) \leq L_1\delta^\beta$, $\omega_1(g, \delta) \leq L_2\delta^\beta$, $0 < \beta \leq 1$, $L_1, L_2 > 0$, $\forall\, \delta > 0$. Then*
i)

$$|D_X(f, g)| \leq \frac{L_1 L_2}{2} \left(\int_a^b \left(\int_a^b (x - y)^{2\beta}dF(y) \right) dF(x) \right).$$
(25.44)

Additionally suppose that F is such that $\exists\ \lambda > 0$ with

$$|F(x) - F(y)| \leq \lambda|x - y|, \ \forall\ x, y \in [a, b].$$

Then
   ii)

$$|D_X(f, g)| \leq \frac{L_1 L_2 \lambda^2 (b - a)^{2(1+\beta)}}{2(1 + \beta)(1 + 2\beta)}. \tag{25.45}$$

Proof. By Theorem 25.4. $\qquad\qquad\square$

Corollary 25.10. *For $0 < \delta \leq b - a$ it holds*

$$|D_X(f, g)| \leq \frac{1}{2}\left[\int_a^b \left(\int_a^b \left\lceil \frac{|x - y|}{\delta} \right\rceil^2 dF(y)\right) dF(x)\right] \omega_1(f, \delta)\omega_1(g, \delta). \tag{25.46}$$

Proof. By Theorem 25.5. $\qquad\qquad\square$

Corollary 25.11. *It holds*
   i)

$$|D_X(f, g)| \leq 2\omega_1\left(f, \left(\int_a^b \left(\int_a^b (x - y)^2 dF(y)\right) dF(x)\right)^{1/2}\right)$$

$$\times \omega_1\left(g, \left(\int_a^b \left(\int_a^b (x - y)^2 dF(y)\right) dF(x)\right)^{1/2}\right), \tag{25.47}$$

and
   ii)

$$|D_X(f, f)| = |E(f^2(X)) - (Ef(X))^2| = Var(f(X)) \tag{25.48}$$

$$\leq 2\omega_1^2\left(f, \left(\int_a^b \left(\int_a^b (x - y)^2 dF(y)\right) dF(x)\right)^{1/2}\right). \tag{25.49}$$

Proof. By Theorem 25.7. $\qquad\qquad\square$

Chapter 26

Chebyshev-Grüss Type and Difference of Integral Means Inequalities Using the Stieltjes Integral

In this chapter we of the establish tight Chebychev-Grüss type and comparison of integral means inequalities involving the Stieltjes integral. The estimates are with respect to $\| \cdot \|_p$, $1 \le p \le \infty$. At the end of the chapter we provide applications to Probability and comparison of means for specific weights. We give also a side result regarding the integral annihilating property of generalized Peano kernel. This treatment follows [51].

26.1 Background

The result that inspired the most this chapter follows.

Theorem 26.1. (Chebychev, 1882, [129]) *Let $f, g : [a, b] \to \mathbb{R}$ absolutely continuous functions. If $f', g' \in L_\infty([a, b])$, then*

$$\left| \frac{1}{b-a} \int_a^b f(x)g(x)dx - \frac{1}{(b-a)^2} \left(\int_a^b f(x)dx \right) \left(\int_a^b g(x)dx \right) \right|$$

$$\le \frac{1}{12}(b-a)^2 \|f'\|_\infty \|g'\|_\infty.$$

Next we mention another great motivational result.

Theorem 26.2. (G. Grüss, 1935, [205]) *Let f, g integrable functions from $[a, b] \to \mathbb{R}$, such that $m \le f(x) \le M$, $\rho \le g(x) \le \sigma$, for all $x \in [a, b]$, where $m, M, \rho, \sigma \in \mathbb{R}$. Then*

$$\left| \frac{1}{b-a} \int_a^b f(x)g(x)dx - \frac{1}{(b-a)^2} \left(\int_a^b f(x)dx \right) \left(\int_a^b g(x)dx \right) \right| \le \frac{1}{4}(M-m)(\sigma-\rho).$$

Also, we were inspired a lot by the important and interesting article by S.S. Dragomir [162].

From [32] we use a lot in the proofs the next result.

355

Theorem 26.3. *Let $f : [a,b] \to \mathbb{R}$ be differentiable on $[a,b]$. The derivative $f' : [a,b] \to \mathbb{R}$ is integrable on $[a,b]$. Let $g : [a,b] \to \mathbb{R}$ be of bounded variation and such that $g(a) \neq g(b)$. Let $x \in [a,b]$. We define*

$$P(g(x), g(t)) := \begin{cases} \dfrac{g(t) - g(a)}{g(b) - g(a)}, & a \leq t \leq x, \\[3mm] \dfrac{g(t) - g(b)}{g(b) - g(a)}, & x < t \leq b. \end{cases} \tag{26.1}$$

Then

$$f(x) = \frac{1}{(g(b) - g(a))} \int_a^b f(t) dg(t) + \int_a^b P(g(x), g(t)) f'(t) dt. \tag{26.2}$$

26.2   Main Results

We present the first main result of Chebyshev-Grüss type.

Theorem 26.4. *Let $f_1, f_2 : [a,b] \to \mathbb{R}$ be differentiable on $[a,b]$. The derivatives $f_1', f_2' : [a,b] \to \mathbb{R}$ are integrable on $[a,b]$. Let $g : [a,b] \to \mathbb{R}$ be of bounded variation and such that $g(a) \neq g(b)$. Here P is as in (26.1).*
Denote by

$$\theta := \frac{1}{(g(b) - g(a))} \int_a^b f_1(x) f_2(x) dg(x)$$

$$- \left(\frac{1}{(g(b) - g(a))} \int_a^b f_1(x) dg(x) \right) \left(\frac{1}{(g(b) - g(a))} \int_a^b f_2(x) dg(x) \right) \tag{26.3}$$

where g^* is the total variation function of g, i.e. $g^*(x) := V_g[a, x]$, for $x \in (a, b]$, where $g^*(a) := 0$.
1) If $f_1', f_2' \in L_\infty([a, b])$, then

$$|\theta| \leq \frac{1}{2|g(b) - g(a)|} \left[\|f_1\|_\infty \|f_2'\|_\infty + \|f_2\|_\infty \|f_1'\|_\infty \right]$$

$$\times \left(\int_a^b \left(\int_a^b |P(g(z), g(x))| dx \right) dg^*(z) \right) =: M_1. \tag{26.4}$$

2) Let $p, q > 1$: $\dfrac{1}{p} + \dfrac{1}{q} = 1$. Suppose $f_1', f_2' \in L_p([a, b])$. Then

$$|\theta| \leq \frac{1}{2|g(b) - g(a)|} \left[\|f_2'\|_p \|f_1\|_\infty + \|f_1'\|_p \|f_2\|_\infty \right]$$

$$\times \left(\int_a^b \|P(g(z), g(x))\|_{q,x} dg^*(z) \right) =: M_2. \tag{26.5}$$

When $p = q = 2$, then

$$|\theta| \leq \frac{1}{2|g(b) - g(a)|} [\|f_2'\|_2 \|f_1\|_\infty + \|f_1'\|_2 \|f_2\|_\infty]$$

$$\times \left(\int_a^b \|P(g(z), g(x))\|_{2,x} dg^*(z) \right) =: M_3. \qquad (26.6)$$

3) Also we have

$$|\theta| \leq \frac{1}{2|g(b) - g(a)|} [\|f_2'\|_1 \|f_1\|_\infty + \|f_1'\|_1 \|f_2\|_\infty]$$

$$\times \left(\int_a^b \|P(g(z), g(x))\|_{\infty,x} dg^*(z) \right) =: M_4. \qquad (26.7)$$

Proof. By Theorem 26.3 we have

$$f_1(z) = \frac{1}{(g(b) - g(a))} \int_a^b f_1(x) dg(x) + \int_a^b P(g(z), g(x)) f_1'(x) dx,$$

and

$$f_2(z) = \frac{1}{(g(b) - g(a))} \int_a^b f_2(x) dg(x) + \int_a^b P(g(z), g(x)) f_2'(x) dx,$$

all $z \in [a, b]$.

Consequently we get

$$f_1(z) f_2(z) = \frac{f_2(z)}{(g(b) - g(a))} \int_a^b f_1(x) dg(x) + f_2(z) \int_a^b P(g(z), g(x)) f_1'(x) dx, \quad (26.8)$$

and

$$f_1(z) f_2(z) = \frac{f_1(z)}{(g(b) - g(a))} \int_a^b f_2(x) dg(x) + f_1(z) \int_a^b P(g(z), g(x)) f_2'(x) dx, \quad (26.9)$$

all $z \in [a, b]$.

Hence by integrating the last two identities we obtain

$$\int_a^b f_1(x) f_2(x) dg(x) = \frac{1}{(g(b) - g(a))} \left(\int_a^b f_1(x) dg(x) \right) \left(\int_a^b f_2(x) dg(x) \right)$$

$$+ \int_a^b f_2(z) \left(\int_a^b P(g(z), g(x)) f_1'(x) dx \right) dg(z) \qquad (26.10)$$

and

$$\int_a^b f_1(x) f_2(x) dg(x) = \frac{1}{(g(b) - g(a))} \left(\int_a^b f_1(x) dg(x) \right) \left(\int_a^b f_2(x) dg(x) \right)$$

$$+ \int_a^b f_1(z) \left(\int_a^b P(g(z), g(x)) f_2'(x) dx \right) dg(z). \tag{26.11}$$

Therefore we find

$$\int_a^b f_1(x) f_2(x) dg(x) - \frac{1}{(g(b) - g(a))} \left(\int_a^b f_1(x) dg(x) \right) \left(\int_a^b f_2(x) dg(x) \right)$$

$$= \int_a^b f_2(z) \left(\int_a^b P(g(z), g(x)) f_1'(x) dx \right) dg(z)$$

$$= \int_a^b f_1(z) \left(\int_a^b P(g(z), g(x)) f_2'(x) dx \right) dg(z). \tag{26.12}$$

So we derive that

$$\theta := \frac{1}{(g(b) - g(a))} \int_a^b f_1(x) f_2(x) dg(x)$$

$$- \left(\frac{1}{(g(b) - g(a))} \int_a^b f_1(x) dg(x) \right) \left(\frac{1}{(g(b) - g(a))} \int_a^b f_2(x) dg(x) \right)$$

$$= \frac{1}{2(g(b) - g(a))} \left[\int_a^b f_1(z) \left(\int_a^b P(g(z), g(x)) f_2'(x) dx \right) dg(z) \right.$$

$$\left. + \int_a^b f_2(z) \left(\int_a^b P(g(z), g(x)) f_1'(x) dx \right) dg(z) \right]. \tag{26.13}$$

1) Estimate with respect to $\| \cdot \|_\infty$. We observe that

$$|\theta| \le \frac{1}{2|g(b) - g(a)|} [\|f_1\|_\infty \|f_2'\|_\infty + \|f_2\|_\infty \|f_1'\|_\infty]$$

$$\times \left(\int_a^b \left(\int_a^b |P(g(z), g(x))| dx \right) dg^*(z) \right), \tag{26.14}$$

where R.H.S. (26.14) is finite and makes sense. That is proving the claim (26.4).

2) Estimate with respect to $\| \cdot \|_p$, $p, q > 1$ such that $\frac{1}{p} + \frac{1}{q} = 1$.
We do have

$$|\theta| \le \frac{1}{2|g(b) - g(a)|} \left[\|f_2'\|_p \int_a^b |f_1(z)| \| P(g(z), g(x)) \|_{q,x} dg^*(z) \right.$$

$$\left. + \|f_1'\|_p \int_a^b |f_2(z)| \| P(g(z), g(x)) \|_{q,x} dg^*(z) \right] \tag{26.15}$$

$$\leq \frac{1}{2|g(b)-g(a)|}[\|f_2'\|_p\|f_1\|_\infty + \|f_1'\|_p\|f_2\|_\infty]\left(\int_a^b \|P(g(z),g(x))\|_{q,x}dg^*(z)\right),$$

$$(26.16)$$

where R.H.S. (26.16) is finite and makes sense. That is proving the claim (26.5).

3) Estimate with respect to $\|\cdot\|_1$. We see that

$$|\theta| \leq \frac{1}{2|g(b)-g(a)|}\left[\|f_2'\|_1\int_a^b |f_1(z)|\|P(g(z),g(x))\|_{\infty,x}dg^*(z)\right.$$

$$\left. +\|f_1'\|_1\int_a^b |f_2(z)|\|P(g(z),g(x))\|_{\infty,x}dg^*(z)\right] \qquad (26.17)$$

$$\leq \frac{1}{2|g(b)-g(a)|}[\|f_2'\|_1\|f_1\|_\infty + \|f_1'\|_1\|f_2\|_\infty]\left(\int_a^b \|P(g(z),g(x))\|_{\infty,x}dg^*(z)\right),$$

$$(26.18)$$

where R.H.S. (26.18) is finite and makes sense. That is proving the claim (26.7).

Justification of the existence of the integrals in R.H.S. of (26.14), (26.16), (26.18).

i) We see that

$$I(z) := \int_a^b |P(g(z),g(x))|dx$$

$$= \int_a^z |g(x)-g(a)|dx + \int_z^b |g(x)-g(b)|dx =: I_1(z) + I_2(z). \qquad (26.19)$$

Both integrals $I_1(z)$, $I_2(z)$ exist as Riemann integrals of bounded variation functions which are bounded functions. Thus $I_1(z)$, $I_2(z)$ are continuous functions in $z \in [a,b]$, hence $I(z)$ is continuous function in z. Consequently

$$\int_a^b \left(\int_a^b |P(g(z),g(x))|dx\right)dg^*(z) = \int_a^b I(z)dg^*(z) \qquad (26.20)$$

is an existing Riemann-Stieltjes integral.

ii) We notice that ($q > 1$)

$$I_q(z) := \|P(g(z),g(x))\|_{q,x} = \left(\int_a^b |P(g(z),g(x))|^q dx\right)^{1/q}$$

$$= \left[\int_a^z |g(x)-g(a)|^q dx + \int_z^b |g(x)-g(b)|^q dx\right]^{1/q} =: [I_{1,q}(z)+I_{2,q}(z)]^{1/q}. \qquad (26.21)$$

Again both integrals $I_{1,q}(z)$, $I_{2,q}(z)$ exist as Riemann integrals of bounded variation functions which are bounded functions.

Clearly here by g being a function of bounded variation, $|g(x)-g(a)|$, $|g(x)-g(b)|$ are of bounded variation, and $|g(x)-g(a)|^q$, $|g(x)-g(b)|^q$ are of bounded variation too. Thus $I_{1,q}(z)$, $I_{2,q}(z)$ are continuous functions in $z \in [a,b]$, hence $I_q(z)$ is a continuous function in z. Finally

$$\int_a^b \|P(g(z), g(x))\|_{q,x} dg^*(z) = \int_a^b I_q(z) dg^*(z) \tag{26.22}$$

is an existing Riemann-Stieltjes integral.

iii) We observe that

$$h(z) := \|P(g(z), g(x))\|_{\infty, x \in [a,b]} \tag{26.23}$$

$$= \max\{\|g(x) - g(a)\|_{\infty, x \in [a,z]}, \|g(x) - g(b)\|_{\infty, x \in [z,b]}\} \tag{26.24}$$

$$=: \max\{\|h_1(x)\|_{\infty, x \in [a,z]}, \|h_2(x)\|_{\infty, x \in [z,b]}\} =: \max\{A_1(z), A_2(z)\}, \tag{26.25}$$

for any $z \in [a,b]$.

Here A_1 is increasing in z and A_2 is decreasing in z. That is, A_1, A_2 are functions of bounded variation, hence $h(z)$ is a function of bounded variation.

Consequently

$$\int_a^b \|P(g(z), g(x))\|_{\infty, x} dg^*(z) = \int_a^b h(z) dg^*(z) \tag{26.26}$$

is an existing Lebesgue-Stieltjes integral.

The proof of the theorem is now complete. $\square$

We give

Corollary 26.5. *(1) All as in the basic assumptions and terms of Theorem 26.4. If f_1', $f_2' \in L_\infty([a,b])$, then*

$$|\theta| \le \min\{M_1, M_2, M_3, M_4\}. \tag{26.27}$$

(2) All as in the basic assumptions of Theorem 26.4 with g being increasing. Then

$$|\theta| \le \frac{1}{2|g(b) - g(a)|} [\|f_1\|_\infty \|f_2'\|_1 + \|f_2\|_\infty \|f_1'\|_1]$$

$$\times \left(\int_a^b \max(g(z) - g(a), g(b) - g(z)) dg(z) \right). \tag{26.28}$$

Proof. By Theorem 26.4. $\square$

Next we present a general comparison of averages main result.

Theorem 26.6. *Let $f : [a,b] \to \mathbb{R}$ be differentiable on $[a,b]$. The derivative $f' : [a,b] \to \mathbb{R}$ is integrable on $[a,b]$. Let $g_1, g_2 : [a,b] \to \mathbb{R}$ be of bounded variation and such that $g_1(a) \ne g_1(b)$, $g_2(a) \ne g_2(b)$. Let $x \in [a,b]$. We define*

$$P(g_i(x), g_i(t)) := \begin{cases} \dfrac{g_i(t) - g_i(a)}{g_i(b) - g_i(a)}, & a \le t \le x, \\[2ex] \dfrac{g_i(t) - g_i(b)}{g_i(b) - g_i(a)}, & x < t \le b, \end{cases} \tag{26.29}$$

for $i = 1, 2$.

Let $[c, d] \subseteq [a, b]$.

We define

$$\Delta := \frac{1}{(g_2(d) - g_2(c))} \int_c^d f(x) dg_2(x) - \frac{1}{(g_1(b) - g_1(a))} \int_a^b f(x) dg_1(x). \quad (26.30)$$

Here g_2^* is the total variation of g_2.

Then

i) If $f' \in L_\infty([a, b])$, it holds

$$|\Delta| \leq \frac{1}{|g_2(d) - g_2(c)|} \left(\int_c^d \left(\int_a^b |P(g_1(x), g_1(t))| dt \right) dg_2^*(x) \right) \|f'\|_\infty =: E_1.$$
$$(26.31)$$

ii) Let $p, q > 1$: $\dfrac{1}{p} + \dfrac{1}{q} = 1$. Suppose $f' \in L_p([a, b])$. Then

$$|\Delta| \leq \frac{1}{|g_2(d) - g_2(c)|} \left[\int_c^d \|P(g_1(x), g_1(t))\|_{q,t,[a,b]} dg_2^*(x) \right] \|f'\|_{p,[a,b]} =: E_2.$$
$$(26.32)$$

When $p = q = 2$, it holds

$$|\Delta| \leq \frac{1}{|g_2(d) - g_2(c)|} \left[\int_c^d \|P(g_1(x), g_1(t))\|_{2,t,[a,b]} dg_2^*(x) \right] \|f'\|_{2,[a,b]} =: E_3.$$
$$(26.33)$$

iii) Also we have

$$|\Delta| \leq \frac{1}{|g_2(d) - g_2(c)|} \left(\int_c^d \|P(g_1(x), g_1(t))\|_{\infty,t,[a,b]} dg_2^*(x) \right) \|f'\|_{1,[a,b]} =: E_4.$$
$$(26.34)$$

Proof. We have by Theorem 26.3 that

$$f(x) = \frac{1}{(g_1(b) - g_1(a))} \int_a^b f(t) dg_1(t)$$

$$+ \int_a^b P(g_1(x), g_1(t)) f'(t) dt, \ \forall \ x \in [a, b]. \quad (26.35)$$

Here $[c, d] \subseteq [a, b]$ and by integrating the last (26.35) we obtain

$$\int_c^d f(x) dg_2(x) = \left(\frac{g_2(d) - g_2(c)}{g_1(b) - g_1(a)} \right) \int_a^b f(t) dg_1(t)$$

$$+ \int_c^d \left(\int_a^b P(g_1(x), g_1(t)) f'(t) dt \right) dg_2(x). \quad (26.36)$$

Thus

$$\Delta := \frac{1}{(g_2(d) - g_2(c))} \int_c^d f(x) dg_2(x) - \frac{1}{(g_1(b) - g_1(a))} \int_a^b f(t) dg_1(t)$$

$$= \frac{1}{(g_2(d) - g_2(c))} \left(\int_c^d \left(\int_a^b P(g_1(x), g_1(t)) f'(t) dt \right) dg_2(x) \right). \qquad (26.37)$$

Next we estimate Δ.

1) Estimate with respect to $\| \cdot \|_\infty$.

By (26.37) we find

$$|\Delta| \leq \frac{1}{|g_2(d) - g_2(c)|} \left(\int_c^d \left(\int_a^b |P(g_1(x), g_1(t))| dt \right) dg_2^*(x) \right) \|f'\|_\infty. \qquad (26.38)$$

2) Estimate with respect to $\| \cdot \|_p$, here $p, q > 1$: $\dfrac{1}{p} + \dfrac{1}{q} = 1$.

We have by (26.37) that

$$|\Delta| \leq \frac{1}{|g_2(d) - g_2(c)|} \left[\int_c^d \|P(g_1(x), g_1(t))\|_{q,t,[a,b]} dg_2^*(x) \right] \|f'\|_{p,[a,b]}. \qquad (26.39)$$

When $p = q = 2$ we derive

$$|\Delta| \leq \frac{1}{|g_2(d) - g_2(c)|} \left[\int_c^d \|P(g_1(x), g_1(t))\|_{2,t,[a,b]} dg_2^*(x) \right] \|f'\|_{2,[a,b]}. \qquad (26.40)$$

3) Estimate with respect to $\| \cdot \|_1$.

We observe by (26.37) that

$$|\Delta| \leq \frac{1}{|g_2(d) - g_2(c)|} \left(\int_c^d \|P(g_1(x), g_1(t))\|_{\infty,t,[a,b]} dg_2^*(x) \right) \|f'\|_{1,[a,b]}. \qquad (26.41)$$

Similarly as in the proof of Theorem 26.4, we establish that

$$\int_c^d \left(\int_a^b |P(g_1(x), g_1(t))| dt \right) dg_2^*(x) \qquad (26.42)$$

is an existing Riemann-Stieltjes integral. Also

$$\int_c^d \|P(g_1(x), g_1(t))\|_{q,t,[a,b]} dg_2^*(x) \qquad (26.43)$$

is an existing Riemann-Stieltjes integral.

Finally

$$\int_c^d \|P(g_1(x), g_1(t))\|_{\infty,t,[a,b]} dg_2^*(x) \qquad (26.44)$$

is an existing Lebesgue-Stieltjes integral.

Therefore the upper bounds (26.38), (26.39), (26.40), (26.41) make sense and exist.

The proof of the theorem is now complete. $\qquad\square$

We give

Corollary 26.7. *All as in the basic assumptions and terms of Theorem 26.6. If $f' \in L_\infty([a,b])$, then*

$$|\Delta| \le \min\{E_1, E_2, E_3, E_4\}. \tag{26.45}$$

Proof. By Theorem 26.6. $\qquad\square$

We also give

Corollary 26.8. *All as in the basic assumptions of Theorem 26.6 with g_1, g_2 being increasing. Then*

$$|\Delta| \le \frac{1}{(g_2(d) - g_2(c))} \left(\int_c^d \max(g_1(x) - g_1(a), g_1(b) - g_1(x)) dg_2(x) \right) \|f'\|_{1,[a,b]}. \tag{26.46}$$

Proof. By (26.34). $\qquad\square$

We make on Theorem 26.6 the following

Remark 26.9. When $[c,d] = [a,b]$ we make the next comments. In that case

$$\Delta^* := \Delta = \frac{1}{(g_2(b) - g_2(a))} \int_a^b f(t) dg_2(t) - \frac{1}{(g_1(b) - g_1(a))} \int_a^b f(t) dg_1(t)$$

$$= \frac{1}{(g_2(b) - g_2(a))} \left(\int_a^b \left(\int_a^b P(g_1(x), g_1(t)) f'(t) dt \right) dg_2(x) \right), \tag{26.47}$$

and the results of Theorem 26.6 can be modified accordingly.

Alternatively, by Theorem 26.3 we have

$$f(x) = \frac{1}{(g_2(b) - g_2(a))} \int_a^b f(t) dg_2(t) + \int_a^b P(g_2(x), g_2(t)) f'(t) dt, \ \forall \ x \in [a,b]. \tag{26.48}$$

Thus

$$\int_a^b f(x) dg_1(x) = \left(\frac{g_1(b) - g_1(a)}{g_2(b) - g_2(a)} \right) \int_a^b f(t) dg_2(t)$$

$$+ \int_a^b \left(\int_a^b P(g_2(x), g_2(t)) f'(t) dt \right) dg_1(x), \tag{26.49}$$

and

$$\frac{1}{(g_1(b) - g_1(a))} \int_a^b f(t) dg_1(t) - \frac{1}{(g_2(b) - g_2(a))} \int_a^b f(t) dg_2(t)$$

$$= \frac{1}{(g_1(b) - g_1(a))} \left(\int_a^b \left(\int_a^b P(g_2(x), g_2(t)) f'(t) dt \right) dg_1(x) \right). \tag{26.50}$$

Finally we derive that

$$\Delta^* = -\frac{1}{(g_1(b) - g_1(a))} \left(\int_a^b \left(\int_a^b P(g_2(x), g_2(t)) f'(t) dt \right) dg_1(x) \right). \tag{26.51}$$

Reasoning as in Theorem 26.6 we have established

Theorem 26.10. *Here all as in Theorem 26.6.*
Case $[c, d] = [a, b]$. We define

$$\Delta^* := \frac{1}{(g_2(b) - g_2(a))} \int_a^b f(t) dg_2(t) - \frac{1}{(g_1(b) - g_1(a))} \int_a^b f(t) dg_1(t). \tag{26.52}$$

Here g_1^* is the total variation function of g_1.
Then
i) If $f' \in L_\infty([a, b])$, it holds

$$|\Delta^*| \leq \frac{1}{|g_1(b) - g_1(a)|} \left(\int_a^b \left(\int_a^b |P(g_2(x), g_2(t))| dt \right) dg_1^*(x) \right) \|f'\|_\infty =: H_1. \tag{26.53}$$

ii) Let $p, q > 1$: $\dfrac{1}{p} + \dfrac{1}{q} = 1$. Suppose $f' \in L_p([a, b])$. Then

$$|\Delta^*| \leq \frac{1}{|g_1(b) - g_1(a)|} \left[\int_a^b \|P(g_2(x), g_2(t))\|_{q,t,[a,b]} dg_1^*(x) \right] \|f'\|_{p,[a,b]} =: H_2. \tag{26.54}$$

When $p = q = 2$, it holds

$$|\Delta^*| \leq \frac{1}{|g_1(b) - g_1(a)|} \left[\int_a^b \|P(g_2(x), g_2(t))\|_{2,t,[a,b]} dg_1^*(x) \right] \|f'\|_{2,[a,b]} =: H_3. \tag{26.55}$$

iii) Also we have

$$|\Delta^*| \leq \frac{1}{|g_1(b) - g_1(a)|} \left(\int_a^b \|P(g_2(x), g_2(t))\|_{\infty,t,[a,b]} dg_1^*(x) \right) \|f'\|_{1,[a,b]} =: H_4. \tag{26.56}$$

We give

Corollary 26.11. *Here all as in Theorem 26.6.*
Case $[c, d] = [a, b]$, Δ^* as in (26.52). Then
i) If $f' \in L_\infty([a, b])$, it holds

$$|\Delta^*| \leq \min\{E_1, H_1\}, \tag{26.57}$$

where E_1 as in (26.31), H_1 as in (26.53).

ii) Let $p, q > 1$: $\frac{1}{p} + \frac{1}{q} = 1$. Suppose $f' \in L_p([a,b])$. Then

$$|\Delta^*| \leq \min\{E_2, H_2\}, \tag{26.58}$$

where E_2 as in (26.32), H_2 as in (26.54).
   When $p = q = 2$, it holds

$$|\Delta^*| \leq \min\{E_3, H_3\}, \tag{26.59}$$

where E_3 as in (26.33), H_3 as in (26.55).
   iii) Also we have

$$|\Delta^*| \leq \min\{E_4, H_4\}, \tag{26.60}$$

where E_4 as in (26.34), H_4 as in (26.56).

Proof. By Theorems 26.6, 26.10. $\qquad\square$

   We continue with

Remark 26.12. Motivated by Remark 26.9 we expand as follows.
   Let here $f \in C([a,b])$, g, g_1, g_2 of bounded variation on $[a,b]$.
   We put

$$P^*(g(x), g(t)) := \begin{cases} g(t) - g(a), \, a \leq t \leq x, \\ \\ g(t) - g(b), \, x < t \leq b, \end{cases} \tag{26.61}$$

$x \in [a,b]$.
   Similarly are defined $P^*(g_i(x), g_i(t))$, $i = 1, 2$. Using integration by parts we observe that

$$\int_a^x (g(t) - g(a))df(t) = f(t)(g(t) - g(a))\Big|_a^x - \int_a^x f(t)dg(t)$$

$$= f(x)(g(x) - g(a)) - \int_a^x f(t)dg(t), \tag{26.62}$$

and

$$\int_x^b (g(t) - g(b))df(t) = (g(t) - g(b))f(t)\Big|_x^b - \int_x^b f(t)dg(t)$$

$$= (g(b) - g(x))f(x) - \int_x^b f(t)dg(t). \tag{26.63}$$

   Adding (26.62), (26.63) we get

$$\int_a^x (g(t) - g(a))df(t) + \int_x^b (g(t) - g(b))df(t) = f(x)(g(b) - g(a)) - \int_a^b f(t)dg(t). \tag{26.64}$$

   That is

$$\int_a^b P^*(g(x), g(t))df(t) = f(x)(g(b) - g(a)) - \int_a^b f(t)dg(t). \tag{26.65}$$

Hence

$$\int_a^b \left(\int_a^b P^*(g(x), g(t)) df(t) \right) dg(x)$$

$$= \left(\int_a^b f(x) dg(x) \right) (g(b) - g(a)) - \left(\int_a^b f(t) dg(t) \right) (g(b) - g(a)) = 0, \quad (26.66)$$

that is proving

$$\int_a^b \left(\int_a^b P^*(g(x), g(t)) df(t) \right) dg(x) = 0. \tag{26.67}$$

If $f \in T([a,b]) := \{ f \in C([a,b]) : f'(x)$ exists and is finite for all but a countable set of x in (a,b) and that $f' \in L_1([a,b]) \}$ or $f \in AC([a,b])$ (absolutely continuous), then

$$\int_a^b \left(\int_a^b P^*(g(x), g(t)) f'(t) dt \right) dg(x) = 0. \tag{26.68}$$

Additionally if $g = id$, then

$$\int_a^b \left(\int_a^b P^*(x, t) f'(t) dt \right) dx = 0. \tag{26.69}$$

If $g = f = id$, then

$$\int_a^b \left(\int_a^b P^*(x, t) dt \right) dx = 0, \tag{26.70}$$

where

$$P^*(x, t) = \begin{cases} t - a, \ a \le t \le x \\ \\ t - b, \ x < t \le b. \end{cases} \tag{26.71}$$

Equality (26.70) can be proved directly by plugging in (26.71).
Similarly to (26.65) we have

$$\int_a^b P^*(g_1(x), g_1(t)) df(t) = f(x)(g_1(b) - g_1(a)) - \int_a^b f(t) dg_1(t), \tag{26.72}$$

and

$$\int_a^b P^*(g_2(x), g_2(t)) df(t) = f(x)(g_2(b) - g_2(a)) - \int_a^b f(t) dg_2(t). \tag{26.73}$$

Integrating (26.72) with respect to g_2 and (26.73) with respect to g_1 we have

$$\int_a^b \left(\int_a^b P^*(g_1(x), g_1(t)) df(t) \right) dg_2(x)$$

$$= \left(\int_a^b f(x) dg_2(x) \right) (g_1(b) - g_1(a)) - \left(\int_a^b f(t) dg_1(t) \right) (g_2(b) - g_2(a)), \quad (26.74)$$

and

$$\int_a^b \left(\int_a^b P^*(g_2(x), g_2(t)) df(t) \right) dg_1(x)$$

$$= \left(\int_a^b f(x) dg_1(x) \right) (g_2(b) - g_2(a)) - \left(\int_a^b f(t) dg_2(t) \right) (g_1(b) - g_1(a)). \quad (26.75)$$

By adding (26.74) and (26.75) we derive the interesting identity

$$\int_a^b \left(\int_a^b P^*(g_1(x), g_1(t)) df(t) \right) dg_2(x)$$

$$+ \int_a^b \left(\int_a^b P^*(g_2(x), g_2(t)) df(t) \right) dg_1(x) = 0. \quad (26.76)$$

If $f \in T([a,b])$ or $f \in A\,C([a,b])$, then

$$\int_a^b \left(\int_a^b P^*(g_1(x), g_1(t)) f'(t) dt \right) dg_2(x)$$

$$+ \int_a^b \left(\int_a^b P^*(g_2(x), g_2(t)) f'(t) dt \right) dg_1(x) = 0. \quad (26.77)$$

Next consider $h \in L_1([a,b])$ and define

$$H(x) := \int_a^x h(t) dt, \ \text{i.e. } H(a) = 0. \quad (26.78)$$

Then by [312], Theorem 9, p.103 and Theorem 13, p.106 we get that $H'(x) = h(x)$ a.e. in $[a,b]$ and H is an absolutely continuous function. That is H is continuous. Therefore by (26.67) and (26.68) we have

$$\int_a^b \left(\int_a^b P^*(g(x), g(t)) h(t) dt \right) dg(x) = 0. \quad (26.79)$$

Also by (26.76) and (26.77) we derive

$$\int_a^b \left(\int_a^b P^*(g_1(x), g_1(t)) h(t) dt \right) dg_2(x)$$

$$+ \int_a^b \left(\int_a^b P^*(g_2(x), g_2(t)) h(t) dt \right) dg_1(x) = 0. \quad (26.80)$$

From the above we have established the following result

Theorem 26.13. *Consider* $h \in L_1([a,b])$, *and* g, g_1, g_2 *of bounded variation on* $[a,b]$. *Put*

$$P^*(g(x), g(t)) := \begin{cases} g(t) - g(a), \ a \leq t \leq x, \\ \\ g(t) - g(b), \ x < t \leq b, \end{cases} \qquad (26.81)$$

$\forall \ x \in [a,b]$.
　　Similarly define $P^*(g_i(x), g_i(t))$, $i = 1, 2$. *Then*

$$\int_a^b \left(\int_a^b P^*(g(x), g(t)) h(t) dt \right) dg(x) = 0, \qquad (26.82)$$

and

$$\int_a^b \left(\int_a^b P^*(g_1(x), g_1(t)) h(t) dt \right) dg_2(x)$$

$$+ \int_a^b \left(\int_a^b P^*(g_2(x), g_2(t)) h(t) dt \right) dg_1(x) = 0. \qquad (26.83)$$

Clearly, for $g_1 = g_2 = g$ *identity (26.83) implies identity (26.82).*

We hope to explore more the merits of Theorem 26.13 elsewhere.

26.3　Applications

1) Let $(\Omega, \mathcal{A}, P)$ be a probability space, $X : \Omega \to [a,b]$, a.e., be a random variable with distribution function F. Let $f_1, f_2 : [a,b] \to \mathbb{R}$ be differentiable on $[a,b]$, f_1', f_2' are integrable on $[a,b]$. Here P as in (26.1). Denote by

$$\theta_X(f_1, f_2) := \int_a^b f_1(x) f_2(x) dF(x)$$

$$- \left(\int_a^b f_1(x) dF(x) \right) \left(\int_a^b f_2(x) dF(x) \right). \qquad (26.84)$$

That is,

$$\theta_X(f_1, f_2) = E(f_1(X) f_2(X)) - E(f_1(X)) E(f_2(X)). \qquad (26.85)$$

In particular we have

$$\theta_X(f_1, f_1) = E(f_1^2(X)) - (E(f_1(X)))^2 = Var(f_1(X)) \geq 0. \qquad (26.86)$$

Then by Theorem 26.4 we derive

Corollary 26.14. *It holds*
　　i) If $f_1', f_2' \in L_\infty([a,b])$, *then*

$$|\theta_X(f_1, f_2)| \leq \frac{1}{2} [\|f_1\|_\infty \|f_2'\|_\infty + \|f_2\|_\infty \|f_1'\|_\infty]$$

$$\times \left(\int_a^b \left(\int_a^b |P(F(z), F(x))| dx \right) dF(z) \right) =: \overline{M}_1. \tag{26.87}$$

ii) Let $p, q > 1$: $\dfrac{1}{p} + \dfrac{1}{q} = 1$. Suppose $f_1', f_2' \in L_p([a, b])$. Then

$$|\theta_X(f_1, f_2)| \le \frac{1}{2}[\|f_2'\|_p \|f_1\|_\infty + \|f_1'\|_p \|f_2\|_\infty]$$

$$\times \left(\int_a^b \|P(F(z), F(x))\|_{q,x} dF(z) \right) =: \overline{M}_2. \tag{26.88}$$

When $p = q = 2$, then

$$|\theta_X(f_1, f_2)| \le \frac{1}{2}[\|f_2'\|_2 \|f_1\|_\infty + \|f_1'\|_2 \|f_2\|_\infty]$$

$$\times \left(\int_a^b \|P(F(z), F(x))\|_{2,x} dF(z) \right) =: \overline{M}_3. \tag{26.89}$$

iii) Also we have

$$|\theta_X(f_1, f_2)| \le \frac{1}{2}[\|f_2'\|_1 \|f_1\|_\infty + \|f_1'\|_1 \|f_2\|_\infty]$$

$$\left(\int_a^b \|P(F(z), F(x))\|_{\infty,x} dF(z) \right) =: \overline{M}_4. \tag{26.90}$$

Corollary 26.15. *If $f_1', f_2' \in L_\infty([a, b])$, then*

$$|\theta_X(f_1, f_2)| \le \min\{\overline{M}_1, \overline{M}_2, \overline{M}_3, \overline{M}_4\}. \tag{26.91}$$

Proof. By Corollary 26.14. $\qquad\square$

Next we estimate $Var(f(X))$.

Corollary 26.16. *It holds*
i) If $f' \in L_\infty([a, b])$, then

$$Var(f(X)) \le \|f\|_\infty \|f'\|_\infty \left(\int_a^b \left(\int_a^b |P(F(z), F(x))| dx \right) dF(z) \right) =: M_1^*. \tag{26.92}$$

ii) Let $p, q > 1$: $\dfrac{1}{p} + \dfrac{1}{q} = 1$. Suppose $f' \in L_p([a, b])$. Then

$$Var(f(X)) \le \|f\|_\infty \|f'\|_p \left(\int_a^b \|P(F(z), F(x))\|_{q,x} dF(z) \right) =: M_2^*. \tag{26.93}$$

When $p = q = 2$, then

$$Var(f(X)) \leq \|f\|_\infty \|f'\|_2 \left(\int_a^b \|P(F(z), F(x))\|_{2,x} dF(z) \right) =: M_3^*. \qquad (26.94)$$

iii) Also we have

$$Var(f(X)) \leq \|f\|_\infty \|f'\|_1 \left(\int_a^b \|P(F(z), F(x))\|_{\infty,x} dF(z) \right) =: M_4^*. \qquad (26.95)$$

Proof. By Corollary 26.14. $\qquad\qquad\qquad\qquad\qquad\qquad\qquad\qquad\qquad\qquad\square$

Corollary 26.17. *If $f' \in L_\infty([a,b])$, then*

$$Var(f(X)) \leq \min\{M_1^*, M_2^*, M_3^*, M_4^*\}. \qquad (26.96)$$

Proof. By Corollary 26.16. $\qquad\qquad\qquad\qquad\qquad\qquad\qquad\qquad\qquad\qquad\square$

2) Here we apply Theorem 26.6 for $g_1(x) = e^x$, $g_2(x) = 3^x$, both continuous and increasing.

We have $dg_1(x) = e^x dx$ and $dg_2(x) = 3^x \ln 3 dx$.

Also $f : [a, b] \to \mathbb{R}$ is differentiable and $f' \in L_1([a,b])$. Define

$$P(e^x, e^t) := \begin{cases} \dfrac{e^t - e^a}{e^b - e^a}, & a \leq t \leq x, \\[3mm] \dfrac{e^t - e^b}{e^b - e^a}, & x < t \leq b, \end{cases} \qquad (26.97)$$

and

$$P(3^x, 3^t) := \begin{cases} \dfrac{3^t - 3^a}{3^b - 3^a}, & a \leq t \leq x, \\[3mm] \dfrac{3^t - 3^b}{3^b - 3^a}, & x < t \leq b. \end{cases} \qquad (26.98)$$

Let $[c, d] \subseteq [a, b]$.

Set

$$\Delta(e^x, 3^x) := \frac{\ln 3}{(3^d - 3^c)} \int_c^d f(x) 3^x dx - \frac{1}{(e^b - e^a)} \int_a^b f(x) e^x dx. \qquad (26.99)$$

We give

Corollary 26.18. *It holds*

i) If $f' \in L_\infty([a,b])$, then

$$|\Delta(e^x, 3^x)| \leq \frac{(\ln 3) \|f'\|_\infty}{(e^b - e^a)(3^d - 3^c)}$$

$$\left\{ -\left(\frac{1}{(\ln 3)^2(1+ln3)} \right) [2\{(3e)^c - (3e)^d\}(\ln 3)^2 \right.$$

$$+ e^a(1+\ln 3)\{3^c - 3^d - 3^c \ln 3 + 3^d \ln 3 + (3^c - 3^d) + a\ln 3 - 3^c c \ln 3 + 3^d d \ln 3\}$$

$$\left. + e^b(1+\ln 3)\{3^c - 3^d - 3^c \ln 3 + 3^d \ln 3 + (3^c - 3^d)b\ln 3 - 3^c c \ln 3 + 3^d d \ln 3\}] \right\}$$

$$=: \overline{E}_1. \tag{26.100}$$

ii) Let $p,q > 1$: $\dfrac{1}{p} + \dfrac{1}{q} = 1$. *Suppose* $f' \in L_p([a,b])$. *If* $q \notin \mathbb{N}$, *then*

$$|\Delta(e^x, 3^x)| \le \frac{(\ln 3)\|f'\|_{p,[a,b]}}{(e^b - e^a)(3^d - 3^c)} \times$$

$$\left\{ \int_c^d 3^x \left\{ \frac{e^{qx}(\,_2F_1[-q,-q,1-q,e^{a-x}] + (-1)^{q+1}\,_2F_1[-q,-q,1-q,e^{b-x}])}{q} \right. \right.$$

$$\left. \left. + \pi CSC(\pi q)((-1)^q e^{qb} - e^{qa})\right\}^{1/q} dx \right\}$$

$$= \overline{E}_2. \tag{26.101}$$

When $p = q = 2$, *it holds*

$$|\Delta(e^x, 3^x)| \le \frac{(\ln 3)\|f'\|_{2,[a,b]}}{\sqrt{2}(e^b - e^a)(3^d - 3^c)} \times$$

$$\left\{ \int_c^d 3^x \left(\sqrt{-4e^{a+x} + 4e^{b+x} + e^{2b}(-3 + 2b - 2x) + e^{2a}(3 - 2a + 2x)} \right) dx \right\}$$

$$=: \overline{E}_3. \tag{26.102}$$

When $q, n \in \mathbb{N} - \{1,2\}$, *then*

$$|\Delta(e^x, 3^x)| \le \frac{(\ln 3)\|f'\|_{\frac{n}{n-1},[a,b]}}{(e^b - e^a)(3^d - 3^c)} \times$$

$$\left\{ \int_c^d 3^x \left\{ (-1)^n e^{an} \left((x-a) + \sum_{k=1}^{n} \binom{n}{k}(-1)^{k-1}\left(\frac{1 - e^{k(x-a)}}{k} \right) \right) \right. \right.$$

$$\left. \left. + e^{bn}\left((b-x) + \sum_{k=0}^{n-1} \binom{n}{k}(-1)^{n-k}\left(\frac{1 - e^{(n-k)(x-b)}}{n-k} \right) \right) \right\}^{1/n} dx \right\} = \overline{E}_4 \tag{26.103}$$

iii) Also we have

$$|\Delta(e^x, 3^x)| \leq \frac{(\ln 3)\|f'\|_{1,[a,b]} \times \Phi}{(e^b - e^a)(3^d - 3^c)} =: \overline{E}_5.$$ (26.104)

Here

$$\Phi := \begin{cases} \Phi_1, & if \ln\left(\dfrac{e^a + e^b}{2}\right) \geq d \\[2ex] \Phi_2, & if \ln\left(\dfrac{e^a + e^b}{2}\right) \leq c \\[2ex] \Phi_3, & if \, c \leq \ln\left(\dfrac{e^a + e^b}{2}\right) \leq d \end{cases}$$ (26.105)

where

$$\Phi_1 := \frac{((3e)^c - (3e)^d)(\ln 3) - (3^c - 3^d)e^b(1 + \ln 3)}{\ln 3(1 + \ln 3)}$$ (26.106)

$$\Phi_2 := \frac{(-(3e)^c + (3e)^d)\ln 3 + (3^c - 3^d)e^a(1 + \ln 3)}{\ln 3(1 + \ln 3)}$$ (26.107)

$$\Phi_3 := \frac{1}{\ln 3(1 + \ln 3)}[e^{a - (\ln 2)(\ln 3) + (\ln 3)\ln(e^a + e^b)} + e^{b - (\ln 2)(\ln 3) + (\ln 3)(\ln(e^a + e^b))}$$
$$+((3e)^c + (3e)^d)(\ln 3) - 3^d e^a(1 + \ln 3) - 3^c e^b(1 + \ln 3)].$$ (26.108)

Proof. By Theorem 26.6 and (26.46). More precisely we have:

i) If $f' \in L_\infty([a,b])$, then

$$|\Delta(e^x, 3^x)| \leq \frac{\ln 3}{(3^d - 3^c)}\left(\int_c^d \left(\int_a^b |P(e^x, e^t)|dt\right) 3^x dx\right)\|f'\|_\infty.$$ (26.109)

ii) Let $p, q > 1$: $\dfrac{1}{p} + \dfrac{1}{q} = 1$. Suppose $f' \in L_p([a,b])$. Then

$$|\Delta(e^x, 3^x)| \leq \frac{\ln 3}{(3^d - 3^c)}\left[\int_c^d \|P(e^x, e^t)\|_{q,t,[a,b]} 3^x dx\right]\|f'\|_{p,[a,b]}.$$ (26.110)

When $p = q = 2$, it holds

$$|\Delta(e^x, 3^x)| \leq \frac{\ln 3}{(3^d - 3^c)}\left[\int_c^d \|P(e^x, e^t)\|_{q,t,[a,b]} 3^x dx\right]\|f'\|_{2,[a,b]}.$$ (26.111)

iii) Also we have

$$|\Delta(e^x, 3^x)| \leq \frac{\ln 3}{(3^d - 3^c)}\left(\int_c^d \max(e^x - e^a, e^b - e^x)3^x dx\right)\|f'\|_{1,[a,b]}.$$ (26.112)

Integrals are calculated mostly by Mathematica 5.2. $\square$

We finally give

Corollary 26.19. *If $f' \in L_\infty([a,b])$, then*
$$|\Delta(e^x, 3^x)| \leq \min\{\overline{E}_1, \overline{E}_2, \overline{E}_3, \overline{E}_4, \overline{E}_5\}.$$ (26.113)

Proof. By Corollary 26.18. $\square$

Chapter 27

An Expansion Formula

In this chapter Taylor like expansion formula is given. Here the Riemann–Stieltjes integral of a function is expanded into a finite sum form which involves the derivatives of the function evaluated at the right end point of the interval of integration. The error of the approximation is given in an integral form involving the nth derivative of the function. Implications and applications of the formula follow. This treatment relies on [36].

27.1 Results

Here we give the first and main result of the chapter.

Theorem 27.1. *Let g_0 be a Lebesgue integrable and of bounded variation function on $[a, b]$, $a < b$. We set*

$$g_1(x) := \int_a^x g_0(t)dt, \ldots \tag{27.1}$$

$$g_n(x) := \int_a^x \frac{(x-t)^{n-1}}{(n-1)!} g_0(t)dt, \quad n \in \mathbb{N}, \ x \in [a, b]. \tag{27.2}$$

Let f be such that $f^{(n-1)}$ is a absolutely continuous function on $[a, b]$. Then

$$\int_a^b f \, dg_0 = \sum_{k=0}^{n-1} (-1)^k f^{(k)}(b) g_k(b) - f(a) g_0(a)$$

$$+ (-1)^n \int_a^b g_{n-1}(t) f^{(n)}(t) dt. \tag{27.3}$$

Proof. We apply integration by parts repeatedly (see [91], p. 195):

$$\int_a^b f \, dg_0 = f(b) g_0(b) - f(a) g_0(a) - \int_a^b g_0 f' \, dt,$$

and

$$\int_a^b g_0 f' dt = \int_a^b f' dg_1 = f'(b)g_1(b) - f'(a)g_1(a) - \int_a^b g_1 f'' dt$$

$$= f'(b)g_1(b) - \int_a^b g_1 f'' dt.$$

Furthermore

$$\int_a^b f'' g_1 dt = \int_a^b f'' dg_2 = f''(b)g_2(b) - f''(a)g_2(a) - \int_a^b g_2 f''' dt$$

$$= f''(b)g_2(b) - \int_a^b g_2 f''' dt.$$

So far we have obtained

$$\int_a^b f dg_0 = f(b)g_0(b) - f(a)g_0(a) - f'(b)g_1(b)$$

$$+ f''(b)g_2(b) - \int_a^b g_2 f''' dt.$$

Similarly we find

$$\int_a^b g_2 f''' dt = \int_a^b f''' dg_3 = f'''(b)g_3(b) - \int_a^b g_3 f^{(4)} dt.$$

That is,

$$\int_a^b f dg_0 = f(b)g_0(b) - f(a)g_0(a) - f'(b)g_1(b)$$

$$+ f''(b)g_2(b) - f'''(b)g_3(b) + \int_a^b g_3 f^{(4)} dt.$$

The validity of (27.3) is now clear. $\qquad\square$

On Theorem 27.1 we have

Corollary 27.2. *Additionally suppose that $f^{(n)}$ exists and is bounded. Then*

$$\left| \int_a^b f dg_0 - \sum_{k=0}^{n-1} (-1)^k f^{(k)}(b)g_k(b) + f(a)g_0(a) \right|$$

$$= \left| \int_a^b g_{n-1}(t) f^{(n)}(t) dt \right| \le \|f^{(n)}\|_\infty \int_a^b |g_{n-1}(t)| dt. \tag{27.4}$$

As a continuation of Corollary 27.2 we have

Corollary 27.3. *Assuming that g_0 is bounded we derive*

$$\left| \int_a^b f dg_0 - \sum_{k=0}^{n-1} (-1)^k f^{(k)}(b)g_k(b) + f(a)g_0(a) \right|$$

$$\le \|f^{(n)}\|_\infty \|g_0\|_\infty \frac{(b-a)^n}{n!}. \tag{27.5}$$

Proof. Here we estimate the right-hand side of inequality (27.4).

We observe that

$$|g_{n-1}(x)| = \left| \int_a^x \frac{(x-t)^{n-2}}{(n-2)!} g_0(t)dt \right|$$

$$\leq \int_a^x \frac{(x-t)^{n-2}}{(n-2)!} |g_0(t)|dt$$

$$\leq \frac{\|g_0\|_\infty}{(n-2)!} \int_a^x (x-t)^{n-2}dt$$

$$= \frac{\|g_0\|_\infty}{(n-1)!}(x-a)^{n-1}.$$

That is

$$|g_{n-1}(t)| \leq \|g_0\|_\infty \frac{(t-a)^{n-1}}{(n-1)!}, \quad \text{all } t \in [a,b].$$

Consequently,

$$\int_a^b |g_{n-1}(t)|dt \leq \|g_0\|_\infty \frac{(b-a)^n}{n!}. \qquad \square$$

A refinement of (27.5) follows in

Corollary 27.4. *Here the assumptions are as in Corollary 27.3. We additionally suppose that*

$$\|f^{(n)}\|_\infty \leq K, \quad \forall n \geq 1,$$

i.e., f possess infinitely many derivatives and all are uniformly bounded. Then

$$\left| \int_a^b f dg_0 - \sum_{k=0}^{n-1}(-1)^k f^{(k)}(b)g_k(b) + f(a)g_a(a) \right|$$

$$\leq K\|g_0\|_\infty \frac{(b-a)^n}{n!}, \quad \forall n \geq 1. \tag{27.6}$$

A consequence of (27.6) comes next.

Corollary 27.5. *Same assumptions as in Corollary 27.4. For some $0 < r < 1$ it holds that*

$$O(r^n) = K\|g_0\|_\infty \frac{(b-a)^n}{n!} \to 0, \quad \text{as } n \to +\infty. \tag{27.7}$$

That is

$$\lim_{n \to +\infty} \sum_{k=0}^{n-1}(-1)^k f^{(k)}(b)g_k(b) = \int_a^b f dg_0 + f(a)g_0(a). \tag{27.8}$$

Proof. Call $A := b - a > 0$, we want to prove that $\frac{A^n}{n!} \to 0$ as $n \to +\infty$. Call

$$x_n := \frac{A^n}{n!}, \quad n = 1, 2, \dots .$$

Notice that

$$x_{n+1} = \frac{A}{n+1} \cdot x_n, \quad n = 1, 2, \dots .$$

But there exists $n_0 \in \mathbb{N}$: $n > A - 1$, i.e., $A < n_0 + 1$, that is, $r := \frac{A}{n_0+1} < 1$ (in fact take $n_0 := \lceil A - 1 \rceil + 1$, where $\lceil \cdot \rceil$ is the ceiling of the number). Thus

$$x_{n_0+1} = rc,$$

where

$$c := x_{n_0} > 0.$$

Therefore

$$x_{n_0+2} = \frac{A}{n_0+2} x_{n_0+1} = \frac{A}{n_0+2} rc < r^2 c,$$

i.e.,

$$x_{n_0+2} < r^2 c.$$

Likewise we get

$$x_{n_0+3} < r^3 c.$$

And in general we obtain

$$0 < x_{n_0+k} < r^k \cdot c = c^* \cdot r^{n_0+k}, \quad k \in \mathbb{N}$$

where $c^* := \frac{c}{r^{n_0}}$. That is

$$0 < x_N < c^* r^N, \quad \forall N \geq n_0 + 1.$$

Since $r^N \to 0$ as $N \to +\infty$, we derive that $x_N \to 0$ as $N \to +\infty$. That is, $\frac{(b-a)^n}{n!} \to 0$, as $n \to +\infty$. $\qquad\square$

Remark 27.6. (On Theorem 27.1) Furthermore we observe that (here $f \in C^n([a,b])$)

$$\left| \int_a^b f \, dg_0 + f(a) g_0(a) \right|$$

$$\overset{(27.3)}{=} \left| \sum_{k=0}^{n-1} (-1)^k f^{(k)}(b) g_k(b) + (-1)^n \int_a^b g_{n-1}(t) f^{(n)}(t) dt \right|$$

$$\leq \sum_{k=0}^{n-1} \| f^{(k)} \|_\infty |g_k(b)| + \| f^{(n)} \|_\infty \int_a^b |g_{n-1}(t)| dt.$$

Set

$$L := \max\{ \| f \|_\infty, \| f' \|_\infty, \dots, \| f^{(n)} \|_\infty \}. \tag{27.9}$$

Then

$$\left| \int_a^b f \, dg_0 + f(a) g_0(a) \right|$$

$$\leq L \cdot \left\{ \sum_{k=0}^{n-1} |g_k(b)| + \int_a^b |g_{n-1}(t)| dt \right\}, \quad n \in \mathbb{N} \text{ fixed.} \tag{27.10}$$

27.2 Applications

(I) Let $\{\mu_m\}_{m\in\mathbb{N}}$ be a sequence of Borel finite signed measures on $[a,b]$. Consider the functions $g_{o,m}(x) := \mu_m[a,x]$, $x \in [a,b]$, $m \in \mathbb{N}$, which are of bounded variation and Lebesgue integrable. Clearly (here $f \in C^n[a,b]$)

$$\int_a^b f d\mu_m = \int_a^b f dg_{o,m} + g_{o,m}(a) \cdot f(a). \tag{27.11}$$

We would like to study the weak convergence of μ_m to zero, as $m \to +\infty$. From (27.10) and (27.11) we obtain

$$\left| \int_a^b f d\mu_m \right| \le L \cdot \left\{ \sum_{k=0}^{n-1} |g_{k,m}(b)| + \int_a^b |g_{n-1,m}(t)| dt \right\}, \quad \forall m \in \mathbb{N}. \tag{27.12}$$

Here we suppose that

$$g_{o,m}(b) \to 0,$$

and

$$\int_a^b |g_{o,m}(t)| dt \to 0, \quad \text{as } m \to +\infty. \tag{27.13}$$

Let $k \in \mathbb{N}$, then

$$g_{k,m}(b) = \int_a^b \frac{(b-t)^{k-1}}{(k-1)!} g_{o,m}(t) dt$$

by (27.2).

   Hence

$$|g_{k,m}(b)| \le \left(\int_a^b |g_{o,m}(t)| dt \right) \frac{(b-a)^{k-1}}{(k-1)!} \to 0.$$

That is,

$$|g_{k,m}(b)| \to 0, \quad \forall k \in \mathbb{N}, \; m \to +\infty. \tag{27.14}$$

Next we have

$$g_{n-1,m}(x) = \int_a^x \frac{(x-t)^{n-2}}{(n-2)!} g_{o,m}(t) dt.$$

Thus

$$|g_{n-1,m}(x)| \le \int_a^x \frac{(x-t)^{n-2}}{(n-2)!} |g_{o,m}(t)| dt$$

$$\le \frac{(x-a)^{n-2}}{(n-2)!} \int_a^x |g_{o,m}(t)| dt$$

$$\le \frac{(b-a)^{n-2}}{(n-2)!} \int_a^b |g_{o,m}(t)| dt.$$

Consequently we obtain

$$\int_a^b |g_{n-1,m}(x)|dx \leq \frac{(b-a)^{n-1}}{(n-2)!} \int_a^b |g_{o,m}(t)|dt \to 0,$$

as $m \to +\infty$. That is,

$$\int_a^b |g_{n-1,m}(t)|dt \to 0, \quad \text{as } m \to +\infty. \tag{27.15}$$

Finally from (27.12), (27.13), (27.14) and (27.15) we derive that

$$\int_a^b fd\mu_m \to 0, \quad \text{as } m \to +\infty.$$

That is, μ_m converges weakly to zero as $m \to +\infty$.

The last result was first proved (case of $n = 0$) in [216] and then in [219].

(II) Formula (27.3) is expected to have applications to Numerical Integration and Probability.

(III) Let $[a, b] = [0, 1]$ and $g_0(t) = t$. Let f be such that $f^{(n-1)}$ is a absolutely continuous function on $[0, 1]$. Then by Theorem 27.1 we find $g_n(x) = \frac{x^{n+1}}{(n+1)!}$, all $n \in \mathbb{N}$. Furthermore we get

$$\int_0^1 fdt = \sum_{k=0}^{n-1}(-1)^k \frac{f^{(k)}(1)}{(k+1)!} + \frac{(-1)^n}{n!} \int_0^1 t^n f^{(n)}(t)dt. \tag{27.16}$$

One can derive other formulas like (27.16) for various basic g_0's.

Chapter 28

Integration by Parts on the Multidimensional Domain

In this chapter various integrations by parts are presented for very general type multivariate functions over boxes in $\mathbb{R}^m$, $m \geq 2$. These are given under special and general boundary conditions, or without these conditions for $m = 2, 3$. Also are established other related multivariate integral formulae and results. Multivariate integration by parts is hardly touched in the literature, so this is an attempt to set the matter into the right perspective. This treatment relies on [34] and is expected to have applications in Numerical Analysis and Probability.

28.1   Results

We need the following result which by itself has its own merit.

Theorem 28.1. *Let $f(x_1, \ldots, x_m)$ be a Lebesgue integrable function on $\times_{i=1}^{m}[a_i, b_i]$, $m \in \mathbb{N}$; $a_i < b_i$, $i = 1, 2, \ldots, m$. Consider*

$$F(x_1, x_2, \ldots, x_{m-1}, y) := \int_{a_m}^{y} f(x_1, x_2, \ldots, x_{m-1}, t)dt,$$

for all $a_m \leq y \leq b_m$, $\overrightarrow{x} := (x_1, x_2, \ldots, x_{m-1}) \in \times_{i=1}^{m-1}[a_i, b_i]$.
Then $F(x_1, x_2, \ldots, x_m)$ is a Lebesgue integrable function on $\times_{i=1}^{m}[a_i, b_i]$.

Proof. One can write

$$F(\overrightarrow{x}, y) = \int_{a_m}^{y} f(\overrightarrow{x}, t)dt.$$

Clearly $f(\overrightarrow{x}, \cdot)$ is an integrable function. Let $y \to y_0$, then $\forall \varepsilon > 0$, $\exists \delta := \delta(\varepsilon) > 0$:

$$|F(\overrightarrow{x}, y) - F(\overrightarrow{x}, y_0)| = \left| \int_{a_m}^{y} f(\overrightarrow{x}, t)dt - \int_{a_m}^{y_0} f(\overrightarrow{x}, t)dt \right|$$

$$= \left| \int_{y}^{y_0} f(\overrightarrow{x}, t)dt \right| < \varepsilon, \quad \text{with } |y - y_0| < \delta,$$

by exercise 6, p. 175, [11]. Therefore $F(\overrightarrow{x}, y)$ is continuous in y.

379

Next we prove that F is a Lebesgue measurable function with respect to $\overrightarrow{x}$. Because $f = f^+ - f^-$, it is enough to establish the above for $f(\overrightarrow{x}, t) \geq 0$, for all $(\overrightarrow{x}, t) \in \times_{i=1}^m [a_i, b_i]$. By Theorem 17.7, p. 131, [11], since f is Lebesgue measurable there exists a sequence $\{\phi_n\}$ of simple functions such that $0 \leq \phi_n(\overrightarrow{x}, t) \uparrow f(\overrightarrow{x}, t)$ for all $(\overrightarrow{x}, t) \in \times_{i=1}^m [a_i, b_i]$. Indeed here $0 \leq \phi_n(\overrightarrow{x}, t) \leq \phi_{n+1}(\overrightarrow{x}, t) \leq f(\overrightarrow{x}, t)$. These are defined as follows:

For each n let

$$A_n^i := \{ (\overrightarrow{x}, t) \in \times_{i=1}^m [a_i, b_i] : (i-1)2^{-n} \leq f(\overrightarrow{x}, t) < i 2^{-n} \}$$

for $i = 1, 2, \ldots, n2^n$, and note that $A_n^i \cap A_n^j = \phi$ if $i \neq j$. Since f is Lebesgue measurable all A_n^i are Lebesgue measurable sets. Here

$$\phi_n(\overrightarrow{x}, t) := \sum_{i=1}^{n2^n} 2^{-n}(i-1)\chi_{A_n^i}(\overrightarrow{x}, t),$$

where χ stands for the characteristic function.

Let $A \subseteq X \times Y$ in general and $x \in X$, then the *x-section* A_x of A is the subset of Y defined by

$$A_x := \{ y \in Y : (x, y) \in A \}.$$

Let $\theta : X \times Y \to \mathbb{R}$ be a function in general. Then for fixed $x \in X$ we define $\theta_x(y) := \theta(x, y)$ for all $y \in Y$. Let $E \subseteq X \times Y$. Then it holds that $\chi_{E_x} = (\chi_E)_x$, see p. 266, [312]. Also it holds for $A_i \subseteq X \times Y$ that

$$\left(\bigcap_{i \in I} A_i \right)_x = \bigcap_{i \in I} (A_i)_x.$$

Clearly here we have that $(\phi_n)_{\overrightarrow{x}} \uparrow f_{\overrightarrow{x}}$. Furthermore we obtain

$$(\phi_n)_{\overrightarrow{x}}(t) \cdot \chi_{[a_m, y]}(t) \uparrow f_{\overrightarrow{x}}(t) \cdot \chi_{[a_m, y]}(t).$$

Hence by the Monotone convergence theorem we find that

$$\int_{a_m}^{b_m} (\phi_n)_{\overrightarrow{x}}(t) \cdot \chi_{[a_m, y]}(t) dt \to \int_{a_m}^{b_m} f_{\overrightarrow{x}}(t) \cdot \chi_{[a_m, y]}(t) dt.$$

That is, we have

$$\int_{a_m}^{y} (\phi_n)_{\overrightarrow{x}}(t) dt \to \int_{a_m}^{y} f_{\overrightarrow{x}}(t) dt = F(\overrightarrow{x}, y).$$

By linearity we get

$$(\phi_n)_{\overrightarrow{x}}(t) = \sum_{i=1}^{n2^n} 2^{-n}(i-1)(\chi_{A_n^i})_{\overrightarrow{x}}(t) = \sum_{i=1}^{n2^n} 2^{-n}(i-1)\chi_{(A_n^i)_{\overrightarrow{x}}}(t).$$

Here $(A_n^i)_{\overrightarrow{x}}$ are Lebesgue measurable subsets of $[a_m, b_m]$ for all $\overrightarrow{x}$, see Theorem 6.1, p. 135, [235]. In particular we have that

$$\sum_{i=1}^{n2^n} 2^{-n}(i-1) \int_{a_m}^{y} \chi_{(A_n^i)_{\overrightarrow{x}}}(t) dt \to F(\overrightarrow{x}, y),$$

and

$$\sum_{i=1}^{n2^n} 2^{-n}(i-1) \int_{a_m}^{b_m} \chi_{[a_m,y]}(t) \cdot \chi_{(A_n^i)_{\overrightarrow{x}}}(t)dt \to F(\overrightarrow{x},y).$$

That is,

$$\sum_{i=1}^{n2^n} 2^{-n}(i-1) \int_{a_m}^{b_m} \chi_{[a_m,y] \cap (A_n^i)_{\overrightarrow{x}}}(t) \cdot dt \to F(\overrightarrow{x},y).$$

More precisely we find that

$$\sum_{i=1}^{n2^n} 2^{-n}(i-1)\lambda_1\big((A_n^i)_{\overrightarrow{x}} \cap [a_m,y]\big) \to F(\overrightarrow{x},y),$$

as $n \to +\infty$.

Here λ_1 is the Lebesgue measure on $\mathbb{R}$. Notice that

$$\big((\times_{i=1}^{m-1}[a_i,b_i]) \times [a_m,y]\big)_{\overrightarrow{x}} = [a_m,y].$$

Therefore we have

$$\big(A_n^i \cap [\times_{i=1}^{m-1}[a_i,b_i] \times [a_m,y]]\big)_{\overrightarrow{x}} = (A_n^i)_{\overrightarrow{x}} \cap [a_m,y].$$

Here $A_n^i \cap [\times_{i=1}^{m-1}[a_i,b_i] \times [a_m,y]]$ is a Lebesgue measurable set in $\times_{i=1}^{m}[a_i,b_i]$.

Clearly $(A_n^i)_{\overrightarrow{x}} \cap [a_m,y]$ is a Lebesgue measurable subset of $[a_m,b_m]$. And by Theorem 6.4, p. 143, [235], we have that

$$\gamma(\overrightarrow{x}) := \lambda_1\big((A_n^i)_{\overrightarrow{x}} \cap [a_m,y]\big)$$

is a Lebesgue measurable function defined on $\times_{i=1}^{m-1}[a_i,b_i]$. Consequently $F(\overrightarrow{x},y)$ is a Lebesgue measurable function in $\overrightarrow{x} \in \times_{i=1}^{m-1}[a_i,b_i]$ as the limit of Lebesgue measurable functions. Clearly we have established so far that $F(\overrightarrow{x},y)$ is a Caratheodory function. Hence by Theorem 20.15, p. 156, [11], $F(\overrightarrow{x},y)$ is a Lebesgue measurable function jointly in $(\overrightarrow{x},y)$ on $\times_{i=1}^{m}[a_i,b_i]$.

Next we notice that

$$|F(\overrightarrow{x},y)| \le \int_{a_m}^{y} |f(\overrightarrow{x},t)|dt \le \int_{a_m}^{b_m} |f(\overrightarrow{x},t)|dt.$$

And we see that

$$\int_{\times_{i=1}^{m-1}[a_i,b_i]} |F(\overrightarrow{x},y)|d\overrightarrow{x} \le \int_{\times_{i=1}^{m-1}[a_i,b_i]} \left(\int_{a_m}^{b_m} |f(\overrightarrow{x},t)|dt\right) d\overrightarrow{x}$$

$$\le M < +\infty, \quad M > 0,$$

by f being Lebesgue integrable on $\times_{i=1}^{m}[a_i,b_i]$. Hence

$$\int_{a_m}^{b_m} \left(\int_{\times_{i=1}^{m-1}[a_i,b_i]} |F(\overrightarrow{x},y)|d\overrightarrow{x}\right) dy \le M(b_m - a_m) < +\infty.$$

Therefore by Tonelli's theorem, p. 213, [11] we get that $F(x_1,\ldots,x_m)$ is a Lebesgue integrable function on $\times_{i=1}^{m}[a_i,b_i]$. $\qquad\square$

As a related result we give

Proposition 28.2. *Let $f(x_1, \ldots, x_m)$ be a Lebesgue integrable function on $\times_{i=1}^m [a_i, b_i]$, $m \in \mathbb{N}$; $a_i < b_i$, $i = 1, 2, \ldots, m$. Consider*

$$F(x_1, x_2, \ldots, x_r, \ldots, x_m) := \int_{a_r}^{x_r} f(x_1, x_2, \ldots, t, \ldots, x_m) dt,$$

for all $a_r \le x_r \le b_r$, $r \in \{1, \ldots, m\}$. Then $F(x_1, x_2, \ldots, x_m)$ is a Lebesgue integrable function on $\times_{i=1}^m [a_i, b_i]$.

Proof. Similar to Theorem 28.1. $\qquad\square$

Above Theorem 28.1 and Proposition 28.2 are also valid in a Borel measure setting.

Next we give the first main result.

Theorem 28.3. *Let g_0 be a Lebesgue integrable function on $\times_{i=1}^m [a_i, b_i]$, $m \in \mathbb{N}$; $a_i < b_i$, $i = 1, 2, \ldots, m$. We put*

$$g_{1,r}(x_1, \ldots, x_r, \ldots, x_m) := \int_{a_r}^{x_r} g_0(x_1, \ldots, t, \ldots, x_m) dt, \qquad (28.1)$$

$$\cdots$$

$$g_{k,r}(x_1, \ldots, x_r, \ldots, x_m) := \int_{a_r}^{x_r} g_{k-1,r}(x_1, \ldots, t, \ldots, x_m) dt, \qquad (28.2)$$

$k = 1, \ldots, n \in \mathbb{N}$; $r \in \{1, \ldots, m\}$ be fixed. Let $f \in C^{n-1}(\times_{i=1}^m [a_i, b_i])$ be such that

$$\frac{\partial^j f}{\partial x_r^j}(x_1, \ldots, b_r, \ldots, x_m) = 0, \quad for \ j = 0, 1, \ldots, n-1. \qquad (28.3)$$

Additionally suppose that:

$$\begin{cases} \dfrac{\partial^{n-1} f}{\partial x_r^{n-1}} \ is \ absolutely \ continuous \ function \\[20pt] with \ respect \ to \ the \ variable \ x_r, \ and \ \dfrac{\partial^n f}{\partial x_r^n} \ is \ integrable. \end{cases} \qquad (28.4)$$

Then

$$\int_{\times_{i=1}^m [a_i, b_i]} f g_0 d\overrightarrow{x} = (-1)^n \int_{\times_{i=1}^m [a_i, b_i]} g_{n,r}(\overrightarrow{x}) \frac{\partial^n f(\overrightarrow{x})}{\partial x_r^n} d\overrightarrow{x}. \qquad (28.5)$$

Proof. Here we use integration by parts repeatedly and we find:

$$\int_{a_r}^{b_r} f(x_1, \ldots, x_r, \ldots, x_m) g_0(x_1, \ldots, x_r, \ldots, x_m) dx_r$$

$$= \int_{a_r}^{b_r} f(x_1, \ldots, x_r, \ldots, x_m) \frac{\partial g_{1,r}}{\partial x_r}(x_1, \ldots, x_r, \ldots, x_m) dx_r$$

$$= f(x_1, \ldots, b_r, \ldots, x_m) g_{1,r}(x_1, \ldots, b_r, \ldots, x_m)$$

$$\quad - f(x_1, \ldots, a_r, \ldots, x_m) g_{1,r}(x_1, \ldots, a_r, \ldots, x_m)$$

$$\quad - \int_{a_r}^{b_r} g_{1,r}(x_1, \ldots, x_r, \ldots, x_m) f_{x_r}(x_1, \ldots, x_r, \ldots, x_m) dx_r$$

$$= -\int_{a_r}^{b_r} g_{1,r}(x_1, \ldots, x_r, \ldots, x_m) f_{x_r}(x_1, \ldots, x_r, \ldots, x_m) dx_r.$$

That is,

$$\int_{a_r}^{b_r} f(x_1,\ldots,x_r,\ldots,x_m)g_0(x_1,\ldots,x_r,\ldots,x_m)dx_r$$

$$= -\int_{a_r}^{b_r} g_{1,r}(x_1,\ldots,x_r,\ldots,x_m)f_{x_r}(x_1,\ldots,x_r,\ldots,x_m)dx_r. \qquad (28.6)$$

Similarly we have

$$\int_{a_r}^{b_r} g_{1,r}(x_1,\ldots,x_r,\ldots,x_m)f_{x_r}(x_1,\ldots,x_r,\ldots,x_m)dx_r$$

$$= \int_{a_r}^{b_r} f_{x_r}(x_1,\ldots,x_r,\ldots,x_m)\frac{\partial g_{2,r}}{\partial x_r}(x_1,\ldots,x_r,\ldots,x_m)dx_r$$

$$= f_{x_r}(x_1,\ldots,b_r,\ldots,x_m)g_{2,r}(x_1,\ldots,b_r,\ldots,x_m)$$

$$- f_{x_r}(x_1,\ldots,a_r,\ldots,x_m)g_{2,r}(x_1,\ldots,a_r,\ldots,x_m)$$

$$- \int_{a_r}^{b_r} g_{2,r}(x_1,\ldots,x_r,\ldots,x_m)f_{x_r x_r}(x_1,\ldots,x_r,\ldots,x_m)dx_r$$

$$= -\int_{a_r}^{b_r} g_{2,r}(x_1,\ldots,x_r,\ldots,x_m)f_{x_r x_r}(x_1,\ldots,x_r,\ldots,x_m)dx_r.$$

That is, we find that

$$\int_{a_r}^{b_r} f(x_1,\ldots,x_r,\ldots,x_m)g_0(x_1,\ldots,x_r,\ldots,x_m)dx_r$$

$$= (-1)^2 \int_{a_r}^{b_r} g_{2,r}(x_1,\ldots,x_r,\ldots,x_m)f_{x_r x_r}(x_1,\ldots,x_r,\ldots,x_m)dx_r. \qquad (28.7)$$

Continuing in the same manner we get

$$\int_{a_r}^{b_r} f(x_1,\ldots,x_r,\ldots,x_m)g_0(x_1,\ldots,x_r,\ldots,x_m)dx_r$$

$$= (-1)^n \int_{a_r}^{b_r} g_{n,r}(x_1,\ldots,x_r,\ldots,x_m)\frac{\partial^n f}{\partial x_r^n}(x_1,\ldots,x_r,\ldots,x_m)dx_r. \qquad (28.8)$$

Here by Proposition 28.2 $g_{n,r}(x_1,\ldots,x_m)$ is Lebesgue integrable. Thus integrating (28.8) with respect to $x_1,\ldots,x_{r-1},x_{r+1},\ldots,x_m$ and using Fubini's theorem we establish (28.5). $\qquad\square$

Corollary 28.4 (to Theorem 28.3). *Additionally suppose that the assumptions (28.3) and (28.4) are true for each $r = 1,\ldots,m$. Then*

$$\int_{\times_{i=1}^m [a_i,b_i]} f g_0 d\overrightarrow{x} = \frac{(-1)^n}{m}\sum_{r=1}^m \int_{\times_{i=1}^m [a_i,b_i]} g_{n,r}(\overrightarrow{x})\frac{\partial^n f}{\partial x_r^n}(\overrightarrow{x})d\overrightarrow{x}. \qquad (28.9)$$

Proof. Clear. $\qquad\square$

Corollary 28.5 (to Corollary 28.4). *Additionally assume that* $\|g_0\| < +\infty$. *Put*

$$M := \frac{\left(\max_{r=1,\ldots,m}\left\|\frac{\partial f^n}{\partial x_r^n}\right\|_\infty\right)\|g_0\|_\infty}{(n+1)!}. \tag{28.10}$$

Then

$$\left|\int_{\times_{i=1}^m [a_i,b_i]} f g_0 d\overrightarrow{x}\right| \le \frac{M}{m}\left\{\sum_{r=1}^m \left[(b_r - a_r)^{n+1}\left(\prod_{\substack{j=1\\j\neq r}}^m (b_j - a_j)\right)\right]\right\}. \tag{28.11}$$

Proof. Notice that

$$g_{n,r}(x_1,\ldots,x_r,\ldots,x_m) = \int_{a_r}^{x_r} \frac{(x_r - t)^{n-1}}{(n-1)!} g_0(x_1,\ldots,t,\ldots,x_m)dt.$$

Thus

$$|g_{n,r}(x_1,\ldots,x_r,\ldots,x_m)| \le \left(\int_{a_r}^{x_r} \frac{(x_r - t)^{n-1}}{(n-1)!}dt\right)\|g_0\|_\infty$$

$$= \frac{(x_r - a_r)^n}{n!}\|g_0\|_\infty.$$

That is,

$$|g_{n,r}(\overrightarrow{x})| \le \frac{(x_r - a_r)^n}{n!}\|g_0\|_\infty, \quad \forall \overrightarrow{x} \in \times_{i=1}^m [a_i,b_i].$$

Consequently we obtain

$$\left|\int_{\times_{i=1}^m [a_i,b_i]} g_{n,r}(\overrightarrow{x})\frac{\partial^n f(\overrightarrow{x})}{\partial x_r^n}d\overrightarrow{x}\right|$$

$$\le \frac{\|\partial^n f/\partial x_r^n\|_\infty\|g_0\|_\infty}{n!}\int_{\times_{i=1}^m [a_i,b_i]} (x_r - a_r)^n dx_1\cdots dx_m$$

$$= \frac{\|\partial f^n/\partial x_r^n\|_\infty\|g_0\|_\infty}{(n+1)!}(b_r - a_r)^{n+1}\prod_{\substack{j=1\\j\neq r}}^m (b_j - a_j).$$

Finally we have

$$\left|\sum_{r=1}^m \int_{\times_{i=1}^m [a_i,b_i]} g_{n,r}(\overrightarrow{x})\frac{\partial^n f(\overrightarrow{x})}{\partial x_r^n}d\overrightarrow{x}\right| \le M\left(\sum_{r=1}^m (b_r - a_r)^{n+1}\left(\prod_{\substack{j=1\\j\neq r}}^m (b_j - a_j)\right)\right).$$

That is establishing (28.11).   □

Some interesting observations follow in

Remark 28.6 (I). Let f, g be Lebesgue integrable functions on $\times_{i=1}^m [a_i,b_i]$, $a_i < b_i$, $i = 1,\ldots,m$. Let f, g be coordinatewise absolutely continuous functions. Also all partial derivatives f_{x_i}, g_{x_i} are suppose to be Lebesgue integrable functions, $i = 1, \ldots, m$. Furthermore assume that

$$f(x_1, \ldots, a_i, \ldots, x_m) = f(x_1, \ldots, b_i, \ldots, x_m) = 0, \tag{28.12}$$

for all $i = 1, \ldots, m$, i.e., f is zero on the boundary of $\times_{i=1}^{m}[a_i, b_i]$. By applying integration by parts we obtain

$$\int_{a_i}^{b_i} g_{x_i} f dx_i = g(x_1, \ldots, b_i, \ldots, x_m) f(x_1, \ldots, b_i, \ldots, x_m)$$
$$- g(x_1, \ldots, a_i, \ldots, x_m) f(x_1, \ldots, a_i, \ldots, x_m)$$
$$- \int_{a_i}^{b_i} f_{x_i} g dx_i = - \int_{a_i}^{b_i} f_{x_i} g dx_i.$$

That is, we have

$$\int_{a_i}^{b_i} g_{x_i} f dx_i = - \int_{a_i}^{b_i} f_{x_i} g dx_i \tag{28.13}$$

for all $i = 1, \ldots, m$.

Integrating (28.13) over

$$[a_1, b_i] \times \cdots \times \widehat{[a_i, b_i]} \times \cdots \times [a_m, b_m];$$

$\widehat{[a_i, b_i]}$ means this entity is absent, and using Fubini's theorem, we get

$$\int_{\times_{i=1}^{m}[a_i,b_i]} g_{x_i} f d\overrightarrow{x} = - \int_{\times_{i=1}^{m}[a_i,b_i]} f_{x_i} g d\overrightarrow{x}, \tag{28.14}$$

true for all $i = 1, \ldots, m$. One consequence is

$$\int_{\times_{i=1}^{m}[a_i,b_i]} f \cdot \left(\sum_{i=1}^{m} g_{x_i} \right) d\overrightarrow{x} = - \int_{\times_{i=1}^{m}[a_i,b_i]} g \cdot \left(\sum_{i=1}^{m} f_{x_i} \right) d\overrightarrow{x}. \tag{28.15}$$

Another consequence: multiply (28.14) by the unit vectors in $\mathbb{R}^m$ and add, we obtain

$$\int_{\times_{i=1}^{m}[a_i,b_i]} f \cdot (\nabla g) d\overrightarrow{x} = - \int_{\times_{i=1}^{m}[a_i,b_i]} g \cdot (\nabla f) d\overrightarrow{x}, \tag{28.16}$$

true under the above assumptions, where ∇ here stands for the gradient.

(II) Next we suppose that all the first order partial derivatives of g exist and they are coordinatewise absolutely continuous functions. Furthermore we assume that all the second order partial derivatives of g are Lebesgue integrable on $\times_{i=1}^{m}[a_i, b_i]$. Clearly one can see that g is as in (I); see Proposition 28.2, etc. We still take f as in (I). Here we apply (28.13) for $g := g_{x_j}$ to get

$$\int_{a_i}^{b_i} (g_{x_j})_{x_i} f dx_i = - \int_{a_i}^{b_i} f_{x_i} g_{x_j} dx_i.$$

And hence by Fubini's theorem we derive

$$\int_{\times_{i=1}^{m}[a_i,b_i]} g_{x_j x_i} f d\overrightarrow{x} = - \int_{\times_{i=1}^{m}[a_i,b_i]} f_{x_i} g_{x_j} d\overrightarrow{x}, \tag{28.17}$$

for all $i = 1, \ldots, m$. Consequently we have

$$\int_{\times_{i=1}^{m}[a_i,b_i]} f \cdot \left(\sum_{i=1}^{m} g_{x_j x_i}\right) d\overrightarrow{x} = -\int_{\times_{i=1}^{m}[a_i,b_i]} g_{x_j} \left(\sum_{i=1}^{m} f_{x_i}\right) d\overrightarrow{x}. \qquad (28.18)$$

Similarly as before we find

$$\int_{\times_{i=1}^{m}[a_i,b_i]} f \cdot (\nabla g_{x_j}) d\overrightarrow{x} = -\int_{\times_{i=1}^{m}[a_i,b_i]} g_{x_j} \cdot (\nabla f) d\overrightarrow{x}, \qquad (28.19)$$

for all $j = 1, \ldots, m$.

From (28.13) again, if $i = j$ we have

$$\int_{a_i}^{b_i} g_{x_i x_i} f dx_i = -\int_{a_i}^{b_i} f_{x_i} g_{x_i} dx_i, \qquad (28.20)$$

and consequently it holds

$$\int_{\times_{i=1}^{m}[a_i,b_i]} g_{x_i x_i} f d\overrightarrow{x} = -\int_{\times_{i=1}^{m}[a_i,b_i]} f_{x_i} g_{x_i} d\overrightarrow{x}. \qquad (28.21)$$

Furthermore by summing up we obtain

$$\int_{\times_{i=1}^{m}[a_i,b_i]} \left(\sum_{i=1}^{m} g_{x_i x_i}\right) f d\overrightarrow{x} = -\int_{\times_{i=1}^{m}[a_i,b_i]} \left(\sum_{i=1}^{m} f_{x_i} g_{x_i}\right) d\overrightarrow{x}.$$

That is, we have established that

$$\int_{\times_{i=1}^{m}[a_i,b_i]} (\nabla^2 g) f d\overrightarrow{x} = -\int_{\times_{i=1}^{m}[a_i,b_i]} \left(\sum_{i=1}^{m} f_{x_i} g_{x_i}\right) d\overrightarrow{x} \qquad (28.22)$$

true under the above assumptions, where ∇^2 is the Laplacian. If g is harmonic, i.e., when $\nabla^2 g = 0$ we find that

$$\sum_{i=1}^{m} \int_{\times_{i=1}^{m}[a_i,b_i]} f_{x_i} g_{x_i} d\overrightarrow{x} = 0. \qquad (28.23)$$

(III) Next we suppose that all the first order partial derivatives of f exist and they are coordinatewise absolutely continuous functions. Furthermore we assume that all the second order partial derivatives of f are Lebesgue integrable on $\times_{i=1}^{m}[a_i, b_i]$. Clearly one can see that f is as in (I), see Proposition 28.2, etc.

Here we take g as in (I) and we suppose that

$$g(x_1, \ldots, a_i, \ldots, x_m) = g(x_1, \ldots, b_i, \ldots, x_m) = 0, \qquad (28.24)$$

for all $i = 1, \ldots, m$. Then, by symmetry, we also get that

$$\int_{\times_{i=1}^{m}[a_i,b_i]} (\nabla^2 f) g d\overrightarrow{x} = -\int_{\times_{i=1}^{m}[a_i,b_i]} \left(\sum_{i=1}^{m} f_{x_i} g_{x_i}\right) d\overrightarrow{x}. \qquad (28.25)$$

So finally, under the assumptions (28.12), (28.24) and with f, g having all of their first order partial derivatives existing and coordinatewise absolutely continuous functions, along with their second order single partial derivatives Lebesgue integrable on $\times_{i=1}^{m}[a_i, b_i]$ (hence f, g are as in (I)), we have proved that

$$\int_{\times_{i=1}^{m}[a_i,b_i]} (\nabla^2 f)g\,d\overrightarrow{x} = \int_{\times_{i=1}^{m}[a_i,b_i]} f(\nabla^2 g)\,d\overrightarrow{x} = -\sum_{i=1}^{m} \int_{\times_{i=1}^{m}[a_i,b_i]} f_{x_i} g_{x_i}\,d\overrightarrow{x}.$$

$$(28.26)$$

From [36] we need the following

Theorem 28.7. *Let g_0 be a Lebesgue integrable and of bounded variation function on $[a,b]$, $a < b$. We call*

$$g_1(x) := \int_a^x g_0(t)dt, \ldots$$

$$g_n(x) := \int_a^x \frac{(x-t)^{n-1}}{(n-1)!} g_0(t)dt, \quad n \in \mathbb{N}, \ x \in [a,b].$$

Let f be such that $f^{(n-1)}$ is a absolutely continuous function on $[a,b]$. Then

$$\int_a^b f\,dg_0 = \sum_{k=0}^{n-1} (-1)^k f^{(k)}(b)g_k(b) - f(a)g_0(a)$$

$$+ (-1)^n \int_a^b g_{n-1}(t)f^{(n)}(t)dt. \tag{28.27}$$

We use also the next result which motivates the following work, it comes from [91], p. 221.

Theorem 28.8. *Let $\mathcal{R} = \{(x,y) \mid a \le x \le b, \ c \le y \le d\}$. Suppose that α is a function of bounded variation on $[a,b]$, β is a function of bounded variation on $[c,d]$, and f is continuous on $\mathcal{R}$. If $(x,y) \in \mathcal{R}$, define*

$$F(y) = \int_a^b f(x,y)d\alpha(x), \quad G(x) = \int_c^d f(x,y)d\beta(y).$$

Then $F \in R(\beta)$ on $[c,d]$, $G \in R(\alpha)$ on $[a,b]$, and we have

$$\int_c^d F(y)d\beta(y) = \int_a^b G(x)d\alpha(x).$$

Here $R(\beta)$, $R(\alpha)$ mean, respectively, that the functions are Riemann–Stieltjes integrable with respect to β and α. That is, we have true

$$\int_a^b \left(\int_c^d f(x,y)d\beta(y) \right) d\alpha(x) = \int_c^d \left(\int_a^b f(x,y)d\alpha(x) \right) d\beta(y). \tag{28.28}$$

We state and prove

Theorem 28.9. *Let g_0 be a Lebesgue integrable and of bounded variation function on $[a,b]$, $a < b$. We set*

$$g_1(x) := \int_a^x g_0(t)dt, \ldots \tag{28.29}$$

$$g_n(x) := \int_a^x \frac{(x-t)^{n-1}}{(n-1)!} g_0(t)dt, \quad n \in \mathbb{N}, \ x \in [a,b].$$

Let r_0 be a Lebesgue integrable and of bounded variation function on $[c,d]$, $c < d$. We form

$$r_1(y) = \int_c^y r_0(t)dt, \ldots \tag{28.30}$$

$$r_n(y) = \int_c^y \frac{(y-s)^{n-1}}{(n-1)!} r_0(s)ds, \quad n \in \mathbb{N}, \ y \in [c,d].$$

Let $f : \mathcal{R} = [a,b] \times [c,d] \to \mathbb{R}$ be a continuous function. We suppose for $k = 0, \ldots, n-1$ that $f_{xy}^{(k,n-1)}(x, \cdot)$ is an absolutely continuous function on $[c,d]$ for any $x \in [a,b]$. Also we assume for $\ell = 0, 1, \ldots, n$ that $f_{xy}^{(n,\ell)}(x,y)$ is continuous on $\mathcal{R}$. Then it holds

$$\int_c^d \left(\int_a^b f(x,y)dg_0(x) \right) dr_0(y)$$

$$= \sum_{k=0}^{n-1} \sum_{\ell=0}^{n-1} (-1)^{k+\ell} f_{xy}^{(k,\ell)}(b,d) g_k(b) r_\ell(d) + r_0(c) \left(\sum_{k=0}^{n-1} (-1)^{k+1} f_x^{(k)}(b,c) g_k(b) \right)$$

$$+ g_0(a) \cdot \left[\sum_{\ell=0}^{n-1} (-1)^{\ell+1} f_y^{(\ell)}(a,d) r_\ell(d) + f(a,c) r_0(c) \right]$$

$$+ \sum_{k=0}^{n-1} (-1)^{k+n} g_k(b) \left(\int_c^d r_{n-1}(y) f_{xy}^{(k,n)}(b,y)dy \right)$$

$$+ (-1)^{n+1} g_0(a) \int_c^d r_{n-1}(y) f_y^{(n)}(a,y)dy$$

$$+ \sum_{\ell=0}^{n-1} (-1)^{\ell+n} r_\ell(d) \left(\int_a^b g_{n-1}(x) f_{xy}^{(n,\ell)}(x,d)dx \right)$$

$$+ (-1)^{n+1} r_0(c) \left(\int_a^b g_{n-1}(x) f_x^{(n)}(x,c)dx \right)$$

$$+ \int_c^d \left(\int_a^b g_{n-1}(x) r_{n-1}(y) f_{xy}^{(n,n)}(x,y)dx \right) dy. \tag{28.31}$$

Proof. Here first we apply (28.27) for $f(\cdot, y)$, to find

$$\int_a^b f(x,y)dg_0(x) = \sum_{k=0}^{n-1} (-1)^k f_x^{(k)}(b,y) g_k(b)$$

$$- f(a,y)g_0(a) + (-1)^n \int_a^b g_{n-1}(x) f_x^{(n)}(x,y)dx,$$

where $f_x^{(k)}$ means the kth single partial derivative with respect to x. Therefore

$$\int_c^d \left(\int_a^b f(x,y) dg_0(x) \right) dr_0(y)$$

$$= \sum_{k=0}^{n-1} (-1)^k g_k(b) \cdot \int_c^d f_x^{(k)}(b,y) dr_0(y) - g_0(a) \int_c^d f(a,y) dr_0(y)$$

$$+ (-1)^n \int_c^d \left(\int_a^b g_{n-1}(x) f_x^{(n)}(x,y) dx \right) dr_0(y) =: \otimes.$$

Consequently we obtain by (28.27) that

$$\otimes = \sum_{k=0}^{n-1} (-1)^k g_k(b) \left\{ \sum_{\ell=0}^{n-1} (-1)^\ell f_{xy}^{(k,\ell)}(b,d) r_\ell(d) \right.$$

$$\left. - f_x^{(k)}(b,c) r_0(c) + (-1)^n \int_c^d r_{n-1}(y) f_{xy}^{(k,n)}(b,y) dy \right\}$$

$$- g_0(a) \left\{ \sum_{\ell=0}^{n-1} (-1)^\ell f_y^{(\ell)}(a,d) r_\ell(d) - f(a,c) r_0(c) + (-1)^n \int_c^d r_{n-1}(y) f_y^{(n)}(a,y) dy \right\}$$

$$+ (-1)^n \cdot \left\{ \sum_{\ell=0}^{n-1} (-1)^\ell \left(\int_a^b g_{n-1}(x) f_x^{(n)}(x,d) dx \right)_y^{(\ell)} r_\ell(d) \right.$$

$$- \left(\int_a^b g_{n-1}(x) f_x^{(n)}(x,c) dx \right) r_0(c)$$

$$\left. + (-1)^n \cdot \left(\int_c^d r_{n-1}(y) \left(\int_a^b g_{n-1}(x) f_x^{(n)}(x,y) dx \right)_y^{(n)} dy \right) \right\}.$$

Finally (28.31) is established by the use of Theorem 9-37, p. 219, [91] for differentiation under the integral sign. Some integrals in (28.31) exist by Theorem 28.8.

$\square$

As a related result on bivariate integration by parts we give

Proposition 28.10. *Let g_0 and r_0 be as in Theorem 28.9. Let $f : \mathcal{R} :=$ $[a,b] \times [c,d] \to \mathbb{R}$ be a continuous function. We suppose that $f(x, \cdot)$ is a absolutely continuous function on $[c,d]$ for any $x \in [a,b]$. Also we assume that f_x, f_{xy}*

are existing and continuous functions on $\mathcal{R}$. Then

$$\int_c^d \int_a^b f(x,y)dg_0(x)dr_0(y)$$

$$= f(b,d)g_0(b)r_0(d) - f(b,c)g_0(b)r_0(c) + f(a,c)g_0(a)r_0(c)$$

$$- f(a,d)g_0(a)r_0(d) + g_0(a)\int_c^d r_0(y)f_y(a,y)dy - g_0(b)\int_c^d r_0(y)f_y(b,y)dy$$

$$- r_0(d)\left(\int_a^b g_0(x)f_x(x,d)dx\right) + r_0(c)\left(\int_a^b g_0(x)f_x(x,c)dx\right)$$

$$+ \int_c^d \left(\int_a^b g_0(x)r_0(y)f_{xy}(x,y)dx\right)dy. \tag{28.32}$$

Proof. Apply (28.31) for $n = 1$, then $k = \ell = 0$, etc. $\qquad\square$

The next and final result of the chapter deals with integration by parts on $\mathbb{R}^3$.

Theorem 28.11. *Let g_i be Lebesgue integrable and of bounded variation functions on $[a_i, b_i]$, $a_i < b_i$, $i = 1,2,3$. Let $f : K = \times_{i=1}^3 [a_i, b_i] \to \mathbb{R}$ be a continuous function. We suppose that f_{x_i}, $i = 1,2,3$; $\frac{\partial^2 f}{\partial x_2 \partial x_1}$, $\frac{\partial^2 f}{\partial x_3 \partial x_2}$, $\frac{\partial^2 f}{\partial x_3 \partial x_1}$, $\frac{\partial^3 f}{\partial x_3 \partial x_2 \partial x_1}$ are all existing and continuous functions on K. Then*

$$\int_{a_1}^{b_1}\int_{a_2}^{b_2}\int_{a_3}^{b_3} f(x_1,x_2,x_3)dg_1(x_1)dg_2(x_2)dg_3(x_3)$$

$$= g_1(b_1)g_2(b_2)g_3(b_3)f(b_1,b_2,b_3) - g_1(a_1)g_2(b_2)g_3(b_3)f(a_1,b_2,b_3)$$

$$- g_1(b_1)g_2(a_2)g_3(b_3)f(b_1,a_2,b_3) + g_1(a_1)g_2(a_2)g_3(b_3)f(a_1,a_2,b_3)$$

$$- g_1(b_1)g_2(b_2)g_3(a_3)f(b_1,b_2,a_3) + g_1(a_1)g_2(b_2)g_3(a_3)f(a_1,b_2,a_3)$$

$$- g_1(a_1)g_2(a_2)g_3(a_3)f(a_1,a_2,a_3) + g_1(b_1)g_2(a_2)g_3(a_3)f(b_1,a_2,a_3)$$

$$- g_2(b_2)g_3(b_3)\int_{a_1}^{b_1} g_1(x_1)\frac{\partial f(x_1,b_2,b_3)}{\partial x_1}dx_1$$

$$+ g_2(a_2)g_3(b_3)\int_{a_1}^{b_1} g_1(x_1)\frac{\partial f(x_1,a_2,b_3)}{\partial x_1}dx_1$$

$$- g_1(b_1)g_3(b_3)\int_{a_2}^{b_2} g_2(x_2)\frac{\partial f(b_1,x_2,b_3)}{\partial x_2}dx_2$$

$$+ g_1(a_1)g_3(b_3)\int_{a_2}^{b_2} g_2(x_2)\frac{\partial f(a_1,x_2,b_3)}{\partial x_2}dx_2$$

$$+ g_2(b_2)g_3(a_3)\int_{a_1}^{b_1} g_1(x_1)\frac{\partial f(x_1,b_2,a_3)}{\partial x_1}dx_1$$

$$- g_2(a_2)g_3(a_3)\int_{a_1}^{b_1} g_1(x_1)\frac{\partial f(x_1,a_2,a_3)}{\partial x_1}dx_1$$

$$+ g_1(b_1)g_3(a_3)\int_{a_2}^{b_2} g_2(x_2)\frac{\partial f(b_1,x_2,a_3)}{\partial x_2}dx_2$$

$$- g_1(a_1)g_3(a_3) \int_{a_2}^{b_2} g_2(x_2) \frac{\partial f(a_1, x_2, a_3)}{\partial x_2} dx_2$$

$$- g_1(b_1)g_2(b_2) \int_{a_3}^{b_3} g_3(x_3) \frac{\partial f(b_1, b_2, x_3)}{\partial x_3} dx_3$$

$$+ g_1(a_1)g_2(b_2) \int_{a_3}^{b_3} g_3(x_3) \frac{\partial f(a_1, b_2, x_3)}{\partial x_3} dx_3$$

$$+ g_1(b_1)g_2(a_2) \int_{a_3}^{b_3} g_3(x_3) \frac{\partial f(b_1, a_2, x_3)}{\partial x_3} dx_3$$

$$- g_1(a_1)g_2(a_2) \int_{a_3}^{b_3} g_3(x_3) \frac{\partial f(a_1, a_2, x_3)}{\partial x_3} dx_3$$

$$- g_3(a_3) \int_{a_1}^{b_1} \int_{a_2}^{b_2} g_1(x_1)g_2(x_2) \frac{\partial^2 f(x_1, x_2, a_3)}{\partial x_2 \partial x_1} dx_2 dx_1$$

$$+ g_2(b_2) \int_{a_1}^{b_1} \int_{a_3}^{b_3} g_1(x_1)g_3(x_3) \frac{\partial^2 f(x_1, b_2, x_3)}{\partial x_3 \partial x_1} dx_3 dx_1$$

$$+ g_3(b_3) \int_{a_1}^{b_1} \int_{a_2}^{b_2} g_1(x_1)g_2(x_2) \frac{\partial^2 f(x_1, x_2, b_3)}{\partial x_2 \partial x_1} dx_2 dx_1$$

$$- g_2(a_2) \int_{a_1}^{b_1} \int_{a_3}^{b_3} g_1(x_1)g_3(x_3) \frac{\partial^2 f(x_1, a_2, x_3)}{\partial x_3 \partial x_1} dx_3 dx_1$$

$$+ g_1(b_1) \int_{a_2}^{b_2} \int_{a_3}^{b_3} g_2(x_2)g_3(x_3) \frac{\partial^2 f(b_1, x_2, x_3)}{\partial x_3 \partial x_2} dx_3 dx_2$$

$$- g_1(a_1) \int_{a_2}^{b_2} \int_{a_3}^{b_3} g_2(x_2)g_3(x_3) \frac{\partial^2 f(a_1, x_2, x_3)}{\partial x_3 \partial x_2} dx_3 dx_2$$

$$- \int_{a_1}^{b_1} \int_{a_2}^{b_2} \int_{a_3}^{b_3} g_1(x_1)g_2(x_2)g_3(x_3) \frac{\partial^3 f(x_1, x_2, x_3)}{\partial x_3 \partial x_2 \partial x_1} dx_3 dx_2 dx_1. \qquad (28.33)$$

Similarly one can get such formulae on $\times_{i=1}^{m} [a_i, b_i]$, *for any* $m > 3$.

Proof. Here we apply integration by parts repeatedly, see Theorem 9-6, p. 195, [91]. Set

$$I := \int_{a_1}^{b_1} \int_{a_2}^{b_2} \int_{a_3}^{b_3} f(x_1, x_2, x_3) dg_1(x_1) dg_2(x_2) dg_3(x_3).$$

The existence of I and other integrals in (28.33) follows from Theorem 28.8. We observe that

$$\int_{a_3}^{b_3} f(x_1, x_2, x_3) dg_3(x_3) = f(x_1, x_2, b_3)g_3(b_3) - f(x_1, x_2, a_3)g_3(a_3)$$

$$- \int_{a_3}^{b_3} g_3(x_3) \frac{\partial f(x_1, x_2, x_3)}{\partial x_3} dx_3.$$

Thus

$$\int_{a_2}^{b_2} \left(\int_{a_3}^{b_3} f(x_1, x_2, x_3) dg_3(x_3) \right) dg_2(x_2)$$

$$= g_3(b_3) \int_{a_2}^{b_2} f(x_1, x_2, b_3) dg_2(x_2) - g_3(a_3) \int_{a_2}^{b_2} f(x_1, x_2, a_3) dg_2(x_2)$$

$$- \int_{a_2}^{b_2} \left(\int_{a_3}^{b_3} g_3(x_3) \frac{\partial f(x_1, x_2, x_3)}{\partial x_3} dx_3 \right) dg_2(x_2)$$

$$= g_3(b_3) \left\{ f(x_1, b_2, b_3) g_2(b_2) - f(x_1, a_2, b_3) g_2(a_2) - \int_{a_2}^{b_2} g_2(x_2) \frac{\partial f(x_1, x_2, b_3)}{\partial x_2} dx_2 \right\}$$

$$- g_3(a_3) \left\{ f(x_1, b_2, a_3) g_2(b_2) - f(x_1, a_2, a_3) g_2(a_2) - \int_{a_2}^{b_2} g_2(x_2) \frac{\partial f(x_1, x_2, a_3)}{\partial x_2} dx_2 \right\}$$

$$- \left\{ \left(\int_{a_3}^{b_3} g_3(x_3) \frac{\partial f(x_1, b_2, x_3)}{\partial x_3} dx_3 \right) g_2(b_2) - \left(\int_{a_3}^{b_3} g_3(x_3) \frac{\partial f(x_1, a_2, x_3)}{\partial x_3} dx_3 \right) g_2(a_2) \right.$$

$$\left. - \int_{a_2}^{b_2} g_2(x_2) \left(\int_{a_3}^{b_3} g_3(x_3) \frac{\partial^2 f(x_1, x_2, x_3)}{\partial x_3 \partial x_2} dx_3 \right) dx_2 \right\}.$$

Therefore we obtain

$$I = g_3(b_3) \left\{ g_2(b_2) \left(\int_{a_1}^{b_1} f(x_1, b_2, b_3) dg_1(x_1) \right) - g_2(a_2) \left(\int_{a_1}^{b_1} f(x_1, a_2, b_3) dg_1(x_1) \right) \right.$$

$$\left. - \int_{a_1}^{b_1} \left(\int_{a_2}^{b_2} g_2(x_2) \frac{\partial f(x_1, x_2, b_3)}{\partial x_2} dx_2 \right) dg_1(x_1) \right\}$$

$$- g_3(a_3) \left\{ g_2(b_2) \int_{a_1}^{b_1} f(x_1, b_2, a_3) dg_1(x_1) - g_2(a_2) \int_{a_1}^{b_1} f(x_1, a_2, a_3) dg_1(x_1) \right.$$

$$\left. - \int_{a_1}^{b_1} \left(\int_{a_2}^{b_2} g_2(x_2) \frac{\partial f(x_1, x_2, a_3)}{\partial x_2} dx_2 \right) dg_1(x_1) \right\}$$

$$- \left\{ g_2(b_2) \left(\int_{a_1}^{b_1} \left(\int_{a_3}^{b_3} g_3(x_3) \frac{\partial f(x_1, b_2, x_3)}{\partial x_3} dx_3 \right) dg_1(x_1) \right) \right.$$

$$- g_2(a_2) \left(\int_{a_1}^{b_1} \left(\int_{a_3}^{b_3} g_3(x_3) \frac{\partial f(x_1, a_2, x_3)}{\partial x_3} dx_3 \right) dg_1(x_1) \right)$$

$$\left. - \int_{a_1}^{b_1} \left(\int_{a_2}^{b_2} g_2(x_2) \left(\int_{a_3}^{b_3} g_3(x_3) \frac{\partial^2 f(x_1, x_2, x_3)}{\partial x_3 \partial x_2} dx_3 \right) dx_2 \right) dg_1(x_1) \right\}.$$

Consequently we get

$$
\begin{aligned}
I = g_3(b_3)\Bigg\{ & g_2(b_2)\bigg[f(b_1,b_2,b_3)g_1(b_1) - f(a_1,b_2,b_3)g_1(a_1) \\
& - \int_{a_1}^{b_1} g_1(x_1)\frac{\partial f(x_1,b_2,b_3)}{\partial x_1}dx_1 \bigg] - g_2(a_2)\bigg[f(b_1,a_2,b_3)g_1(b_1) \\
& - f(a_1,a_2,b_3)g_1(a_1) - \int_{a_1}^{b_1} g_1(x_1)\frac{\partial f(x_1,a_2,b_3)}{\partial x_1}dx_1 \bigg] \\
& - \bigg[\left(\int_{a_2}^{b_2} g_2(x_2)\frac{\partial f(b_1,x_2,b_3)}{\partial x_2}dx_2 \right) g_1(b_1) \\
& - \left(\int_{a_2}^{b_2} g_2(x_2)\frac{\partial f(a_1,x_2,b_3)}{\partial x_2}dx_2 \right) g_1(a_1) \\
& - \int_{a_1}^{b_1} g_1(x_1) \left(\int_{a_2}^{b_2} g_2(x_2)\frac{\partial^2 f(x_1,x_2,b_3)}{\partial x_2 \partial x_1}dx_2 \right) dx_1 \bigg]\Bigg\} \\
& - g_3(a_3)\Bigg\{ g_2(b_2)\bigg[f(b_1,b_2,a_3)g_1(b_1) - f(a_1,b_2,a_3)g_1(a_1) \\
& - \int_{a_1}^{b_1} g_1(x_1)\frac{\partial f(x_1,b_2,a_3)}{\partial x_1}dx_1 \bigg] \\
& - g_2(a_2)\bigg[f(b_1,a_2,a_3)g_1(b_1) - f(a_1,a_2,a_3)g_1(a_1) \\
& - \int_{a_1}^{b_1} g_1(x_1)\frac{\partial f(x_1,a_2,a_3)}{\partial x_1}dx_1 \bigg] \\
& - \bigg[\left(\int_{a_2}^{b_2} g_2(x_2)\frac{\partial f(b_1,x_2,a_3)}{\partial x_2}dx_2 \right) g_1(b_1) \\
& - \left(\int_{a_2}^{b_2} g_2(x_2)\frac{\partial f(a_1,x_2,a_3)}{\partial x_2}dx_2 \right) g_1(a_1) \\
& - \int_{a_1}^{b_1} g_1(x_1) \left(\int_{a_2}^{b_2} g_2(x_2)\frac{\partial^2 f(x_1,x_2,a_3)}{\partial x_2 \partial x_1}dx_2 \right) dx_1 \bigg]\Bigg\} \\
& - \Bigg\{ g_2(b_2)\bigg[\left(\int_{a_3}^{b_3} g_3(x_3)\frac{\partial f(b_1,b_2,x_3)}{\partial x_3}dx_3 \right) g_1(b_1) \\
& - \left(\int_{a_3}^{b_3} g_3(x_3)\frac{\partial f(a_1,b_2,x_3)}{\partial x_3}dx_3 \right) g_1(a_1) \\
& - \int_{a_1}^{b_1} g_1(x_1) \left(\int_{a_3}^{b_3} g_3(x_3)\frac{\partial^2 f(x_1,b_2,x_3)}{\partial x_3 \partial x_1}dx_3 \right) dx_1 \bigg] \\
& - g_2(a_2)\bigg[\left(\int_{a_3}^{b_3} g_3(x_3)\frac{\partial f(b_1,a_2,x_3)}{\partial x_3}dx_3 \right) g_1(b_1)
\end{aligned}
$$

$$-\left(\int_{a_3}^{b_3} g_3(x_3)\frac{\partial f(a_1,a_2,x_3)}{\partial x_3}dx_3\right)g_1(a_1)$$

$$\left.-\int_{a_1}^{b_1} g_1(x_1)\left(\int_{a_3}^{b_3} g_3(x_3)\frac{\partial^2 f(x_1,a_2,x_3)}{\partial x_3\partial x_1}dx_3\right)dx_1\right]$$

$$-\left[\left(\int_{a_2}^{b_2}\int_{a_3}^{b_3} g_2(x_2)g_3(x_3)\frac{\partial^2 f(b_1,x_2,x_3)}{\partial x_3\partial x_2}dx_3 dx_2\right)g_1(b_1)\right.$$

$$-\left(\int_{a_2}^{b_2}\int_{a_3}^{b_3} g_2(x_2)g_3(x_3)\frac{\partial^2 f(a_1,x_2,x_3)}{\partial x_3\partial x_2}dx_3 dx_2\right)g_1(a_1)$$

$$\left.\left.-\int_{a_1}^{b_1} g_1(x_1)\left(\int_{a_2}^{b_2}\int_{a_3}^{b_3} g_2(x_2)g_3(x_3)\frac{\partial^3 f(x_1,x_2,x_3)}{\partial x_3\partial x_2\partial x_1}dx_3 dx_2\right)dx_1\right]\right\}.$$

Above we used repeatedly the Dominated Convergence theorem to establish continuity of integrals, see Theorem 5.8, p. 121, [235]. Also we used repeatedly Theorem 9-37, p. 219, [91], for differentiation under the integral sign. $\qquad\square$

Bibliography

1. M. Abramowitz, I. A. Stegun (Eds.), Handbook of mathematical functions with formulae, graphs and mathematical tables, National Bureau of Standards, Applied Math. Series 55, 4th printing, Washington, 1965.
2. G. Acosta and R. G. Durán, An Optimal Poincaré inequality in L_1 for convex domains, Proc. A.M.S., Vol. 132 (1) (2003), 195-202.
3. W. Adamski, An integral representation theorem for positive bilinear forms, Glasnik Matematicki, Ser III 26 (46) (1991), No. 1-2, 31-44.
4. R. P. Agarwal, An integro-differential inequality, General Inequalities III (E.F. Beckenbach and W. Walter, eds.), Birkhäuser, Basel, 1983, 501-503.
5. Ravi. P. Agarwal, Difference equations and inequalities. Theory, methods and applications. Second Edition, Monographs and Textbooks in Pure and Applied Mathematics, 228. Marcel Dekker, Inc., New York, 2000.
6. R. P. Agarwal, N.S. Barnett, P. Cerone, S.S. Dragomir, A survey on some inequalities for expectation and variance, Comput. Math. Appl. 49 (2005), No. 2-3, 429-480.
7. Ravi P. Agarwal and Peter Y. H. Pang, Opial Inequalities with Applications in Differential and Difference Equations, Kluwer Academic Publishers, Dordrecht, Boston, London, 1995.
8. Ravi P. Agarwal, Patricia J.Y. Wong, Error inequalities in polynomial interpolation and their applications, Mathematics and its Applications, 262, Kluwer Academic Publishers, Dordrecht, Boston, London, 1993.
9. N.L. Akhiezer, The classical moment problem, Hafner, New York, 1965.
10. Sergio Albeverio, Frederik Herzberg, The moment problem on the Wiener space, Bull. Sci. Math. 132 (2008), No. 1, 7-18.
11. Charalambos D. Aliprantis and Owen Burkinshaw, Principles of Real Analysis, third edition, Academic Press, Boston, New York, 1998.
12. A. Aglic Aljinovic, Lj. Dedic, M. Matic and J. Pecaric, On Weighted Euler harmonic Identities with Applications, Mathematical Inequalities and Applications, 8, No. 2 (2005), 237-257.
13. A. Aglic Aljinovic, M. Matic and J. Pecaric, Improvements of some Ostrowski type inequalities, J. of Computational Analysis and Applications, 7, No. 3 (2005), 289-304.
14. A. Aglic Aljinovic and J. Pecaric, The weighted Euler Identity, Mathematical Inequalities and Applications, 8, No. 2 (2005), 207-221.
15. A. Aglic Aljinovic and J. Pecaric, Generalizations of weighted Euler Identity and Ostrowski type inequalities, Adv. Stud. Contemp. Math. (Kyungshang) 14 (2007), No. 1, 141-151.
16. A. Aglic Aljinovic, J. Pecaric and A. Vukelic, The extension of Montgomery Identity

via Fink Identity with Applications, Journal of Inequalities and Applications, Vol. 2005, No. 1 (2005), 67-80.

17. Noga Alon, Assaf Naor, Approximating the cut-norm via Grothendieck's inequality, SIAM J. Comput. 35 (2006), No. 4, 787-803 (electronic).

18. G.A. Anastassiou, The Levy radius of a set of probability measures satisfying basic moment conditions involving $\{t, t^2\}$, Constr. Approx. J., 3 (1987), 257–263.

19. G.A. Anastassiou, Weak convergence and the Prokhorov radius, J. Math. Anal. Appl., 163 (1992), 541–558.

20. G.A. Anastassiou, Moments in Probability and Approximation Theory, Pitman Research Notes in Math., 287, Longman Sci. & Tech., Harlow, U.K., 1993.

21. G.A. Anastassiou, Ostrowski type inequalities, Proc. AMS 123 (1995), 3775-3781.

22. G.A. Anastassiou, Multivariate Ostrowski type inequalities, Acta Math. Hungarica, 76, No. 4 (1997), 267-278.

23. G.A. Anastassiou, Optimal bounds on the average of a bounded off observation in the presence of a single moment condition, V. Benes and J. Stepan (eds.), Distributions with given Marginals and Moment Problems, 1997, Kluwer Acad. Publ., The Netherlands, 1-13.

24. G.A. Anastassiou, Opial type inequalities for linear differential operators, Mathematical Inequalities and Applications, 1, No. 2 (1998), 193-200.

25. G.A. Anastassiou, General Fractional Opial type Inequalities, Acta Applicandae Mathematicae, 54, No. 3 (1998), 303-317.

26. G.A. Anastassiou, Opial type inequalities involving fractional derivatives of functions, Nonlinear Studies, 6, No. 2 (1999), 207-230.

27. G.A. Anastassiou, Quantitative Approximations, Chapman & Hall/CRC, Boca Raton, New York, 2001.

28. G.A. Anastassiou, General Moment Optimization problems, Encyclopedia of Optimization, eds. C. Floudas, P. Pardalos, Kluwer, Vol. II (2001), 198-205.

29. G.A. Anastassiou, Taylor integral remainders and moduli of smoothness, Y.Cho, J.K.Kim and S.Dragomir (eds.), Inequalities Theory and Applications, 1 (2001), Nova Publ., New York, 1-31.

30. G.A. Anastassiou, Multidimensional Ostrowski inequalities, revisited, Acta Mathematica Hungarica, 97 (2002), No. 4, 339-353.

31. G.A. Anastassiou, Probabilistic Ostrowski type inequalities, Stochastic Analysis and Applications, 20 (2002), No. 6, 1177-1189.

32. G.A. Anastassiou, Univariate Ostrowski inequalities, revisited, Monatshefte Math. 135 (2002), No. 3, 175-189.

33. G.A. Anastassiou, Multivariate Montgomery identities and Ostrowski inequalities, Numer. Funct. Anal. and Opt. 23 (2002), No. 3-4, 247-263.

34. G.A. Anastassiou, Integration by parts on the Multivariate domain, Anals of University of Oradea, fascicola mathematica, Tom IX, 5-12 (2002), Romania, 13-32.

35. G.A. Anastassiou, On Grüss Type Multivariate Integral Inequalities, Mathematica Balkanica, New Series Vol.17 (2003) Fasc.1-2, 1-13.

36. G.A. Anastassiou, A new expansion formula, Cubo Matematica Educational, 5 (2003), No. 1, 25-31.

37. G.A. Anastassiou, Basic optimal approximation of Csiszar's f−divergence, Proceedings of 11th Internat. Conf. Approx. Th, editors C.K.Chui et al, 2004, 15-23.

38. G.A. Anastassiou, A general discrete measure of dependence, Internat. J. of Appl. Math. Sci., 1 (2004), 37-54, www.gbspublisher.com.

39. G.A. Anastassiou, The Csiszar's $f−$ divergence as a measure of dependence, Anal. Univ. Oradea, fasc. math., 11 (2004), 27-50.

40. G.A. Anastassiou, Holder-Like Csiszar's type inequalities, Int. J. Pure & Appl. Math. Sci., 1 (2004), 9-14, www.gbspublisher.com.

41. G.A. Anastassiou, Higher order optimal approximation of Csiszar's f−divergence, Nonlinear Analysis, 61 (2005), 309-339.

42. G.A. Anastassiou, Fractional and other Approximation of Csiszar's f−divergence, (Proc. FAAT 04-Ed. F. Altomare), Rendiconti Del Circolo Matematico Di Palermo, Serie II, Suppl. 76 (2005), 197-212.

43. G.A. Anastassiou, Representations and Estimates to Csiszar's f- divergence, Panamerican Mathematical Journal, Vol. 16 (2006), No.1, 83-106.

44. G.A. Anastassiou, Applications of Geometric Moment Theory Related to Optimal Portfolio Management, Computers and Mathematics with Appl., 51 (2006), 1405-1430.

45. G.A. Anastassiou, Optimal Approximation of Discrete Csiszar f−divergence, Mathematica Balkanica, New Series, 20 (2006), Fasc. 2, 185-206.

46. G.A. Anastassiou, Csiszar and Ostrowski type inequalities via Euler-type and Fink identities, Panamerican Math. Journal, 16 (2006), No. 2, 77-91.

47. G.A. Anastassiou, Multivariate Euler type identity and Ostrowski type inequalities, Proceedings of the International Conference on Numerical Analysis and Approximation theory, NAAT 2006, Cluj-Napoca (Romania), July 5-8, 2006, 27-54.

48. G.A. Anastassiou, Difference of general Integral Means, in Journal of Inequalities in Pure and Applied Mathematics, 7 (2006), No. 5, Article 185, pp. 13, http://jipam.vu.edu.au.

49. G.A. Anastassiou, Opial type inequalities for Semigroups, Semigroup Forum, 75 (2007), 625-634.

50. G.A. Anastassiou, On Hardy-Opial Type Inequalities, J. of Math. Anal. and Approx. Th., 2 (1) (2007), 1-12.

51. G.A. Anastassiou, Chebyshev-Grüss type and Comparison of integral means inequalities for the Stieltjes integral, Panamerican Mathematical Journal, 17 (2007), No. 3, 91-109.

52. G.A. Anastassiou, Multivariate Chebyshev-Grüss and Comparison of integral means type inequalities via a multivariate Euler type identity, Demonstratio Mathematica, 40 (2007), No. 3, 537-558.

53. G.A. Anastassiou, High order Ostrowski type Inequalities, Applied Math. Letters, 20 (2007), 616-621.

54. G.A. Anastassiou, Opial type inequalities for Widder derivatives, Panamer. Math. J., 17 (2007), No. 4, 59-69.

55. G.A. Anastassiou, Grüss type inequalities for the Stieltjes Integral, J. Nonlinear Functional Analysis & Appls., 12 (2007), No. 4, 583-593.

56. G.A. Anastassiou, Taylor-Widder Representation Formulae and Ostrowski, Grüss, Integral Means and Csiszar type inequalities, Computers and Mathematics with Applications, 54 (2007), 9-23.

57. G.A. Anastassiou, Multivariate Euler type identity and Optimal multivariate Ostrowski type inequalities, Advances in Nonlinear Variational Inequalities, 10 (2007), No. 2, 51-104.

58. G.A. Anastassiou, Multivariate Fink type identity and multivariate Ostrowski, Comparison of Means and Grüss type Inequalities, Mathematical and Computer Modelling, 46 (2007), 351-374.

59. G.A. Anastassiou, Optimal Multivariate Ostrowski Euler type inequalities, Studia Univ. Babes-Bolyai Mathematica, Vol. LII (2007), No. 1, 25-61.

60. G.A. Anastassiou, Chebyshev-Grüss Type Inequalities via Euler Type and Fink identities, Mathematical and Computer Modelling, 45 (2007), 1189-1200.

61. G.A. Anastassiou, Ostrowski type inequalities over balls and shells via a Taylor-Widder formula, JIPAM. J. Inequal. Pure Appl. Math. 8 (2007), No. 4, Article 106, 13 pp.

62. G.A. Anastassiou, Representations of Functions, accepted, Nonlinear Functional Analysis and Applications, 13 (2008), No. 4, 537-561.

63. G.A. Anastassiou, Poincaré and Sobolev type inequalities for vector valued functions, Computers and Mathematics with Applications, 56 (2008), 1102-1113.

64. G.A. Anastassiou, Opial type inequalities for Cosine and Sine Operator functions, Semigroup Forum, 76 (2008), No. 1, 149-158.

65. G.A. Anastassiou, Chebyshev-Grüss Type Inequalities on $\mathbb{R}^N$ over spherical shells and balls, Applied Math. Letters, 21 (2008), 119-127.

66. G.A. Anastassiou, Ostrowski type inequalities over spherical shells, Serdica Math. J. 34 (2008), 629-650.

67. G.A. Anastassiou, Poincare type Inequalities for linear differential operators, CUBO, 10 (2008), No. 3, 13-20.

68. G.A. Anastassiou, Grothendieck type inequalities, accepted, Applied Math. Letters, 21 (2008), 1286-1290.

69. G.A. Anastassiou, Poincaré Like Inequalities for Semigroups, Cosine and Sine Operator functions, Semigroup Forum, 78 (2009), 54-67.

70. G.A. Anastassiou, Distributional Taylor Formula, Nonlinear Analysis, 70 (2009), 3195-3202.

71. G.A. Anastassiou, Opial type inequalities for vector valued functions, Bulletin of the Greek Mathematical Society, Vol. 55 (2008), pp. 1-8.

72. G.A. Anastassiou, Fractional Differential Inequalities, Springer, NY, 2009.

73. G.A. Anastassiou, Poincaré and Sobolev type inequalities for Widder derivatives, Demonstratio Mathematica, 42 (2009), No. 2, to appear.

74. G.A. Anastassiou and S. S. Dragomir, On Some Estimates of the Remainder in Taylor's Formula, J. of Math. Analysis and Appls., 263 (2001), 246-263.

75. G.A. Anastassiou and J. Goldstein, Ostrowski type inequalities over Euclidean domains, Rend. Lincei Mat. Appl. 18 (2007), 305–310.

76. G.A. Anastassiou and J. Goldstein, Higher order Ostrowski type inequalities over Euclidean domains, Journal of Math. Anal. and Appl. 337 (2008), 962–968.

77. G.A. Anastassiou, Gisèle Ruiz Goldstein and J. A. Goldstein, Multidimensional Opial inequalities for functions vanishing at an interior point, Atti Accad. Lincei Cl. Fis. Mat. Natur. Rend. Math. Acc. Lincei, s. 9, v. 15 (2004), 5–15.

78. G.A. Anastassiou, Gisèle Ruiz Goldstein and J. A. Goldstein, Multidimensional weighted Opial inequalities, Applicable Analysis, 85 (May 2006), No. 5, 579–591.

79. G.A. Anastassiou and V. Papanicolaou, A new basic sharp integral inequality, Revue Roumaine De Mathematiques Pure et Appliquees, 47 (2002), 399-402.

80. G.A. Anastassiou and V. Papanicolaou, Probabilistic Inequalities and remarks, Applied Math. Letters, 15 (2002), 153-157.

81. G.A. Anastassiou and J. Pecaric, General weighted Opial inequalities for linear differential operators, J. Mathematical Analysis and Applications, 239 (1999), No. 2, 402-418.

82. G.A. Anastassiou and S. T. Rachev, How precise is the approximation of a random queue by means of deterministic queueing models, Computers and Math. with Appl., 24 (1992), No. 8/9, 229-246.

83. G.A. Anastassiou and S. T. Rachev, Moment problems and their applications to characterization of stochastic processes, queueing theory and rounding problems, Proceedings of 6th S.E.A. Meeting, Approximation Theory, edited by G. Anastassiou,

Proceedings of 6th S.E.A. Meeting, Approximation Theory, edited by G. Anastassiou, 1-77, Marcel Dekker, Inc., New York, 1992.

84. G.A. Anastassiou and S. T. Rachev, Solving moment problems with application to stochastics, Monografii Matematice, University of the West from Timisoara, Romania, No. 65/1998, pp. 77.

85. G.A. Anastassiou and T. Rychlik, Moment Problems on Random Rounding Rules Subject to Two Moment Conditions, Comp. and Math. with Appl., 36 (1998), No. 1, 9–19.

86. G.A. Anastassiou and T. Rychlik, Rates of uniform Prokhorov convergence of probability measures with given three moments to a Dirac one, Comp. and Math. with Appl. 38 (1999), 101–119.

87. G.A. Anastassiou and T. Rychlik, Refined rates of bias convergence for generalized L-statistics in the I.I.D. case, Applicationes Mathematicae, (Warsaw) 26 (1999), 437–455.

88. G.A. Anastassiou and T. Rychlik, Prokhorov radius of a neighborhood of zero described by three moment constraints, Journal of Global Optimization 16 (2000), 69–75.

89. G.A. Anastassiou and T. Rychlik, Moment Problems on Random Rounding Rules Subject to One Moment Condition, Communications in Applied Analysis, 5 (2001), No. 1, 103–111.

90. G.A. Anastassiou and T. Rychlik, Exact rates of Prokhorov convergence under three moment conditions, Proc. Crete Opt. Conf., Greece, 1998, Combinatorial and Global Optimization, pp. 33–42, P. M. Pardalos et al ets. World Scientific (2002).

91. T. M. Apostol, Mathematical Analysis, Second Edition, Addison-Wesley Publ. Co., 1974.

92. S. Aumann, B. M. Brown, K. M. Schmidt, A Hardy-Littlewood-type inequality for the p-Laplacian, Bull. Lond. Math. Soc., 40 (2008), No. 3, 525-531.

93. Laith Emil Azar, On some extensions of Hardy-Hilbert's Inequality and applications, J. Inequal. Appl., 2008, Art. ID 546829, 14 pp.

94. M. A. K. Baig, Rayees Ahmad Dar, Upper and lower bounds for Csiszar f-divergence in terms of symmetric J-divergence and applications, Indian J. Math., 49 (2007), No. 3, 263-278.

95. Drumi Bainov, Pavel Simeonov, Integral Inequalities and applications, translated by R. A. M. Hoksbergen and V. Covachev, Mathematics and its Applications (East European Series), 57, Kluwer Academic Publishers Group, Dordrecht, 1992.

96. Andrew G. Bakan, Representation of measures with polynomial denseness in $L_p(\mathbb{R}, d\mu)$, $0 < p < \infty$, and its application to determinate moment problems, Proc. Amer. Math. Soc. 136 (2008), No. 10, 3579-3589.

97. Dominique Bakry, Franck Barthe, Patrick Cattiaux, Arnaud Guilin, A simple proof of the Poincaré inequality for a large class of probability measures including the log-concave case, Electron. Commun. Probab., 13 (2008), 60-66.

98. Gejun Bao, Yuming Xing, Congxin Wu, Two-weight Poincaré inequalities for the projection operator and A-harmonic tensors on Riemannian manifolds, Illinois J. Math., 51 (2007), No. 3, 831-842.

99. N. S. Barnett, P. Cerone, S. S. Dragomir, Ostrowski type inequalities for multiple integrals, pp. 245-281, Chapter 5 from Ostrowski Type Inequalities and Applications in Numerical Integration, edited by S. S. Dragomir and T. Rassias, on line: http://rgmia.vu.edu.au/monographs, Melbourne (2001).

100. N. S. Barnett, P. Cerone, S. S. Dragomir, Inequalities for Distributions on a Finite Interval, Nova Publ., New York, 2008.

101. N. S. Barnett, P. Cerone, S. S. Dragomir and J. Roumeliotis, Approximating Csiszar's f-divergence via two integral identities and applications, (paper #13, pp. 1-15), in Inequalities for Csiszar f-Divergence in Information Theory, S.S.Dragomir (ed.), Victoria University, Melbourne, Australia, 2000. On line: http://rgmia.vu.edu.au

102. N. S. Barnett, P. Cerone, S. S. Dragomir and A. Sofo, Approximating Csiszar's f-divergence by the use of Taylor's formula with integral remainder, (paper #10, pp.16), in Inequalities for Csiszar's f-Divergence in Information Theory, S.S.Dragomir (ed.), Victoria University, Melbourne, Australia, 2000. On line: http://rgmia.vu.edu.au

103. N. S. Barnett, P. Cerone, S. S. Dragomir and A. Sofo, Approximating Csiszar's f-divergence via an Ostrowski type identity for n time differentiable functions, article #14 in electronic monograph "Inequalities for Csiszar's f-Divergence in Information Theory", edited by S.S.Dragomir, Victoria University, 2001, http://rgmia.vu.edu.au/monographs/csiszar.htm

104. N. S. Barnett, P. Cerone, S. S. Dragomir and A. Sofo, Approximating Csiszar's f-divergence via an Ostrowski type identity for n time differentiable functions. Stochastic analysis and applications. Vol. 2, 19-30, Nova Sci. Publ., Huntington, NY, 2002.

105. N. S. Barnett, S. S. Dragomir, An inequality of Ostrowski's type for cumulative distribution functions, RGMIA (Research Group in Mathematical Inequalities and Applications), Vol. 1, No. 1 (1998), pp. 3-11, on line: http://rgmia.vu.edu.au

106. N. S. Barnett and S. S. Dragomir, An Ostrowski type inequality for double integrals and applications for cubature formulae, RGMIA Research Report Collection, 1 (1998), 13-22.

107. N. S. Barnett and S. S. Dragomir, An Ostrowski type inequality for a random variable whose probability density function belongs to $L_\infty[a, b]$, RGMIA (Research Group in Mathematical Inequalities and Applications, Vol. 1, No. 1, (1998), pp. 23-31, on line: http://rgmia.vu.edu.au

108. N. S. Barnett and S. S. Dragomir, An Ostrowski type inequality for double integrals and applications for cubature formulae, Soochow J. Math. 27 (2001), No. 1, 1-10.

109. N. S. Barnett, S. S. Dragomir, On the weighted Ostrowski inequality. JIPAM. J. Inequal. Pure Appl. Math. 8 (2007), No. 4, Article 96, pp. 10.

110. N. S. Barnett, S. S. Dragomir, Advances in Inequalities from Probability Theory and Statistics, edited volume, Nova Publ., New York, 2008.

111. N. S. Barnett, S. S. Dragomir, R. P. Agarwal, Some inequalities for probability, expectation and variance of random variables defined over a finite interval, Comput. Math. Appl. 43 (2002), No. 10-11, 1319-1357.

112. R.G. Bartle, The Elements of Real Analysis, Second Edition, Wiley, New York, 1976.

113. P. R. Beesack, On an integral inequality of Z. Opial, Trans. Amer. Math. Soc., 104 (1962), 479-475.

114. V. Benes and J. Stepan (Eds.), Distributions with given marginals and moment problems, Kluwer, London, 1997.

115. Alexandru Bica, A probabilistic inequality and a particular quadrature formula, An. Univ. Oradea Fasc. Mat. 9 (2002), 47-60.

116. Martin Bohner, Thomas Matthews, Ostrowski inequalities on time scales, JIPAM. J. Inequal. Pure Appl. Math. 9 (2008), no. 1, Article 6, pp.8.

117. I. Brnetic and J. Pecaric On an Ostrowski type inequality for a random variable, Mathematical Inequalities and Applications, Vol. 3, No.1 (2000), 143-145.

118. V. I. Burenkov, Sobolev's integral representation and Taylor's formula, (Russian), Studies in the theory of differentiable functions of several variables and its applications, V. Trudy Mat. Inst. Steklov. 131 (1974), 33-38.

119. V. I. Burenkov, Sobolev space on domains, Teubner-Texte zur Mathematik [Teubner Texts in Mathematics], 137. B. G. Teubner Verlagsgesellschaft mbtt, Stuttgart, 1998, pp. 312.

120. P. L. Butzer and H. Berens, Semi-Groups of Operators and Approximation, Springer-Verlag, New York, 1967.

121. P. L. Butzer and H. G. Tillmann, Approximation Theorems for Semi-groups of Bounded Linear Transformations, Math. Annalen 140 (1960), 256-262.

122. José A. Canavati, The Riemann-Liouville integral, Nieuw Archief Voor Wiskunde, 5, No. 1 (1987), 53-75.

123. P. Cerone, Difference between weighted integral means, Demonstratio Mathematica, Vol. 35, No. 2, 2002, pp. 251-265.

124. P. Cerone, Approximate multidimensional integration through dimension reduction via the Ostrowski functional, Nonlinear Funct. Anal. Appl. 8 (2003), No. 3, 313-333.

125. P. Cerone and S. S. Dragomir, Trapezoidal-type rules from an inequalities point of view, Chapter 3 in Handbook of Analytical-Computational Methods in Applied Mathematics, edited by George Anastassiou, Chapman & Hall/CRC, 2000, pp. 65–134.

126. P. Cerone and S. S. Dragomir, Midpoint type rules from an inequality point of view, Chapter 4 in Handbook of Analytical-Computational Methods in Applied Mathematics, edited by George Anastassiou, Chapman & Hall/CRC, 2000, pp. 135–200.

127. Pietro Cerone, Sever S. Dragomir and Ferdinand Österreicher, Bounds on extended f-divergences for a. variety of classes, Kybernetika (Prague) 40 (2004), No. 6, 745-756.

128. P. Cerone and S. S. Dragomir, Advances in Inequalities for special Functions, edited volume, Nova Publ., New York, 2008.

129. P.L. Chebyshev, Sur les expressions approximatives des integrales definies par les autres prises entre les mêmes limites, Proc. Math. Soc. Charkov, 2 (1882), 93-98.

130. Wing-Sum Cheung, Chang-Jian Zhao, On a discrete Opial-type inequality, JIPAM. J. Inequal. Pure Appl. Math. 8 (2007), No. 4, Article 98, pp. 4.

131. Wing-Sum Cheung, Chang-Jian Zhao, On Opial-type integral inequalities, J. Inequal. Appl. 2007, Art. ID 38347, pp. 15.

132. Wing-Sum Cheung, Zhao Dandan, Josip Pecaric, Opial-type inequalities for differential operators, Nonlinear Anal. 66 (2007), No. 9, 2028-2039.

133. Seng-Kee Chua and R. L. Wheeden, A note on sharp 1-dimensional Poincaré inequalities, Proc. AMS, Vol. 134, No. 8 (2006), 2309-2316.

134. D.-K. Chyan, S.-Y. Shaw and S. Piskarev, On Maximal regularity and semivariation of Cosine operator functions, J. London Math. Soc. (2) 59 (1999), 1023-1032.

135. Andrea Cianchi, Sharp Morrey-Sobolev inequalities and the distance from extremals, Trans. Amer. Math. Soc. 360 (2008), No. 8, 4335-4347.

136. Andrea Cianchi, Adele Ferone, A strengthened version of the Hardy-Littlewood inequality, J. Lond. Math. Soc. (2) 77 (2008), No. 3, 581-592.

137. A. Civljak, Lj. Dedic and M. Matic, On Ostrowski and Euler-Grüss type inequalities involving measures, J. Math. Inequal. 1 (2007), No. 1, 65-81.

138. Hong Rae Cho, Jinkee Lee, Inequalities for the integral means of holomorphic functions in the strongly pseudoconvex domain, Commun. Korean Math. Soc. 20 (2005), No. 2, 339-350.

139. I. Csiszar, Eine Informationstheoretische Ungleichung und ihre Anwendung auf den Beweis der Ergodizität von Markoffschen Ketten, Magyar Tud. Akad. Mat. Kutato Int. Közl. 8 (1963), 85-108.

140. I. Csiszar, Information-type measures of difference of probability distributions and indirect observations, Studia Math. Hungarica 2 (1967), 299-318.

141. I. Csiszar, On topological properties of f-divergences, Studia Scientiarum Mathematicarum Hungarica, 2 (1967), 329-339.

142. Raul E. Curto, Lawrence A. Fialkow, Michael Moeller, The extremal truncated moment problem, Integral Equations Operator Theory 60 (2008), No. 2, 177-200.

143. Lj. Dedic, M. Matic and J. Pecaric, On generalizations of Ostrowski inequality via some Euler- type identities, Mathematical Inequalities and Applications Vol. 3, No. 3(2000), 337-353.

144. Lj. Dedic, M. Matic, J. Pecaric and A. Vukelic, On generalizations of Ostrowski inequality via Euler harmonic identities, Journal of Inequalities and Applications Vol. 7, No. 6(2002), 787-805.

145. Cristina Draghici, Inequalities for integral means over symmetric sets, J. Math. Anal. Appl. 324 (2006), No. 1, 543-554.

146. S. S. Dragomir, New estimation of the remainder in Taylor's formula using Grüss' type inequalities and applications, Math. Ineq. Appl. 2(2) (1999), 183-193.

147. S. S. Dragomir, Ostrowski's inequality for mappings of bounded variation and applications, Math. Ineq. & Appl., 4 (1) (2001), 59-66, and on-line in (http://rgmia.vu.edu.au); RGMIA Res. Rep. Coll., 2 (1) (1999), 103-110.

148. S. S. Dragomir, A survey on Cauchy-Buniakowsky-Schwartz Type Discrete Inequalities, RGMIA Monographs, Victoria University, 2000. (on-line http://rgmia.vu.edu.au/monographs/).

149. S. S. Dragomir, Semi-Inner Products and Applications, RGMIA Monographs, Victoria University, 2000. (on-line http://rgmia.vu.edu.au/monographs/).

150. S. S. Dragomir, Some Gronwall Type Inequalities and Applications, RGMIA Monographs, Victoria University, 2000. (on-line http://rgmia.vu.edu.au/monographs/).

151. S. S. Dragomir, Some inequalities of Grüss type, Indian J. Pure Appl. Math., 31 (2000), 397-415.

152. S. S. Dragomir (ed.), Inequalities for Csiszar f-Divergence in Information Theory, Victoria University, Melbourne, Australia, 2000. On-line http://rgmia.vu.edu.au

153. S. S. Dragomir, Upper and lower bounds for Csiszar's f-divergence in terms of the Kullback–Leibler distance and applications, paper #5 from [152], 2000.

154. S. S. Dragomir, New estimates of the Čebyšev functional for Stieltjes integrals and applications, RGMIA 5(2002), Supplement Article 27, electronic, http://rgmia.vu.edu.au

155. S. S. Dragomir, Sharp bounds of Čebyšev functional for Stieltjes integrals and applications, RGMIA 5(2002), Supplement Article 26, electronic, http://rgmia.vu.edu.au

156. S. S. Dragomir, A Generalized f-Divergence for Probability Vectors and Applications, Panamer. Math. J. 13 (2003), No. 4, 61-69.

157. S. S. Dragomir, Some Gronwall Type Inequalities and Applications, Nova Science Publishers, Inc., Hauppauge, NY, 2003.

158. S. S. Dragomir, A weighted Ostrowski type inequality for functions with values in Hilbert spaces and Applications, J. Korean Math. Soc., 40 (2003), 207-224.

159. S. S. Dragomir, Bounds for f-divergences under likelihood ratio constraints, Appl. Math. 48 (2003), No. 3, 205-223.

160. S. S. Dragomir, Semi-Inner Products and Applications, Nova Science Publishers, Inc., Hauppauge, NY, 2004.

161. S. S. Dragomir, Discrete Inequalities of the Cauchy–Bunyakovsky–Schwarz Type, Nova Science Publishers, Inc., Hauppauge, NY, 2004.

162. S. S. Dragomir, Inequalities of Grüss type for the Stieltjes integral and applications, Kragujevac J. Math., 26 (2004), 89–122.

163. S. S. Dragomir, Advances in inequalities of the Schwarz, Grüss and Bessel Type

in Inner Product Spaces, RGMIA Monographs, Victoria University, 2004. (on-line http://rgmia.vu.edu.au/monographs/).

164. S. S. Dragomir, Advances in Inequalities of the Schwarz, Triangle and Heisenberg Type in Inner Product Spaces, RGMIA Monographs, Victoria University, 2005. (on-line http://rgmia.vu.edu.au/monographs/).

165. S. S. Dragomir, Advances in inequalities of the Schwarz, Grüss and Bessel Type in Inner Product Spaces, Nova Science Publishers, Inc., Hauppauge, NY, 2005.

166. S. S. Dragomir, Advances in Inequalities of the Schwarz, Triangle and Heisenberg Type in Inner Product Spaces, Nova Science Publishers, Inc., Hauppauge, NY, 2007.

167. S. S. Dragomir, Grüss type discrete inequalities in inner product spaces, revisited. Inequality theory and applications. Vol. 5 (2007), Nova Sci. Publ., NY, 61-69.

168. S. S. Dragomir, Sharp Grüss-type inequalities for functions whose derivatives are of bounded variation, JIPAM J. Inequal. Pure Appl. Math. 8 (2007), No. 4, Article 117, pp. 13.

169. S. S. Dragomir and N. S. Barnett, An Ostrowski type inequality for mappings whose second derivatives are bounded and applications, RGMIA Research Report Collection, Vol.1 , No. 2, 1998, 69-77, on line http://rgmia.vu.edu.au

170. S. S. Dragomir and N.S. Barnett, An Ostrowski type inequality for mappings whose second derivatives are bounded and applications, J. Indian Math. Soc. (S.S.), 66 (1-4) (1999), 237-245.

171. S. S. Dragomir, N. S. Barnett and P. Cerone, An Ostrowski type inequality for double integrals in terms of L_p -norms and applications in numerical integration, RGMIA Research Report Collection, Vol.1 , No. 2, 1998, 79-87, on line http://rgmia.vu.edu.au

172. S. S. Dragomir, N. S. Barnett and P. Cerone, An n-dimension Version of Ostrowski's Inequality for mappings of the Holder Type, RGMIA Res. Rep. Collect, 2 (1999), 169-180.

173. S. S. Dragomir, N. S. Barnett and P. Cerone, An n-Dimensional Version of Ostrowski's Inequality for Mappings of the Holder Type, Kyungpook Math. J. 40 (2000), No. 1, 65-75.

174. S. S. Dragomir, N. S. Barnett and S. Wang, An Ostrowski type inequality for a random variable whose probability density function belongs to $L_p[a, b], p > 1$, Mathematical Inequalities and Applications, Vol 2, No. 4 (1999), 501–508.

175. S. S. Dragomir, P. Cerone, N. S. Barnett and J. Roumeliotis, An inequality of the Ostrowski type for double integrals and applications to cubature formulae, RGMIA Research Report Collection, 2, 1999, 781-796.

176. S. S. Dragomir, P. Cerone, N. S. Barnett and J. Roumeliotis, An inequality of the Ostrowski type for double integrals and applications to Cubature formulae, Tamsui Oxford Journal of Mathematical Sciences, 16 (2000), 1-16.

177. S. S. Dragomir, N.T. Diamond, A Discrete Grüss Type Inequality and Applications for the Moments of Random Variables and Guessing Mappings, Stochastic analysis and applications, Vol. 3, 21-35, Nova Sci. Publ., Hauppauge, NY, 2003.

178. S. S. Dragomir and V. Gluscevic, Approximating Csiszar f-Divergence via a Generalized Taylor Formula, pp. 1-11, in S.S. Dragomir (ed.), Inequalities for Csiszar f-Divergence in Information Theory, Victoria University, Melbourne, Australia, 2000, on line: http://rgmia.vu.edu.au

179. S. S. Dragomir and Anca C. Gosa, A Generalization of an Ostrowski Inequality in Inner Product Spaces, Inequality theory and applications, Vol. 4, 61-64, Nova Sci. Publ., New York, 2007.

180. S. S. Dragomir and C.E.M. Pearce, Selected Topics on Hermite-Hadamard Inequal-

ities and Applications, RGMIA Monographs, Victoria University, 2000. (on-line http://rgmia.vu.edu.au/monographs/).

181. S. S. Dragomir and J. Pečarić, Refinements of some inequalities for isotonic functionals,. Anal. Num. Theor. Approx., 11 (1989), 61-65.

182. S. S. Dragomir, J. E. Pečarić and B. Tepes, Note on Integral Version of the Grüss Inequality for Complex Functions, Inequality theory and applications, Vol. 5, 91-96, Nova Sci. Publ., New York, 2007.

183. S. S. Dragomir and T. M. Rassias, Ostrowski Type Inequalities and Applications in Numerical Integration, RGMIA Monographs, Victoria University, 2000. (on-line http://rgmia.vu.edu.au/monographs/).

184. Didier Dubois, Laurent Foulloy, Gilles Mauris and Henri Prade, Probability-Possibility Transformations, Triangular Fuzzy Sets, and Probabilistic Inequalities, Reliab. Comput. 10 (2004), No. 4, 273-297.

185. R. M. Dudley, Real Analysis and Probability, Wadsworth &. Brooks/Cole, Pacific Grove, CA, 1989.

186. J. Duoandikoetxea, A unified approach to several inequalities involving functions and derivatives, Czechoslovak Mathematical Journal, 51 (126) (2001), 363-376.

187. A. L. Durán, R. Estrada and R. P. Kanwal, Pre-asymptotic Expansions, J. of Math. Anal. and Appls., 202 (1996), 470-484.

188. R. Estrada and R. P. Kanwal, A Distributional Theory for Asymptotic Expansions, Proceedings of the Royal Society of London, Series A, Mathematical and Physical Sciences, Vol. 428, No. 1875 (1990), 399-430.

189. R. Estrada and R. P. Kanwal, The Asymptotic Expansion of Certain Multi-Dimensional Generalized Functions, J. of Math. Analysis and Appls. 163 (1992), 264-283.

190. R. Estrada and R. P. Kanwal, Taylor Expansions for Distributions, Mathematical Methods in the Applied Sciences, Vol. 16 (1993), 297–304.

191. R. Estrada and R. P. Kanwal, A Distributional Approach to Asymptotics, Theory and Applications, 2nd Edition, Birkhäuser, Boston, Basel, Berlin, 2002.

192. L. C. Evans, Partial Differential Equations, Graduate Studies in Mathematics, vol 19, American Mathematical Society, Providence, RI, 1998.

193. H. O. Fattorini, Ordinary differential equations in linear topological spaces, I. J. Diff. Equations, 5 (1968), 72-105.

194. W. Feller, An Introduction to Probability Theory and its Applications, Vol. II, 2nd ed., J. Wiley, New York, 1971.

195. A. M. Fink, Bounds on the deviation of a function from its averages, Czechoslovak Mathematical Journal, 42 (117) (1992), No. 2, 289-310.

196. Wendell Fleming, Functions of Several Variables, Springer-Verlag, New York, Berlin, Undergraduate Texts in Mathematics, 2nd edition, 1977.

197. A. Flores-Franulič and H. Román-Flores, A Chebyshev type inequality for fuzzy integrals, Appl. Math. Comput. 190 (2007), No. 2, 1178-1184.

198. Peng Gao, Hardy-type inequalities via auxiliary sequences, J. Math. Anal. Appl. 343 (2008), No. 1, 48-57.

199. I. Gavrea, On Chebyshev type inequalities involving functions whose derivatives belong to L_p spaces via isotonic functionals, JIPAM. J. Inequal. Pure Appl. Math. 7 (2006), No. 4, Article 121, 6 pp. (electronic).

200. I. M. Gel'fand and G. E. Shilov, Generalized Functions, Vol. I., 1964, Academic Press, New York, London.

201. Gustavo L. Gilardoni, On the minimum f-divergence for given total variation, C. R. Math. Acad. Sci. Paris 343 (2006), No. 11-12, 763-766.

202. J. A. Goldstein, Semigroups of Linear Operators and Applications, Oxford Univ. Press, Oxford, 1985.

203. S. P. Goyal, Manita Bhagtani and Kantesh Gupta, Integral Mean Inequalities for Fractional Calculus Operators of Analytic Multivalent Function, Bull. Pure Appl. Math. 1 (2007), No. 2, 214-226.

204. A. Grothendieck, Résumé de la théorie métrique des produits tensoriels topologiques, Bol. Soc. Matem., São Paolo 8 (1956), 1-79.

205. G. Grüss, Über das Maximum des absoluten Betrages von $\left[\left(\frac{1}{b-a}\right)\int_a^b f(x)g(x)dx - \left(\frac{1}{(b-a)^2}\int_a^b f(x)dx \int_a^b g(x)dx\right)\right]$, Math. Z. 39 (1935), pp. 215-226.

206. H. Özlem Güney and Shigeyoshi Owa, Integral Means Inequalities for Fractional Derivatives of a Unified Subclass of Prestarlike Functions with Negative Coefficients, J. Inequal. Appl. 2007, Art. ID 97135, 9 pp.

207. Alexander A. Gushchin, Denis A. Zhdanov, A Minimax Result for f-Divergences, From stochastic calculus to mathematical finance, Springer, Berlin, 2006, 287-294.

208. James Guyker, An inequality for Chebyshev connection coefficients, JIPAM. J. Inequal. Pure Appl. Math. 7 (2006), No. 2, Article 67, 11 pp. (electronic).

209. U. Haagerup, A new upper bound for the complex Grothendieck constant, Israeli J. Math. 60 (1987), 199-224.

210. Jan Haluska, Ondrej Hutnik, Some inequalities involving integral means, Tatra Mt. Math. Publ. 35 (2007), 131-146.

211. Junqiang Han, A class of improved Sobolev-Hardy inequality on Heisenberg groups, Southeast Asian Bull. Math. 32 (2008), No.3, 437-444.

212. G. H. Hardy, J. E. Littlewood, G. Polya, Inequalities, Reprint of the 1952 edition, Cambridge Mathematical Library, Cambridge University Press, Cambridge, 1988.

213. Petteri Harjulehto, Ritva Hurri-Syrjanen, On a (q,p)-Poincaré inequality, J. Math. Anal. Appl. 337 (2008), No. 1, 61-68.

214. E. Hewitt and K. Stromberg, Real and Abstract Analysis, Springer-Verlag, #25, New York, Berlin, 1965.

215. E. Hille and R. S. Phillips, Functional Analysis and Semigroups, revised edition, pp. XII a. 808, Amer. Math. Soc. Colloq. Publ., Vol. 31, Amer. Math. Soc., Providence, RI, 1957.

216. G. Hognas, Characterization of weak convergence of signed measures on [0,1], Math. Scand. 41 (1977), 175-184.

217. K. Isii, The extreme of probability determined by generalized moments (I): bounded random variables, Annals Inst. Math. Statist. 12 (1960), 119-133.

218. K. C. Jain, Amit Srivastava, On symmetric information divergence measures of Csiszar's f-divergence class, Far East J. Appl. Math. 29 (2007), No. 3, 477-491.

219. J. Johnson, An elementary characterization of weak convergence of measures, The American Math. Monthly, 92, No. 2 (1985), 136-137.

220. N. L. Johnson and C. A. Rogers, The moment problems for unimodal distributions, Annals Math. Statist. 22 (1951), 433-439.

221. Marius Junge, Embedding of the operator space OH and the logarithmic little Grothendieck inequality, Invent. Math. 161 (2005), No. 2, 225-286.

222. S. Karlin and L. S. Shapley, Geometry of moment spaces, Memoirs Amer. Math. Soc., No. 12, Providence, 1953.

223. S. Karlin and W. J. Studden, Tchebycheff systems: with applications in Analysis and Statistics, Interscience, New York, 1966.

224. Y. Katznelson, An introduction to Harmonic Analysis, Dover, New York, 1976.

225. Stephen Keith, Xiao Zhong, The Poincaré inequality is an open ended condition, Ann. of Math. (2) 167 (2008), No. 2, 575-599.
226. H. G. Kellerer, Verteilungsfunktionen mit gegebenen Marginalverteilungen, Zeitschr. fur Warsch. 3 (1964), 247-270.
227. H. G. Kellerer, Duality theorems for marginal problems, Zeitschr. fur Warsch. 67 (1984), 399-432.
228. J. H. B. Kemperman, On the sharpness of Tchebysheff type inequalities, Indagationes Math. 27 (1965), 554-601.
229. J. H. B. Kemperman, The general moment problem, a geometric approach, Annals Math. Statist. 39 (1968), 93-122.
230. J. H. B. Kemperman, Moment problems with convexity conditions, Optimizing methods in statistics (J.S. Rustagi, ed.) 1971, Academic Press, New York, 115-178.
231. J. H. B. Kemperman, On a class of moment problems, Proc. Sixth Berkeley Symposium on Math. Statist. and Prob. 2 (1972), 101-126.
232. J. H. B. Kemperman, Moment Theory, Class notes, University of Rochester, NY, Spring 1983.
233. J. H. B. Kemperman, On the role of duality in the theory of moments, Semi-infinite programming and applications (A.V. Fiacco and K.O. Kortanck, eds.), Lecture Notes in Economics and Mathematical Systems, Vol. 215 (1983), Springer-Verlag, New York, 63-92.
234. J. H. B. Kemperman, Geometry of the moment problem, Moment in Mathematics, American Mathematical Society, Lecture Notes, presented in San Antonio, TX, January 20-22, 1987, pp. 1-34.
235. J. Kingman and S. Taylor, Introduction to Measure and Probability, Cambridge University Press, Cambridge, NY, 1966.
236. P. P. Korovkin, Inequalities, Translated from the Russian by Sergei Vrubel, Reprint of the 1975 edition, Little Mathematics Library, Mir, Moscow, distributed by Imported Publications, Chicago, IL, 1986.
237. D. Kreider, R. Kuller, D. Ostberg and F. Perkins, An Introduction to Linear Analysis, Addison-Wesley Publishing Company, Inc., Reading, Mass., USA, 1966.
238. M. G. Krein, The ideas of P.L. Cebysev and A.A. Markov in the theory of limiting values of integrals and their further development, Uspehi Mat. Nauk 6 (1951), 3-130 (Russian), English translation: Amer. Math. Soc. Translations, Series 2, 12 (1959), 1-121.
239. M. G. Krein and A. A. Nudel'man, The Markov moment problem and external problems, Amer. Math. Soc., Providence, Rhode Island, 1977.
240. I. L. Krivine, Sur la constante de Grothendieck, C.R.A.S. 284 (1977), 445-446.
241. V. I. Krylov, Approximate Calculation of Integrals, Macmillan, New York, London, 1962.
242. Jonas Kubilius, On some inequalities in the probabilistic number theory. New trends in probability and statistics, Vol. 4 (Palanga, 1996), 345-356, VSP, Utrecht, 1997.
243. S. Kullback, Information Theory and Statistics, Wiley, New York, 1959.
244. S. Kullback and R. Leibler, On information and sufficiency, Ann. Math. Statist., 22 (1951), 79-86.
245. Man Kam Kwong, On an Opial inequality with a boundary condition, JIPAM. J. Inequal. Pure Appl. Math. 8 (2007), No. 1, Article 4, 6 pp. (electronic).
246. V. Lakshmikantham and S. Leela, Differential and integral inequalities: Theory and applications. vol. I: Ordinary differential equations, Mathematics in Science and Engineering, Vol. 55-I, Academic Press, New York - London, 1969.
247. V. Lakshmikantham and S. Leela, Differential and integral inequalities: Theory and

applications. vol. II: Functional, partial, abstract, and complex differential equations, Mathematics in Science and Engineering, Vol. 55-II, Academic Press, New York - London, 1969.

248. Alberto Lastra, Javier Sanz, Linear continuous operators for the Stieltjes moment problem in Gelfand-Shilov spaces, J. Math. Anal. Appl. 340 (2008), No. 2, 968-981.

249. Juha Lehrbäck, Pointwise Hardy inequalities and uniformly fat sets, Proc. Amer. Math. Soc. 136 (2008), No. 6, 2193-2200.

250. N. Levinson, On an inequality of Opial and Beesack, Proc. Amer. Math. Soc. 15 (1964), 565-566.

251. Wei-Cheng Lian, Shiueh-Ling Yu, Fu-Hsiang Wong, Shang-Wen Lin, Nonlinear type of Opial's inequalities on time scales, Int. J. Differ. Equ. Appl. 10 (2005), No. 1, 101-111.

252. E. H. Lieb, M. Loss, Analysis, 2nd Edition, American Math. Soc., 2001, Providence, R. I.

253. C. T. Lin and G. S. Yang, A generalized Opial's inequality in two variables, Tamkang J. Math., 15 (1983), 115-122.

254. Wenjun Liu, Jianwei Dong, On new Ostrowki type inequalities, Demonstratio Math. 41 (2008), No. 2, 317-322.

255. Zheng Liu, Notes on a Grüss type inequality and its application, Vietnam J. Math. 35 (2007), No. 2, 121-127.

256. Françoise Lust-Piquard and Quanhua Xu, The little Grothendieck theorem and Khintchine inequalities for symmetric spaces of measurable operators, J. Funct. Anal. 244 (2007), No. 2, 488-503.

257. Yong-Hua Mao, General Sobolev type inequalities for symmetric forms, J. Math. Anal. Appl. 338 (2008), No. 2, 1092-1099.

258. A. Markov, On certain applications of algebraic continued fractions, Thesis, St. Petersburg, 1884.

259. M. Matic, J. Pecaric and N. Ujevic, Weighted version of multivariate Ostrowski type inequalities, Rocky Mountain J. Math. 31 (2001), No. 2, 511-538.

260. Janina Mihaela Mihaila, Octav Olteanu and Constantin Udriste, Markov-type and operator-valued multidimensional moment problems, with some applications, Rev. Roumaine Math. Pures Appl. 52 (2007), No. 4, 405-428.

261. Janina Mihaela Mihaila, Octav Olteanu and Constantin Udriste, Markov-type moment problems for arbitrary compact and some non-compact Borel subsets of $\mathbb{R}^n$, Rev. Roumaine Math. Pures Appl. 52 (2007), No. 6, 655-664.

262. R. von Mises, The limits of a distribution function if two expected values are given, Ann. Math. Statist. 10 (1939), 99-104.

263. D. S. Mitrinovic, J. E. Pecaric and A. M. Fink, Inequalities involving functions and their integrals and derivatives, Mathematics and its Applications (East European Series), 53, Kluwer Academic Publishers Group, Dordrecht, 1991.

264. D. S. Mitrinovic, J. E. Pecaric and A. M. Fink, Classical and New Inequalities in Analysis, Kluwer Academic, Dordrecht, 1993.

265. D. S. Mitrinovic, J. E. Pecaric and A. M. Fink, Inequalities for Functions and their Integrals and Derivatives, Kluwer Academic Publishers, Dordrecht, 1994.

266. B. Mond, J. Pecaric, I. Peric, On reverse integral mean inequalities, Houston J. Math. 32 (2006), No. 1, 167-181 (electronic).

267. Davaadorjin Monhor, A Chebyshev inequality for multivariate normal distribution, Probab. Engrg. Inform. Sci. 21 (2007), No. 2, 289-300.

268. Cristinel Mortici, A new simple proof for an inequality of Cebyshev type, An. Stiint. Univ. Ovidius Constanta Ser. Mat. 13 (2005), No. 2, 39-43.

269. H. Mulholland and C. Rogers, Representation theorems for distribution functions, Proc. London Math. Soc., 8 (1958), 177-223.

270. S. V. Nagaev, Probabilistic and moment inequalities for dependent random variables, (Russian) Teor. Veroyatnost. i Primenen. 45 (2000), No. 1, 194-202, Translation in Theory Probab. Appl. 45 (2001), No. 1, 152-160.

271. B. Nagy, On cosine operator functions in Banach spaces, Acta Scientiarum Mathematicarum Szeged, 36 (1974), 281-289.

272. B. Nagy, Approximation theorems for Cosine operator functions, Acta Mathematica Academiae Scientiarum Hungaricae, 29 (1-2) (1977), 69-76.

273. I. D. Necaev, Integral inequalities with gradients and derivatives, Soviet. Math. Dokl., 22 (1973), 1184-1187.

274. Tim Netzer, An elementary proof of Schmudgen's theorem on the moment problem of closed semi-algebraic sets, Proc. Amer. Math. Soc. 136 (2008), No. 2, 529-537 (electronic).

275. C. Olech, A simple proof of a certain result of Z. Opial, Ann. Polon. Math., 8 (1960), 61-63.

276. Z. Opial, Sur une inegalite, Ann. Polon. Math., 8 (1960), 29-32.

277. A. Ostrowski, Uber die Absolutabweichung einer differentiebaren Funktion von ihrem Integralmittelwert, Comment. Math. Helv. 10 (1938), 226-227.

278. B. G. Pachpatte, On an inequality of Ostrowski type in three independent variables, J. Math. Anal. Appl. 249 (2000), No. 2, 583-591.

279. B. G. Pachpatte, An inequality of Ostrowski type in n independent variables, Facta Univ. Ser. Math. Inform. No. 16 (2001), 21-24.

280. B. G. Pachpatte, On multivariate Ostrowski type inequalities, JIPAM. J. Inequal. Pure. Appl. Math. 3 (2002), No. 4, Article 58, 5 pp. (electronic).

281. B. G. Pachpatte, On Grüss type inequalities for double integrals, J. Math. Anal. & Appl., 267 (2002), 454-459.

282. B. G. Pachpatte, New weighted multivariate Grüss type inequalities, electronic Journal: Journal of Inequalities in Pure and Applied Mathematics, http://jipam.vu.edu.au/, Volume 4, Issue 5, Article 108, 2003.

283. B. G. Pachpatte, Inequalities similar to Opial's inequality involving higher order derivatives, Tamkang J. Math. 36 (2005), No. 2, 111-117.

284. B. G. Pachpatte, A note on Opial type finite difference inequalities, Tamsui Oxf. J. Math. Sci. 21 (2005), No. 1, 33-39.

285. B. G. Pachpatte, Mathematical inequalities, North Holland Mathematical Library, 67, Elsevier B. V., Amsterdam, 2005.

286. B. G. Pachpatte, Integral and finite difference inequalities and applications, North-Holland Mathematics Studies, 205, Elsevier Science B. V., Amsterdam, 2006.

287. B. G. Pachpatte, On Chebyshev-Grüss type inequalities via Pecaric's extension of the Mongomery identity, Journal of Inequalities in Pure and Applied Mathematics, http://jipam.vu.edu.au/, Vol 7, Issue 1, Article 11, 2006.

288. B. G. Pachpatte, Some new Ostrowski and Grüss type inequalities, Tamkang J. Math. 38 (2007), No. 2, 111-120.

289. B. G. Pachpatte, New inequalities of Ostrowski-Grüss type, Fasc. Math., No. 38 (2007), 97-104.

290. B. G. Pachpatte, On a new generalization of Ostrowski type inequality, Tamkang J. Math. 38 (2007), No. 4, 335-339.

291. B. G. Pachpatte, New generalization of certain Ostrowski type inequalities, Tamsui Oxf. J. Math. Sci. 23 (2007), No. 4, 393-403.

292. B. G. Pachpatte, New discrete Ostrowski-Grüss like inequalities, Facta. Univ. Ser. Math. Inform. 22 (2007), No. 1, 15-20.

293. B. G. Pachpatte, A note on Grüss type inequalities via Cauchy's mean value theorem, Math. Inequal. Appl. 11 (2008), No. 1, 75-80.

294. J. E. Pecaric, On the Cebyshev inequality, Bul. Sti. Tehn. Inst. Politehn. "Traian Vuia" Timisoara, 25 (39) (1980), No. 1, 5-9.

295. J. Pecaric and I. Peric, A multidimensional generalization of the Lupas-Ostrowski inequality, Acta Sci. Math. (Szeged), 72 (2006), 65-72.

296. Josip E. Pecaric, Frank Proschan, Y. L. Tong, Convex functions, partial orderings, and statistical applications, Mathematics in Science and Engineering, 187, Academic Press, Inc., Boston, MA, 1992.

297. J. Pecaric, A. Vukelic, Milovanovic-Pecaric-Fink inequality for difference of two integral means, Taiwanese J. Math. 10 (2006), No. 4, 933-947.

298. Antonio M. Peralta, Angel Rodriguez Palacios, Grothendieck's inequalities revisited. Recent progress in functional analysis (Valencia, 2000), 409-423, North-Holland Math. Stud., 189, North-Holland, Amsterdam, 2001.

299. J. Pitman, Probability, Springer-Verlag, New York, Berlin, 1993.

300. E. A. Plotnikova, Integral representations and the generalized Poincaré inequality on Carnot groups, (Russian) Sibirsk. Mat. Zh. 49 (2008), No. 2, 420-436.

301. G. Polya, G. Szego, Isoperimetric Inequalities in Mathematical Physics, Annals of Mathematics Studies, No. 27, Princeton University Press, Princeton, N. J., 1951.

302. Svetlozar T. Rachev, Probability Metrics and the Stability of Stochastic Models, John Wiley & Sons, NY, 1991.

303. Svetlozar T. Rachev and Stefan Mittnik, Stable Paretian Models in Finance, John Wiley & Sons, NY, 1999.

304. Arif Rafig, Fiza Zafar, New bounds for the first inequality of Ostrowski-Grüss type and applications in numerical integration, Nonlinear Funct. Anal. Appl. 12 (2007), No. 1, 75-85.

305. A. Rényi, On measures of dependence, Acta Math. Acad. Sci. Hungar. 10 (1959), 441-451.

306. A. Rényi, On measures of entropy and information, Proceedings of the 4th Berkeley Symposium on Mathematical Statistics and Probability, I, Berkeley, CA, 1960, 547-561.

307. H. Richter, Parameterfreie Abschätzung und Realisierung von Erwartungswerten, Blätter der Deutschen Gesellschaft für Versicherungsmathematik, 3 (1957), 147-161.

308. F. Riesz, Sur certaines systèmes singuliers d'équations integrales, Ann. Sci. Ecole Norm. Sup. 28 (1911), 33-62.

309. Leonard James Rogers, An extension of a certain theorem in inequalities, Messenger of Math. 17 (1888), 145-150.

310. W. Rogosinsky, Moments of non-negative mass, Proc. Roy. Soc. London, Ser. A, 245 (1958), 1-27.

311. W. W. Rogosinsky, Non-negative linear functionals, moment problems, and extremum problems in polynomial spaces, Studies in mathematical analysis and related topics, 316-324, Stanford Univ. Press, 1962.

312. H. L. Royden, Real Analysis, Second edition, Macmillan, (1968), New York, London.

313. T. Rychlik, The complete solution of a rounding problem under two moment conditions, Proc. 3rd Conf. on Distributions with Given Marginals and Moment Problems, Prague, 1996 (V. Benes, J. Stepan, eds.), Kluwer Acad, Publ. Dordrecht, 1997, 15-20.

314. Tomasz Rychlik, Upper bounds for the expected Jefferson rounding under mean-variance-skewness conditions, Probab. Math. Statist. 20 (2000), No. 1, Acta Univ. Wratislav. No. 2246, 1-18.

315. Tomasz Rychlik, Sharp evaluation of the Prokhorov distance to zero under general moment conditions, J. Comput. Anal. Appl. 6 (2004), No. 2, 165-172.

316. Mohammad Sababheh, Two-sided probabilistic versions of Hardy's inequality, J. Fourier Anal. Appl. 13 (2007), No. 5, 577-587.

317. L. Schumaker, Spline functions basic theory, 1981, Wiley, New York.

318. L. Schwartz, Analyse Mathematique, Paris, 1967.

319. H. L. Selberg, Zwei Ungleichungen zur Ergänzung des Tchebycheffschen Lemmas, Skand. Aktuarietidskrift, 23 (1940), 121-125.

320. Zhongwei Shen, Peihao Zhao, Uniform Sobolev inequalities and absolute continuity of periodic operators, Trans. Amer. Math. Soc. 360 (2008), No. 4, 1741-1758.

321. G. Shilov, Elementary Functional Analysis, The MIT Press Cambridge, Massachusetts, 1974.

322. J. A. Shohat, J. D. Tamarkin, The problem of moments, Math. Surveys 1, Amer. Math. Soc., Providence, 1943.

323. R. M. Shortt, Strassen's marginal problem in two or more dimensions, Zeitsch. fur Wahrsch. 64 (1983), 313-325.

324. G. M. Skylar, S. Yu. Ignatovich, Development on the Markov moment problem approach in the optimal control theory, Methods Funct. Anal. Topology, 13 (2007), No. 4, 386-400.

325. A. Sofo, Integral inequalities of the Ostrowski type, JIPAM. J. Inequal. Pure Appl. Math. 3 (2002), No. 2, Article 21, 39 pp. (electronic).

326. 326. M. Sova, Cosine operator functions, Rozprawy Matematyczne XLIX (Warszawa, 1966)

327. M. R. Spiegel, Advances Calculus, Schaum's Outline Series, McGraw-Hill Book Co., NY, 1963.

328. V. Stakenas, On some inequalities of probabilistic number theory, Liet. Mat. Rink, 46 (2006), No. 2, 256-266, translation in Lithuanian Math. J. 46 (2006), No. 2, 208-216.

329. B. Stankovic, Taylor Expansion for Generalized Functions, J. of Math. Analysis and Appl., 203, 31-37 (1996).

330. Stevo Stević, Area type inequalities and integral means of harmonic functions on the unit ball, J. Math. Soc. Japan 59 (2007), No. 2, 583-601.

331. W. T. Sulaiman, On a reverse of Hardy-Hilbert's integral inequality in its general form, Int. J. Math. Anal. (Ruse) 2 (2008), No. 1-4, 67-74.

332. Inder Jeet Taneja, Pranesh Kumar, Relative information of type s, Csiszar's f-divergence, and information inequalities, Inform. Sci. 166 (2004), No. 1-4, 105-125.

333. W. C. Troy, On the Opial-Olech-Beesack inequalities, USA-Chile Workshop on Nonlinear Analysis, Electron. J. Diff. Eqns., Conf., 06 (2001), 297-301. http://ejde.math.swt.edu or http://ejde.math.unt.edu

334. Kuei-Lin Tseng, Shiow Ru Hwang, S. S. Dragomir, Generalizations of weighted Ostrowski type inequalities for mappings of bounded variation and their applications, Comput. Math. Appl. 55 (2008), No. 8, 1785-1793.

335. K. Watanabe, Y. Kametaka, A. Nagai, K. Takemura, H. Yamagishi, The best constant of Sobolev inequality on a bounded interval, J. Math. Anal. Appl. 340 (2008), No. 1, 699-706.

336. Jia Jin Wen, Chao Bang Gao, The best constants of Hardy type inequalities for p=-1, J. Math. Res. Exposition 28 (2008), No. 2, 316-322.

337. Jia Jin Wen, Wan-Ian Wang, Chebyshev type inequalities involving permanents and their applications, Linear Algebra Appl., 422 (2007), No. 1, 295-303.

338. E. T. Whittaker and G. N. Watson, A Course in Modern Analysis, Cambridge University Press, 1927.

339. D. V. Widder, A Generalization of Taylor's Series, Transactions of AMS, 30, No. 1 (1928), 126-154.

340. D. Willett, The existence - uniqueness theorems for an nth order linear ordinary differential equation, Amer. Math. Monthly, 75 (1968), 174-178.

341. Fu-Hsiang Wong, Wei-Cheng Lian, Shiueh-Ling Yu, Cheh-Chih Yeh, Some generalizations of Opial's inequalities on time scales, Taiwanese J. Math. 12 (2008), No. 2, 463-471.

342. Quanhua Xu, Operator-space Grothendieck inequalities for noncommutative L_p-space, Duke Math. J. 131 (2006), No. 3, 525-574.

343. G. S. Yang, Inequality of Opial-type in two variables, Tamkang J. Math., 13 (1982), 255-259.

344. Cheh-Chih Yeh, Ostrowski's inequality on time scales, Appl. Math. Lett. 21 (2008), No. 4, 404-409.

345. Qi S. Zhang, A uniform Sobolev inequality for Ricci flow with surgeries and applications, C. R. Math. Acad. Sci. Paris 346 (2008), No. 9-10, 549-552.

346. Chang-Jian Zhao, Wing-Sum Cheung, On multivariate Grüss inequalities, J. Inequal. Appl. 2008, Art. ID 249438, 8 pp.

347. Bogdan Ziemian, Taylor formula for distributions, Dissertationes Math. (Rozprawy Mat.), 264 (1988), 56 pp.

348. Bogdan Ziemian, The Melin transformation and multidimensional generalized Taylor expansions of singular functions, J. Fac. Sci. Univ. Tokyo Sect. IA Math. 36 (1989), No. 2, 263-295.

349. Bogdan Ziemian, Generalized Taylor expansions and theory of resurgent functions of Jean E'calle, Generalized functions and convergence (Katowice, 1988), 285-295, World Sci. Publ., Teaneck, NJ, 1990.

List of Symbols

Index